The Handbook of Environmental Chemistry

Founded by Otto Hutzinger

Editors-in-Chief: Damià Barceló • Andrey G. Kostianoy

Volume 67

Advisory Board:
Jacob de Boer, Philippe Garrigues, Ji-Dong Gu,
Kevin C. Jones, Thomas P. Knepper, Alice Newton,
Donald L. Sparks

More information about this series at http://www.springer.com/series/698

Applications of Advanced Oxidation Processes (AOPs) in Drinking Water Treatment

Volume Editors: Antonio Gil · Luis Alejandro Galeano · Miguel Ángel Vicente

With contributions by

M. Antonopoulou · M. Brienza · B. R. Contesini ·
A. P. B. R. de Freitas · L. V. de Freitas · B. M. Esteves ·
P. Fernández-Ibáñez · L. A. Galeano · A.-M. García · A. Gil ·
F. M. Gomes · M. Guerrero-Flórez · J. Hofman · R. Hofman-Caris ·
H. J. Izário Filho · S. Jafarinejad · H. K. Karapanagioti · K. Katsanou ·
I. Konstantinou · G. Li Puma · C. C. A. Loures · L. M. Madeira ·
B. K. Mayer · S. Nahim-Granados · A. M. Nasser · C. B. Özkal ·
A. Pintar · M. I. Polo-López · J. H. Ramírez · C. E. R. Reis ·
C. S. D. Rodrigues · D. R. Ryan · C.-A. Sánchez · M. B. Silva ·
S. Sorlini · T. Tišler · R. A. Torres-Palma · M. Á. Vicente

Editors
Antonio Gil
Department of Sciences
Public University of Navarra
Pamplona, Spain

Luis Alejandro Galeano
Department of Chemistry
University of Nariño
Pasto, Nariño
Colombia

Miguel Ángel Vicente
Department of Inorganic Chemistry
University of Salamanca
Salamanca, Spain

ISSN 1867-979X   ISSN 1616-864X   (electronic)
The Handbook of Environmental Chemistry
ISBN 978-3-319-76881-6   ISBN 978-3-319-76882-3 (eBook)
https://doi.org/10.1007/978-3-319-76882-3

Library of Congress Control Number: 2018946685

© Springer International Publishing AG, part of Springer Nature 2019
This work is subject to copyright. All rights are reserved by the Publisher, whether the whole or part of the material is concerned, specifically the rights of translation, reprinting, reuse of illustrations, recitation, broadcasting, reproduction on microfilms or in any other physical way, and transmission or information storage and retrieval, electronic adaptation, computer software, or by similar or dissimilar methodology now known or hereafter developed.
The use of general descriptive names, registered names, trademarks, service marks, etc. in this publication does not imply, even in the absence of a specific statement, that such names are exempt from the relevant protective laws and regulations and therefore free for general use.
The publisher, the authors, and the editors are safe to assume that the advice and information in this book are believed to be true and accurate at the date of publication. Neither the publisher nor the authors or the editors give a warranty, express or implied, with respect to the material contained herein or for any errors or omissions that may have been made. The publisher remains neutral with regard to jurisdictional claims in published maps and institutional affiliations.

Printed on acid-free paper

This Springer imprint is published by the registered company Springer International Publishing AG part of Springer Nature.
The registered company address is: Gewerbestrasse 11, 6330 Cham, Switzerland

Editors-in-Chief

Prof. Dr. Damià Barceló
Department of Environmental Chemistry
IDAEA-CSIC
C/Jordi Girona 18–26
08034 Barcelona, Spain
and
Catalan Institute for Water Research (ICRA)
H20 Building
Scientific and Technological Park of the
  University of Girona
Emili Grahit, 101
17003 Girona, Spain
dbcqam@cid.csic.es

Prof. Dr. Andrey G. Kostianoy
P.P. Shirshov Institute of Oceanology
Russian Academy of Sciences
36, Nakhimovsky Pr.
117997 Moscow, Russia
kostianoy@gmail.com

Advisory Board

Prof. Dr. Jacob de Boer
IVM, Vrije Universiteit Amsterdam, The Netherlands

Prof. Dr. Philippe Garrigues
University of Bordeaux, France

Prof. Dr. Ji-Dong Gu
The University of Hong Kong, China

Prof. Dr. Kevin C. Jones
University of Lancaster, United Kingdom

Prof. Dr. Thomas P. Knepper
University of Applied Science, Fresenius, Idstein, Germany

Prof. Dr. Alice Newton
University of Algarve, Faro, Portugal

Prof. Dr. Donald L. Sparks
Plant and Soil Sciences, University of Delaware, USA

The Handbook of Environmental Chemistry Also Available Electronically

The Handbook of Environmental Chemistry is included in Springer's eBook package *Earth and Environmental Science*. If a library does not opt for the whole package, the book series may be bought on a subscription basis.

For all customers who have a standing order to the print version of *The Handbook of Environmental Chemistry*, we offer free access to the electronic volumes of the Series published in the current year via SpringerLink. If you do not have access, you can still view the table of contents of each volume and the abstract of each article on SpringerLink (www.springerlink.com/content/110354/).

You will find information about the

– Editorial Board
– Aims and Scope
– Instructions for Authors
– Sample Contribution

at springer.com (www.springer.com/series/698).

All figures submitted in color are published in full color in the electronic version on SpringerLink.

Aims and Scope

Since 1980, *The Handbook of Environmental Chemistry* has provided sound and solid knowledge about environmental topics from a chemical perspective. Presenting a wide spectrum of viewpoints and approaches, the series now covers topics such as local and global changes of natural environment and climate; anthropogenic impact on the environment; water, air and soil pollution; remediation and waste characterization; environmental contaminants; biogeochemistry; geoecology; chemical reactions and processes; chemical and biological transformations as well as physical transport of chemicals in the environment; or environmental modeling. A particular focus of the series lies on methodological advances in environmental analytical chemistry.

Series Preface

With remarkable vision, Prof. Otto Hutzinger initiated *The Handbook of Environmental Chemistry* in 1980 and became the founding Editor-in-Chief. At that time, environmental chemistry was an emerging field, aiming at a complete description of the Earth's environment, encompassing the physical, chemical, biological, and geological transformations of chemical substances occurring on a local as well as a global scale. Environmental chemistry was intended to provide an account of the impact of man's activities on the natural environment by describing observed changes.

While a considerable amount of knowledge has been accumulated over the last three decades, as reflected in the more than 70 volumes of *The Handbook of Environmental Chemistry*, there are still many scientific and policy challenges ahead due to the complexity and interdisciplinary nature of the field. The series will therefore continue to provide compilations of current knowledge. Contributions are written by leading experts with practical experience in their fields. *The Handbook of Environmental Chemistry* grows with the increases in our scientific understanding, and provides a valuable source not only for scientists but also for environmental managers and decision-makers. Today, the series covers a broad range of environmental topics from a chemical perspective, including methodological advances in environmental analytical chemistry.

In recent years, there has been a growing tendency to include subject matter of societal relevance in the broad view of environmental chemistry. Topics include life cycle analysis, environmental management, sustainable development, and socio-economic, legal and even political problems, among others. While these topics are of great importance for the development and acceptance of *The Handbook of Environmental Chemistry*, the publisher and Editors-in-Chief have decided to keep the handbook essentially a source of information on "hard sciences" with a particular emphasis on chemistry, but also covering biology, geology, hydrology and engineering as applied to environmental sciences.

The volumes of the series are written at an advanced level, addressing the needs of both researchers and graduate students, as well as of people outside the field of "pure" chemistry, including those in industry, business, government, research establishments, and public interest groups. It would be very satisfying to see these volumes used as a basis for graduate courses in environmental chemistry. With its high standards of scientific quality and clarity, *The Handbook of Environmental Chemistry* provides a solid basis from which scientists can share their knowledge on the different aspects of environmental problems, presenting a wide spectrum of viewpoints and approaches.

The Handbook of Environmental Chemistry is available both in print and online via www.springerlink.com/content/110354/. Articles are published online as soon as they have been approved for publication. Authors, Volume Editors and Editors-in-Chief are rewarded by the broad acceptance of *The Handbook of Environmental Chemistry* by the scientific community, from whom suggestions for new topics to the Editors-in-Chief are always very welcome.

Damià Barceló
Andrey G. Kostianoy
Editors-in-Chief

Foreword

Safe and readily available water is crucial for public health, and thus, the purification of drinking water is of such vital importance. Advanced oxidation processes (AOPs) are among the most promising methods to replace or integrate with conventional drinking water purification technologies. AOPs are rarely applied alone, but in combination with other treatment methods to obtain optimal removal rates.

All AOPs involve the generation of highly reactive oxygen species. These methods are of interest to public health, as they have the potential for complete mineralization of recalcitrant compounds, but at high concentrations of organic compounds, their application is more limited due to the high energy and oxidant consumption.

AOPs are efficient in natural organic matter (NOM) reduction and mitigation of disinfection by-product (DBP) formation, though the use of AOPs for disinfection is less studied in comparison to chemical degradation. Also, they can be used for the removal of taste and odor causing components from water as well as for the elimination of the emerging contaminants, such as pharmaceuticals. Importantly, the selection of a specific AOP depends strongly on the physicochemical properties of the water to be treated.

Full-scale applications of AOPs are still limited. This book offers major developments in recent research to put AOPs in practice. It provides a unique, holistic perspective on basic and applied research issues regarding the application of AOPs in drinking water purification. The reader finds up-to-date information and solutions on AOPs from many leading experts in the field. The 15 chapters of the book form a most useful, structured, and timely contribution to our understanding of the application of AOPs in drinking water purification.

Lappeenranta University of Technology Mika Sillanpää
Lappeenranta, Finland

Florida International University
Miami, FL, USA

Preface

Drinking water treatment plants worldwide are mostly based on simple physicochemical and disinfecting technologies very well known from over a century. Perhaps this is why recent challenges coming from more variable and complex water supplies raised along the past 30 years still remain unsolved. Although developed countries have in general adopted highly effective solutions, they are usually energy expensive and then not cost-effective for application in the real context of developing and more populated countries. On the other hand, within the same timeframe, advanced oxidation processes (AOPs) have emerged as alternative technologies for the treatment of quite dissimilar polluted streams; though undoubtedly stressed in wastewaters, drinking water has remained somewhat ignored. This has motivated this revision about most up-to-date studies dedicated to devise most promising process conditions, like AOPs, for simultaneous oxidation of chemicals and inactivation of microorganisms, as well as types of water supplies feasible for production of drinking water. Moreover, special attention has also been paid to most critical concerns that have prevented more decided application of AOPs in this field, including potential formation of disinfection by-products (DBPs), formation of intermediates of different molecular sizes and polar character, challenging analytical techniques in recording NOM substrates and by-products, pH close to neutral in most available supplies, presence of inorganic ions acting as radical scavengers, and simultaneous role as disinfection technologies, among others.

Accordingly, this book series volume in *The Handbook of Environmental Chemistry* reviews the most typical sources of drinking water, namely, ground and surface waters, not only from a purely technical approach but also taking into account some influencing social factors. It is followed by a revision about the most critical limitations of conventional technologies in the treatment of the water supplies currently available. Second big topic is covered by the chapters "Natural Organic Matter: Characterization and Removal by AOPs to Assist Drinking Water Facilities," "Natural Organic Matter Removal by Heterogeneous

Catalytic Wet Peroxide Oxidation (CWPO)," "Separation and Characterization of NOM Intermediates Along AOPs Oxidation," and "Photo (Catalytic) Oxidation Processes for the Removal of Natural Organic Matter and Contaminants of Emerging Concern from Water" devoted to natural organic matter (NOM) separation, fractionation, characterization, and removal by several AOPs including photo-assisted and other catalyzed processes under either homogeneous or heterogeneous regimes. Special attention has been given to analytic challenges that should be faced in the short term in order to get more specific insights regarding molecular size distributions and polar nature along a plethora of AOPs, whereas the most important expected limitations for every given family of technologies are also stressed. Just on the border of this and a novel and exciting field, the removal of taste and odor compounds by AOPs is comprehensibly depicted in the chapter "AOPs Methods for the Removal of Taste and Odor Compounds," where the most critical aspects to get it more feasible in the future are openly established.

Third even more promising but also challenging topic is the inactivation/disinfection of a diversity of microorganisms that could be seriously enhanced by AOPs in drinking water facilities in comparison to more conventional chlorine or UV-based techniques. This part starts with those more thoroughly studied homogeneous Fenton and photo-Fenton variants, where a gradient of microorganism's susceptibilities, main involved oxidizing species, and interferences are clearly stated (chapter "Homogeneous Fenton and Photo-Fenton Disinfection of Surface and Ground Water"), followed by a more comprehensive chapter devoted to revise AOP's mechanisms of inactivation aimed at reporting foremost and more hazardous waterborne pathogens as well as particular interactions with other AOPs including heterogeneous solid (photo)-catalysts and electrocatalysts (chapter "Disinfection by Chemical Oxidation Methods"). Finally, the actual potential of AOPs treating more resistant to disinfection protozoan *Cryptosporidium* gets enclosed in the chapter "Inactivation of Cryptosporidium by Advanced Oxidation Processes". Last but not least, this interesting section finishes with the chapters "Impact on Disinfection Byproducts Using Advanced Oxidation Processes for Drinking Water Treatment" and "Evolution of Toxicity and Estrogenic Activity Throughout AOP's Surface and Drinking Water Treatment" dedicated to analyze the impact of AOPs on disinfection by-products as target final pollutants of special concern in the context of drinking water and the evolution of toxicity along the treatments, respectively. Finally, within a kind of miscellaneous topics, cost-effective materials for heterogeneous AOPs (chapter "Cost-Effective Catalytic Materials for AOP-Treatment Units"), regime change of heterogeneous Fenton-like processes toward continuous reactors (chapter "Wastewater Treatment by Heterogeneous Fenton-Like Processes in Continuous Reactors"), and application of chemometric methods, namely, statistical tools of process optimization, are also critically exposed in the volume.

We hope this work to encourage not only scientists but also decision makers at industrial and public health levels to embrace AOPs in the ultimate goal of practically applying such a helpful family of technological alternatives supplementing drinking water treatment plants in the short future.

Pamplona, Spain
Pasto, Colombia
Salamanca, Spain

Antonio Gil
Luis Alejandro Galeano
Miguel Ángel Vicente

Contents

Surface Water and Groundwater Sources for Drinking Water 1
Konstantina Katsanou and Hrissi K. Karapanagioti

Limitations of Conventional Drinking Water Technologies
in Pollutant Removal 21
Roberta Hofman-Caris and Jan Hofman

Natural Organic Matter: Characterization and Removal by AOPs
to Assist Drinking Water Facilities 53
S. Sorlini

Natural Organic Matter Removal by Heterogeneous Catalytic Wet
Peroxide Oxidation (CWPO) 69
José Herney Ramírez and Luis Alejandro Galeano

Separation and Characterization of NOM Intermediates Along AOP
Oxidation 99
Ana-María García, Ricardo A. Torres-Palma, Luis Alejandro Galeano,
Miguel Ángel Vicente, and Antonio Gil

Photo(Catalytic) Oxidation Processes for the Removal
of Natural Organic Matter and Contaminants of Emerging
Concern from Water 133
Monica Brienza, Can Burak Özkal, and Gianluca Li Puma

Homogeneous Fenton and Photo-Fenton Disinfection
of Surface and Groundwater 155
María Inmaculada Polo-López, Samira Nahim-Granados,
and Pilar Fernández-Ibáñez

AOPs Methods for the Removal of Taste and Odor Compounds 179
M. Antonopoulou and I. Konstantinou

Wastewater Treatment by Heterogeneous Fenton-Like Processes in Continuous Reactors 211
Bruno M. Esteves, Carmen S.D. Rodrigues, and Luis M. Madeira

Disinfection by Chemical Oxidation Methods 257
Luis-Alejandro Galeano, Milena Guerrero-Flórez, Claudia-Andrea Sánchez, Antonio Gil, and Miguel-Ángel Vicente

Inactivation of *Cryptosporidium* by Advanced Oxidation Processes 297
Abidelfatah M. Nasser

Cost-Effective Catalytic Materials for AOP Treatment Units 309
Shahryar Jafarinejad

Impact on Disinfection Byproducts Using Advanced Oxidation Processes for Drinking Water Treatment 345
Brooke K. Mayer and Donald R. Ryan

Evolution of Toxicity and Estrogenic Activity Throughout AOP's Surface and Drinking Water Treatment 387
Tatjana Tišler and Albin Pintar

Chemometric Methods for the Optimization of the Advanced Oxidation Processes for the Treatment of Drinking and Wastewater ... 405
Messias Borges Silva, Cristiano Eduardo Rodrigues Reis, Fabrício Maciel Gomes, Bruno dal Rovere Contesini, Ana Paula Barbosa Rodrigues de Freitas, Hélcio José Izário Filho, Leandro Valim de Freitas, and Carla Cristina Almeida Loures

Index 423

Surface Water and Groundwater Sources for Drinking Water

Konstantina Katsanou and Hrissi K. Karapanagioti

Abstract Raw water is groundwater, surface water, or rainwater that has not received any treatment in order to be suitable for drinking. Its quality must be good enough to produce when treated a safe and acceptable drinking water, and it must come from a source that can consistently provide sufficient required quantity.

Polluted or contaminated water sources can contain chemical or microbiological hazards which can lead to sickness and require treatment before consumption. In many cases it is better to protect the quality of the raw water providing sustainable management than to treat it after it has become deteriorated.

Keywords Aquifer, Artificial recharge, Municipal water supply, Protection, Quality deterioration, Water treatment

Contents

1 Introduction 2
2 Water Cycle and Climate Change 3
3 Drinking Water Sources 4
   3.1 Surface Water 4
   3.2 Groundwater 6
4 Deterioration Issues 8
   4.1 Surface Water 11
   4.2 Groundwater 11
5 Treatment 13
   5.1 Surface Water 13
   5.2 Groundwater 14

K. Katsanou
Laboratory of Hydrogeology, Department of Geology, University of Patras, Patras, Greece
e-mail: katsanou@upatras.gr

H.K. Karapanagioti (✉)
Department of Chemistry, University of Patras, Patras, Greece
e-mail: Karapanagioti@upatras.gr

6	Water Resource Management and Protection	15
	6.1 European Regulations-Legislation and Protection Measures	15
	6.2 Artificial Recharge	17
7	Discussion and Conclusions	17
References		17

1 Introduction

The saline water of the oceans comprises approximately 97% of the global water supply. The remaining 3% is freshwater, out of which nearly 69% is captured in glaciers, 30% is hosted in karst and porous aquifers, and the rest is in surface water reservoirs such as lakes, rivers, streams, and marshes.

Freshwater is the basic substance of life on earth and is increasingly in short supply. According to [1], 75% of the European and 33% of the global population use groundwater as their primary source of drinking water. Yet, in many countries, both the quantity and quality of this resource have been compromised by human activities. Nowadays, water scarcity affects 88 developing countries that are home to half of the world's population [2].

Surface water is any source of water that is open to the atmosphere and is subject to run off from the land. Hence, it is very likely to contain microorganisms that can cause sickness and in some cases more serious, even fatal, illnesses. In some areas, a substantial portion of the surface drinking water is derived from bank filtration that carries a diverse chemicals' and pathogens' load [3] and requires purification.

On the other hand, groundwater is covered by soils and sediments and is considered to be less vulnerable than surface water. Its abstraction though requires drilling and pumping equipment that is not always available or sustainable especially in developing countries.

As the population increases, the groundwater abstraction is expected to rise in the coming century, while available sites for surface reservoirs become limited.

The most important step in providing an area with safe drinking water is the selection of the best available source water. The more protected source waters are the easier and the cheaper to be transformed into safe drinking water [4].

The availability of freshwater varies both spatially and temporarily. The renewable fraction of the earth freshwater is usually found in the form of surface water and displays an uneven distribution. Groundwater is more evenly distributed, though much of it is nonrenewable, fossil water.

The water use in a particular region is determined not only by the natural groundwater availability but also by the population and the land use of an area, as well as economical factors.

In the developed countries, the municipality is obliged by law to supply the consumers with high-quality water, while in the developing countries, this is not always valid. Hence, the economy of a community defines attitudes and funding toward water development and treatment.

The local climate also plays a significant role due to influences on evaporation rates and practices such as lawn watering and cooling requirements. Additionally, cultural values, actions, policies, and laws of national governments also have an impact on water use. Finally, the issue of ownership of the resource which is linked to government influences can be an important factor.

Municipal water supply systems include facilities for storage, transmission, treatment, and distribution. The design of these facilities depends on the quality of the water, the particular needs of the user or consumer, and the quantities of water that must be processed. In certain cases, seawater can also be used as a drinking water source through the process of desalination.

The objective of this chapter is to present the drinking water sources, their characteristics, conventional treatment schemes, and protection with respect to sustainable use.

2 Water Cycle and Climate Change

The only naturally renewable source of freshwater globally is precipitation (about 110,000 km^3/year). Out of the precipitation occurring over land, a large fraction (70,000 km^3/year) moves back to the atmosphere through evaporation and transpiration from plants [5]. Infiltration rates vary depending on land use, the character, and the moisture content of the soil, as well as the intensity and duration of precipitation.

In case that the rate of precipitation exceeds the rate of infiltration, overland flow occurs. About 26% of this part of the cycle (18,000 km^3/year) can be used by humans.

Worldwide, the total annual runoff, including soil infiltration and groundwater replenishment, is estimated to be 41,000 km^3/year. Geographical remoteness and seasonal issues such as floods limit the total annual accessible runoff to 12,500 km^3/year [6].

Therefore, it is estimated that about 54% of the accessible runoff and 23% of the total renewable resource (precipitation over land) are currently appropriate for human use in some form.

Although globally groundwater is not very significant in volume, it is a critical source of water that can cover the human needs, because it is part of the limited budget of freshwater. From the perspective of long-term sustainability, it is the renewable resource that is most critical. However, it can also be viewed as a nonrenewable resource because the rate that may be withdrawn can be higher than the rate that it is replenished.

Climate change leads to atmosphere warming, which in turn alters the hydrologic cycle. This, depending on the area, results in changes to the amount, timing, form, and intensity of precipitation, the water flow in watersheds, as well as the quality of aquatic and marine environments.

Climate change can also change groundwater level and temperature and poses another risk to groundwater and water supply that is not yet well understood, though it needs attention to find solutions for adaptation.

Although several scenarios have been developed to predict the potential impact of climate change, there is still no general consensus on the quantitative effects. However, it is broadly accepted that many semi- or even arid areas will become drier, resulting in less groundwater recharge. In humid zones, recharge is expected to increase due to a higher number of extreme precipitation events. Efforts are being undertaken so as to better understand the climate change mechanisms on a regional and even local scale and hence to be able to determine mitigating measures.

3 Drinking Water Sources

The freshwater drinking sources are summarized in Fig. 1. Although water is a single resource, to understand the quality and the accessibility of the different formations, each geological water occurrence will be discussed separately. Thus, surface water will be discussed as lake water and river water and groundwater as springwater and well water.

3.1 Surface Water

Surface water is accumulated on the ground or in a stream, river, lake, reservoir, or ocean. The total land area that contributes surface runoff to a lake or river is called catchment area (Fig. 2). The volume of water depends mostly on the amount of rainfall but also on the size of the watershed, the slope of the ground, the soil type and vegetation, and the land use. Any changes in the water level of a lake are controlled by the difference between the input and output compared to the total volume of the lake.

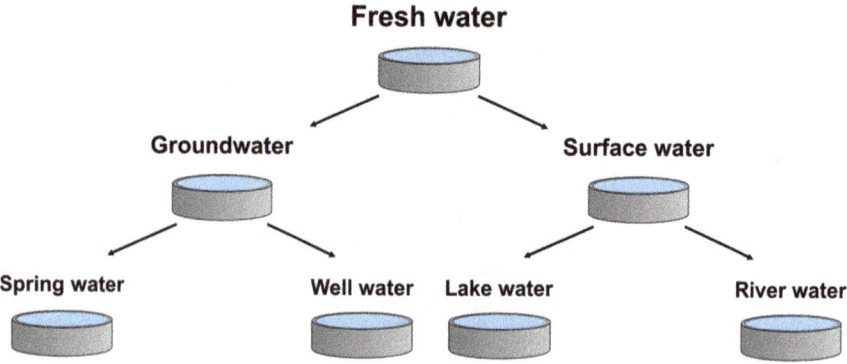

Fig. 1 Freshwater drinking sources

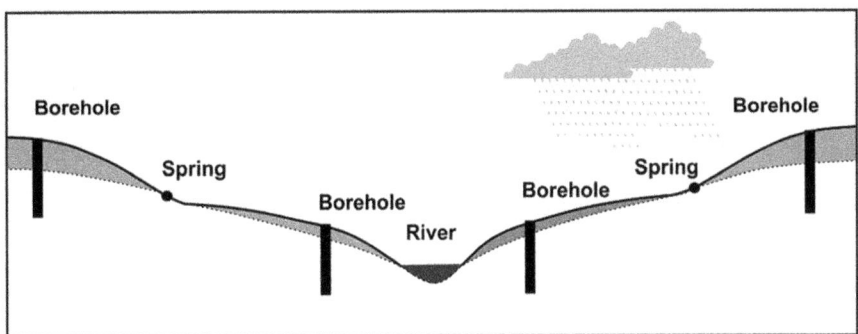

Fig. 2 The occurrence of rivers, springs, and boreholes in a catchment area

Surface water is often used for large urban water supply systems, as rivers and lakes can supply a large, regular volume of water. For small community supplies, other forms of water supply, such as wells or spring-fed gravity systems, are generally preferred to surface water. This is because the cost of treatment and delivery of surface water is likely to be high and operation and maintenance less reliable.

The advantages of the use of surface water as a resource for domestic water supply are many. Surface water, among others, is easy to be abstracted by direct pumping and can be treated after use and put back into a river. However, surface water is seasonal and will always need treatment.

3.1.1 Lakes

Lakes have numerous features, such as catchment area, inflow and outflow, nutrient content, dissolved oxygen, pollutants, pH, sedimentation, type, etc. The most significant inputs are precipitation onto the lake surface, the runoff carried by streams from the lake catchment area, the aquifers, and artificial sources from outside the catchment area. The most significant outputs are evaporation from the lake surface, surface water and groundwater flows, and any water extraction for human activities. The water level of a lake displays fluctuations related to climate and water abstraction variations.

The majority of lakes globally contain freshwater. Their water composition varies depending on many factors. Freshwater lakes are important natural resources that, among others, also serve as drinking water sources. However, they face various water-quality deterioration problems due to the impact of human activities, i.e., pollution, agriculture and fishery activities, and climate change [7, 8].

Based on their type, lakes can be classified into tectonic, landslide, saline, volcanic, glacial, and others such as karst lakes, while based on their nutrient content, they can be classified into oligotrophic, mesotrophic, eutrophic, and hypertrophic.

Oligotrophic lakes are characterized by clear waters that display low concentrations of plant life, while hypertrophic ones are characterized by waters excessively enriched in nutrients, poor clarity, and algal blooms due to an over-enrichment of nutrients. Such lakes are of little use to humans and have a poor ecosystem due to decreased dissolved oxygen. Variations in nutrient enrichment are influenced by environmental changes and anthropogenic activities.

3.1.2 Rivers and Streams

Rivers are part of the hydrological cycle. Their flow is a function of many factors including precipitation, runoff, interflow, groundwater flow, and pumped inflow and outflow. The discharge of a river varies seasonally and among the years. In many countries, rivers and streams have a wide seasonal variation in flow which affects their water quality.

The chemical composition of the river water is complex and depends on the inputs from the atmosphere, the geology of the catchment area, and the human activities. It has a large effect on the ecology, and it also affects the uses of the river water. In order to determine the river water chemistry, a well-designed sampling and analysis is required.

During dry seasons and drought periods, springs feeding small watercourses in river headwater areas are the main source of downstream waters [9]. When the average discharge of a river cannot serve the water supply throughout the year, a dam is built to block the flow of water. Gradually with time, an artificial lake is formed. The construction of a dam for its abstraction is expensive and environmentally damaging that may trigger earthquakes. Moreover it requires sufficient precipitation and large river catchment while such reservoirs will eventually silt up.

3.2 *Groundwater*

Groundwater is a major source of drinking water worldwide and is hosted in aquifers. Hydrological recharge of aquifers hugely varies geographically and strongly depends, among other factors, on climate, geology, soil type, vegetation, and land use [10]. Groundwater is recharged from precipitation, which is complemented by natural infiltration by surface water or by artificial recharge. On a global scale, 20% of the irrigation water and 40% of the water used in industry are derived from groundwater [11].

Groundwater within an unconsolidated rock moves only a few centimeters a day, i.e., about 10 m/year. Its velocity largely depends on the steepness of the aquifer slopes and the permeability of the rocks. In consolidated rocks the water velocity can be many times higher. A typical example is the karst formations.

Groundwater occurs in aquifers under two different conditions, the unconfined and the confined. In an unconfined aquifer, the water only fills the aquifer partly, and its upper surface of the saturated zone may rise and decline. On the other hand in a confined aquifer, the water completely fills the aquifer that is overlain by a confining bed. The recharge of the saturated zone occurs by percolation of water from the land surface through the unsaturated zone.

The advantages of the use of groundwater as a resource for domestic water supply are many. In most inhabited parts of the world, there is a large amount of groundwater, and despite that the abstracted volumes are huge, they are often readily supplemented. Another advantage is that the upper soil layers act as a filter against physical, chemical, and biological deterioration which is effective both in terms of quality and cost. Finally, groundwater use often brings great economic benefits per unit volume compared to surface water because of ready local availability, high drought reliability, and a generally good quality requiring only minimal treatment [12].

The movement of pollutants in an aquifer is defined by the hydraulic characteristics of the hosting soils and rocks. Substances dissolved in water move along with it unless they are tied up or delayed by adsorption. Thus, the movement of pollutants tends to be through the most permeable zones; the farther their point of origin from a groundwater discharge area, the deeper they penetrate into the groundwater system and the larger the area that is ultimately affected [13].

Groundwater exploration utilizes hydrogeological mapping, hydrogeophysical prospecting, investigation drilling, pumping tests, and groundwater regime observation, but also remote sensing, isotopic studies, shallow seismic prospecting, and velocity logging are also being applied in groundwater exploration [14].

3.2.1 Springs

The physical outlets of an aquifer are known as springs. Springs occur mainly in mountainous or hilly terrains, in locations where the water table meets the ground surface (Fig. 2).

In the past, the community water supplies were often based on springs, and they still comprise a source of water, because springwater usually displays a high natural quality and its intake is relatively easy (Fig. 3). Springs can be classified as artesian, gravity, perennial, intermittent, tubular, seepage, and thermal springs.

3.2.2 Boreholes and Wells

Groundwater can be obtained from an aquifer by drilling a well or a borehole below the water table. The water level in boreholes or wells drilled into unconfined aquifers stands in the same position as the water table in the surrounding aquifer.

The water level in boreholes drilled into confined aquifers stands at some height above the top of the aquifer but not necessarily above the land surface. If the water level in a borehole stands above the land surface, the well displays artesian flow, where water rises to the surface without pumping (Fig. 4). Pumping water out of an aquifer lowers the water level near the well.

Fig. 3 Intake of water from a spring in a remote mountainous area in Greece

The "safe yield" of an aquifer is the volume of water that can be withdrawn without depleting the aquifer, i.e., the water that is renewable. In case that a higher volume is withdrawn, a number of undesirable effects can occur (see Sect. 4). A groundwater source of drinking water, such as a borehole, can be used as safe water supply after no or very little treatment. Even a dug well that pumps shallow water and is subject to weather can be treated effectively with relatively simple equipment.

4 Deterioration Issues

As already mentioned, freshwater ecosystems are important natural resources for the survival of living organisms of the biosphere and are under pressure from a range of stressors, including changes in land use, pollution, agriculture, climate change, and human activities [15, 16]. In many regions, the increasing economic development and human migration also stress water resources and water quality.

More than half of the world's population live adjacent to water bodies and carry out activities that increase aquatic stressors such as anthropogenic eutrophication and algal blooms [17]. The problems most frequently encountered in the operation of supply wells are related either to declines in yield or to deterioration in the quality of water (Fig. 5).

Fig. 4 A borehole with artesian flow

Groundwater overexploitation has caused large and continuous decline of groundwater levels, which in many cases results in the arrival of water from different directions to a well or borehole that may contain a large concentration of any substance.

In urban areas, groundwater overexploitation leads to land subsidence. In coastal areas, the intensive groundwater use is not simply limited to subsidence. It can also lead to the deterioration of groundwater quality. When freshwater heads are declined by withdrawals through boreholes, the freshwater-saltwater interface shifts until a new balance is established. Seawater can migrate kilometers toward the inland [18]. Since seawater intrusion takes a long time to remedy and often is irreversible, immediate actions should be taken for its prevention or mitigation.

Intensive overexploitation of groundwater globally has already created serious environmental problems. Therefore, the analysis of the influence of exploited groundwater on the environment and the allowable withdrawal of a well field should be determined beforehand.

In aquifers, water moves slower than in surface water systems, and the time for contaminant infiltration depends on the type of the contaminant, the characteristics of the aquifer, and the length of the flow path. This can last from a number of years up to centuries. As a consequence, there is a great deal of time for intimate

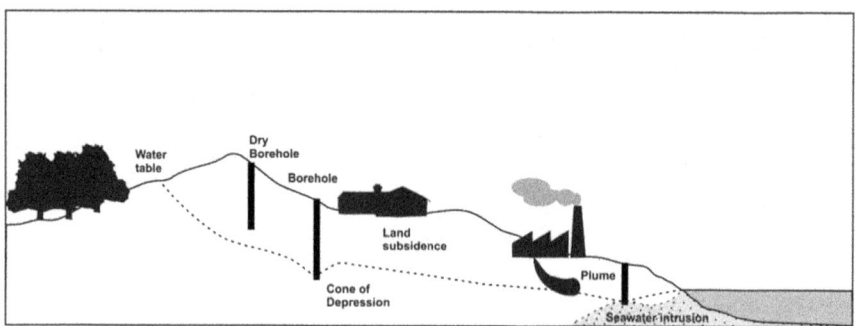

Fig. 5 Deterioration issues

contact between the water and soil and rock material, which results in changes of the chemical composition of water that may involve almost any substance soluble in water.

The arrival of water containing dissolved chemicals, which are naturally occurring or human-introduced ones, in an undesirably large concentration will deteriorate the water quality of a borehole. The most commonly observed increases in concentration involve NaCl and NO_3^-. Additionally, the growing number of chemicals that is applied in manufacturing processes increases water-related risks.

Hazardous substances include herbicides, pesticides, and inorganic and complex organic substances. Deterioration in physical quality has to do with changes in appearance, taste, and temperature of water that may originate from rock particles of variable sizes.

Deterioration in biological quality is related to the appearance of bacteria and/or viruses associated with human, animal, and food processing wastes. It normally indicates a connection between the land and the water surface.

Each water source has a unique set of contaminants. Groundwater contains pesticides, chemicals, and nitrates, while surface water mostly contains bacteria and other microorganisms as well as other suspended particles. Groundwater and surface water recharge each other, and this interconnection between the two sources of drinking water may lead to exchange of contaminants.

Human health can be harmed in case that contaminants and pathogens end up in drinking water. The World Health Organization has estimated that contaminated drinking water causes 502,000 diarrheal deaths each year [19]. The deterioration in water quality does not only impact human health but also has economic consequences [20].

Groundwater from deep aquifers is protected from pathogen contamination by the covering soil layers. Although groundwater is better protected than surface water, shallower groundwater sources, or groundwater that can be influenced by surface water, are still vulnerable to fecal contamination [21].

4.1 Surface Water

4.1.1 Lakes

Eutrophication has become a worldwide environmental problem [22, 23]. Recently, studies have shown that lake eutrophication could accelerate in response to the effects of anthropogenic activities and climate change [24]. Water in lakes, bogs, and swamps may gain color, taste, and odor from decaying vegetation and other natural organic matters.

It can also be polluted by (a) runoff of soils and farm chemicals from agricultural lands, (b) waste from cities, (c) discharges from industrial areas, (d) leachate from disposal sites, and (e) direct atmospheric pollutants such as rain, snow, or dust.

The interaction between lakes and groundwater systems has been studied with numerical models and on the field [25], with attention paid to the place of a lake within a local, intermediate, or regional groundwater system.

4.1.2 Rivers and Streams

Chemical composition of rivers in higher topographical areas where the water volume is smaller is affected by fewer rock types and can reflect the local geology, and in particular cases, their quality can be poor. In the absence of any pollution source, the river water quality is similar to the rain and springwater.

However, as a river flows to lower topographic areas, its water accumulates fine soil particles, microbes, organic material, and soluble minerals. In populated areas, the quality of surface water as well as groundwater is directly influenced by land use and human activities. According to Environmental Protection Agency (EPA) [26], the most common source of river pollution is agriculture.

In wet periods, especially at their beginning, there is a high risk of fecal contamination, as feces are washed into the river. However, along with the increase in flow, there is a higher dilution rate, and thus, wastewater will be diluted, so the health risk decreases [4]. In dry periods, the silt load is often lower than in wet ones, but the dissolved solids are much more concentrated.

4.2 Groundwater

Nowadays, groundwater resources are facing qualitative and quantitative problems. In many cases, the groundwater abstraction rates for many large aquifers worldwide significantly exceed the natural renewal rate [27–29].

Growing industrialization, waste deposition, and the exponentially increasing production and use of synthetic chemicals, which are often released into the environment, put groundwater resources under growing pressure [30].

The chemical composition of groundwater also reflects the geology of the area that has been in contact. If the groundwater is in contact with limestone, then, the water is expected to be hard and to contain high concentrations of calcium and carbonates. In general though, the water quality does not differentiate significantly enough to be a concern to humans or any other living organisms and ecosystems.

The human activities are mainly responsible for groundwater quality degradation. Groundwater pollution most often results from overuse of fertilizers and pesticides, improper disposal of wastes on land, industrial and household chemicals, garbage landfills, wastewater from mines, oil field brine pits, leaking of oil storage tanks, sewage sludge, septic systems, etc.

Groundwater can be also contaminated by overpumping in coastal regions that triggers seawater intrusion, a widespread and serious problem in many parts of the world, especially in the Mediterranean region.

There are cases though where pollution sources are absent, while groundwater does not lie within the potable regulations. These waters are mostly found in areas with recent volcanic activity, where groundwater displays higher temperatures than the groundwater of the surrounding watershed area and higher residence time and salt content than the freshwaters. Moreover, in many coastal areas worldwide, particularly in karst areas, natural seawater intrusion may exist.

4.2.1 Springs

The most vulnerable water supplies are those hosted in unconfined and karst aquifers, where pollutants can readily diffuse into groundwater supplies. In case that springs are used for water supply, special attention should be paid to any variation of their water temperature during the day and coloration of their water shortly after rain, which indicate that the aquifer layer is not deep enough or that there is direct infiltration of surface water through the topsoil, and thus, the spring is extremely vulnerable to contamination.

4.2.2 Boreholes and Wells

Deterioration in water quality may result either from changes in the biological, chemical, or physical quality of water in the aquifer or changes in the well. The monitoring of the biological and chemical quality of water of wells that supply domestic needs is important, in order to spot any differentiations in their water quality. Especially before using a new well for public supply, it should be clarified first that its water quality meets the potability standards. Otherwise treatment is required.

5 Treatment

In all cases, drinking water should be free of microbes or be disinfected before it is consumed. In addition, surface water should be treated for the removal of suspended solids and algae. Groundwater may not need additional treatment if it is springwater. Springwaters often do not even need disinfection. However, depending on the formations groundwater passes through, it may need softening or iron and manganese removal or removal of other substances such as arsenic or chromium. The following sections describe the conventional treatment methods that are mostly described in [31]. Recent advances in disinfection are critically reviewed and presented in [32].

5.1 Surface Water

If drinking water originates from surface water, it requires treatment that needs substantial infrastructure, i.e., water treatment plants.

5.1.1 Lake Conventional Treatment

The most common treatment for lake or dam water is a combination of coagulation, flocculation, sedimentation, sand filtration, and disinfection. Coagulation, flocculation, and sedimentation are complimentary processes needed for the removal of stable colloid particles that are light enough to be suspended and will take a long time to settle due to gravity only. If these suspended solids are microalgae, more sophisticated treatment is necessary, and expensive membranes are employed. Coagulation is a chemical process that requires the addition of coagulants which are usually salts in the form of powders. They are used to neutralize the surface of colloids, in other words to destabilize their solution and allow their flocculation. Colloids repulse each other due to common surface charge since they usually originate from the same geological formation. Once their surface charge is neutralized, then, they flocculate due to gravity forces forming bigger particles that settle much faster by gravity. Sand filtration usually follows the sedimentation tank and is used to remove the finer flocs that did not settle or the destabilized particles if no sedimentation tank is present.

Disinfection is used to deactivate microorganisms in water and keep the water supply safe throughout the pipelines of the municipality. In some countries for surface water, sand filtration is mandatory before disinfection since attached microbes to suspended particles are more persistent than suspended microbes. Thus, disinfection would not be as effective in the presence of suspended particles.

Some industries use pressure filters, ion exchange, and ultraviolet radiation to remove solids and alkalinity and to disinfect. This way they avoid the addition of any chemicals that could create a taste and odor problem. However, these processes are more expensive and more energy intensive.

5.1.2 River and Stream Conventional Treatment

River water follows similar treatment schemes as lake water for the removal of suspended solids and for disinfection. Although river water is most commonly pumped from the river sources or from the river banks, it still requires treatment for suspended solids.

5.2 Groundwater

If drinking water originates from groundwater, it is usually treated on site of the point it is collected, e.g., on the well.

5.2.1 Spring Conventional Treatment

If the spring is protected, springwater can be of high quality and may not require any treatment, not even disinfection. However, in order to be safe, usually municipalities disinfect springwater in order to keep it safe during transportation through the pipelines of the distribution system.

5.2.2 Borehole and Well Conventional Treatment

For groundwater originating from boreholes or wells, the most common treatment scheme is a combination of softening, aeration, and disinfection.

Softening is a chemical process used for the removal of hardness that is due to calcium and magnesium carbonate minerals dissolved in groundwater. Hardness is associated with scaling in the pipelines or in machines that heat water.

Aeration is also a chemical process used for oxidation and, thus, formation of less soluble species and precipitation of iron and manganese. It is also possible to use an oxidation medium such as potassium permanganate or chlorine to help metal precipitation.

Disinfection, also a chemical process, is used to deactivate microorganisms in water and keep the water supply safe throughout the pipelines of the municipality. The most common disinfection method is chlorination that requires the mixing of water with chlorine gas or with sodium hypochlorite. Chlorine is hydrolyzed and hypochlorite acid is formed. Chlorine is effective for the disinfection from bacteria and is also persistent in the supply pipeline without requiring the addition of another disinfectant. Ozone and UV are also used and are effective disinfectants. However, they require the addition of chlorine in water before it is transported through pipelines. The use of chlorine as a disinfectant could be a problem since chlorine has strong taste and odor.

Other methods to get water free from salts are distillation, ion exchange, and membrane filtration. Conventional distillation requires high amounts of power and wasting lots of water for cooling. Solar distillation is friendlier to the environment but requires space and sun shining. Ion exchange requires the use of specific resins that remove both anions and cations. They still require water softening to avoid fast consumption of the resins by hardness and frequent regeneration of resins. Membrane filtration requires high pressure to push water through the membrane.

If water is polluted, it may require additional treatment before it can be consumed. Toxic metals present in groundwater due to pollution or to natural minerals can be removed by sorbent materials that will strongly bind them. However, in many countries, it is considered preferable to change water supply than try to treat a polluted one. In places where water supplies are scarce, in-house water filters are commonly used employing ceramic pots and sorbent filtering media (e.g., activated carbon, zeolites, biosorbents, bone chars, etc.).

6 Water Resource Management and Protection

Although water sources, mostly the surface ones, are treated before entering the households, their treatment costs and the risks posed to public health could be reduced by investing on their protection.

In Europe, 20% of surface water is at serious risk from pollution; 60% of the cities overexploit their groundwater resources, while 50% of wetlands are endangered [33]. There is a constantly increased demand for water that has as consequence that nearly half of the European population live in "water-stressed" countries, where abstraction of freshwater is extremely high.

The trends of deterioration of water quality and decrease of urban groundwater supplies are being corrected by water allocation projects and protection measures. For urban water pollution control, urban sewage and industrial wastewater will be treated and reused increasingly.

6.1 European Regulations-Legislation and Protection Measures

In 1998, the Council Directive (98/83/EC) was introduced and included essential potability standards [34]. This Directive requires member states to monitor the quality of water intended for human consumption regularly, by using a "sampling points" method, providing regular information to consumers, and reporting the results every 3 years to the Commission.

In 2000, the European Union adopted the Water Framework Directive (WFD, 2000/60/EC). According to it, member states have had to build river basin management plans (RBMPs), which were finalized after extensive public consultation. They are aiming to protect each 1 of the 110 river basin districts and are valid for 6 years. The overall targets of the RBMPs are to accelerate water-saving techniques to alleviate the crisis of urban water shortage, to protect groundwater resources in the overexploited areas, to improve the coordination between socioeconomic development and environmental improvement, and to seek sustainable utilization of groundwater resources.

According to the WFD, protection zones have been established for both surface water and groundwater resources in important water supply regions. They comprise of three different zones:

- An "immediate" proximity protection area: This area is generally small and encloses the well itself. It belongs to the municipality or the utility and is fenced. Its purpose is to protect the well itself from direct pollution.
- A proximity area: Its purpose is to protect the well, allowing a sufficient reaction time in case of pollution from nearby point source pollution. The criterion of 50-day water transit is generally applied to define the limits of this area. Its size varies according to the hydrogeological characteristics of the area and the volume of abstraction. In this area there are limitations in the use of the soil, such as banning of manure, pesticides, road construction, etc., which are derived by hydrogeological study of the area.
- A distant area: Its purpose is to protect the catchment area of a water source. In this area, whose extent may vary from place to place, the use of soil is only subject to recommendations.

In 2006 a new directive aiming to protect groundwater from contamination (GWD, 2006/118/EC) was published. It stems from the European Water Framework Directive of 2000 [35] and sets Community Quality Standards for groundwater with respect to the major contaminants, nitrates and pesticides. It also creates the framework for setting any other Groundwater Quality Standards.

The Groundwater Directive [36] is based on a mixed regime that is in compliance with Groundwater Quality Standards and measures to prevent hazardous or limit nonhazardous substance emissions to the groundwater.

According to it, the member states are responsible for the establishment of certain Groundwater Quality Standards based on the local or regional conditions [37].

Moreover, in 2007 the Commission launched the Water Information System for Europe (WISE), an instrument for the collection and exchange of data and information within Europe and for the monitoring of pollutants released to surface waters or within the aquatic environment [38].

6.2 Artificial Recharge

The overexploitation of groundwater is controlled by increased infiltration and extensive use of artificial recharge. Artificial recharge can be used to increase the availability of groundwater storage and reduce saltwater intrusion in coastal aquifers, where pumping and droughts have severely affected the groundwater quality. Stored water from artificial recharge can be used not only for drinking but also for industrial and agricultural water supplies.

In coastal areas, it can protect groundwater from saltwater intrusion.

In certain areas, artificial recharge can be used to provide protection from land subsidence caused by drought and aquifer overexploitation [39].

7 Discussion and Conclusions

Groundwater plays an important role for urban water supplies. Nowadays that there is recognition of the vertical dimension of hydrologic connectivity [40] and consent that many surface water ecosystems depend on groundwater at different levels, conservation and management of this linked resource are even more important [41].

For the efficient and sustainable exploitation, management, and remediation of groundwater resources, it is important to know the water and contaminant flow patterns and the way they interact with the geological formations.

Assessment and monitoring of water resources and their quality are generally a routine in developed countries. However, this is not always the case in developing or in remote areas where vulnerable communities are hugely affected by poor water quality [42]. Areas that suffer from limited availability or accessibility of water resources may also be at greater water risk [43] especially if limited water resources are undermined by natural and anthropogenic contamination, climate change, or other factors as poverty, remoteness, insufficient water management, and lack of treatment [44–46].

Certain measures for groundwater protection such as reduction in water consumption, enhancement of artificial recharge use, pollution control, and holistic management of groundwater resources are already taken [47].

References

1. Sampat P (2000) Deep trouble-the hidden threat of groundwater pollution. World Watch Paper 154. Worldwatch Institute, Washington, DC
2. Howard KWF, Gelo KK (2002) Intensive groundwater use in urban areas: the case of megacities. In: Ramon Llamas M, Custodio E (eds) Intensive use of groundwater: challenges and opportunities. A.A. Balkema Publishers, CRC Press, Rotterdam

3. Tufenkji N, Ryan JN, Elimelech M (2002) The promise of bank filtration. A simple technology may inexpensively clean up poor-quality raw surface water. Environ Sci Technol 36:422A–428A
4. Medema GJ, Shaw S, Waite M, Snozzi M, Morreau A, Grabow W (2003) Catchment characterization and source water quality. In: Assessing microbial safety of drinking water; improving approaches and methods. Chapter 4, World Health Organization, London
5. AbuZeid KM (2012) Mediterranean water outlook: perspective on policies and water management in Arab countries. In: Choukr-Allah R, Ragab R, Rodriguez-Clemente R (eds) Integrated water resources management in the Mediterranean region: dialogue towards new strategy. Springer Science & Business Media, Dordrecht
6. Postel SL, Daily GC, Ehrlich PR (1996) Human appropriation of renewable fresh water. Science 271(5250):785–788
7. Mushtaq F, Nee Lala MG (2017) Remote estimation of water quality parameters of Himalayan lake (Kashmir) using Landsat 8 OLI imagery. Geocarto Int 32(3):274–285
8. Zhou Y, Zhang Y, Shi K, Liu X, Niu C (2015) Dynamics of chromophoric dissolved organic matter influenced by hydrological conditions in a large, shallow, and eutrophic lake in China. Environ Sci Pollut R 22:12992–13003
9. Di Matteo L, Dragoni W, Maccari D, Piacentini SM (2017) Climate change, water supply and environmental problems of headwaters: the paradigmatic case of the Tiber, Savio and Marecchia rivers (Central Italy). Sci Total Environ 598:733–748
10. Scanlon B, Healy R, Cook P (2002) Choosing appropriate techniques for quantifying groundwater recharge. Hydrgeol J 10:18–39
11. Millennium Assessment (2005) Millennium ecosystem assessment. Ecosystems and human well-being: wetlands and water synthesis. World Resources Institute, Washington, DC
12. Burke JJ, Moench M (2000) Groundwater and society: resources, tensions and opportunities. United Nations, New York
13. U.S. Environmental Protection Agency (1993) A review of methods for assessing aquifer sensitivity and ground water vulnerability to pesticide contamination. U.S. EPA/813/R-93/002, EPA, Washington, DC
14. Changping Y, Tingfang S, Liangyu J, Tingzuo W, Shaodu L (1993) Exploration and evaluation of groundwater well field. Geological Publishing House, Beijing
15. Rashid I, Romshoo SA (2012) Impact of anthropogenic activities on water quality of Lidder River in Kashmir Himalayas. Environ Monit Assess 185(6):4705–4719
16. Mushtaq F, Pandey AC (2014) Assessment of land use/land cover dynamics vis-à-vis hydrometeorological variability in Wular Lake environs Kashmir Valley, India using multitemporal satellite data. Arab J Geosci 7:4707–4715
17. Torbick N, Hession S, Hagen S, Wiangwang N, Becker B, Qi J (2013) Mapping inland lake water quality across the Lower Peninsula of Michigan using Landsat TM imagery. Int J Remote Sens 34:7607–7624
18. Bear J (1972) Dynamics of fluids in porous media. American Elsevier, New York. https://pdfs.semanticscholar.org/0000/f028e1fe064dd1ae4bd3f9bb19fcbe2f7adf.pdf
19. World Health Organization (2016) Drinking-water – fact sheet. Geneve. http://www.who.int/mediacentre/factsheets/fs391/en/. Accessed 13 Jan 2017
20. Juntunen J, Meriläinen P, Simola A (2017) Public health and economic risk assessment of waterborne contaminants and pathogens in Finland. Sci Total Environ 599–600:873–882
21. Bouchoucha M, Piquet JC, Chavanon F, Dufresne C, Le Guyader FS (2016) Faecal contamination of echinoderms: first report of heavy Escherichia coli loading of sea urchins from a natural growing area. Lett Appl Microbiol 62:105–110
22. Huang J, Zhan J, Yan H, Wu F, Deng X (2013) Evaluation of the impacts of land use on water quality: a case study in the Chaohu Lake Basin. Sci World J 2013:329187
23. Wang L, Liang T (2015) Distribution characteristics of phosphorus in the sediments and overlying water of Poyang Lake. PLoS One 10(5):e0125859

24. Li Y, Zhang Y, Shi K, Zhu G, Zhou Y, Zhang Y, Guo Y (2017) Monitoring spatiotemporal variations in nutrients in a large drinking water reservoir and their relationships with hydrological and meteorological conditions based on Landsat 8 imagery. Sci Total Environ 599–600:1705–1717
25. Roningen JM, Burbey TJ (2012) Hydrogeologic controls on lake level: a case study at Mountain Lake, Virginia, USA. Hydrgeol J 20:1149–1167
26. EPA (2015) Water Quality Assessment and TMDL Information. https://ofmpub.epa.gov/waters10/attains_nation_cy.control
27. Danielopol DL, Greber C, Gunatilaka A, Notenboom J (2003) Present state and future prospects for groundwater ecosystems. Environ Conserv 30(2):104–130
28. Gleeson T, Wada Y, Bierkens MFP, van Beek LPH (2012) Water balance of global aquifers revealed by groundwater footprint. Nature 488:197–200
29. Griebler C, Avramov M (2015) Groundwater ecosystem services: a review. Freshw Sci 34 (1):355–367
30. Zaisheng H (1998) Groundwater for urban water supplies in northern China – an overview. Hydrgeol J 6:416–420
31. American Water Works Association (AWWA) (2011) Water quality and treatment: a handbook on drinking water, 6th edn. McGraw Hill, New York
32. Ngwenya N, Ncube EJ, Parsons J (2013) Recent advances in drinking water disinfection: successes and challenges. In: Whitacre DM (ed) Reviews of environmental contamination and toxicology, vol 222. Springer Science+Business Media, New York, pp 111–170
33. http://ec.europa.eu/environment/pubs/pdf/factsheets/water-framework-directive.pdf
34. EU Council Directive 98/83/EC of 3 November 1998 on the quality of water intended for human consumption. http://eur-lex.europa.eu/LexUriServ/LexUriServ.do?uri=OJ:L:1998:330:0032:0054:EN:PDF
35. Directive 2000/60/EC of the European Parliament and of the Council of 23 October 2000 establishing a framework for Community action in the field of water policy. http://eur-lex.europa.eu/resource.html?uri=cellar:5c835afb-2ec6-4577-bdf8-756d3d694eeb.0004.02/DOC_1&format=PDF
36. Directive 2006/118/EC of 12 December 2006 on the protection of groundwater against pollution and deterioration. http://eur-lex.europa.eu/LexUriServ/LexUriServ.do?uri=OJ:L:2006:372:0019:0031:EN:PDF
37. Quevauviller P (2006) Agreement on new EU groundwater directive. J Soil Sediment 6(4):254
38. WISE (2007) http://water.europa.eu/
39. Masciopinto C, Vurro M, Palmisano VN, Liso IS (2017) A suitable tool for sustainable groundwater management. Water Resour Manag 31(13):4133–4147
40. Pringle C (2003) What is hydrologic connectivity and why is it ecologically important? Hydrol Process 17(13):2685–2689
41. Boulton AJ (2005) Editorial: chances and challenges in the conservation of groundwaters and their dependent ecosystems. Aquat Conserv 15:319–323
42. UNEP (2010) Africa Water Atlas. Division of Early Warning and Assessment (DEWA), United Nations Environment Programme (UNEP), Nairobi, Kenya
43. World Health Organization (2011) Noncommunicable diseases country profiles. Geneve. http://apps.who.int/iris/bitstream/10665/44704/1/9789241502283_eng.pdf
44. Cai WJ, Wang Y, Krest J, Moore WS (2003) The geochemistry of dissolved inorganic carbon in a surficial groundwater aquifer in North Inlet, South Carolina, and the carbon fluxes to the coastal ocean. Geochim Cosmochim Acta 67:631–639
45. Garrett V, Ogutu P, Mabonga P, Ombeki S, Mwaki A, Aluoch G, Phelan M, Quick RE (2008) Diarrhoea prevention in a high-risk rural Kenyan population through point-of-use chlorination, safe water storage, sanitation, and rainwater harvesting. Epidemiol Infect 136(11):1463–1471
46. Hanjra MA, Ferede T, Gutta DG (2009) Reducing poverty in sub-Saharan Africa through investments in water and other priorities. Agr Water Manage 96(7):1062–1070
47. Ferrier RC, Jenkins A (2010) The catchment management concept. Handbook of catchment management. Wiley-Blackwell

Limitations of Conventional Drinking Water Technologies in Pollutant Removal

Roberta Hofman-Caris and Jan Hofman

Abstract This chapter gives an overview of the more traditional drinking water treatment from ground and surface waters. Water is treated to meet the objectives of drinking water quality and standards. Water treatment and water quality are therefore closely connected.

The objectives for water treatment are to prevent acute diseases by exposure to pathogens, to prevent long-term adverse health effects by exposure to chemicals and micropollutants, and finally to create a drinking water that is palatable and is conditioned in such a way that transport from the treatment works to the customer will not lead to quality deterioration.

Traditional treatment technologies as described in this chapter are mainly designed to remove macro parameters such as suspended solids, natural organic matter, dissolved iron and manganese, etc. The technologies have however only limited performance for removal of micropollutants. Advancing analytical technologies and increased and changing use of compounds however show strong evidence of new and emerging threats to drinking water quality. Therefore, more advanced treatment technologies are required.

Keywords Conventional water treatment, Disinfection, Groundwater, Micropollutants, Surface water

R. Hofman-Caris (✉)
KWR Watercycle Research Institute, PO Box 1072, 3430 BB Nieuwegein, The Netherlands
e-mail: Roberta.Hofman@kwrwater.nl

J. Hofman
Water Innovation and Research Centre, Department of Chemical Engineering, University of Bath, Claverton Down, Bath BA2 7AY, UK
e-mail: J.A.H.Hofman@bath.ac.uk

Contents

1 Safe Drinking Water Quality ... 23
  1.1 General Aspects ... 23
  1.2 Microbiological Standards ... 25
  1.3 Chemical Water Quality ... 25
  1.4 Water Quality: Presence and Characterization of Natural Organic Matter ... 28
  1.5 Water Quality: Conditioning ... 30
2 Water Treatment: Drinking Water from Groundwater Sources ... 31
3 Water Treatment: Drinking Water from Surface Water Sources ... 34
  3.1 Coagulation, Floc Formation and Floc Removal ... 35
  3.2 Application of Activated Carbon ... 44
  3.3 Disinfection ... 46
4 Water Treatment: Drinking Water from River Bank Filtrate or Dune Filtrate ... 47
5 Conclusions ... 48
References ... 48

Abbreviations

°D	German degree
AC	Activated carbon
AOC	Assimilable organic carbon
BB	Building block
BP	Biopolymer
DAF	Dissolved air flotation
DOC	Dissolved organic carbon
E. coli	*Escherichia coli*
E2	17-Beta-estradiol
EC	European Commission
EDC	Endocrine-disrupting compound
EE2	17-Alpha-ethinylestradiol
FEEM	Fluorescence excitation emission matrix
GAC	Granular activated carbon
HS	Humic substance
LC-OCD	Liquid chromatography – organic carbon detection
LMw	Low molecular weight
LP	Low pressure (UV lamp, 253.7 nm)
LRV	Logarithmic reduction values
LSI	Langelier saturation index
MP	Medium pressure (UV lamp, 200–300 nm)
Mw	Molecular weight
NOM	Natural organic matter
PAC	Powdered activated carbon
PACl	Polyaluminium chloride
PCCPP	Practical calcium carbonate precipitation potential
PRAM	Polarity rapid assessment method

QMRA	Quantitative microbial risk assessment
REACH	Registration, evaluation, and authorization chemicals
SAX	Strong anion exchanger
SDWA	Safe Drinking Water Act
SEC	Size exclusion chromatography
SI	Saturation index
SMP	Soluble microbiological product
SPE	Solid phase extraction
SUVA	Specific UV absorbance
TCCPP	Theoretical calcium carbonate precipitation potential
TH	Total hardness
TOC	Total organic carbon
TTC	Threshold of toxicological concern
UV	Ultraviolet
UV-A	315–380 nm
UV-B	280–315 nm
UV-C	200–280 nm
VUV	Vacuum ultraviolet (100–200 nm)
WHO	World Health Organization
WWTP	Wastewater treatment plant

1 Safe Drinking Water Quality

1.1 General Aspects

The availability of safe drinking water is of utmost importance for human health, everywhere in the world. For centuries people have been aware of this, although they didn't always realize what caused the problems, and why the methods they applied were effective. Sanskrit texts dating from about 2000 BC indicate that drinking water should be irradiated by sunlight and afterwards filtrated over charcoal. In case the source water was unclean, it should first be boiled, then a piece of copper would have to be immersed for seven times, and finally the water would have to be filtrated. It took until about 1700 AD until Antony van Leeuwenhoek discovered microorganisms, although by that time no one yet understood the importance of this finding for human health. In 1854 during a cholera epidemic in London, John Snow, an English physician, realized that the source of the epidemic was a public water pump on Broad Street. By disabling the pump, the outbreak was stopped. Since this time, the importance of safe drinking water has become more and more clear, and since the end of the nineteenth and beginning of the twentieth century, distribution of safe drinking water has become an important task of authorities. By the same time, measurements were taken to improve sewerage. The combination of these two facts resulted in a significant increase in health and average lifespan of people.

As becomes clear from the above, the microbiological safety of drinking water is a very important parameter in drinking water treatment. However, it also became clear that the chemical composition of the water may play an important role. Some Roman emperors are notorious because of their mental health. This probably was caused by the fact that they were rich enough to afford drinking cups containing lead, as a result of which they obtained a very high dose of lead. Originally, drinking water mains also contained high lead concentrations, but as became clear this may cause problems for public health the mains have been replaced by polyvinyl chloride or polyethylene materials. Nowadays in some areas, like Bangladesh, high arsenic concentrations cause serious health problems. In 1989 in the Netherlands, bentazon, a pesticide, was detected in drinking water. As during the last quarter of the twentieth century analytical techniques were significantly improved, it became clear that sources for drinking water often contained (too) high concentrations of pesticides. And shortly after it also was observed that many other organic micropollutants, like pharmaceuticals, personal care products, flame retardants, solvents and many other industrial chemicals, may be present in sources for drinking water.

Nowadays, the World Health Organization (WHO) forms an authoritative basis for the setting of national regulations and standards for water safety in support of public health. Because of the "precautionary principle", in many cases these regulations and standards are set lower than the WHO guidelines. The objective of the EU Drinking Water Directive (98/83/EC) is to protect human health from adverse effects of any contamination of water intended for human consumption by ensuring that it is wholesome and clean. Member states of the European Union can include additional requirements, e.g. regulate additional substances that are relevant within their territory or set more stringent standards. In the USA drinking water has to comply with the Safe Drinking Water Act (SDWA). Here standards are set and in some cases treatment technology is prescribed. The number of compounds in water seems to increase, partly because of the improvement of analytical techniques, increasing the number of detectable compounds, and partly because the number of chemicals used in practice is increasing. Although there already are numerous standards set, it is likely that the number of standards will further increase. Water treatment techniques aim at safe drinking water, which contains no pathogens or toxic substances. Besides, the water has to be palatable, clear, colourless and odourless. Furthermore, it should not be corrosive (for tanks and pipes) and have a low organic content, in order to prevent biological growth in pipes and tanks. Nowadays, for customer convenience and environmental reasons (decreased use of detergents and energy requirements), water often is softened and conditioned. Finally, all these requirements should be met at low costs, as drinking water is a first necessity for life. In this way it is tried to ensure safe drinking water and to keep customers trust.

1.2 Microbiological Standards

Microbiological standards are important because when drinking water contains pathogens, exposure to these microorganisms can lead to acute infectious diseases and illness, and in severe cases even to death. Symptoms are gastroenteritis, diarrhoea, inflammations, etc. In more severe conditions, waterborne diseases like typhoid fever, cholera or poliomyelitis can occur.

Roughly three categories of pathogens with a high health impact can be distinguished: bacteria, viruses and protozoa. Sometimes a fourth category is added: helminths. Bacteria have in general a low resistance to chlorine or other disinfectants, and persistence in water supply is moderate. Some bacteria may multiply in drinking water and their infectivity is in general low to moderate. Viruses are moderately resistant to chlorine and have a high infectivity. Protozoa are highly infective and very resistant to chlorine [1].

For microbiological standards two approaches are used. The more traditional approach is the use of faecal indicators in drinking water. The standards in the European Drinking Water Directive are 0/100 mL for *E. coli* and *Enterococci* [2]. However, for the more infective viruses and protozoa, this approach is insufficient, and a health-based quantitative microbial risk assessment (QMRA) is required to determine the microbiological safety of drinking water. The QMRA methodology requires information on exposure, expressed as the number of microorganisms ingested, and dose-response models to determine the infection probability. Finally, epidemiological data can be used to determine the disease effects and severity. Using these data, the QMRA can be used to determine the health effects of drinking water at very low concentrations of pathogens. For viruses the acceptable concentrations are extremely low (one microorganism in a few hundred to thousands m^3) and thus are not measurable. Instead, logarithmic reduction values (LRV) for pathogens by individual treatment steps are used. These LRVs can be added for all individual steps in a treatment. If the number of pathogens in the raw water is known, it can be calculated by the LRV what the concentration in the treated water will be and whether this imposes an acceptable health risk.

1.3 Chemical Water Quality

Chemical water quality relates to several organic and inorganic compounds, both from natural and from anthropogenic origin, that may be present in drinking water. Problems caused by the presence of heavy metals have been recognized for quite some time now, and measurements have been taken to prevent these problems. Drinking water mains are no longer made of lead, and the water is conditioned (i.e. the pH is adjusted) in order to prevent dissolution of, e.g. copper or carbonate from cement pipes. However, it is well known that in some parts of the world, sources for drinking water contain too high concentrations of e.g. arsenic or chromium, which still causes a lot of human health problems worldwide. Coagulation/flocculation and adsorption, sometimes in combination with oxidation, can be applied to remove these metals from drinking water, and new techniques are being developed [3, 4]. For arsenic the WHO standard is 10 μg/L, but it is known that this isn't a "safe" concentration, as arsenic is a very toxic metal. However, in some cases it is technically difficult or too expensive to further decrease the arsenic concentration.

Pesticides (Fig. 1) are applied in both municipal and agricultural areas and end up in sources for drinking water either by run-off to surface water or by penetration into groundwater. They are organic compounds which have been designed to be harmful for certain organisms. Therefore, in principle they shouldn't be present in drinking water. However, as analytical techniques are optimized to be able to detect lower and lower concentrations, it is very difficult to ensure that "no" pesticides are present. Because of this often the "threshold of toxicological concern" (TTC) is applied, which gives a guideline for safe concentrations for certain compounds, below which no negative effects on human health are expected. In the Netherlands a standard has been set at 0.1 μg/L for individual pesticides, with a total concentration <0.5 μg/L.

In the past decade, it has become known that sources for drinking water also may contain pharmaceuticals. These too are compounds designed for their effect on living organisms. The major part of these pharmaceuticals after use is excreted in urine and faeces and thus is present in municipal wastewater. However, wastewater treatment plants (WWTPs) in general have not been designed to deal with these

Fig. 1 Several pesticides

compounds; as a result of which, only 60–70% of the total load of pharmaceuticals and their metabolites is removed from WWTP effluent. The rest ends up in surface water, which in turn is used to produce drinking water. Research in the Netherlands has shown that WWTPs significantly contribute to the pharmaceutical concentrations in small surface waters [5], which leads to the conclusion that the presence of pharmaceuticals in surface water may become a problem for drinking water production. At the moment more than 4,000 chemical compounds are being applied as pharmaceuticals, and in a small country like the Netherlands (with approximately 17 million people), yearly 3.5 million kg of pharmaceuticals is used, 140,000 kg of which end up in surface water. It is expected that these amounts will increase in the coming years, as a result of the development of new pharmaceuticals and of ageing of the people [6]. Besides, due to climate change, longer periods of draught are expected to occur, resulting in lower river discharge and thus higher concentrations. It already has been shown that some of these compounds, like diclofenac and hormone disruptors, also have a negative effect on the aquatic environment. Recently, the EU has put some of these compounds (17-beta-estradiol (E2), 17-alpha-ethinylestradiol (EE2), and diclofenac) on a watch list, and it is expected that eventually standards will be set for these compounds [7, 8].

A separate category of pharmaceuticals are the veterinary pharmaceuticals. As these often are excreted with manure, they may enter the environment more diffusely and may finally end up in groundwater [9]. This, however, may take several years; as a result of which, it has to be kept in mind that even after a certain type of pharmaceutical may have been banned, still it may be observed in groundwater for many years.

Apart from the above-mentioned pesticides and pharmaceuticals, also industrial compounds, originating from industrial wastewater treatment plants, can be found in surface waters. Their relevance for drinking water production depends on their concentrations and characteristics, which also determine their behaviour in water treatment processes. Unfortunately, very often little is known about their presence in wastewater or surface water.

Another category of pollutants are (micro)plastics. As in many cases this is particulate matter, other techniques will be required to deal with these materials. They cannot be degraded or adsorbed by common processes, and often the particles are too small (micro- and nanoscale) for removal in regular filtration processes like sand filtration. No standards have been set yet for water.

A recent trend observed, initiated by REACH registration, is that apolar chemicals are replaced by more polar ones [10]. More polar chemicals are better soluble in water, and therefore removal in a treatment process is often more difficult.

1.4 Water Quality: Presence and Characterization of Natural Organic Matter

Natural organic matter (NOM) is present in all surface ground and soil waters. It affects biogeochemical processes (like metal complexation and redox conditions) as well as water treatment processes through several mechanisms. Therefore, it is a key parameter with respect to design and operation of water treatment processes. It is responsible for colour, taste and odour problems and the major part of the coagulant and disinfectant requirements; it hinders the removal of other contaminants (e.g. by competition for adsorption sites in activated carbon, by pore blocking in filters and by interference with photochemical and oxidation processes); it acts a as a precursor for unwanted (disinfection) byproducts during treatment with chlorine and ozone; it contributes to membrane fouling, corrosion and the formation of metal complexes; and it acts as a substrate for bacterial growth, resulting in biologically unstable water and metal complexes [11–14].

NOM is a complex heterogeneous mixture of various organic molecules originating from the natural biological activity in water. Its composition varies from largely aliphatic to highly coloured and aromatic, from highly charged to uncharged, with a wide variety of chemical compositions and molecular weights, depending on its origin [11, 15–17]. NOM originating from plant matter has a high lignin content, with a predominant aromatic fraction [18]. It is suggested that aromatic parts of wood and nonwoody plants are the precursors of soil humic acids [19]. These authors describe that aromatic hydroxyl carboxylic acids and aldehydes are formed during UV and sunlight irradiation of lake and river NOM. Often NOM represents a family of polymeric chains resulting from the condensation of polyphenols [20]. Colour often is caused by the presence of fulvic acids, humic acids and hymatomelanic acids.

The amount and character of NOM in water differ with climate and the hydrological regime as well as with other environmental factors. NOM found in natural waters consists of both hydrophobic and hydrophilic components. Approximately 50% of the total organic carbon (TOC) in water consists of hydrophobic acids: humic acids, fulvic acids and humin. This fraction contains much aromatic carbon, phenolic structures and conjugated double bonds. Hydrophilic NOM contains more aliphatic carbon and nitrogenous compounds like carbohydrates, sugars and amino acids. Dissolved organic carbon (DOC) content and TOC content often are used as indicators for NOM, but they give no information on its composition. The composition of NOM can be characterized in various ways, for example, by dividing it in a soluble (<0.45 μm, also containing cell fragments and macromolecules) and a particulate (suspended) fraction (>0.45 μm). Soluble microbiological products (SMPs) end up in the water during, e.g. biological treatment in municipal wastewater treatment plants. They originate either from the conversion of organic compounds by microorganisms or from dead microorganisms. The SMPs contain humic acids, polysaccharides, proteins, amino acids, antibiotics, extracellular enzymes, parts of microorganisms themselves and conversion products [21, 22].

A common method to characterize organic material is by means of LC-OCD (liquid chromatography-organic carbon detection) techniques [23]. In this case the following classification is applied:

- Biopolymers (BP) with molecular weight (MW) $\gg$20,000
- Humic substances (HS) with MW $\approx$1,000
- "Building blocks" (BB) with MW $\approx$300–500 (These are natural conversion products of humic substances)
- Neutral components with MW <350
- Acidic components (Low MW-acids) with MW <350

Size exclusion chromatography (SEC) often also is applied to determine the molecular weight distribution of the material.

Assimilable organic carbon (AOC) is a mixture of various fractions of organic material, which differ per type of water [24]. Grefte concluded that per type of water, a specific linear relation can be observed between the average AOC concentration and the concentration of LMW acids.

Important parameters in the characterization of NOM are the aromaticity and the hydrophobicity of the material. Both variables are related. In literature, the material often is characterized by its specific UV absorbance (SUVA). Material with a high-SUVA value in general contains much high molecular weight compounds, whereas low SUVA value material contains hydrophilic, low molar mass and low charge density compounds. Moreover, compounds with a high SUVA have a higher aromaticity and more unsaturated carbon bonds. However, size distribution and SUVA value are not necessarily related: it is possible that, e.g. coagulation largely affects the SUVA value but at the same time hardly affects the molecular weight of the DOC. Audenaert et al. [25] used the UV absorption at a wavelength of 310 nm to determine the presence and amount of NOM.

Fluorescence (fluorescence excitation emission matrix; FEEM) too is applied to characterize dissolved organic material from a biological treatment process [26]. These authors studied a method to determine the polarity of various NOM fractions by means of the "polarity rapid assessment method" (PRAM). In this method water is extracted by means of various adsorbents (solid phase extraction, SPE). Apart from the hydrophobic surface of the material (and its aromatic character), also the molecular weight and molecular weight distribution play an important role. Column materials used for this technique are, e.g. C_2, C_8 and C_{18}, which show an increasing capacity for hydrophobic components. The most important parameter to characterize the various fractions is the difference in hydrophobic surface of the various components. Furthermore, dipole interactions and hydrogen bridging are used for characterization, for example, by applying anion exchangers with NH_2 (a weak anion exchanger) and SAX (a strong anion exchanger). This method is affected by the pH and ionic strength [26–30].

The PRAM method differs from the commonly applied extraction using a XAD resin, as in the XAD method, a low pH is applied, and separation of the fractions is carried out in series instead of in a parallel execution. In XAD different fractions can be isolated and analysed. Thus, it is possible to determine a mass balance based on the XAD method, whereas this cannot be done using the PRAM results [26, 29].

1.5 Water Quality: Conditioning

Carbon dioxide and carbonate, and as a result pH, play a very important role in the characteristics of (drinking) water and water treatment.

Carbon dioxide dissolves in water according to Eq. (1):

$$CO_2 + H_2O \rightarrow H_2CO_3 \tag{1}$$

Subsequently, carbonic acid dissociates into bicarbonate and carbonate, according to Eqs. (2) and (3).

$$H_2CO_3 + H_2O \rightarrow HCO_3^- + H_3O^+ \tag{2}$$

$$HCO_3^- + H_2O \rightarrow CO_3^{2-} + H_3O^+ \tag{3}$$

The dissociation constant for reaction (2) (K_1) is 4.5×10^{-7} mol/L, and the dissociation constant for reaction (3) (K_2) is 4.7×10^{-11} mol/L at 25°C. From this it follows that at equilibrium:

$$pK_1 = pH + \log\left(\frac{[H_2CO_3]}{[HCO_3^-]}\right) = 6.35 \tag{4}$$

$$pK_2 = pH + \log\left(\frac{[HCO_3^-]}{[CO_3^{2-}]}\right) = 10.33 \tag{5}$$

Ca^{2+} may react with CO_3^{2-}, forming $CaCO_3$, which has a very limited solubility:

$$K_s = [Ca^{2+}] \cdot [CO_3^{2-}] = 3.8 \times 10^{-9} \text{ at } 25°C \tag{6}$$

The degree of super or subsaturation of calcium carbonate is expressed as the saturation index SI (also Langelier saturation index (LSI)) and is defined as (Eq. 7)

$$SI = \log\left(\frac{[Ca^{2+}] \cdot [CO_3^{2-}]}{K_s}\right) = pH - pH_s \tag{7}$$

In Eq. (7) $[Ca^{2+}]$ is the Ca^{2+} concentration, $[CO_3^{2-}]$ is the carbonate concentration, and K_s is the solubility constant for $CaCO_3$. pH_s is the equilibrium pH of water containing identical concentrations of Ca^{2+} and HCO_3^-.

$$pH_s = pK_2 - pK_s - \log([Ca^{2+}][CO_3^-]) \tag{8}$$

In general, in the Netherlands it is strongly recommended that the pH of drinking water should be above 7.4 in order to prevent the dissolution of zinc from brass in taps and of copper and iron from pipe materials. In this way also the dissolution of lead will be decreased in places where still lead piping is applied. Furthermore, the SI should be > -0.2 in order to prevent the dissolution of calcium carbonate from cement. This recommendation sometimes contradicts the pH requirements mentioned above [31]. For conditioning marble filtration may be applied. As many organic micropollutants are salts, the pH controls their degree of dissociation and thus not only their solubility in water but also their behaviour in adsorption, photolysis and oxidation processes.

A way to describe the precipitation of calcium carbonate is by applying the total hardness (TH), the theoretical calcium carbonate precipitation potential at 90°C ($TCCPP_{90}$) and the practical calcium carbonate precipitation potential (PCCPP). The TH is defined as the total concentration of calcium and magnesium ions in water (in mmol/L or in German degrees (°D), 1 mmol/L corresponding to 5.6°D). In the Netherlands it is recommended to apply a TH <1.8 mmol/L, in order to increase customer comfort and decrease the environmental impact (use of detergents and energy requirements). Besides, in the Netherlands a minimum calcium concentration of 1 mmol/L is required, based on the fact that sufficient calcium uptake is essential for human health [32]. However, there is no evidence that the presence of calcium in drinking water would positively contribute to this [33–36].

Furthermore, it is recommended that TCCPP <0.6 mmol/L and PCCPP <0.4 mmol/L, in order to prevent precipitation of calcium carbonate, which results in higher maintenance costs and lower lifespan for apparatus like laundry machines and dish washers.

Under these conditions most organic micropollutants are soluble; as a result of which, they are relatively difficult to remove from the water. Another point of attention are heavy metals. By applying milk of lime for softening or marble filtration for conditioning, small concentrations of heavy metals may be introduced into drinking water. Precipitation of $CaCO_3$ does not contribute to the removal of e.g. organic micropollutants.

2 Water Treatment: Drinking Water from Groundwater Sources

In many countries, groundwater is used as a source for drinking water production. Groundwater is in general relatively clean, although due to biogeochemical processes a wide range of compounds can be dissolved. The compounds present in the groundwater depend largely on the soil composition of the aquifer. Also, the well or borehole conditions can have influence on the water quality of the abstracted water. In specific circumstances, groundwater can be under influence of surface water. This can be the case in karstic aquifers where cracks in rocks result in a direct shortcut between the surface water and the groundwater. Sometimes, groundwater is deliberately put under influence of surface waters by application of river bed filtration.

In principle, groundwater can be aerobic, slightly anaerobic and deeply anaerobic. Aerobic groundwater is open to the atmosphere and thus contains oxygen. Most dissolved compounds will be present in an oxidized state. Under these conditions iron from the soil doesn't dissolve in the water. In order to produce drinking water, even aerobic groundwater is aerated to increase the oxygen concentration and simultaneously decrease the CO_2 concentration. The most important parameters that will have to be adjusted in this case are pH, calcium content, bicarbonate concentration and saturation index (SI). This is called "conditioning".

When the groundwater originates from sandy soil, it often is lime aggressive; as a result of which, cement pipe material may be dissolved (SI < −0.2) or enhanced corrosion of drinking water mains may occur. Due to several degradation processes, carbon dioxide is present in the water, and as there is hardly any calcium present, the CO_2 concentration may be (much) higher than its equilibrium concentration. This is reflected in a large negative value of the SI. Air stripping will remove CO_2, but the pH and HCO_3^- concentration still will have to be adjusted. For example, marble filtration may be applied for this purpose, dissolving the marble pellets and adding some hardness to bring the water near equilibrium.

Sometimes the groundwater contains a relatively high calcium concentration (in calcium-rich aquifers). Hard waters are no threat for public health. Nevertheless, hardness can be a nuisance as warm water devices have a shorter life and require more maintenance due to scaling at higher temperatures (calcium carbonate solubility decreases with increasing temperatures), and, for instance, higher concentrations of detergents will be required. For these reasons water in many cases is softened.

As mentioned before, there are three types of groundwater: aerobic, slightly anaerobic and deeply anaerobic [37]. The treatment requirements for the three types differ considerably. Figs. 2–12 shows block schemes for typical groundwater treatment schemes.

The aerobic groundwater doesn't contain dissolved iron, but the anaerobic groundwater does. Besides, it contains ammonium and manganese. For slightly anaerobic groundwater, aeration and stripping result in the removal of CO_2 but also in the oxidation of Fe^{2+} to Fe^{3+}, of NH_4^+ to NO_3^- by biological processes and of Mn^{2+} to MnO_2, partly by chemical and partly by biological processes [38]. Fe^{3+} reacts with hydroxyl ions, forming $Fe(OH)_3$ flocs, which slowly converts into iron oxide (Fe_2O_3). Bacteria, *Nitrosomonas* and *Nitrobacter* are responsible for the biological conversion of ammonium, which requires a relatively high amount of oxygen.

Deeply anaerobic groundwater may contain high concentrations of iron, manganese, ammonium, hydrogen sulphide and methane. Sometimes also chlorinated compounds, originating from industrial contaminations and spills, are found. Aeration and stripping are applied to remove the gasses and to oxidize iron, manganese and ammonium. For the nitrate formation, "dry filtration" has to be carried out, in

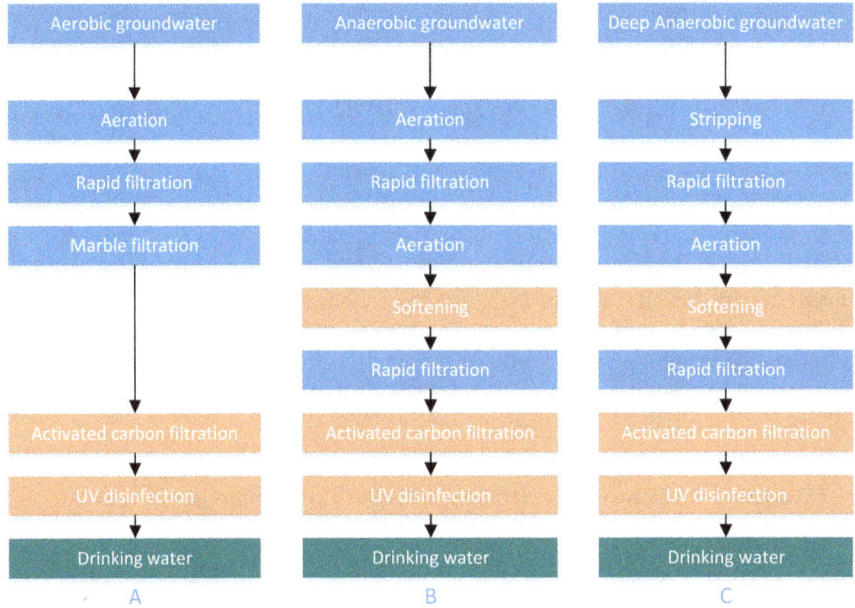

Fig. 2 Typical groundwater treatment schemes: (**a**) aerobic groundwater; (**b**) anaerobic groundwater, (**c**) deeply anaerobic groundwater. *Orange blocks* are optional, depending on the water quality: softening for hard water, activated carbon filtration and UV for the presence of organic micropollutants and disinfection

order to provide sufficient oxygen for the oxidation process. The removal of gasses can be described by applying Henry's law (Eq. 9):

$$C_w = K_H \times C_g \qquad (9)$$

In which C_w is the equilibrium concentration of a gas in water, C_g is the equilibrium concentration of a gas in air and K_H is Henry's constant or distribution coefficient. Stripping may also be an effective method to remove volatile contaminants, like vinyl chloride and 1,2-dichloro ethane, which sometimes are encountered in groundwater due to industrial pollution. However, most organic micropollutants are soluble salts, which cannot be removed in this way.

Simultaneously, iron and manganese are oxidized, forming precipitates. Colloidal particles, with a size <1 μm, are kept floating by the Brownian motion. In case of neutral particles, stirring will result in collisions of the particles, upon which London-van der Waals forces will keep them together, forming larger flocs. These flocs may also include other species, like heavy metals (arsenic) and NOM/DOC. They precipitate the rate depending on their size and composition. Rapid sand filtration then can be applied to remove the flocs from the water. Thus, also colour, which originates from DOC, may be removed from the water.

Organic micropollutants, like pesticides and pharmaceuticals, resemble NOM in certain ways and thus also may be included into the flocs. However, often these compounds are more hydrophilic and polar; as a result of which, flocculation is not a very efficient process for the removal of these compounds. During the last decades, more and more filtration over activated carbon (AC) has been applied to remove these compounds. AC is very effective for the removal of organic and preferably hydrophobic compounds. As a result, serious competition is observed by NOM in the removal of micropollutants. Other problems that are encountered are pore blocking by NOM; as a result of which, micropollutants cannot reach the pores anymore, and the effective surface for adsorption is reduced, and displacement of already adsorbed compounds by less soluble and more hydrophobic NOM compounds occurs.

3 Water Treatment: Drinking Water from Surface Water Sources

Surface water has a much more dynamic behaviour than groundwater. Water quality can vary quite rapidly, and river discharge flows can vary, in bigger rivers sometimes up to three orders of magnitude. Moreover, surface water contains a much wider variety of pollutants, as it is under direct anthropogenic influence by discharging of municipal and industrial wastewaters, surface run-off, atmospheric deposition and incidental spills and calamities. It is therefore also impossible to control water quality. This puts additional pressure on a treatment system to produce safe drinking water.

Surface water treatment exists in a wide variety of systems, and in recent years innovation in this area is accelerating. Nevertheless, a number of general treatment objectives have to be met: removal of suspended solids and disinfection, removal of taste and odour compounds and removal of a wide variety of micropollutants. Figure 3 shows a number of more traditional surface water treatment schemes.

Suspended solids are mostly removed by application of coagulation, flocculation and sedimentation. Often these steps are followed by a rapid filtration. Disinfection traditionally was done chemically by adding chlorine. However, because of the detrimental health effects of the byproducts, these systems are mostly abandoned nowadays. Other disinfection systems and multi-barrier approaches have been introduced. Many surface water treatment works also contain improved barriers against micropollutants often in combined processes for oxidation and disinfection like ozone and activated carbon filtration. Below the removal of suspended solids, disinfection and activated carbon filtration will be discussed.

Limitations of Conventional Drinking Water Technologies in Pollutant Removal

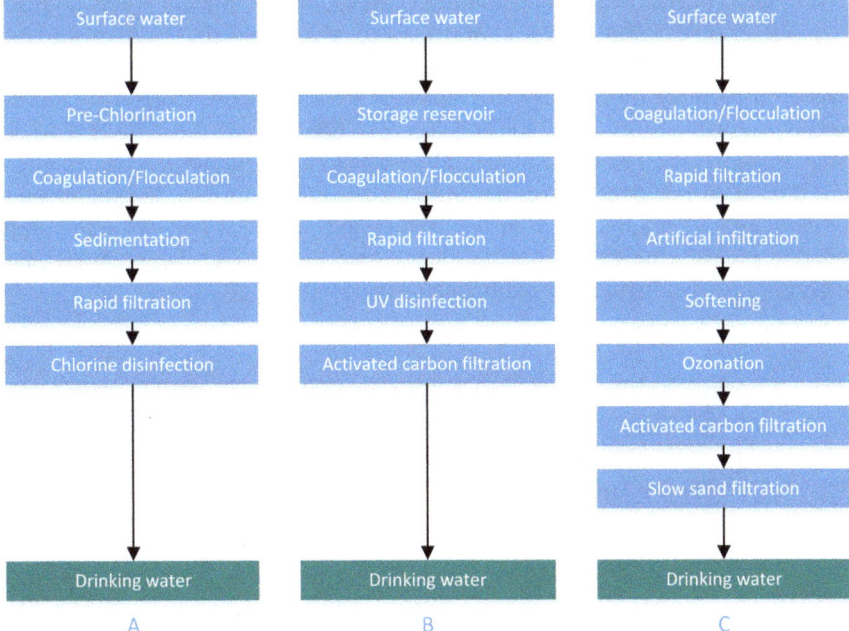

Fig. 3 Typical surface water treatment schemes. (**a**) Traditional, (**b**) advanced, (**c**) advanced including soil passage

3.1 Coagulation, Floc Formation and Floc Removal

An important step in producing clean water is the removal of suspended matter. Suspended matter can consist of a large variety of materials and a wide range of particle sizes, e.g. clay and silt, organic debris, plastics and engineered nanoparticles. Part of the suspended matter may be stable in water, whilst other fractions may settle or float. This depends on the water chemistry, the density of the particulate matter relative to water, the temperature and viscosity of the water, flow and mixing, etc. These conditions often vary hourly, daily and over the seasons.

Suspended matter causes turbidity in water, which is not only an aesthetic problem but also leads to poor treatment and disinfection performance. When the suspended matter content is relatively low, the particles can often be removed by filtration. However, if the concentration increases or if the particles are too small, a filtration step is in many cases insufficient or hindered by operational problems such as filter blocking. In that case, it is necessary to utilize other particle removal techniques like coagulation, flocculation and clarification. The suspended solid concentration that determines whether filtration or coagulation should be applied depends largely on the characteristics of the particulate matter and the water composition and is therefore difficult to predict.

Although coagulation, flocculation and clarification are different processes from a physico-chemical perspective, in practice they always are connected and taking place in the same unit operation. Coagulation is a process of destabilizing the suspended particles by reducing the repulsive forces between them. This can be done by adding a salt or metal ion solution. The effect is that the particles start to form agglomerates. The agglomerate now starts to grow further into larger flocs. This process, called flocculation, needs gentle stirring to create collisions between the particles. Often a flocculant or flocculation aid, such as a polymer solution, is added to increase the stability of the flocs. Once the flocs are large enough, the suspension can be clarified by sedimentation or floatation.

3.1.1 Colloid Stability

Suspended particles in water can span a large size range. The smallest particles are often a few nanometres and can have a natural origin (minerals, clays) or can be man-made (e.g. TiO_2, n-Ag, ZnO, n-C_{60}), whilst the larger particles can be a few hundred micrometres. The latter can be of organic origin or larger sand grains. The smaller size particles or colloids are often quite stable in water, which means that the particles remain in suspension if the water is stagnant. They have a typical diameter of a few micrometres or less. Only larger particles or the heavier ones will sink in a stagnant water body. Particles with a lower density than water, e.g. organic materials, oils and fats, will float on the surface.

Colloids in water can be stabilized by two independent physical principles. The first one is steric stabilization (Fig. 4). Steric stabilization occurs if the water

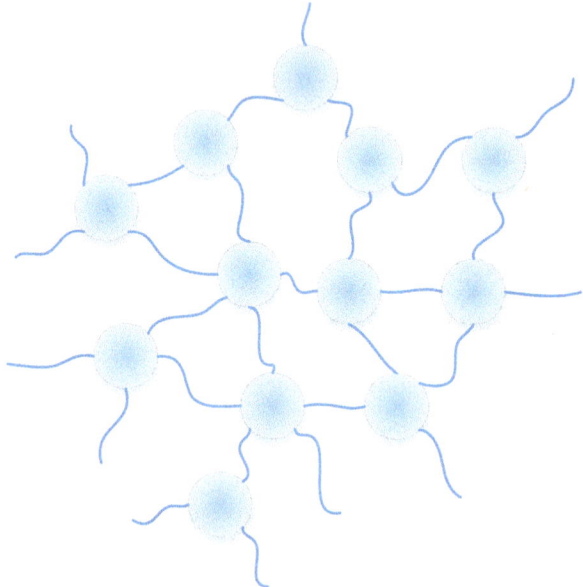

Fig. 4 Steric stabilization

contains water soluble polymers or large molecules like humic substances. These materials can adsorb on the particle surfaces. The length of the polymers prevents that the particles can approach each other at a sufficient short distance to form an agglomerate. However, depending on the polymer chain length and/or particle concentrations, the materials can start to form a stable network in the water.

The second form of stabilization can occur if the particle surface is charged. If a surface charge is present at the particle surface, an electrical double layer will occur around the particle (Fig. 5). This is caused by adsorption of ions from the water phase or ionization of molecules at the particle surface. Ions of an opposite charge are attracted by this charge, thus forming an electric double layer [39]. In most cases particles in water, partly being clay minerals, are negatively charged. Only at very acidic pH positive surface charges may occur.

Also significant parts of the NOM, the humic acids, are negatively charged. Upon mixing, the colloidal particles and compounds will repel each other; as a result of which coagulation will not occur. In this case the electrolyte content of the water can be increased. In this way, the double layer surrounding the colloidal particles may be compressed, decreasing the zeta potential of the surface [40]. It is also possible to include positively charged particles in the flocs by adsorption of highly charged counter ions, which also decrease the zeta potential. Thus, the repelling Coulomb forces can be decreased until they become smaller than the London-van der Waals forces.

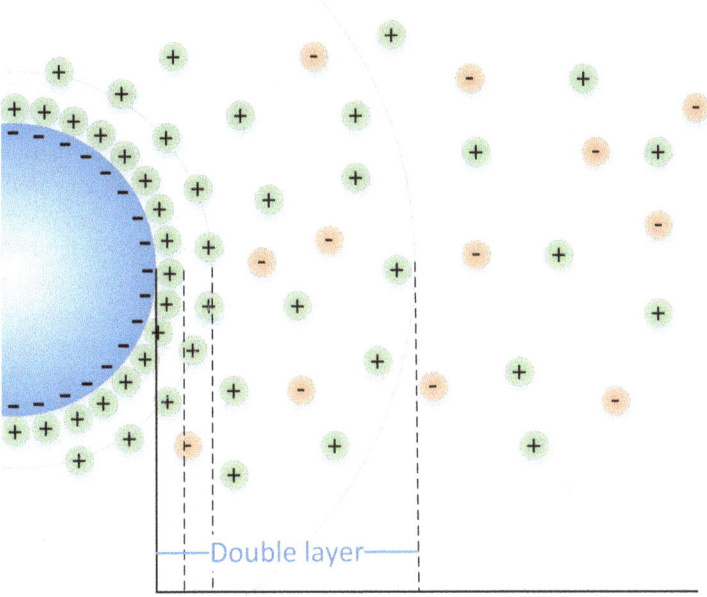

Fig. 5 Electric double layer

3.1.2 Destabilization and Floc Formation

Suspended colloids can be destabilized by reducing the repulsive forces between the particles. Adding positively charged ions will reduce the electrostatic repulsive forces between the particles, because the positive ions will interact with negatively charged particle surface. Divalent ions appear to be about eight times as effective as a monovalent ion in causing coagulation, whereas a trivalent ion is even 600 times as effective [41]. For this purpose, in practice often Fe(III) or Al(III) salts are added to the water.

All metal cations are hydrated to some extend in water. In the primary hydration, shell water molecules are in direct contact with the metal ion. The bonding of the water in the secondary hydration shell is more loose. For Al^{3+} and Fe^{3+} ions, the primary hydration shell consists of six water molecules in an octahedral coordination. Due to the high charge on the metal ion, the water molecules in this shell are polarized. This may lead to a loss of one or more protons, depending on the solution pH. As a result, the water molecules in the hydration shell are progressively replaced by OH ions, lowering the positive charge, according to Eq. (10):

$$Me^{3+} \rightarrow Me(OH)^{2+} \rightarrow Me(OH)_2^+ \rightarrow Me(OH)_3 \rightarrow Me(OH)_4^- \qquad (10)$$

In Eq. (10) Me is the metal (iron of aluminium). However, Eq. (10) is a simplified scheme, as in practice various polynuclear forms are observed [42]. Examples are $Al_2(OH)_2^{4+}$ and $Al_3(OH)_4^{5+}$, but equivalent species can be found for iron. Furthermore, also polynuclear hydrolysis products exist, like $Al_{13}O_4(OH)_{24}^{7+}$. This is known as polyaluminium (PACl), and its chloride salt is a very effective flocculant.

At about neutral pH, both Al(III) and Fe(III) salts show a limited solubility. As a result, amorphous hydroxide precipitates, which plays a very important role in the coagulation and flocculation processes. Positively charged precipitate particles may deposit on colloidal particles (heterocoagulation), resulting in charge neutralization and thus destabilization [42]. In order to obtain effective coagulation, a very short mixing time is beneficial, which means often high mixing intensity is required. This can be realized by, e.g. dosing in a cascade (Fig. 6) or utilizing a high-shear turbine mixer. In the Netherlands mainly Fe-salts are applied, whereas in other countries Al-salts are preferred, sometimes combined with active silica [43]. An aspect which requires some attention is the pH of the salt solutions. Often they have a pH value ≤ 2. By administering the solutions at one injection point, pH gradients may occur, with locally very low pH values. At these values certain hydrates can be formed, which are relatively stable, are ineffective adsorbents and cause turbidity in the water. A high mixing rate is required to prevent the formation of such hydrates, and addition should take place at different locations. In practice the process starts at a relatively high mixing rate, which later is reduced in order not to destroy the already formed flocs by too high-shear rates (tapered coagulation).

Fig. 6 Dosing of ferric chloride in a cascade

Another way to form larger particles is by adsorption and bridging of macromolecules and polyelectrolytes. This process is known as "flocculation". For this purpose both natural and synthetic flocculants (mainly polyelectrolytes) can be applied. Anionic, nonionic and cationic polymer can be used. When the concentration of flocculant is too low, bridging cannot occur, and the floc size remains small. At too high flocculant concentrations, the flocculant will cover the total particle surface; as a result of which bridging also cannot occur. In practice the addition of Fe^{3+} or Al^{3+} is often combined with the addition of a polyelectrolyte, thus enhancing flocculation. The use of polymers results in the formation of larger and stronger flocs with an open structure. The higher the molecular weight of the polymer, the more effective the flocculant will be. However, in this case too care should be taken not to apply a too high polymer concentration, as this would shield the colloidal particles or may cause so much steric hindrance that bridging of particles would be hindered. Sometimes addition of a salt will increase the adsorption of polymers by decreasing the electrical double layer. Addition of more particles, like clay particles, may also enhance flocculation, as it facilitates collisions between particles. Examples of flocculation tanks are shown in Fig. 7.

Fig. 7 *Top*, Tapered flocculation tanks; *Centre*, mixing device in flocculation tank; *Bottom*, flocculation tank in Wuppertal (Germany)

When this process occurs at a high rate, other compounds or contaminants may be included in the flocs, which is called "sweep coagulation". This process is more effective than for contaminant removal than only applying charge destabilization. Characteristics of the sweep coagulation process are that flocs form more rapidly and can become much larger. Figure 8 shows the working areas for pH and metal ion dose for the different modes of coagulation, such as sweep coagulation, adsorption destabilization and charge neutralization.

When excess coagulant is used, more than would be required for baseline coagulation, this is called enhanced coagulation. The effect can be increased by changing the pH, the order of chemical addition or by using alternative coagulant chemicals. In this way TOC and NOM removal may be improved [11].

The time required to transport colloidal particles and flocculants is in the order of 10^{-4} s, whereas very high molecular weight polymers would require a few seconds. As a result good mixing is required to obtain optimum flocculation. pH may be an important parameter in this respect. Often it is found that within a certain pH range, in general pH ≈ 8, a minimum flocculant dosage is required. In case this is a narrow range, it is difficult to maintain the right pH during the process. This also depends on the colour of the water: in general it is found that the more colour there is present, the lower the optimum pH will be. In order to remove fulvic acids, a higher dose of flocculant is required than for the removal of humic acids, which probably is caused by the lower molecular mass of the fulvic acids.

During coagulation and flocculation iron(hydr)oxides are formed, which precipitate (see Fig. 9). It is known that these compounds can react with phosphate. As a result, during flocculation also phosphate is removed from the aqueous phase. The same principle is encountered for arsenic, which also can (partly) be removed by flocculation in this way. However, for other heavy metals, the method is less effective.

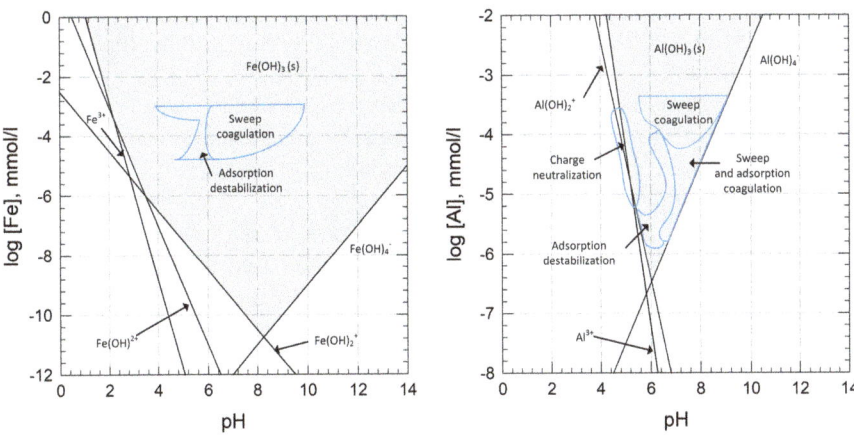

Fig. 8 Coagulation diagrams alum and ferric (adapted from [44] and equilibrium constants from [45])

Fig. 9 Horizontal sedimentation

3.1.3 Pollutant Removal

Organic micropolllutants may be removed by flocculation, but only if they have been adsorbed to some other, high molecular weight compounds. This will, amongst others, depend on pH, which determines the charge of the molecules and thus their ability to be adsorbed. In general, removal of organic micropollutants by flocculation is very limited, as was demonstrated by van der Horst et al. [46] for pharmaceuticals and endocrine disrupting compounds (EDC's) and by Saraiva Soares et al. [47] for pesticides and their metabolites (endosulfan, ethylenethiourea (ETU) and 1,2,4-triazole). This mainly is due to the solubility and hydrophilicity of the micropollutants.

Microorganisms like bacteria and algae may be removed by sweep coagulation and by bridging. For viruses removal may occur by complexation with aluminium and iron. However, the level of removal in general is insufficient to obtain safe drinking water [48].

Recently, in a case of an industrial contamination of river water, used for the production of drinking water in the Netherlands, it has been found that in a rapid sand filter (Fig. 10), micropollutants may be removed by means of biodegradation. However, the effectiveness of this biodegradation strongly depends on circumstances (like temperature) and the presence of certain types of microorganisms.

Fig. 10 Rapid sand filtration in the Netherlands (*left*) and in Germany (*right*)

3.1.4 Floc Removal

Once the flocs are formed, they have to be separated from the water stream. In most cases this is done by sedimentation or gravity settling. In practice, many different forms of sedimentation tanks exist. The simplest form is the long rectangular sedimentation pond as shown in Fig. 9. The length of the tanks should be such that under the low flow conditions, the residence time is sufficient for particles to settle to the bottom of the tank. Periodically the sludge is removed from the tanks by dredging. As these horizontal tanks need a large surface area, other engineering solutions such as lamellae separators and circular clarifiers have been designed. Instead of applying sedimentation to remove suspended matter, also flotation can be applied, especially in case of low-density particles or, e.g. oil droplets or algae (dissolved air flotation; DAF; see Fig. 11). In this case air is dissolved under pressure, and then by means of nozzles, it is released in a tank at atmospheric pressure. As a result small bubbles (10–100 μm, on the average 40 μm) will be formed. In order to ensure the formation of such small bubbles, the pressure differences should be 400–600 kPa [49]. These bubbles may adsorb surfactants and/or NOM and thus obtain a negative charge. Bubbles may adsorb hydrophobic particles. By dosing a coagulant like polyaluminium chloride or ferric chloride, particles will be destabilized, promoting the adherence of the bubbles to the particles, causing them to float to the surface, where they can be removed by means of skimming. Care should be taken not to overdose the coagulant, as this will result in charge restabilization of the positively charged particles and bubbles. It is possible that organic micropollutants also will be removed in this way, but the method is not very effective, as these compounds often have a high solubility.

As a final step to remove residual suspended solids but also to remove excess coagulants, a rapid filter step is applied. These filters also contribute to a multi-barrier disinfection.

Fig. 11 Dissolved air flotation in Norway

3.2 Application of Activated Carbon

Activated carbon often is used to improve the water quality by removing natural organic matter and organic micropollutants. Two types of activated carbon can be applied:

- Powdered activated carbon (PAC)
  The particle size of this material is <0.05 mm. It is added to the water, where it can adsorb organic compounds, like NOM but also organic micropollutants. As it has a high surface area, due to its small particle diameter, filtration requires a short contact time of about 10–15 min. Afterwards the activated carbon has to be removed by means of filtration.
- Granular activated carbon (GAC)
  The granules have a diameter of 0.3–3 mm and have a porous character. The pores contribute to the adsorption capacity of the material. However, pore size (micropores <1 nm, mesopores 1–25 nm and macropores >25 nm) determines which compounds can enter the pores. Another aspect of the pores is that they may be blocked by NOM which adsorbs at the surface. After the GAC has been loaded, it can be reactivated, by heating under an inert atmosphere, followed by activation, e.g. by an acid. GAC is applied in filter beds (Fig. 12).

The adsorption properties of the activated carbon strongly depend on the physical characteristics of the carbon (pore volume and size, surface area) and the chemical characteristics (surface composition, which depends on the activation

Fig. 12 GAC filter at Leiden (the Netherlands)

method applied). Furthermore, in GAC filters microorganisms may develop; as a result of which, biodegradation also will take place inside the column. This improves the removal of organic compounds by the filter and results in a longer time to reactivation of the GAC.

Whether organic micropollutants can be efficiently removed by activated carbon strongly depends on the micropollutant characteristics. In general, it can be assumed that the more hydrophobic the compounds are, the higher the adsorption capacity of the carbon will be. However, NOM may act as a competitor. Other parameters that may affect the effectiveness of activated carbon are temperature, pH and the presence of salts. Small, hydrophilic compounds in general are very difficult to remove by means of activated carbon.

After some time, the activated carbon will be saturated and adsorption will stop. For powdered activated carbon, an equilibrium between the adsorbed and dissolved compounds will occur quickly after dosing it. The PAC can therefore be used only once and has to be removed after dosing. For granular activated carbon, the adsorbing compounds will gradually saturate the filter column. During operation, a mass transfer zone will exist in the filter that slowly moves down in the flow direction in the filter bed. After some time, the compound will break through and the filter column has to be taken out of operation and the activated carbon has to be regenerated.

3.3 Disinfection

In the nineteenth century, it was discovered that some compounds, like chlorine, can be used as a disinfectant. Since the beginning of the twentieth century, disinfectants were applied by drinking water companies. However, since the 1970s it became clear that by adding chlorine (toxic) disinfection, byproducts may be formed by the reaction of chlorine with humic acids [50]. Because of this ozone became more popular as a disinfectant. However, also ozone appeared to give harmful byproducts, as it reacts with bromide to form bromate.

Membrane filtration also can be applied but also appeared to have some disadvantages: fouling of the membrane surface may occur and sometimes leakage is observed, decreasing the disinfection effectiveness.

In Marseille in 1901, UV was applied for disinfection purposes, but it took about 50 years until it was applied on a larger scale in Switzerland, Austria and Norway. Since the last quarter of the twentieth century, application of UV for water disinfection has become quite common.

The photochemical active part of the electromagnetic spectrum can be divided into four regions: vacuum UV (VUV; 100–200 nm), UV-C (200–280 nm), UV-B (280–315 nm) and UV-A (315–380 nm). This division is based on physical, biological or medical parameters.

UV-C radiation appears to be absorbed by DNA and RNA in cells of organisms. This may result in the death of these cells or will prevent their reproduction. The term "inactivation" can be applied to either a cell or its single subsystem. UV irradiation has been shown to be a powerful tool in inactivation of both microorganisms and cells such as bacteria, viruses, protozoan parasites, some spores, living cells and subsystems such as enzymes, amino acids and lipids [51]. One of the main advantages of UV inactivation is that no chemicals have to be added, since irradiation is a physical process. The germicide effect of UV light on bacteria and viruses is primarily due to the formation of thymine, thymine-cytosine (pyrimidine) and cytosine dimers in polynucleotide chains of DNA (they are listed in prevalence of order). It seems that radiation at wavelengths in the range of 200–295 nm (so-called bactericide or germicide range) exerts the most effective action. The absorption spectrum of DNA of viruses and bacteria shows a maximum at about 260 nm. However, also at wavelengths between 115 and 160 nm, the absorption coefficient of DNA is high [52].

For killing microorganisms it is not always necessary to change the DNA of the cells. It may also be enough to damage the cell membrane, allowing other compounds to enter the cell and damage or kill it. (V)UV photons with a wavelength below 275 nm can break C–C (3.8 eV) or C–H (4.5 eV) bonds, thus damaging the cell membrane or proteins in the cell. Furthermore, photo desorption producing species eroding the outer coat of the spore may lead to cell death during germination. Cells can be destroyed by etching [53]. Sosnin et al. also distinguish between two different disinfection methods: the inactivation of microorganisms by UV irradiation or their total VUV-induced photo mineralization.

Bacterial spores, which are the most resistant form of living microorganisms, often are used as model microorganisms in studies [52]. The reason that spores are 10–50 times more resistant to 254 nm UV light than their corresponding growing cells is the presence of a unique UV photoproduct of the spore DNA (called SP) [54]. Finally, spore DNA is protected by multiple layers, which surround the core, namely, a germ cell wall, cortex, inner and outer spore coats and sometimes an exosporium. In B subtilis spores, these form a 150–200-nm-thick proteinous barrier, which can shield the core from the effects of (V)UV photons. It is possible that UV radiation can also kill spores by modifying these outer layers of the spores, but the importance of this second pathway is not yet fully known.

Templeton et al. [55] have shown that humic acid flocs and particles <2 µm may shield viruses from UV radiation, thus decreasing the disinfection effectiveness of a UV process.

In order to inactivate microorganisms, it has been established that it is not necessary to kill them. As a result, UV doses applied for disinfection have been decreased in the past years, in order to decrease the energy demand of the process. At the moment in general, a UV dose of 20–70 mJ/cm^2 is applied in order to obtain effective disinfection of drinking water. In principle, photolysis by UV irradiation can also degrade organic micropollutants. This depends on the wavelength, i.e. the energy of the UV photon that can be absorbed by the molecule. In practice two types of mercury containing UV lamps are being used: low pressure (LP) UV lamps, which emit a single wavelength of 253.7 nm, and medium pressure (MP) UV lamps, which emit a much broader spectrum between 200 and 300 nm. Obviously, MP lamps will be more effective in causing photolysis of micropollutants because of this broader spectrum. However, it has been shown that for effective photolysis, much higher doses will be required than is common for disinfection purposes [56, 57]. Therefore, it can be concluded that UV disinfection in general will contribute very little to the degradation of organic micropollutants.

4 Water Treatment: Drinking Water from River Bank Filtrate or Dune Filtrate

Sometimes river bank filtrate is used instead of groundwater. By filtration through soil, the water quality can significantly be improved, partly by adsorption to soil particles, partly as a result of biodegradation by subsoil microorganisms. In this way also some micropollutants may be removed, depending on the local conditions (like the type of soil and water, presence of microorganisms, molecular properties of the micropollutants, etc.). However, not all micropollutants appear to be removed in this way. Because of this the treatment process may be simpler than the treatment process required for surface water treatment. In general aeration, in order to increase the oxygen content, sand filtration, conditioning and disinfection will be applied, but, depending on the water quality, other techniques, like filtration over activated carbon, also may be necessary.

A type of treatment typical for the Netherlands is dune filtration. This resembles river bank filtration, but in general the residence time is longer. Dune filtration significantly contributes to the removal of pathogens but also of other compounds. To make dune filtration possible, the high water quality is already required to protect the dune ecosystem. Especially organic micropollutants have to be removed before infiltration. For the final treatment, similar processes as in case of river bank filtrate will be necessary.

5 Conclusions

Water is treated to meet the objectives of drinking water quality and standards. These objectives are to prevent acute diseases by exposure to pathogens, to prevent long-term adverse health effects by exposure to chemicals and micropollutants and finally to create a drinking water that is palatable and is conditioned in such a way that transport from the treatment works to the customer will not lead to quality deterioration.

Traditional treatment technologies are mainly designed to remove macro parameters such as suspended solids, natural organic matter, dissolved iron and manganese, etc. Which kinds of technologies are required depends on the drinking water source. For groundwater in general aeration, filtration, conditioning and disinfection will be required, whereas for surface water coagulation/flocculation, sedimentation, several filtration processes, conditioning and disinfection will be necessary. These technologies have however only limited performance for removal of micropollutants. Advancing analytical technologies and increased and changing use of compounds show strong evidence of new and emerging threats to drinking water quality. Therefore, more advanced treatment technologies are required, in order to guarantee the production of safe drinking water.

References

1. WHO (2017) Guidelines for drinking-water quality4th edn. WHO, Gutenberg. First addendum to the fourth edition
2. European Commission (1998) Drinking Water Directive; Council Directive 98/83/EC on the quality of water intended for human consumption. http://eur-lex.europa.eu/LexUriServ/LexUriServ.do?uri=OJ:L:1998:330:0032:0054:EN:PDF
3. Ahmad A, Kools S, Schriks M, Stuyfzand P, Hofs B (2015) Arsenic and chromium concentrations and their speciation in groundwater resources and drinking water supply in the Netherlands. Nieuwegein, KWR Watercycle Research Institute
4. Ahmad A, Richards LA, Bhattacharya P (2016) Arsenic remediation of drinking water: an overview. In: Bhattacharya P, Jovanovich D, Polya D (eds) Best practice guide on control of arsenic in drinking water. IWA Publishing, London, pp 120–138

5. Coppens LJC, van Gils JAG, ter Laak TL, Raterman BW, van Wezel AP (2015) Towards spatially smart abatement of human pharmaceuticals in surface waters: defining impact of sewage treatment plants on susceptible functions. Water Res 81:356–365
6. Van Der Aa NGFM, Kommer GJ, Van Montfoort JE, Versteegh JFM (2011) Demographic projections of future pharmaceutical consumption in the Netherlands. Water Sci Technol 63 (4):825–831
7. European Commission (2015) Commission implementing decision (EU) 2015/495 of 20 March 2015 establishing a watch list of substances for union-wide monitoring in the field of water policy pursuant to Directive 2008/105/EC of the European Parliament and of the Council. http://eur-lex.europa.eu/legal-content/EN/TXT/PDF/?uri=CELEX:32015D0495&from=EN
8. Carvalho RN, Ceriani L, Ippolito A, Lettieri T (2015) Development of the first watch list under the environmental quality standards directive. EC Joint Research Centre, Ispra
9. ter Laak TL, van der Aa M, Houtman CJ, Stoks PG, van Wezel AP (2010) Relating environmental concentrations of pharmaceuticals to consumption: a mass balance approach for the river Rhine. Environ Int 36(5):403–409
10. Schoeps P, Schriks M (2010) The effect of REACH on the log Kow distribution of drinking water contaminants. KWR Watercycle Research Institute, Nieuwegein
11. Matilainen A, Vepsäläinen M, Sillanpää M (2010) Natural organic matter removal by coagulation during drinking water treatment: a review. Adv Colloid Interf Sci 159(2):189–197
12. Westerhoff P, Mezyk SP, Cooper WJ, Minakata D (2007) Electron pulse radiolysis determination of hydroxyl radical rate constants with Suwannee river fulvic acid and other dissolved organic matter isolates. Environ Sci Technol 41(13):4640–4646
13. Cooper WJ, Mezyk SP, Peller JR, Cole SK, Song W, Mincher BJ, Peake BM (2008) Studies in radiation chemistry: application to ozonation and other advanced oxidation processes. Ozone Sci Eng 30(1):58–64
14. Cooper WJ, Song W, Gonsior M, Kalnina D, Peake BM, Mezyk SP (2008) Recent advances in structure and reactivity of dissolved organic matter in natural waters. Water Sci Technol Water Supply 8:615–623
15. Lakretz A, Ron EZ, Harif T, Mamane H (2011) Biofilm control in water by advanced oxidation process (AOP) pre-treatment: effect of natural organic matter (NOM). Water Sci Technol 64 (9):1876–1884
16. Lamsal R, Walsh ME, Gagnon GA (2011) Comparison of advanced oxidation processes for the removal of natural organic matter. Water Res 45(10):3263–3269
17. Matilainen A, Sillanpää M (2010) Removal of natural organic matter from drinking water by advanced oxidation processes. Chemosphere 80(4):351–365
18. Metz DH, Reynolds K, Meyer M, Dionysiou DD (2011) The effect of UV/H2O2 treatment on biofilm formation potential. Water Res 45(2):497–508
19. Thomson J, Parkinson A, Roddick FA (2004) Depolymerization of chromophoric natural organic matter. Environ Sci Technol 38(12):3360–3369
20. Sarathy SR, Stefan MI, Royce A, Mohseni M (2011) Pilot-scale UV/H2O2 advanced oxidation process for surface water treatment and downstream biological treatment: effects on natural organic matter characteristics and DBP formation potential. Environ Technol 32(15):1709–1718
21. Azami H, Sarrafzadeh MH, Mehrnia MR (2012) Soluble microbial products (SMPs) release in activated sludge systems: a review. J Environ Health Sci Eng 9(1):30
22. Xie WM, Ni BJ, Seviour T, Yu HQ (2013) Evaluating the impact of operational parameters on the formation of soluble microbial products (SMP) by activated sludge. Water Res 47 (3):1073–1079
23. Huber SA, Balz A, Abert M, Pronk W (2011) Characterisation of aquatic humic and non-humic matter with size-exclusion chromatography – organic carbon detection – organic nitrogen detection (LC-OCD-OND). Water Res 45(2):879–885
24. Grefte A (2013) Removal of natural organic matter fractions by anion exchange; impact on drinking water treatment processes and biological stability. Civil engineering. Delft University of Technology, Delft

25. Audenaert WTM, Vermeersch Y, Van Hulle SWH, Dejans P, Dumoulin A, Nopens I (2011) Application of a mechanistic UV/hydrogen peroxide model at full-scale: sensitivity analysis, calibration and performance evaluation. Chem Eng J 171(1):113–126
26. Rosario-Ortiz FL, Snyder S, Suffet IH (2007) Characterization of the polarity of natural organic matter under ambient conditions by the Polarity Rapid Assessment Method (PRAM). Environ Sci Technol 41(14):4895–4900
27. Rosario-Ortiz FL, Mezyk SP, Doud DFR, Snyder SA (2008) Quantitative correlation of absolute hydroxyl radical rate constants with non-isolated effluent organic matter bulk properties in water. Environ Sci Technol 42(16):5924–5930
28. Rosario-Ortiz FL, Mezyk SP, Wert EC, Devin FRD, Singh MK, Xin M, Baik S, Snyder SA (2008) Effect of ozone oxidation on the molecular and kinetic properties of effluent organic matter. J Adv Oxid Technol 11(3):529–535
29. Rosario-Ortiz FL, Snyder SA, Suffet IH (2007) Characterization of dissolved organic matter in drinking water sources impacted by multiple tributaries. Water Res 41(18):4115–4128
30. Rosario-Ortiz FL, Wert EC, Snyder SA (2010) Evaluation of UV/H2O2 treatment for the oxidation of pharmaceuticals in wastewater. Water Res 44(5):1440–1448
31. Slaats PGG, Meerkerk MA, Hofman-Caris CHM (2013) Conditionering: de optimale samenstelling van drinkwater; Kiwa-Mededeling 100 –Update 201
32. WHO (2005) Nutrients in drinking water; water, sanitation and health protection and the human environment. WHO, Geneva
33. Leurs LJ, Schouten LJ, Mons MN, Goldbohm RA, Van Den Brandt PA (2010) Relationship between tap water hardness, magnesium, and calcium concentration and mortality due to ischemic heart disease or stroke in the Netherlands. Environ Health Perspect 118(3):414–420
34. Momeni M, Gharedaghi Z, Amin MM, Poursafa P, Mansourian M (2014) Does water hardness have preventive effect on cardiovascular disease? Int J Prev Med 5(2):159–163
35. Rylander R (2014) Magnesium in drinking water – a case for prevention? J Water Health 12 (1):34–40
36. Rapant S, Fajčíková K, Cvečková V, Ďurža A, Stehlíková B, Sedláková D, Ženišová Z (2015) Chemical composition of groundwater and relative mortality for cardiovascular diseases in the Slovak Republic. Environ Geochem Health 37(4):745–756
37. De Moel PJ, Verberk JQJC, Van Dijk JC (2006) Principles and practices. Sdu Editors, Delft University of Technology, Delft
38. Bruins JH, Petrusevski B, Slokar YM, Huysman K, Joris K, Kruithof JC, Kennedy MD (2015) Biological and physico-chemical formation of Birnessite during the ripening of manganese removal filters. Water Res 69:154–161
39. Verwey EJW, Overbeek JTG, van Nes K (1948) Theory of the stability of lyophobic colloids; the interaction of sol particles having an electric double layer. Elsevier, New York
40. Verwey EJW, Overbeek JTG (1955) Theory of the stability of lyophobic colloids. J Colloid Sci 10(2):224–225
41. Meijers AP (1974) De theorie van de vlokvomring. Rijswijk, Kiwa
42. Duan J, Gregory J (2003) Coagulation by hydrolysing metal salts. Adv Colloid Interf Sci 100–102(Suppl):475–502
43. van Melick MJ (1975) Tweede rapport van de commissie Vlokvomring en vlokverwijdering
44. Amirtharajah A, Mills KM (1982) Rapid-mix design for mechanisms of alum coagulation. J Am Water Works Ass 74(4):210–216
45. Crittenden JC, Rhodes Trussell R, Hand DW, Howe KJ, Tchobanoglous G (2012) Water treatment principles and design3rd edn. Wiley, Hoboken
46. Van Der Horst W, Ijpelaar GF, Scholte-Veenendaal P, Rietveld LC, Van Dijk JC (2006) Occurrence and removal of pharmaceuticals and Endocrine Disrupting Compounds (EDCs) from drinking water. In: American Water Works Association – water quality technology conference and exposition 2006: taking water quality to new heights
47. Saraiva Soares AF, Leão MMD, Vianna Neto MR, Da Costa EP, De Oliveira MC, Amaral NB (2013) Efficiency of conventional drinking water treatment process in the removal of endosulfan, ethylenethiourea, and 1,2,4-triazole. J Water Supply Res Technol 62(6):367–376

48. Gale P, Pitchers R, Gray P (2002) The effect of drinking water treatment on the spatial heterogeneity of micro-organisms: implications for assessment of treatment efficiency and health risk. Water Res 36(6):1640–1648
49. Edzwald JK (1995) Principles and applications of dissolved air flotation. Water Sci Technol 31 (3–4):1–23
50. Rook JJ (1974) Formation of haloforms during chlorination of natural water. Water Treat Exams 23(2):234–243
51. Sosnin EA, Oppenländer T, Tarasenko VF (2006) Applications of capacitive and barrier discharge excilamps in photoscience. J Photochem Photobiol C: Photochem Rev 7(4):145–163
52. Lerouge S, Fozza AC, Wertheimer MR, Marchand R, Yahia LH (2000) Sterilization by low-pressure plasma: the role of vacuum-ultraviolet radiation. Plasmas Polym 5(1):31–46
53. Halfmann H, Denis B, Bibinov N, Wunderlich J, Awakowicz P (2007) Identification of the most efficient VUV/UV radiation for plasma based inactivation of Bacillus atrophaeus spores. J Phys D Appl Phys 40(19):5907–5911
54. Wang D, Oppenländer T, El-Din MG, Bolton JR (2010) Comparison of the disinfection effects of vacuum-UV (VUV) and UV light on Bacillus subtilis spores in aqueous suspensions at 172, 222 and 254 nm. Photochem Photobiol 86(1):176–181
55. Templeton MR, Andrews RC, Hofmann R (2005) Inactivation of particle-associated viral surrogates by ultraviolet light. Water Res 39(15):3487–3500
56. Wols BA, Harmsen DJH, Beerendonk EF, Hofman-Caris CHM (2014) Predicting pharmaceutical degradation by UV (LP)/H2O2 processes: a kinetic model. Chem Eng J 255:334–343
57. Wols BA, Harmsen DJH, Beerendonk EF, Hofman-Caris CHM (2015) Predicting pharmaceutical degradation by UV (MP)/H2O2 processes: a kinetic model. Chem Eng J 263:336–345

Natural Organic Matter: Characterization and Removal by AOPs to Assist Drinking Water Facilities

S. Sorlini

Abstract The water sources of drinking water generally contain natural organic matter (NOM) as a result of the interactions between the hydrologic cycle and the environment. The amount, character, and properties of NOM vary considerably according to the origins of the waters and depend on the biogeochemical cycles of their surrounding environments. NOM can negatively influence water quality in drinking water supply systems, and it can significantly influence the performance of drinking water treatment processes. Hence, NOM removal is an important issue in order to optimize drinking water treatment operation and to reduce the risks of water alteration in the distribution systems. Several treatment processes can be applied for NOM removal depending on water quality, the nature of NOM, and the treatments already existing in the supply system. Among the most effective conventional solutions coagulation/flocculation, filtration, and carbon adsorption are available. An interest has recently increased toward nonconventional solutions based on membrane filtration and advanced oxidation processes (AOPs). An overview on the AOPs will be presented and discussed. Moreover, the AOP with ozone and UV radiation, with two low pressure UV lamps, at 254 and 185 nm wavelength, was experimented on a surface water in order to study the removal of odorous and pesticide, organic compounds (UV absorbance and THMs precursors) and bromate formation. Different batch tests were performed with ozone concentration up to 10 mg L^{-1}, UV dose up to 14,000 J m^{-2}, and a maximum contact time of 10 min. The main results show that metolachlor can be efficiently removed with ozone alone while for geosmin and MIB a complete removal can be obtained with the advanced oxidation of ozone, with concentration of 1.5–3 mg L^{-1} and contact time of 2–3 min, with UV radiation (with doses of 5,000–6,000 J m^{-2}). As concerns the influence of the organic precursors, all the experimented processes show a medium

---

S. Sorlini (✉)
Department of Civil, Environmental, Architectural Engineering and Mathematics, University of Brescia, Brescia, Italy
e-mail: sabrina.sorlini@unibs.it

removal of about 20–40% for UV absorbance and 15–30% for THMFP (trihalomethane formation potential).

Keywords Advanced oxidation processes, Characterization, Natural organic matter, Removal

Contents

1 Introduction	55
2 NOM Characteristics and Characterization	56
3 Impact of NOM on Drinking Water Supply Systems	57
4 AOPs for the Removal of NOM	57
4.1 Ozone-Based Treatments (O_3/H_2O_2, O_3/UV, and $O_3/H_2O_2/UV$)	58
4.2 UV Light-Based Treatments	59
4.3 Photocatalytic Oxidation	59
4.4 Fenton's Process	60
5 Case Study: Effect of the O_3/UV Process on Organic Matter Removal and Influence on DBPs Formation	60
5.1 Raw Water	60
5.2 Experimental Plant	61
5.3 Analytical Methods	61
5.4 Results: Removal of Organic Matter	62
5.5 Results: Removal of Geosmin, MIB, and Metolachlor	62
5.6 Results: Bromate Formation	65
5.7 Conclusions	65
6 General Considerations About the Application of AOPs for NOM Removal in Drinking Water Treatment Plants	66
References	66

Abbreviations

AOPs	Advanced oxidation processes
DBPFP	Disinfection by-products formation potential
DBPs	Disinfection by-products
DOC	Dissolved organic carbon
FAs	Fulvic acids
GC	Gas chromatography
HAAFP	Haloacetic acid formation potential
HAAs	Haloacetic acids
HAs	Humic acids
HMM	High molar mass
LMM	Low molecular mass
NMR	Nuclear magnetic resonance
NOM	Natural organic matter
SUVA	Specific UV absorbance
THMFP	Trihalomethane formation potential

THMs   Trihalomethanes
TOC   Total organic carbon
TOX   Total organic halide
UV–Vis Ultraviolet and visible
VUV   Vacuum ultraviolet

1 Introduction

Natural organic matter (NOM) is ubiquitous in water as it comes from the degradation of plants, bacteria, algae, and vegetal organisms in general. NOM is defined as a complex matrix of organic materials, present in all natural water. Water systems often have multiple sources of NOM, thus different organic carbon fractions [1]. The amount, character, and properties of NOM vary considerably according to the origins of the waters and depend on the biogeochemical cycles of their surrounding environments [2]. Concentrations of TOC in the ground water are typically in the range of 0.1–2.0 mg L^{-1} and 1–20 mg L^{-1} in surface water [3, 4] although several dozens of mg L^{-1} of TOC are not uncommon [5]. The factors that determine the composition of NOM are location dependent and include the source of organic matter, the water chemistry, temperature, pH, and biological processes. Moreover, the range of organic components in NOM can vary seasonally at the same location [6], for example, due to rainfall, snowmelt runoff, floods, or droughts. NOM found in natural waters consists of both hydrophobic and hydrophilic components, of which the largest fraction is generally hydrophobic acids, making up approximately 50% of the total organic carbon (TOC) in water [7].

NOM has a significant impact on many aspects of water treatment, including the performance of unit processes, necessity for and application of water treatment chemicals, and the biological stability of the water. Once the composition and quantity of NOM in the water source has been examined, suitable methods for efficient NOM removal can be applied. No single process alone can be used to treat NOM due to its high variability. Among the suitable solutions, the following treatment processes can be efficiently applied for NOM removal: adsorption, biological treatments, coagulation/flocculation, electrochemical methods, ion exchange, membrane technology, and AOP [8]. An increasing interest in drinking water treatment has recently been shown for the AOPs due to several advantages: low chemical consumption in some AOPs, complete mineralization of pollutant, oxidation of disinfection by-product (DBP) rapid reaction, unselective oxidants, harmful to microorganisms, and easily implemented in existing water treatment plants.

This chapter presents an overview on AOPs applied for the NOM removal during drinking water treatments. Moreover, the main results of an experimental study on the O_3/H_2O_2 process for NOM and micropollutant removal will be presented.

2 NOM Characteristics and Characterization

NOM compounds are complex mixtures possessing unique combinations of various functional groups, including esteric, phenolic, quinine, carboxylic, hydroxyl, amino, and nitroso, which are usually negatively charged at neutral pH. NOM found in natural waters consists of both hydrophobic and hydrophilic components, of which the largest fraction is generally hydrophobic acids, making up approximately 50% of the total organic carbon (TOC) in water [7]. These hydrophobic acids can be described as humic substances comprising (1) humic acids (HAs), which are soluble in alkali, but insoluble in acid; (2) fulvic acids (FA), which are soluble in both alkali and acid; and (3) humins, which are insoluble in both alkali and acid. FAs constitute a major fraction of these humic substances, which vary in molecular size and functional group content [9]. Hydrophobic NOM is rich in aromatic carbon, phenolic structures, and conjugated double bonds, while hydrophilic NOM contains more aliphatic carbon and nitrogenous compounds, such as carbohydrates, sugars, and amino acids.

The amount of NOM in water can be predicted with parameters including ultraviolet and visible (UV–Vis) detected compounds, TOC, and SUVA. TOC and dissolved organic carbon (DOC) are the most convenient parameters for analyzing the NOM removal of treatment processes. DOC is the organic carbon in a water sample filtered through a 0.45 μm filter [10]. TOC is the sum of the particulate and DOC when existing inorganic carbon is removed by acidification.

UV–Vis absorption spectroscopy is a semiquantitative method to determine humic substances in natural waters. Any wavelength from 220 to 280 nm is appropriate for NOM measurements: absorbance at 220 nm is associated with both the carboxylic and aromatic chromophores, whereas absorbance at 254 nm is typical for aromatic groups, and it has been identified as a potential surrogate measure for DOC [11].

Specific UV absorbance (SUVA) is calculated as the ratio of a UV absorbance at a specific wavelength (e.g., UV254 absorbance) and the TOC concentration. A high SUVA value indicates that the organic matter is largely composed of hydrophobic, high-MM organic material. A low SUVA value indicates that the water contains organic compounds that are mainly hydrophilic, with a low MM and charge density [6].

Several methods can be used for the characterization of NOM including resin adsorption, size exclusion chromatography, nuclear magnetic resonance (NMR) spectroscopy, and fluorescence spectroscopy. More precise methods for determining NOM structures have been developed recently: pyrolysis gas chromatography mass spectrometry, multidimensional NMR techniques, and Fourier transform ion cyclotron resonance mass spectrometry.

A detailed description of the methods used to characterize various features of natural organic matter (NOM) is reported in literature [12].

3 Impact of NOM on Drinking Water Supply Systems

Significance of NOM for drinking water quality and stability is represented by the impact on several aspects of its treatment [13]:

- NOM greatly affect organoleptic properties of the water (taste, color, and odor) [4, 14].
- NOM influence chemical properties in terms of mostly negative reactions and interferences with other chemicals used for oxidation and disinfection by lowering their effectiveness and thus increasing their consumption to achieve the treatment goal [4, 14].
- NOM can form complexes with the toxic heavy metals and synthetic organic chemicals, making them more soluble and more difficult to remove by the treatment [14].
- NOM are involved in formation of undesired and detrimental DBPs with an increase in bioavailability of organic matter for microorganisms in the water supply systems, enabling their proliferation.

NOM can cause several problems in drinking water treatment and distribution systems. It can affect the performance of unit processes, the necessity for application of water treatment chemicals, and the biological stability of the water. NOM increases the reagent dose necessary during water treatment, especially for coagulant and disinfectant. NOM can also interfere with the performance of unit operations, such as biofilm growth on media, causing rapid filter clogging and fast saturation of activated carbon beds. NOM is also responsible for the fouling of membranes.

During water treatment, NOM has generally been considered the main precursor to DBPs, especially hydrophobic and high molar mass (HMM) NOM, with its high aromatic carbon content [15, 16]. It has also been observed that hydrophilic and low molar mass (LMM) NOM play a significant role in DBP formation [15]. Bromine and iodine appear more reactive with hydrophilic and LMM fractions of NOM in the formation of THMs and HAAs (haloacetic acids). Chlorine has been shown to react more readily with HMM and hydrophobic NOM compounds [15].

NOM can impact also the water behavior in the distribution system as it affects corrosion, is a source of nutrients for heterotrophic bacteria, and acts as a substrate for bacterial growth in the pipes [17].

4 AOPs for the Removal of NOM

AOPs are obtained from the combination of several oxidants, radiation, and catalysts: O_3/H_2O_2, UV/H_2O_2, UV/O_3, UV/TiO_2, Fe^{2+}/H_2O_2, $Fe^{2+}/H_2O_2 + h\nu$, vacuum ultraviolet (VUV) radiation, or ionizing radiation. All these processes involve the generation of highly reactive radical intermediates, especially the $OH^{\bullet}$ radical [18] that is one of the most powerful known oxidants. The reaction rate of a compound in OH• radical-mediated oxidation is usually several orders of magnitude higher than that achieved by molecular ozone under the same conditions.

The reaction rate constants between OH• radicals and organic species are in the range of 10^8-10^{10} M^{-1} s^{-1} [19].

OH• radicals are highly nonselective oxidants, enabling a very large number of reactions. Once free radical reaction has been initiated (following photolysis, ozone, hydrogen peroxide, or heat), a series of simple reactions will occur. The reactions of OH• radicals with NOM proceed in three different ways:

- By the addition of OH• radicals to double bonds
- By H-atom abstraction, which yields carbon-centered radicals
- By the OH• radical gaining an electron from an organic substituent

4.1 Ozone-Based Treatments (O_3/H_2O_2, O_3/UV, and $O_3/H_2O_2/UV$)

Ozone reacts with NOM by an electrophilic addition to double bonds. This reaction is very selective. In addition to the direct reaction of ozone with NOM, a nonselective and fast reaction occurs with the OH• radicals formed when ozone decomposes in water. The OH• radical formation potential is much lower during ozonation than in AOPs. The combination of ozone with other systems, like UV light or hydrogen peroxide, can promote OH• formation.

The influence of ozone on THM precursors depends on the kind and structures of the organic material that can have different reactivities toward ozone and chlorine. Some authors observed that ozonation produces a transformation of natural organic matter from more reactive hydrophobic DOC, that reacts easily with chlorine to produce THM, to a hydrophilic fraction, with a consequent lower THM formation [20]. This finding is in agreement with other authors [21, 22].

The combination of O_3 with H_2O_2 increases total THM concentrations than in ozonated samples [23]. Similarly, other authors [24] did not observe any significant gains in THM reduction after adding H_2O_2 or TiO_2 to the ozone treatment.

Other authors [25] observed that TOC, THMFP, and TOX decreased with O_3/UV in comparison with ozone alone. Moreover, O_3/UV results in significant mineralization of DOC, reduction of trihalomethane formation potential (THMFP), and haloacetic acid formation potential (HAAFP), according to [26].

The combined system $O_3/H_2O_2/UV$ was not remarkable more efficient than O_3/UV in HAA decomposition [27].

4.2 UV Light-Based Treatments

Much research has focused on developing applications for UV/H_2O_2; little attempt has been made to evaluate the impact of UV/H_2O_2 on NOM. Past studies have demonstrated that substantial reduction of DBP formation potential (DBPFP) could be achieved using UV/H_2O_2 [28–30]. But all these studies mainly focused on strong advanced oxidation conditions made possible by very long UV exposures (i.e., fluence) and/or high UV/H_2O_2 concentration. Under such conditions NOM is mineralized leading to a reduction in the concentration of NOM [31].

Although the combination of UV irradiation at 254 nm and H_2O_2 treatment increases the NOM reduction and promotes the OH· radical formation, a combination of high UV fluence and high peroxide concentration is required in order to generate significant reduction of NOM [30]. Moreover, under fluence and H_2O_2 concentrations typically applied in drinking water treatment applications, UV standard dose for disinfection is about 400 mJ cm^{-2} [26], and NOM is not removed.

4.3 Photocatalytic Oxidation

The photocatalytic oxidation combines the UV light with a heterogeneous photocatalysis in which several different semiconductors are employed (e.g., TiO_2, ZnO, ZnS, WO_3, $SrTiO_3$) for catalysis.

The TiO_2/UV process was applied for controlling membrane fouling by NOM [32]; even though the rate of TOC removal was relatively low, membrane fouling of both an MF and a UF was completely eliminated after 20 min of treatment due to the changes in NOM molecular characteristics.

The presence of hardness (Ca^{2+}, Mg^{2+}, Fe^{2+}/Fe^{3+}) and the accompanying common anions (e.g., Cl^-, NO_3^-, SO_4^{2-}, and HCO_3^-/CO_3^{2-}) have considerable effects on the degradation kinetics of NOM expressed as DOC or UV254 [33].

Little is known about the by-products formed during photocatalytic oxidation of NOM. Some authors [34] reported that aldehydes and ketones were typical intermediate products formed during photocatalytic oxidation of NOM in surface waters and these compounds would be oxidized to form carboxylic acids as the reaction proceeded. Photocatalytic oxidation may also form by-products that have different reactivity with chlorine disinfectants [35].

NOM plays an important role as inhibitor of oxidation of target micropollutants during AOPs; the degradation rates can decrease by one order of magnitude or more in the presence of backgrounds such as NOM. Authors [36] provided a novel analytical approach to select strategies to enhance the performance of AOPs, even in systems with high levels of NOM or other background constituents.

4.4 Fenton's Process

Fenton's reagent is a catalytic oxidative mixture that contains iron ions and hydrogen peroxide. In this process hydroxyl radicals are produced during the decomposition of hydrogen peroxide in the presence of ferrous salts.

Fenton's reagent, H_2O_2 with Fe(II)/Fe(III) ions, in water produces OH radicals. Some authors [16] observed that UV/Fe(III) treatment was ineffective in NOM removal, while the UV/Fe(III)/H_2O_2 system has the potential to remove organics with a broad range of MMs. Even though high NOM removal rates were detected, the remaining organic compounds appear fairly reactive to chlorine, thus resulting in low reduction of THMFP.

Fenton and photo-Fenton's processes were compared in order to assess their potential to remove NOM from organic-rich waters [19]. The performance of both processes was dependent on pH, Fe:H_2O_2 ratio, as well as Fe^{2+} dose. Under optimum conditions both processes achieved greater than 90% removal of DOC and UV_{254} absorbance. This removal leads to the THMFP of the water being reduced from 140 to below 10 $\mu g\ L^{-1}$.

A comparison of UV/H_2O_2, Fenton, and photo-Fenton for treating an upland catchment reservoir (DOC = 7.5 mg L^{-1}) showed how both Fenton's processes could remove the excess of 90% of DOC and UV-absorbing species and are significantly quicker than UV/H_2O_2 [37].

5 Case Study: Effect of the O_3/UV Process on Organic Matter Removal and Influence on DBPs Formation

This experimental research was addressed to compare the ozone conventional oxidation with the advanced oxidation with ozone and UV radiation, as concerns geosmin, MIB and metolachlor removal, the influence on DBPs organic precursors, and bromate formation [38].

5.1 Raw Water

Raw water, collected from the river Secchia at the treatment plant of Tressano, managed by AGAC of Reggio Emilia, in the North of Italy, showed the characteristics indicated in Table 1: TOC = 1.8 mg L^{-1}; UV absorbance = 0.36 cm^{-1}; THMFP = 70–80 $\mu g\ L^{-1}$; bromide = 30 $\mu g\ L^{-1}$.

During the experimental tests, water was artificially contaminated with 0.5 $\mu g\ L^{-1}$ geosmin (*trans*-1,10-dimethyl-*trans*-9-decalol), 0.2–0.4 $\mu g\ L^{-1}$ MIB (2-methylisoborneol), and 7–10 $\mu g\ L^{-1}$ metolachlor [2-chloro-*N*-(2-ethyl-6-methylphenyl)-*N*-(2-methoxy-1-methylethyl)-acetamide].

Table 1 Raw water characteristics

Parameter	Average	Maximum	Minimum
pH (pH unit)	8.06	8.33	7.6
Alkalinity (mg HCO_3 L^{-1})	173.84	263.2	136.44
Turbidity (NTU)	7.85	38.8	1.7
TOC (mg L^{-1})	1.80	3.35	0.61
UV absorbance 254 nm (cm^{-1})	0.36	0.9	0.058
Transmittance (%)	63.0	51.0	83.0
THMFP (µg L^{-1})	76.05	124.5	18
Bromide (µg L^{-1})	32.96	91	5
Ammonium (mg NH_4 L^{-1})	0.05	0.14	<0.005

5.2 Experimental Plant

The experimental tests were performed in batch conditions on a laboratory-scale plant (Fig. 1) ($Q = 10$ L min^{-1}) formed by a stainless steel reactor (volume of 20 L), a pump, and an in-line ozone injection followed by a static mixer and a low pressure mercury-vapor ultraviolet (UV) lamp. Ozone was generated by means of "Ozonia Triogen Compact Ozone Generator" (Model TOGC2), with a capacity of 8 gO_3 h^{-1}. Two UV lamps (Model TR-65) were applied separately with various nominal wavelengths of maximum light intensity of 254 and 185 nm. Each lamp had about 80% of the radiation around the maximum wavelength, and the intensity was about 25 W m^{-2}.

The experimental conditions tested on water samples were ozone concentrations $=0$–15 mg L^{-1}, ozone contact time $= 0$–14 min, and UV dose $= 0$–14,000 J m^{-2}. Water samples were collected in an outlet from the UV chamber after 2, 4, 6, 8, and 10 min from the beginning of the test (the time required for one complete cycle in the system is 2 min).

5.3 Analytical Methods

Geosmin, MIB, and metolachlor were analyzed with a capillary column gas chromatography-mass spectrometry (GC/MS). Trihalomethane formation potential (THMFP) was determined following [39]; THMs were determined with a gas chromatograph (GC Perkin-Elmer 8600) with the static headspace method. TOC was analyzed with a total carbon monitor (Model 1010), the UV absorbance with an UV–visible spectrophotometer with a 1 cm quartz cell and turbidity with a 2100 AN Hach turbidimeter. Ozone concentration was measured by means of an ozone analyzer (BMT 961) that analyzes the flow of ozone production and that of ozone outlet from the reactor; the difference between the second and the first term gives the amount of ozone effectively transferred to water.

Fig. 1 Experimental plant

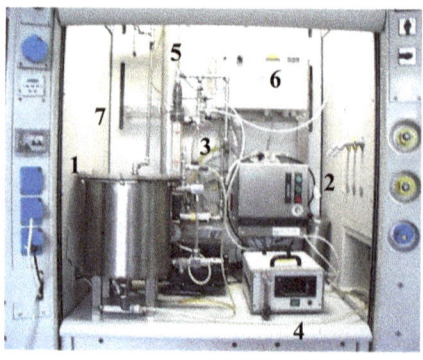

1 steel feed tank
2 ozone generator
3 static mixer
4 BMT 961
5 UV lamp
6 UV generator
7 sampling point

5.4 Results: Removal of Organic Matter

As concerns the influence on organic matter of the experimented treatments, ozone alone and combined with UV radiation reduces the absorption of radiation at the wavelength of 254 nm of about 10–20% with ozone/UV 254 nm and 10% with ozone/UV 185 nm (Fig. 2). This is due to a removal of aromatic structures and double bonds of natural organic matter.

As shown in Fig. 3, the removal of THM precursors with ozone and ozone/UV is very different (from 0 to 40%), and no improvements are observed with increasing ozone and UV dose and contact time. The partial THMFP reduction (10–30%) observed in most of the trials is due to the degradation of humic substances into low molecular weight compounds that are less reactive toward chlorine. However, at the same time, bromide, whose concentration in raw water varies from 5 to 90 $\mu g\ L^{-1}$, is oxidized to hypobromite which further leads to brominated compounds [40], with a consequent higher THM formation.

The combination of ozone with UV does not improve THM precursor removal with respect to ozone alone, according to the results of other authors [25] that found a significant reduction of total organic halide while no differences were shown for chloroform.

5.5 Results: Removal of Geosmin, MIB, and Metolachlor

Molecular ozone has very different reaction rates with organic compounds (Fig. 4): it reacts very fast with metolachlor, while the odorous compounds (MIB and geosmin) are more persistent, and their complete removal can be obtained only with ozone combined with UV radiation (Fig. 5).

The improvement in removal rates for taste and odor compounds obtained by the advanced oxidation process (O_3/UV) can be explained by the action of initiators (UV rays) to introduce the decomposition of the ozone in water, thus generating

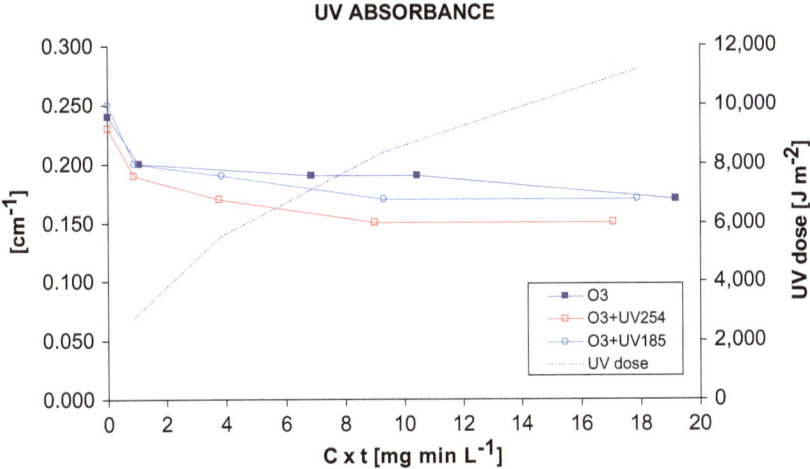

Fig. 2 UV absorbance removal with ozone, ozone/UV 254 nm and ozone/UV 185 nm

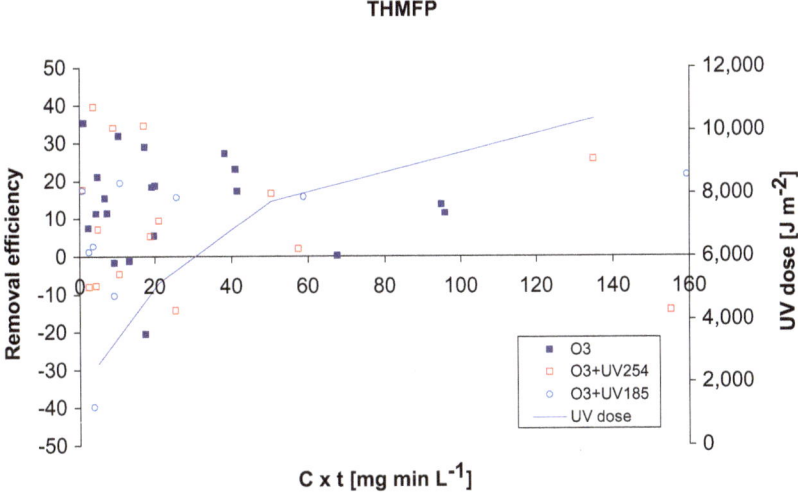

Fig. 3 THMFP removal with ozone, ozone/UV 254 nm and ozone/UV 185 nm

hydroxyl radicals that are very reactive when the water has a low alkalinity (that means low concentration of scavengers like HCO_3^-, CO_3^{2-}, etc.).

The main results show that metolachlor can be efficiently removed with ozone alone with $C \times t = 8\text{--}10$ mg min L^{-1} (ozone concentration = 1 mg L^{-1} and 8–10 min contact time); the same removal can be obtained with $C \times t = 4$ mg min L^{-1} for ozone combined with UV (UV dose = 4,000–6,000 J m^{-2}). This means that, for the same contact time, ozone concentration can be reduced from 1 mg L^{-1} to 0.2–0.4 mg L^{-1}, whereas, for geosmin and MIB, a complete removal can be obtained

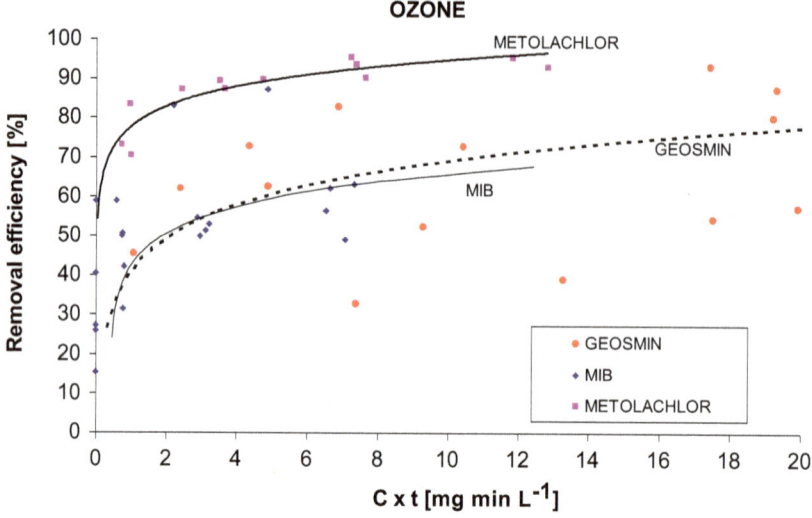

Fig. 4 Geosmin, MIB, and metolachlor removal with ozone

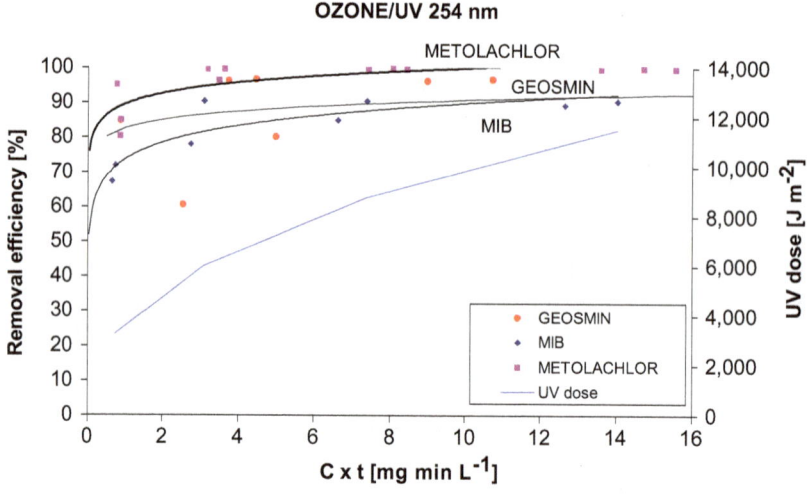

Fig. 5 Geosmin, MIB, and metolachlor removal with ozone/UV 254 nm

only with the combination of ozone (with concentration of 1.5–3 mg L^{-1} and contact time of 2–3 min) with UV radiation, with doses of 5,000–6,000 J m^{-2}.

5.6 Results: Bromate Formation

As concerns bromate formation (Fig. 6), a significant reduction is shown in the AOPs with ozone/UV with respect to conventional oxidation with ozone. UV radiation, in the wavelength range of 180–300 nm, provides energy to reduce bromate to hypobromite ion as intermediate and to bromide and oxygen as end products [41] via complex reactions generated by the primary reaction of photolysis. Ozone combined with UV lamp at 185 nm wavelength is about 10–20% lower than ozone alone; the highest bromate destruction is obtained with ozone combined with UV at 254 nm wavelength, for which bromate is about 40–50% lower than ozone alone and its final concentration is generally lower than the 31/01 Italian Legislative Decree Limit of 10 µg L^{-1}.

5.7 Conclusions

The conventional and advanced oxidation tests performed on water contaminated with geosmin, MIB and metolachlor show that the combination of ozone with UV radiation, both at 254 nm and 185 nm wavelengths, improves the process efficiency and offers a complete removal for all the analyzed contaminants with C × t of 4–10 mg min L^{-1} and UV doses of 4,000–6,000 J m^{-2}. All the experimented treatments offer a good removal of organic precursors, while a significant reduction (below the limit of 10 µg L^{-1} of 31/01 Italian Legislative Decree) of bromate is obtained only with ozone combined with UV 254 nm.

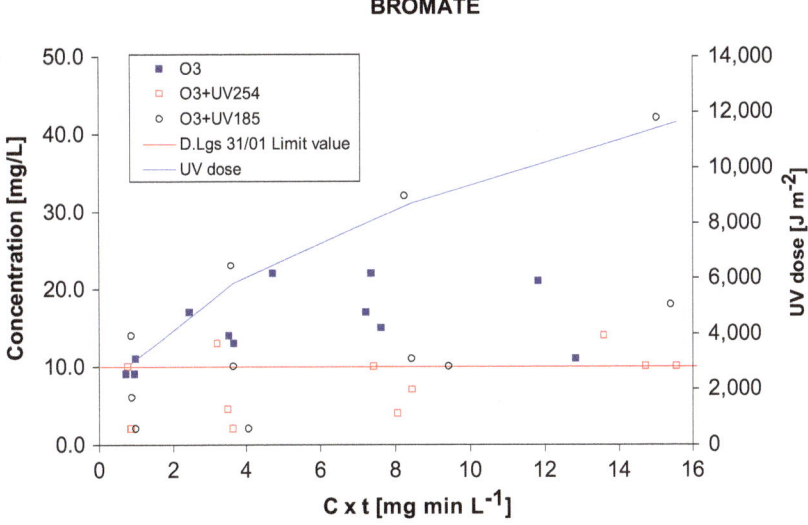

Fig. 6 Bromate formation during ozone, ozone/UV 254 nm and ozone/UV 185 nm

6 General Considerations About the Application of AOPs for NOM Removal in Drinking Water Treatment Plants

AOPs are among the most studied and promising technologies for drinking water purification and disinfection. Although total reduction of NOM has not been achieved with AOPs, several studies have shown efficient NOM reduction and mitigation of DBP formation.

The results of various studies dealing with NOM removal by AOPs are always study specific depending on the water characteristics, such as the amount of organic matter. Therefore, the characterization of NOM in water should be made before the design and optimization of the AOP treatment. Furthermore, in order to assess its influence on the downstream processes, it is important to determine the organic characteristics of the treated water.

The implementation of an AOP process in a drinking water treatment plant can be an interesting solution for the removal of NOM and DBP minimization in drinking water systems. First, the objective of the AOP treatment needs to be defined, whether it is to enhance biodegradability of organic matter or mineralize organic compounds. Therefore, the optimal conditions for each case must be determined according to the type and amount of organic compounds and the interfering background compounds present in the water.

AOPs generally are not applied alone but may find better application in combination with other treatments, thus enhancing their efficiency for NOM removal. For example, coagulation prior to oxidation removes most of the HMM NOM, thus impacting on subsequent AOP treatment. The combination of AOP and BAC treatment has been suggested to offer a more viable option for the reduction of harmful DBPs than the AOP alone.

It should be emphasized that AOPs have site-specific effects, so pilot-scale and full-scale tests must be conducted to define the optimal conditions for the process. Experimental and pilot-scale studies are often conducted under conditions that are not economically feasible in commercial applications. The full-scale application of AOPs in drinking water treatment plants is still limited because of high cost, high level of pretreatment required, a lack of experience, and operational difficulties. Nevertheless, there is an increasing interest for AOPs that will take to an increase of full-scale installation of these processes.

References

1. Rigobello E, Dantas A, Di Bernardo L, Vieira E (2011) Influence of the apparent molecular size of aquatic humic substances on colour removal by coagulation and filtration. Environ Technol 32:1767–1777
2. Fabris R, Chow C, Drikas M, Eikebrokk B (2008) Comparison of NOM character in selected - Australian and Norwegian drinking waters. Water Res 42:4188–4196

3. Rodrigues A, Brito P, Janknecht MF, Proença R, Nogueira R (2009) Quantification of humic acids in surface water: effects of divalent cations, pH, and filtration. J Environ Monit 11:377–382
4. Crittenden JC (2012) MWH's water treatment: principles and design. Wiley, Hoboken
5. Kokorite M, Klavins V, Rodinov G (2012) Trends of natural organic matter concentrations in river waters of Latvia. Environ Monit Assess 184:4999–5008
6. Sharp E, Jarvis P, Parsons S, Jefferson B (2006) Impact of fractional character on the coagulation of NOM. Colloids Surf A 286:104–111
7. Thurman E (1985) Organic geochemistry of natural waters. Martinus Nijhoff/Dr W. Junk Publishers, Dordrecht
8. Shestakova M, Sillanpää M (2013) Removal of dichloromethane from ground and wastewater: a review. Chemosphere 93:1258–1267
9. Snoeyink VL, Jenkins D (1980) Water chemistry. Wiley, New York
10. Danielsson L (1982) On the use of filters for distinguishing between dissolved and particulate fractions in natural waters. Water Res 16:179–182
11. Korshin G, Chow C, Fabris R, Drikas M (2009) Absorbance spectroscopy-based examination of effects of coagulation on the reactivity of fractions of natural organic matter with varying apparent molecular weights. Water Res 43:1541–1548
12. Sillanpää M (2015) Natural organic matter in water. Characterization and treatment methods. Elsevier, New York
13. Cehovin M, Medic A, Scheideler J, Mielcke J, Ried A, Kompare B, Gotvajn AZ (2017) Hydrodynamic cavitation in combination with the ozone, hydrogen peroxide and the UV-based advanced oxidation processes for the removal of natural organic matter from drinking water. Ultrason Sonochem 37:394–404
14. Matilainen A, Sillanpää M (2010) Removal of natural organic matter from drinking water by advanced oxidation processes. Chemosphere 80:351–365
15. Hua G, Reckhow D (2007) Characterization of disinfection byproduct precursors based on hydrophobicity and molecular size. Environ Sci Technol 41:3309–3315
16. Sanly, Lim M, Chiang K, Amal R, Fabris R, Chow C, Drikas M (2007) A study on the removal of humic acid using advanced oxidation process. Sep Sci Technol 42:1391–1404
17. Jacangelo J, DeMarco J, Owen D, Randtke S (1995) Selected processes for removing NOM: an overview. J Am Water Works Assoc 87(1):64–77
18. Glaze W, Kang J, Chapin D (1987) The chemistry of water treatment processes involving ozone, hydrogen peroxide and ultraviolet radiation. Ozone Sci Eng 9:335–352
19. Murray CA, Parsons SA (2004) Removal of NOM from drinking water: Fenton's and photo-Fenton's processes. Chemosphere 54(7):1017–1023
20. Galapate R, Baes A, Okada M (2001) Transformation of dissolved organic matter during ozonation: effects on trihalomethane formation potential. Water Res 35:2201–2206
21. Treguer R, Tatin R, Couvert A, Wolbert D, Tazi-Pain A (2010) Ozonation effect on natural organic matter adsorption and biodegradation – application to a membrane bioreactor containing activated carbon for drinking water production. Water Res 44:781–788
22. Molnar J, Agbaba J, Dalmacija B, Tubić A, Krčmar D, Maletić S, Tomašević D (2013) The effects of matrices and ozone dose on changes in the characteristics of natural organic matter. Chem Eng J 222:435–443
23. Irabelli A, Jasim S, Biswas N (2008) Pilot-scale evaluation of ozone vs. peroxone for trihalomethane formation. Ozone Sci Eng 30:356–366
24. Mosteo R, Miguel N, Martin-Muniesa S, Ormad M, Ovelleiro J (2009) Evaluation of trihalomethane formation potential in function of oxidation processes used during the drinking water production process. J Hazard Mater 172:661–666
25. Kusakabe K, Aso S, Hayashi J, Isomura K, Morooka S (1990) Decomposition of humic acid and reduction of trihalomethane formation potential in water by ozone with UV irradiation. Water Res 24:781–785

26. Chin A, Bérube PR (2005) Removal of disinfection by-product precursors with ozone-UV advanced oxidation process. Water Res 39(10):2136–2144
27. Wang GS, Liao CH, Chen HW, Yang HC (2006) Characteristics of natural organic matter degradation in water by UV/H_2O_2 treatment. Environ Technol 27(3):277–287
28. Kleiser G, Frimmel F (2000) Removal of precursors for disinfection by-products (DBPs) – differences between ozone- and OH-radical-induced oxidation. Sci Total Environ 256:1–9
29. Liu W, Andrews SA, Sharpless C, Stefan M, Linden KG, Bolton JR (2002) Bench-scale investigations into comparative evaluation of DBP formation from different UV/H2O2 technologies. In: Proceedings of the AWWA water quality technology conference, Seattle
30. Toor R, Mohseni M (2007) UV/H_2O_2 based AOP and its integration with biological activated carbon treatment for DBP reduction in drinking water. Chemosphere 66:2087–2095
31. Sarathy SR, Mohseni M (2009) UV/H2O2 treatment of drinking water: impacts on NOM characteristics, vol 11. IUVA News
32. Huang X, Leal M, Li Q (2008) Degradation of natural organic matter by TiO_2 photocatalytic oxidation and its effect on fouling of low-pressure membranes. Water Res 42(4–5):1142–1150
33. Uyguner CS, Bekbolet M (2009) Application of photocatalysis for the removal of natural organic matter in simulated surface and ground waters. J Adv Oxid Technol 12(1):2371–1175
34. Liu S, Lim M, Fabris R, Chow C, Drikas M, Amal R (2008) TiO_2 Photocatalysis of natural organic matter in surface water: impact on trihalomethane and haloacetic acid formation potential. Environ Sci Technol 42:6218–6223
35. Pichat P (2013) In: Pichat P (ed) Photocatalysis and water purification: from fundamentals to recent applications. Wiley, New York
36. Brame J, Long M, Li Q, Alvarez P (2015) Inhibitory effect of natural organic matter or other background constituents on photocatalytic advanced oxidation processes: mechanistic model development and validation. Water Res 84:1–10
37. Parsons S, Byrne A (2004) Water treatment applications. In: Parsons S (ed) Advanced oxidation processes for water and wastewater treatment. IWA Publishing, London, pp 329–34638
38. Collivignarelli C, Sorlini S (2004) AOPs with ozone and UV radiation in drinking water: contaminants removal and effects on disinfection byproducts formation. Water Sci Technol 48(4):51–56
39. Standard methods for the examination of water and wastewater (1998) Front cover, 20th edn. APHA – American Public Health Association
40. Camel V, Bermond A (1998) The use of ozone and associated oxidation processes in drinking water treatment. Water Res 32(11):3208–3222
41. Siddiqui MS, Amy GL, McCollum LJ (1996) Bromate destruction by UV irradiation and electric arc discharge. Ozone Sci Eng 18:271–290

Natural Organic Matter Removal by Heterogeneous Catalytic Wet Peroxide Oxidation (CWPO)

José Herney Ramírez and Luis Alejandro Galeano

Abstract NOM usually reaches drinking water supply sources through metabolic reactions and soil leaching. It has been, in general, considered that NOM is still one of the most problematic contaminants present in this kind of influents. Therefore, in the present chapter, most relevant technologies used for removal of NOM and its constituents from water have been examined, emphasizing in the past few years. An overview of the recent research studies dealing the NOM removal by catalytic wet peroxide oxidation and other closely related heterogeneous Fenton-like AOPs is presented. As revealed from recent literature reports, heterogeneous Fenton processes including CWPO are still emerging, promising catalytic technologies for NOM removal from water. A wide variety of catalytic solids reported within the past few years has been examined focusing on their potential in the NOM removal from water. Main findings offered by several types of catalysts like zeolites, Fe-functionalized activated carbons, carbon nanotubes, but mainly pillared and other clay minerals have been critically discussed emphasizing on the NOM removal by CWPO.

Keywords Al/Fe-pillared clay, Catalytic wet peroxide oxidation, Natural organic matter, Water treatment

---

J.H. Ramírez (✉)
Departamento de Ingeniería Química y Ambiental, Facultad de Ingeniería, Universidad Nacional de Colombia, Bogotá, DC, Colombia
e-mail: jhramirezfra@unal.edu.co

L.A. Galeano
Grupo de Investigación en Materiales Funcionales y Catálisis (GIMFC), Departamento de Química, Facultad de Ciencias Exactas y Naturales, Universidad de Nariño, Pasto-Nariño, Colombia
e-mail: alejandrogaleano@udenar.edu.co

Contents

1 Introduction 71
2 Conventional Techniques Used in NOM Removal from Water 71
    2.1 Adsorption 71
    2.2 Coagulation/Flocculation 72
    2.3 Advanced Oxidation Processes (AOPs) 73
    2.4 Ion Exchange 75
3 NOM Removal from Water by Heterogeneous CWPO and Other Related AOPs 75
    3.1 CWPO by Pillared and Other Related Clay Catalysts 77
    3.2 NOM Removal by CWPO and Other AOPs 84
    3.3 Other Solid Catalysts Used for NOM Removal 88
4 Conclusions 92
References 93

Abbreviations

AC	Activated carbon
AERs	Anion exchange resins
AMR	Atomic metal ratio
AOC	Assimilable organic carbon
AOPs	Advanced oxidation processes
BAC	Biologically activated carbon
BDD	Boron-doped diamond
BDOC	Biodegradable dissolved organic carbon
CNTs	Carbon nanotubes
CWPO	Catalytic wet peroxide oxidation
CEC	Cationic exchange capacity
CC	Chemical coagulation
COD	Chemical oxygen demand
4-CP	4-Chlorophenol
DBPs	Disinfection by-products
DOC	Dissolved organic carbon
EC	Electrocoagulation
EO	Electro-oxidation
GAC	Granular activated carbon
HC	Hydrodynamic cavitation
HDC	Hydro-dechlorination
HMW	High molecular weight
HO$^\bullet$	Hydroxyl radical
IMW	Intermediate molecular weight
LC-OCD	Liquid chromatography-organic carbon detector
LMW	Low molecular weight
NA	Nano-adsorbents
NOM	Natural organic matter
PCMC	P-chloro-m-cresol
PILCs	Pillared clays
ROS	Radical oxygen species
THMs	Trihalomethanes
TOC	Total organic carbon

1 Introduction

Waste and surface waters are currently strongly requiring development and application of novel, cost-effective treatments. A number of techniques such as chemical, physical, biological, incineration, etc. and their combinations are available, but each process has its inherent limitations in applicability, effectiveness, and cost. There are many water systems not suitable for biological treatments, mainly due to the presence of some highly refractory and toxic pollutants. Their treatments by conventional chemical processes may have several drawbacks in terms of efficiency and/or cost.

NOM may cause many problems in wastewater treatment processes, for example, undesirable color, taste, and odor, while reacting with common disinfectants to produce a variety of toxic DBPs [1]. Residual NOM can also promote bacterial regrowth and pipe corrosion in drinking water distribution systems [2]. NOM also reduces the overall efficiency of water treatment plants through increased chemical dosages, interference with the removal of other contaminants and filter fouling [3], as well as increased levels of complexed heavy metals and adsorbed organic pollutants [4], among others.

Changes in NOM loading and composition have a significant influence on the selection, design, and operation of water treatment processes. No single process alone can be used to treat NOM due to its very high variability.

2 Conventional Techniques Used in NOM Removal from Water

Diverse processes have been investigated to remove NOM from water such as adsorption, coagulation/flocculation, advanced oxidation processes, biological and electrochemical methods, ion exchange, and membrane technology. These processes are the most common and economically feasible to remove NOM. In the following sections, the removal of NOM by some conventional techniques has been presented and discussed.

2.1 Adsorption

The adsorption process is generally considered as one of the best water treatment technologies because of its convenience, ease of operation, and simplicity of design. The mechanism of NOM adsorption is mainly ligand exchange with the hydroxyl groups of the mineral surface [5]. Different kinds of adsorbents are used to eliminate NOM, including AC [6]. Li et al. [7] showed that constriction of internal pores of the carbon caused a reduction in the diffusion rate of the targeted compounds. By comparison of two carbons, the extent of reduction in the diffusion coefficient caused by adsorption of the same concentration of a pore-blocking compound, mg/g of AC, was found to be less pronounced for the carbon with larger volume of mesopores. AC is often used in drinking water treatment plants to remove or control unpleasant taste and odor, organic compounds, and NOM. Columns are used in municipal treatments to produce potable water as well as in disposable cartridges at industrial, commercial, and residential installations.

AC has been also used in the treatment of municipal wastewater either as a secondary or a tertiary process, with the advantage over other materials of being able to control odor in the water. The carbon may be either powdered or granular, the former being added as slurry into the water before chlorination, whereas the latter used in conventional gravity sand filters alone or together with sand.

GAC is good choice for removing NOM as well as taste and odor compounds. The removal of NOM by GAC is through reversible and irreversible physical adsorption caused by nonspecific mechanisms, such as van der Waals forces, dipole interactions, and hydrophobic interactions [8]. There are two options for locating GAC units in water treatment plants, i.e., (1) post-filtration adsorption, where the GAC unit is located after the conventional filtration process, and (2) filtration-adsorption, in which either all or a fraction of the filter granular media is replaced by GAC. Compared with filter adsorbents, the post-filtration provides higher flexibility for both handling GAC and design of specific conditions of adsorption and thus often allows for lower operational costs.

Many bench-, pilot-, and full-scale studies have shown GAC as a promising method to effectively remove NOM [9, 10]. However, a major constraint in operating GAC contactors is the cost of routinely replacing the GAC media due to the loss of adsorption capacity that occurs once GAC saturates.

2.2 Coagulation/Flocculation

The most common and economically feasible process available to remove NOM is coagulation and flocculation followed by sedimentation/flotation and filtration. Most of the NOM can be removed by coagulation, although the hydrophilic, LMW fractions of NOM are apparently removed less efficiently than the hydrophobic, HMW compounds [11]. Coagulation/flocculation in water/wastewater treatment plants involves the addition of chemicals to alter the physical state of dissolved and suspended solids and facilitate their removal by sedimentation [12]. As chemical products, coagulants react with the suspended and colloidal particles in the water, causing them to bind together and thus allowing for their removal in the subsequent treatment processes [13]. The aggregation mechanisms through which particles and colloids are removed include a combination of charge neutralization, entrapment, adsorption, and complexation with coagulant ions into insoluble masses [14].

Coagulation treatment has been employed to decrease turbidity and color and also to remove pathogens [15]. It is well established through the long and large literature that the coagulation process efficiency is highly dependent on hydrophilic and hydrophobic properties (Table 1) of NOM and dissolved organics [16–18].

Coagulation can be induced using chemical salts, such as ferric chloride or alum, or via electrocoagulation (EC) which uses sacrificial electrodes to provide a pure source of cations. Unlike chemical coagulation (CC), EC is not a commonly used water treatment technology. Nevertheless, EC has successfully treated a diverse variety of water types at the bench-scale, including municipal, textile dye, and petroleum refinery wastewaters [20–22].

2.3 Advanced Oxidation Processes (AOPs)

AOPs are based on the production of highly reactive, short-lived hydroxyl radicals, which react with organic contaminants with high reaction rate constants.

AOPs destroy the organic molecules, even the more stable, hard-to-degrade compounds, including carcinogens and mutagens, by means of the generation of highly reactive species which oxidize organic matter; thus, AOPs may be of great interest for public health and turned into a promising study field due to its almost total degradation potential of soluble organic contaminants in waters and soils, some of them under reasonably mild temperature and pressure conditions [23–26].

In AOPs, chemical reactions, electron beams, UV light, or ultrasound pulses are used to obtain high oxidation rates, thanks to the generation of free radicals (mainly hydroxyl radicals). Indeed, highly reactive HO$^\bullet$ are traditionally thought to be the main active species responsible for the destruction of the contaminants, including NOM. Their high standard reduction potential of 2.8 V in acidic media enables these radicals to oxidize almost all organic compounds to carbon dioxide and water,

Table 1 NOM fractions and chemical groups (adapted from [19])

Fraction	Chemical groups
Hydrophobic	
Strong acids	Humic and fulvic acids, HMW alkyl monocarboxylic and dicarboxylic acids, aromatic acids
Weak acids	Phenols, tannins, IMW alkyl monocarboxylic and dicarboxylic acids
Bases	Proteins, aromatic amines, HMW alkyl amines
Neutral	Hydrocarbons, aldehydes, HMW methyl ketones and alkyl alcohols, ethers, furans, pyrrole
Hydrophilic	
Acids	Hydroxy acids, sugars, sulfonics, LMW alkyl monocarboxylic and dicarboxylic acids
Bases	Amino acids, purines, pyrimidine, LMW alkyl amines
Neutrals	Polysaccharides, LMW alky alcohols, aldehydes, and ketones

excepting some of the most simple organic compounds, such as acetic, maleic, and oxalic acids, acetone, or simple chloride derivatives as chloroform [27]; however, just these species are of interest because they are typically oxidation products of larger molecules, being continuously generated by chemical, photochemical, or electrochemical reactions.

Among the more used AOPs, photo-Fenton and photocatalysis have been of special interest [28]. The photo-Fenton reaction is well known as an efficient and inexpensive method for wastewater treatment. Photo-Fenton improves the efficiency of dark Fenton or Fenton-like reagents, respectively, by means of the interaction of radiation (UV or Vis) with Fenton reagents. Light exposure increases the rate of HO$^\bullet$ formation by photoreactions of H_2O_2 ($\lambda < 360$ nm) and/or Fe^{3+} either producing HO$^\bullet$ straightforward or regenerating Fe^{2+} [29].

Photocatalysis is defined as catalytic activation of a given reaction via a mechanism that only proceeds if the system is lighted up. Photocatalytic reactions may operate via a number of different mechanisms. The main well-known mechanisms are (1) photolysis of adsorbates, (2) reactions of adsorbed species with photoelectrons or holes (photo-electrochemical reaction), and (3) injection of electrons from an excited adsorbate into a semiconducting mineral [30].

The efficiency of NOM removal using AOPs strongly depends on water characteristics, including the concentration of organic matter. Therefore, characterization of the NOM in water should necessarily anticipate the design and optimization of any AOP treatment. The characteristics of the organics in the treated water must be also determined to assess their influence on downstream processes.

The reactions of HO$^\bullet$ with NOM proceed at least by means of three different ways: (1) HO$^\bullet$ addition on double bonds; (2) H-atom abstraction, which yields carbon-centered radicals; and (3) the HO$^\bullet$ gaining an electron from an organic substituent. The carbon-centered radicals then react very rapidly with oxygen to form organic peroxyl radicals. The mutual reactions of peroxyl radicals can lead to the production of ketones or aldehydes and/or carbon dioxide. The rate of oxidation depends on radical, oxygen, and pollutant concentrations. Other factors affecting formation of the radicals are pH, temperature, the presence of ions, the type of pollutant, and the presence of scavengers such as bicarbonate ions [31].

Finally, HO$^\bullet$ can also be produced without using chemicals by means of cavitation. Generally it is a phenomenon of formation, cyclic growth, and rarefaction with the terminal implosive collapse of vapor bubbles in the liquid phase. Cavitation, caused by pressure gradients of water due to the influence of geometry, is called hydrodynamic cavitation (HC). Recent studies show beneficial effectiveness of HC for degradation of various recalcitrant organic contaminants (pharmaceuticals, synthetic and persistent organic chemicals, micropollutants, etc.) from water and wastewater [32].

2.4 Ion Exchange

Important properties of AERs include polymer composition, porosity, and charged functional groups. Performances of AERs for NOM removal are influenced by the inner characteristics of the resins (strong- or weak-base AER), the water quality (pH, ionic strength, hardness, etc.), and the nature of organic compounds (molecular weight, charge density, and polarity). The composition of most resins is either polystyrene or polyacrylic. Polystyrene resins are more hydrophobic than polyacrylic resins; as a result, polyacrylic resins tend to have more open structure and higher water content. The porosity of resins is defined as either macroporous or gel. Macroporous resins are highly porous solids, while gel resins do not contain any pores [30]. Strong-base resins typically contain either type I ($-N+(CH_3)_3$) or type II ($-N^+(CH_3)_2(C_2H_4OH)$) quaternary ammonium functional groups. Due to their ethanolic content, type II resins are more hydrophilic than type I resins [33]. Most strong-base anion exchange resins are used in the chloride form.

The advantages and disadvantages of NOM removal by using ion exchange treatment are presented in Table 2. As is evident this technology is very interesting since it has proven to be highly efficient, while exhibiting very low formation of DBP; only one disadvantage can be observed for this technology regarding the necessity of an additional treatment stage since the NOM removal is not complete.

3 NOM Removal from Water by Heterogeneous CWPO and Other Related AOPs

AOPs are gaining more and more interest as potential solutions in the field of water treatment; the use of appropriate catalysts can substantially decrease the energy consumption of oxidation processes, such as wet air and wet peroxide oxidation of refractory organic compounds. The main limitation of the conventional Fenton process is the dissolved homogeneous catalyst that cannot be easily recovered throughout the process, leading to additional pollution [34]. Besides, another outstanding issue that has deserved a number of works in the past few years regarding drawbacks of the homogeneously activated Fenton process has been the very narrow range of pH values ($<\sim 4.0$) [35] under which it is operable as efficient catalytic process itself (Fig. 1); it associates high operating costs for industrial-scale

Table 2 Advantages and disadvantages of NOM using ion exchange treatment

Advantages	Disadvantages
Proven technology	Additional stage of treatment required
Potential highly efficient	
Very low formation of DBPs	
Efficient in treating the transphilic fraction of NOM	

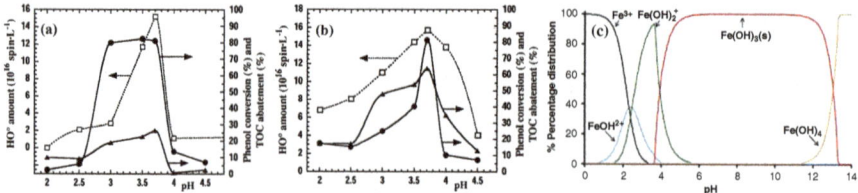

Fig. 1 Effect of pH on the catalytic performance of the Fenton process activated under either homogeneous or heterogeneous conditions: (**a**) homogeneous system ([Fe] = 5 mg/L) and (**b**) heterogeneous catalyst (Al/Fe-PILC FAZA, 5 g/L), amount of DMPO/HO adduct, phenol conversion (after 2 h of reaction), and TOC abatement (after 4 h of reaction) in function of the pH: (*open square*) DMPO/HO. Adduct amount (*dotted line*), phenol conversion (*filled circle*), TOC abatement (*filled triangle*) (reprinted from [35], with permission of Springer). (**c**) Speciation diagram of ferric hydroxyl species as a function of pH for a solution containing 1.0×10^{-5} M of Fe(III) at 25 °C obtained by MINEQL+ software (reprinted from [34], with permission of Elsevier)

applications of other variants of the Fenton process, including photo-Fenton. As it can be seen, although hydroxyl radical concentration becomes significantly affected in both Fenton systems over pH = 4.0, the homogeneous one seems much more susceptible (Fig. 1a, b), according to the speciation undergone by Fe(III) as a function of pH (Fig. 1c). In this sense, several works have been recently revised by Clarizia et al. [34], examining and comparing the most relevant papers dealing with photo-Fenton processes at neutral pH available in the literature. The main strategy to overcome this practical disadvantage so far (most or the real contaminated waters do not match such a range of pH values, including surface waters as common supply sources to produce drinking water) has been to adopt different types of chelating agents. Each iron(III)-ligand complex features a particular mechanism of photolysis, speciation pattern, light absorption properties, quantum yield, biodegradation, and toxicity that must be considered for choosing proper chelating agent and operating conditions in the photo-Fenton degradation under circumneutral pH values.

Therefore, one of the AOPs lately attracting more attention has been the so-called CWPO; it employs hydrogen peroxide as oxidizing agent in the presence of a solid Fenton or Fenton-like catalyst. The redox properties of the transition metals immobilized in solid catalysts (usually either Fe^{2+}/Fe^{3+} or Cu^{2+}) allow the generation of hydroxyl radicals in the presence of hydrogen peroxide under very mild conditions of temperature and pressure [36]. In addition, it enables easy recovery of the catalytic species, so available for use through a long-term, extended number of catalytic cycles. Hence, important efforts have been focused on finding heterogeneous catalysts with adequate catalytic activity and convenient stability even at pHs of reaction exceeding 4.0, taking into account the acidic conditions under which the reaction takes place more efficiently. Another important advantage linked to the Fenton process activated in heterogeneous phase is to avoid expensive extra steps to recover both the excess of dissolved metals and the sludge formed upon application under near-neutral pH conditions, typical of most water streams [37].

Several teams have reported different materials as possible carriers of active sites catalyzing CWPO: alumina-supported Fe [38], Fe-containing zeolites [39, 40], activated and other functionalized carbonaceous supports [37, 41], and pillared clays [42, 43] are the most common reported catalysts for the CWPO of NOM and other organics. The main limitations of the most suggested materials could be summarized as either low effectiveness under near-neutral pH, low stability against active metal leaching, and/or high cost. The first approach still thoroughly investigated by several groups has been the use of supported catalysts of the Fenton-active transition metals on typical hosts like alumina [44, 45] or silica [46, 47]. However, the most significant drawback of this kind of materials has been definitely poor chemical stability to leaching of the active metal in the reaction mixture in comparison to other solid, functional materials. After the pillared and other modified clay materials, activated carbons and synthetic zeolites have been perhaps the most studied heterogeneous carriers in the past few years because of their high effectiveness, and then these are more specifically analyzed hereafter.

3.1 CWPO by Pillared and Other Related Clay Catalysts

The catalytic potential of the clay-modified materials activating the CWPO reaction has been widely investigated along the past two decades, mainly emphasizing on Al/Fe-, Al/Cu-, and Al/Fe-Cu-pillared clays [48–53]. The high performance displayed by Fe, Cu-modified pillared clays in this reaction mainly relies on the high rates of elimination of several model toxic compounds in terms of both contaminant's depletion and both TOC and COD elimination. Besides, this type of catalysts can be obtained from inexpensive raw materials, increasing the cost-effective potential scaling-up in order to solve real problems in water treatment. Several studies have pointed out that smectites once pillared with mixed Al-based polycations containing iron or copper are materials displaying pretty high performance catalyzing the CWPO reaction [51, 52]. Timofeeva et al. [51] reported the effect of some synthesis variables like hydrolysis ratio of the pillaring solution (OH/(Al + Fe)), temperature of calcination in the preparation of Al/Fe-PILCs, as well as the atomic ratio Fe/Cu in Fe, Cu, Al-clays on the catalytic properties of the resulting materials in the CWPO reaction. Presence of Cu or Fe in isolated sites within alumina pillars, that is, truly mixed interlayering Al/M-polycations, has been believed from a time ago [54] to be closely related to the high catalytic response displayed by this kind of layered materials in the CWPO degradation of several types of compounds, as more recently strongly supported also by H_2-TPR measurements in the case of the Al/Fe-PILCs [52]. In this sense, this and other studies have evidenced that higher loadings of Fe in Al/Fe-pillaring solutions lead to higher fractions of the transition metal fixed in the inorganic host, but not necessarily improving the degradation of organics by a purely catalytic pathway of response in heterogeneous phase [52, 55]. Although higher fractions of Fe in the interlayering solutions clearly promote increased amounts of iron incorporated into the PILCs as isolated octahedrally coordinated Fe^{3+} cations (the most active in the CWPO reaction), together there is bigger incidence of Fe aggregates, less active but also less stable against leaching under the strongly oxidizing environment of the catalytic reaction [52, 55, 56]. Thus, leaching of the active metal(s) usually makes more difficult to interpret the contaminant removals along this type of reaction, since it implicitly leads to higher contribution of the homogeneously activated Fenton oxidation to the overall activity [52]. In addition, elimination of the contaminants should always account for the fraction adsorbed on the catalyst's surface in order to rule out a significant contribution, avoiding reporting it as a purely catalytic degradation. Of course, it must be more carefully assessed in the case of high-surface solid carriers as, for instance, metal-functionalized activated carbons.

Another important issue of the Fenton solid catalysts useful in the CWPO reaction is the presence, in most of cases, of an induction period before observing faster degradation of the contaminants. It is typical of the Fenton-like variant of reaction and must be mainly ascribed in the case of the iron-functionalized materials, to the balance (Fe^{3+}/Fe^{2+}) in the accessible, active sites of reaction; as higher ratio in this couple of oxidation states, as more prolonged must be expected such induction period. Zhou et al. [57] recently modeled the apparent induction period followed by a rapid oxidation observed during the catalytic wet peroxide oxidation of 4-chlorophenol (4-CP) by using Fermi's kinetic equation ($R^2 = 0.9938$–0.9993). 4-CP oxidation proceeded via 4-chlorocatechol (major) and hydroquinone (minor) pathways, along the formation of main intermediate (5-chloro-1,2,4-benzenetriol). Besides CO_2, H_2O, and Cl^-, two main compounds (I and II) formed, the former identified as 2,4-dioxopentanedioic acid, whereas the latter as ferric-oxalate complex. Finally, it has reported marked structural and active differences between Al/Fe-PILC and Al/Cu-PILC in which compared to the latter, the former possessed higher specific surface area and catalytic activity, but its optimal calcination temperature was lower. The induction period also resulted longer in the case of the Al/Fe-PILC. In addition, compound II accounted for a considerable proportion in Al/Fe-PILC system, whereas compound I was almost only component in Al/Cu-PILC system. Overwhelming advantage of the Al/Fe-PILC on the Al/Cu-PILC system catalyzing CWPO had been also clearly evidenced earlier dealing on degradation of the azo-dye methyl orange [52]. On the basis of elemental analyses, it was found that within the range of 0–10% of atomic metal ratio (AMR), Fe gets incorporated around 15 times more efficiently in Al/M-PILCs than copper, whereas the patterns of compensation of the starting CEC showed that increasing values of $AMR_{(Fe\ and/or\ Cu)}$ led to less and less efficient compensation of the aluminosilicate's layer charge. AMR is then a key and useful parameter to be considered in preparation of Al-mixed PILCs with Fe and/or Cu for CWPO, since different as could be believed by default, higher fractions of these active metals in the mixed interlayering solutions do not necessarily conduce to more efficient clay catalysts. However, in other reports published a little bit later [55, 56], adsorption of tartrazine showed slight rise with increasing content of Fe^{3+} in Al/Fe-PILCs ($AMR_{(Fe)}$ in the range of 1.0–20.0%) [55], probably related to the well-known high external surface featured by iron oxides, inferring that higher loadings of this active metal are necessarily forced to form iron oxide aggregates, not very useful for this particular application. However, apparently this increased incidence of non-pillaring iron aggregates as a function of $AMR_{(Fe)}$ did not lead to lower stability against leaching in the CWPO reaction, since pretty low concentration of Fe was reported for all samples evaluated. According to Mössbauer analyses, the structure of the incorporated iron-containing oxides resembled those of naturally occurring minerals akaganeite (β-FeOOH) or lepidocrocite (γ-FeOOH) [56]. AMR can be calculated as follows [52]:

$$AMR_{(Fe\ and/or\ Cu)} = \frac{(Fe\ and/or\ Cu)}{(Al + (Fe\ and/or\ Cu))} \times 100$$

In related address, Timofeeva et al. [51] focused in studying the effect of the Fe/Cu ratio in the Al/Fe-Cu-containing pillaring solution on structural and catalytic properties, CWPO degradation of phenol, of final catalysts. The increase in copper loading led to decreased total surface area, micropore volume, and interlayer distance, whereas the decrease in Fe/Cu ratio favored the formation of oligomeric iron species. However, according to the authors, the introduction of copper ions also increased the rate of the catalytic reaction, an effect that was interpreted in terms of higher rate of radical generation. Moreover, a little bit later in other study [52], no cooperative effect in the CWPO degradation of methyl orange by the presence of both active metals in the same Al/M-pillared clay catalyst was found, which was explained by a possible competition between both active metals for octahedral sites into the framework of Al_{13} polycations in this three-metal system.

An important issue that has remained still controversial regarding Al/Fe-PILCs is whether there is or no formation of truly mixed pillars of both metals. In spite that several works have suggested the outstanding activity of this kind of materials in the CWPO reaction as being strongly related to the mixed Al/Fe pillars [52, 54], some others have ruled it out while proposing that only iron oxides either decorating the Al_2O_3 pillars or as external aggregates are stabilized in this type of materials [58, 59]. In this sense, Bankovic et al. [56] based on DR UV–Vis spectra, Mössbauer, and FTIR analyses proposed that in the AlFe1–15 PILCs (AMR_{Fe} values from 1.0 to 15.0%), the Fe^{3+} ion probably partially substituted Al^{3+} ions in Keggin ion, thus forming $Al_{13-x}Fe_x$ oxide pillars, whereas in the case of the AlFe20 PILC, other types of pillars might have been formed including those containing separate Al or Fe oxide pillars. Besides, it was shown that increasing iron content of the PILCs resulted in the increasing presence of species with greater clusters of octahedrally coordinated Fe^{3+} ions. The structure of the incorporated iron-containing oxides resembled those of naturally occurring minerals akaganeite (β-FeOOH) or lepidocrocite (γ-FeOOH) (Fig. 2a). It was in similar address to the trend already found as a function of the increased content of Fe in final materials, based in H_2-TPR analyses (Fig. 2b) [52]. Finally, same authors reported interesting behavior of the textural properties in Al/Fe-PILCs as a function of the increasing content in Fe. Particularly, pillaring with Keggin-like ions led to almost monomodal distribution of mesopore diameters. The addition of Fe^{3+} into the

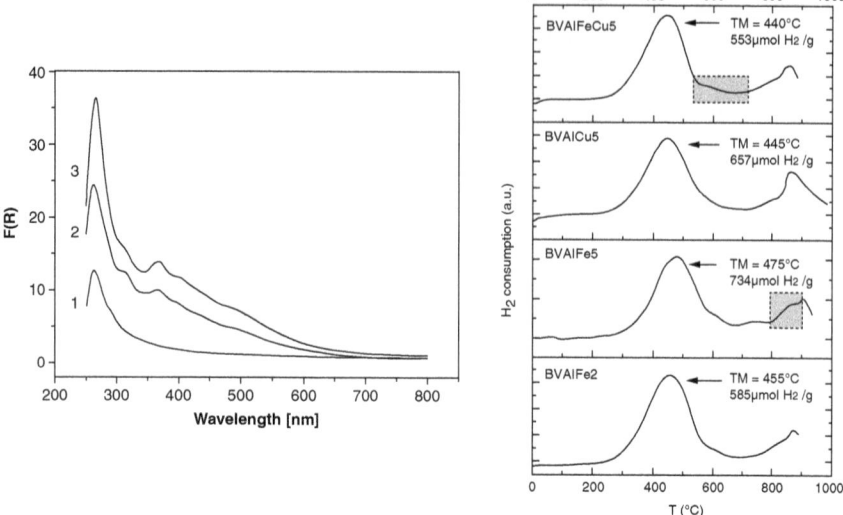

Fig. 2 Physicochemical properties of different Al/Fe- and Al/Cu-PILCs as a function of the AMR_{Fe}: (**a**) DR UV–Vis spectra of (1) Al PILC, (2) AlFe10 PILC (AMR_{Fe} = 10%), and (3) AlFe20 PILC (AMR_{Fe} = 20%) (reprinted from [56], with permission of Elsevier). (**b**) H_2-TPR diagrams of BVAlFe2 (AMR_{Fe} = 2.0%), BVAlFe5 (AMR_{Fe} = 5.0%), BVAlCu5 (AMR_{Cu} = 5.0%), and BVAlFeCu5 ($AMR_{Fe\,+\,Cu}$ = 5.0%) (reprinted from [52], with permission of Elsevier)

pillaring solution resulted in changes in the mesopore diameter distribution. Increasing Fe content led to the broadening of the distribution of mesopore diameters in the sequence Al PILC < AlFe1 PILC < AlFe5 PILC < AlFe10 PILC, which got inverted for increasing AMR_{Fe} values from 10 to 20%. This phenomenon was proposed to be further investigated.

Later, Khankhasaeva et al. [53] studied the degradation of sulfanilamide in water by H_2O_2 in the presence of Fe, Al/M-pillared clays from various cationic forms of the starting layered aluminosilicate (Fe, Al/M-MM, M = Na^+, Ca^{2+}, and Ba^{2+}). The montmorillonitic materials were exchanged with bulky Fe, Al-polyoxocations prepared at Al/Fe = 10/1 (AMR_{Fe} ~9.0%) and OH/(Al + Fe) = 2.0 and then calcinated at 500 °C and from XRD, and chemical analyses found that the rate of crystalline swelling was dependent on the nature of the starting interlayer cations in the clay mineral (Fe, Al–/Na-MM > Fe, Al/Ca-MM > Fe, Al/Ba-MM). In addition, the catalytic properties of Fe, Al/M-MMs depended on the type of exchangeable cations, although only slightly the textural properties. These differences were mainly ascribed to higher iron uptake in the Na-MM cationic form.

Gao et al. [60] recently studied the introduction of nickel in Fe-Al-pillared montmorillonite and its effect in the CWPO degradation of the azo-dye Orange II. The effects of active metal molar loading (AMR_{Fe-Ni}), Fe/Ni molar ratio, and calcination temperature on the pillared clays were measured and discussed. It was proposed the introduction of nickel to the pillaring solution may contribute to better active iron stability and catalytic performance. Iron was found in the clays in the form of FeOOH when nickel was introduced, whose integrity was more easily maintained when a moderate calcination temperature was used. Solid modified with 4/6 Fe/Ni molar ratio and $AMR_{Fe-Ni} = 6\%$, finally heated at 273 °C, displayed better catalytic behavior. Thus, Acid Orange II solution got 72.32% of COD removal after 3 h of reaction: pH = 3.0, $T = 60$ °C, and H_2O_2 dosing = 100% of the theoretical stoichiometric amount. Fe/Ni ratios above 4/6 led to declined catalytic behavior of pillared clays, which was attributed to formation of large iron aggregates.

Some other studies have also attempted combination of properties of Al-, Al/Fe-PILCs with other either transition or precious metals to improve catalytic properties of this kind of materials in water decontamination. For instance, recently a two-step treatment of p-chloro-m-cresol (PCMC) in water by catalytic hydrodechlorination (HDC) followed by CWPO under ambient-like conditions (25 °C, 1 atm) in the presence of catalysts supported on Al-PILCs using Fe and Pd or Rh as active phases was reported [61]. The bimetallic Pd-Fe catalyst showed the best performance, allowing complete dechlorination in less than 1 h reaction time (25 °C, 1 atm, $pH_0 = 6$, $QH_2 = 50$ N mL/min, $[catalyst]_0 = 1$ g/L, $[PCMC]_0 = 0.7$ mmol/L). Meanwhile, CWPO of PCMC with a monometallic Fe catalyst allowed complete conversion of that pollutant but with only about 33% reduction of TOC after 4 h. Then, a two-step approach consisting in HDC of PCMC followed by CWPO has been tested for the first time in two different ways. The first one used a Pd-Fe bimetallic catalyst in both steps, while in the second approach, monometallic Pd-Fe catalysts were used for HDC and CWPO, respectively. The HDC step extended for 30 min was enough to achieve complete dechlorination, the main reaction product being m-cresol (selectivity >85%). After the HDC step, the pH of the reaction medium was adjusted to 3.5, and the stoichiometric H_2O_2 dose was added to start the CWPO step. It allowed complete conversion of m-cresol in 15 min, with 60% TOC reduction after 4 h of reaction time. However, the stability of the catalysts against Fe leaching was more rather poor; the Fe leached from the catalyst in this combined treatment reached 2.5 mg/L in the HDC step and 4.9 mg/L at the end of the experiment. Although it represents only 6.6% of the initial Fe load reported, the recorded CWPO activity was undoubtedly influenced by the typical homogeneous Fenton process. It is a promising combined application of the CWPO process in order to contribute giving overall response to contamination by halogenated organics, but strong efforts should be made in order to significantly increase stability of the active metal in the step of the heterogeneous Fenton process.

Thanks to the highly promising potential exhibited by the Al/Fe-PILCs in the CWPO treatment of a variety of polluted aqueous systems, several efforts have been taking place along the past decade in order to make possible its preparation in both higher to lab scales and starting from concentrated precursors. Some studies have focused on the reduction of the synthesis time, involving the use of either microwaves [59, 62] or ultrasound [63] in the preparation of the metal interlayering precursor but increasing consumption of energy as a drawback. Other approaches have tried to increase concentration of the pillaring precursors, either interlayering polycationic solution or the clay suspension itself [43], where typical concentration of the oligomerized metal precursor was close to tenfold increased from around 0.06 to 0.63 mol/L. It implied the use of lower hydrolysis ratio (OH/(Al + Fe) = 1.6) in comparison to typical lab-scale preparations (~2.0–2.4). In addition, the use of ethanol as suspension medium demonstrated to be advantageous in order to perform the interlayering step (contact of the interlayering, oligomeric metal solution with the clay mineral) on concentrated clay suspensions (up to around 50% w/w). Another important step forward in the same address has been achieved by some researchers [64, 65] by means of the straightforward addition of the raw clays on the interlayering Al or Al/Fe solutions. It has allowed to save time and resources (extra added solvent-free suspension) by working on not previously swollen mineral, but attaining Al- or Al/Fe-pillared materials with basal spacings around 1.8 nm, significant increase in specific surface and pore volume. Moreover, as far as we know right now, there is no complete solution to this issue, and hence in general, reports about the application of this type of clay catalysts at industrial level are not still available. Regarding this issue, as a comparison reference, CWPO treatment in continuous reactors has started to be studied in the presence of other solid catalysts, like for instance zeolites. The CWPO degradation of phenol by a Fe-ZSM-5 catalyst in a fixed bed reactor was recently investigated [66]. The effects of feed flow rate and catalyst bed height were determined by following conversion of phenol, H_2O_2, and TOC concentrations. The Fe-ZSM-5 catalyst achieved the highest activity (99.2% phenol conversion and 77.7% of TOC conversion) at 80 °C, 2 mL/min feed flow rate and 3.8 cm as catalyst bed height. In spite that the authors claim remarkable low iron leaching concentration, since it was around 1.0 mg/L, it may anticipate sustained rapid deactivation of the catalyst under real conditions of treatment. In addition, such a level of concentration has demonstrated to be not negligible catalyzing the CWPO reaction based on a close to purely homogeneous Fenton mechanism. As it was stressed by the authors, further research should be focused on the deactivated mechanism and regeneration strategies of the catalyst. It is big evidence promoting the rapid development of scaled-up preparation of Al/Fe-pillared clays, since the concentrations of iron leached in the CWPO reaction mixture typically do not exceed 0.3 mg/L. The HPLC patterns of the solution at different catalyst bed heights demonstrated that low concentration of quinone was generated and then transformed into low molecular weight organic acids and finally changed into carbon dioxide and water with the increase of residence time. This pattern of degradation has been observed in the presence of other efficient CWPO catalysts.

Finally, some emerging materials attracting interest as CWPO-active clay catalysts along the past 5 years have been some naturally occurring or synthetic clay minerals, whose either structural or added content of Fe or Cu has been investigated as active sites of the catalytic reaction [67–69]. First of all, allophane clay materials with SiO_2/Al_2O_3 ratios between 1.0 and 2.2 were synthesized by a coprecipitation route and further impregnated with iron or copper species, in targeted loadings surrounding 2.0–6.0 w/w % [68]. The lower ratio resembled typical Al-rich soil allophane (AlSi1; BET surface <1.0 m^2/g), whereas the higher one looked like more as a hydrous feldspathoid with a large interspherule surface, thereby exhibiting a high surface area (AlSi$_2$; BET surface ~191 m^2/g; up to 287 c for 8.5 Fe$_2$O$_3$ wt.%). The iron-based AlSi$_2$ catalysts with taillike structure and high surface area proved to be far more active in the CWPO elimination of phenol (pH = 3.7, [phenol]$_0$ = 5 × 10^{-4} mol/L, [catalyst] = 0.4 g/L, [H$_2$O$_2$]$_{added}$ = 0.1 mol/L under constant flow rate of 2.0 mL/h along 4 h for the Fe catalysts and 8 h for the Cu catalysts). The highest catalytic efficiency (94%/1 h of reaction; total organic carbon abatement 63%/4 h of reaction) was obtained at 40 °C for the calcined iron oxide-supported AlSi$_2$ allophane sample (300 °C/1 h), for which very low leaching level of iron species was noticed (0.37 mg/L). By contrast, large differences in terms of catalytic efficiency (conversion rates) and stability were observed for the copper-based counterparts.

Munoz et al. [67] recently explored the potential application of naturally occurring minerals as inexpensive catalysts in heterogeneous Fenton, namely, catalytic wet peroxide oxidation (CWPO). Performance of magnetite, hematite, and ilmenite as CWPO catalysts was tested under different working conditions (25–90 °C, [H$_2$O$_2$] = 250–1,000 mg/L, [catalyst] = 1–4 g/L). In general terms, the use of naturally occurring Fe minerals is attractive because of pretty low cost of this kind of minerals. However, CWPO activated by this kind of materials proceed in fairly longer times of reaction and/or demands for increased temperatures. As expected, in this study the operating temperature showed to play a key role on the rate of H$_2$O$_2$ decomposition, so that in the presence of magnetite, the H$_2$O$_2$ conversion after 4 h increased from 8 to 99% by increasing the temperature from 25 to 90 °C ([phenol]$_0$ = 100 mg/L). Here it is noteworthy that every Celsius degree of higher reaction temperature of course also supposes increased cost of operation of the catalytic system, and then it may rapidly offset the benefit related to the use of a cheaper catalyst. Conversely, as a clear advantage, leaching is not a big deal in the case of the treatment of wastewaters given the very low cost featuring this kind of ferrous minerals. A significant metal leaching may also become not practical in the use of the catalyst in an extended range of pH values, a crucial difference of the CWPO process against the conventional Fenton process. Complete phenol conversion and almost 80% TOC reduction were claimed at 75 °C with a catalyst loading of 2 g/L in the presence of theoretical stoichiometric amount of H$_2$O$_2$ required for complete mineralization of a phenol sample solution (500 mg/L). Among the minerals compared, magnetite (Fe$_3$O$_4$) was particularly attractive, since it showed the highest activity and can be easily separated from the liquid phase given its magnetic properties. Moreover, it must be stressed that such a promising mineral was also the one leaching a higher Fe concentration along three consecutive runs; of course it decreased steadily with the cycles of reuse (over 12 mg/L in the first run, around 4 mg/L after the third run), but it allows to infer quite significant contribution of homogeneous Fenton reaction.

In summary, Fe-containing naturally occurring minerals could be devised as interesting CWPO catalysts in the field of wastewater treatment, mainly on those with significant carbon loadings in order for the exothermal enthalpies of oxidation to contribute decreasing the costs of heating of the catalytic system over around

70 °C. Such a kind of materials, together with other types of synthetic-based supported catalysts, mainly Fe and/or Cu oxides as active sites, has demonstrated as a general drawback the low stability against metal leaching in the oxidizing reaction mixture. Outstanding, worth mentioning exception has been the case of modified allophane clay synthetic materials functionalized by impregnation with Fe, which recently showed proper activity and chemical stability ($Fe_{leached}$ <0.4 mg/L) in phenol oxidation under still middle temperature of reaction (40 °C). Al/M-mixed pillared clays in general have exhibited lower contents of the active metals (Cu and/or mainly Fe) incorporated in final materials. They usually display high performance in contaminant depletion even at room temperatures, with output concentrations of active metals typically below 0.3 mg/L, thanks to the specific location of them in very active sites at the interlayer space of the starting aluminosilicates. Moreover, in order to ensure such a set of desirable properties in the CWPO application, cumbersome steps and parameters of preparation must be carefully followed, which definitely have delayed the scaled-up preparation of this kind of materials and, in turn, their more widely spread use in the treatment of a variety of contaminated waters at industrial scale. Therefore, further work in the short term in these still exciting materials should focus in the development of Al/Fe-PILC preparation from concentrated precursors (significantly intensified process), exhibiting physicochemical properties comparable with products obtained in conventional preparations from diluted precursors as well as under reproducible conditions at higher to lab scales (let's say bench followed by pilot scale).

3.2 NOM Removal by CWPO and Other AOPs

AOPs have been intensively studied about the treatment of wide variety of contaminated waters but predominantly wastewaters. NOM removal by such a set of powerful technologies has attracted interest from more rather short time ago, strongly focused in the improvement of drinking water treatment systems. In this context, effect of the NOM removal to decrease the potential of formation of disinfection by-products, more particularly THMs, has been studied. The presence of NOM causes many problems in drinking water treatment processes, including (1) negative effect on water quality by color, taste, and odor problems, (2) increased coagulant and disinfectant dose requirements (which in turn results in increased sludge and potential harmful disinfection by-product formation), (3) promoted biological growth in distribution system, and (4) increased levels of complexed heavy metals and adsorbed organic pollutants [4, 70]. Among the assessed AOPs have been O_3/H_2O_2, O_3/UV, UV/H_2O_2, TiO_2/UV, H_2O_2/catalyst, Fenton, and photo-Fenton processes as well as ultrasound so far, as reviewed in 2010 by Matilainen et al. [4]. Thus, in this section we focus in reviewing the heterogeneous Fenton-like applications reported on the past few years together with the most strongly related others.

First of all, within this recent period, several efforts have been made in order to more confidently and in depth characterize NOM. This is because NOM is a very complex matrix, with widely distributed compositions, chemical functionalities, polar properties, and molecular weight distributions. In addition, there is a high variability of their properties as a function of the source (surface water, groundwater, place, etc.) and even season of the year. For instance, dissolved organic matter (~6.0–15.0 mg/L as dissolved organic carbon) present in two secondary effluents was recently characterized and monitored through UV/H_2O_2 and ozonation treatments by using LC-OCD technique [71]. Biopolymers, humic substances, building blocks, low molar mass neutrals, and low molar mass acids were the fractions resolved. Monitoring of the organic matter fractions with LC-OCD demonstrated that the reduction of effluent's aromaticity (decreasing in specific UV absorbance – SUVA) was not strictly correlated with the complete depletion of humic substances in the effluents for both advanced treatments. However, the UV/H_2O_2 process led to an effluent with lower biopolymer content together with important increase in low molecular mass fractions, although significant amounts of humic acids still remained after extended oxidation treatments. In spite that both AOPs efficiently removed different fractions of the dissolved organic matter, the final composition of the treated effluents was significantly different between the two processes.

During the treatment of natural sources of water by AOPs to remove micropollutants, NOM gets broken down into smaller species, potentially affecting biostability by increasing AOC and BDOC. Bazri et al. [72] found that by means of the UV/H_2O_2 treatment, both AOC and BDOC increased by about three to four times over the course of treatment, indicating the reduction of biological stability. Although a wide range of organic molecular weights were found responsible for AOC increase, low molecular weight organics seemed to contribute more, which could be a serious drawback in order to apply in general AOPs in drinking water facilities. Accordingly, it can be easily anticipated that higher fraction of the TOC represented in low molecular weight organic compounds in this kind of effluents may lead to either higher incidence of microbiological growth in distribution pipes or higher demand of residual chlorine in the effluents of typical disinfection units. Another very important issue in this field is the pH and alkalinity at which a given AOP can operate efficiently. Typical range of pH values for surface waters is close to neutral, sometimes slightly basic. This condition, together with variable alkalinity, may significantly compete with NOM scavenging $HO^{\bullet}$. A little bit later, Black et al. [73] compared UV/H_2O_2 and ozonation of surface water focused on the performance exhibited by a biofiltration unit (acclimatized biological activated carbon) before and after the AOP treatment, finding significantly different results regarding biodegradability of the final oxidized products. Straight oxidation on source water investigated in this case did not preferentially react with the biodegradable or nonbiodegradable NOM. In addition, the type or dose of oxidation applied did not affect the observed rate of biodegradation. Although the oxidation prior to biofiltration increased the overall removal of organic matter, it did not affect the rate of NOM biodegradation. Moreover, most outstanding conclusions reached in this study regarding NOM and NOM intermediaries recording could be summarized as follows: (1) Ozonation preferentially reacted with higher molecular weight chromophoric NOM. However, the reduction in high molecular weight NOM did not appear to increase the biodegradability of the raw water. (2) Advanced oxidation processes reacted with all molecular weight chromophoric NOM fractions equally. (3) Neither ozonation nor AOP treatment with UV/H_2O_2 preferentially reacted with either the biodegradable or the nonbiodegradable fractions of organic matter. (4) Regardless of the oxidation condition applied, the rate of biodegradation did not change. Therefore, and also according to the same authors, such a set of final statements are very source-depending and contradicted results reported by others, who suggested that AOP oxidation increases biodegradability of NOM.

One of the main aspects speeding up research in NOM removal by advanced oxidation technologies has been certainly to decrease occurrence of DBPs, especially THMs. THM generation can be limited by reducing the levels of NOM prior to the chlorination step. It was recently reported that a solar photo-Fenton system (total intensity of Suntest solar simulator, 600 W/m^2; UV intensity 20–30 W/m^2; experiments performed at room temperature around 25 °C, increased up to 30 °C during irradiation) degraded either humic acid solutions (as NOM model compounds) or NOM contained in a river water, dramatically reducing THM formation during the subsequent chlorination step under close to neutral pH of reaction [74]. Whereas in non-pretreated river water, 100–160 μg/L of THMs was formed upon chlorination, values of 20–60 μg/L were reported for water previously treated through 4 h by neutral photo-Fenton under solar simulator in the presence of 1.0 mg Fe/L as initial concentration. Undoubtedly, such a very low concentration of dissolved iron could be one outstanding advantage of this solar photo-Fenton system, since apparently it would not be too much susceptible against metal precipitation in the strongly chelating environment provided by dissolved NOM. It is noteworthy the peroxide concentration used in this study was around four times the stoichiometric dosage required for full oxidation (calculated on the basis of 2.12 mg H_2O_2/L consumed per 1.0 mg COD/L as theoretical mass ratio reported by Deng et al. for NOM dissolved in leachates of landfills when treated by the Fenton process [75]; together, it must be assumed a ratio 3.33 mg COD/mg TOC corresponding to own experimental results [76] obtained from a synthetic standard resembling NOM composition in most surface and underground waters[1]). Although highly source dependent, the use of a standard theoretical mass ratio COD/TOC for NOM could be very useful in the purpose of unbiased comparison between results

---

[1]Distribution of synthetic fractions in the NOM standard model solution (mass %):

1. Polyacrylic acid (PAA) (transphilic fraction; average MW, 130,000 Da), 20
2. Polystyrene sulfonate (PSS1) (hydrophobic fraction; average MW, 200,000 Da), 12.5
3. Polystyrene sulfonate (PSS2) (hydrophobic fraction; average MW, 1,000,000 Da), 12.5
4. Polygalacturonic acid (PGUA) (hydrophilic fraction; MW, 25,000–50,000 Da), 30
5. Humic acid (HA) (hydrophobic fraction; average MW, 1,000,000 Da), 25

of NOM degradation reported by several Fenton and AOP treatments, in the presence of widely variable concentrations of H_2O_2.

From very different focus, some studies have stressed the role played by dissolved NOM hindering targeted reactions and removal efficiency exhibited by photocatalytic AOPs. Brame et al. [77] recently claimed the development of an analytical model to account for various inhibition mechanisms in catalytic AOPs, including competitive adsorption of inhibitors, scavenging of produced ROS at the catalyst's surface and in solution, and the inner filtering of the excitation illumination, which combine to decrease ROS-mediated degradation, in the case of the photocatalytic processes. Competitive adsorption by NOM and ROS scavenging were found to be the most influential inhibitory mechanisms and should be carefully taken into account in forthcoming studies mainly in heterogeneously activated photocatalytic AOPs. Among this kind of technologies, it was very recently reported the removal of the NOM present in raw drinking water (around 30 mg TOC/L) by coupling an optimized conventional coagulation-flocculation process with heterogeneous photocatalysis (either TiO_2-P25 suspended catalyst or TiO_2-P25/β-SiC foam-supported material). It was claimed that 80% of mineralization of humic substances was achieved after 220 min of irradiation on the clarified water (7.8 mg TOC/L), effluent of the coagulation-flocculation unit. According to the authors, the fraction remaining after the photocatalytic tests was only the hydrophilic fraction of humic acids. The stability tests of the supported catalyst with clarified water collected in treatment plant showed a progressive deactivation due to adsorption of different ions coexisting with humic substances, resulting in the decrease in catalyst efficiency.

Another interesting family of techniques recently reviewed as promising for NOM removal from surface waters has been that of the electrochemical methods like EC and EO [11]. Whereas in EC systems significant increase of NOM removal rates can be achieved when combined with membrane filtration hybrid systems, in EO technology electrolysis efficiency is strongly linked to electrode composition. Efficiency could be increased by changing the reactor design, using commercial electrodes and exploring the semiconducting properties of oxide mixtures. In this purpose, particularly Boron-doped diamond (BDD) anodes have proved to be effective in humic acid removal from aqueous solutions and potentially their total mineralization.

Kasprzyk et al. [38] studied the NOM removal from water using catalytic ozonation in the presence of Al_2O_3. The main purpose of the paper was to show the potential of alumina for longer-term usage. Alumina found application in water treatment technology mainly as an ion exchanger for the removal of inorganic anions such as As, Se, and F. The usage of alumina as a support for active species, mainly metal or metal oxides in the process of catalytic ozonation of several organic compounds was also studied.

As so far showed, the assessment of the Fenton systems activated by either homogeneous or heterogeneous catalysts has been more rather scarcely studied in NOM removal. The Al/Fe-PILC-catalyzed CWPO removal of NOM from a raw surface water employed as supply source in a drinking water treatment plant (pH around 7.5) was investigated in semi-batch lab-scale experiments [37]. It has achieved full color removal in less than 45 min of reaction and 96% of COD removal in 4 h of reaction for the best clay catalyst (5.0 g/L) under room temperature (18 ± 2 °C) and atmospheric pressure (72 kPa). The clay catalyst displayed high chemical stability against iron leaching even under a very high humic/catalyst ratio, around 28 times more concentrated than surface water studied, and longtime reaction (24 h). This kind of clay catalysts is very promising for this application at real scale given their pretty low cost, since more than 90% in weight is constituted by natural, widely available layered clay minerals. However, only one additional study has been published in the past few years regarding the interaction between humic substances and clay minerals [78]. The aim of the research was to investigate the influence of montmorillonite as a representative clay mineral on the TiO_2 photocatalytic removal of humic acids as model compound of NOM. The effect of Mt was found to be proceeding through a dose-independent trend mainly resulting in the humic structural changes rather than an efficient degree of mineralization. The molecular size distribution profiles displayed the formation of lower molecular size fractions through oxidative degradation of higher molecular size fractions. In comparison to the regular decreasing scheme attained for the specific UV–Vis parameters of HA, the presence of Mt significantly altered the spectroscopic properties of the molecular size subfractions of HA. Therefore, in the field of the Al/Fe-pillared clay-catalyzed CWPO degradation of NOM, an interesting chapter in the short term should account for the degree of adsorption as well as to realize if there is any molecular size and/or polar selectivity displayed by this type of functional materials to adsorb typical fractions of NOM.

3.3 Other Solid Catalysts Used for NOM Removal

3.3.1 Zeolites

Zeolite-based materials are extremely versatile, and their main applications include ion exchange resins, catalytic applications in the petroleum industry, separation process, and as an adsorbent for water, carbon dioxide, and hydrogen sulfide. Zeolites are the compounds of aluminosilicates and can be artificially synthesized by reacting sodium aluminate with sodium silicate. The ratio of silica to the alumina determines the type (X or Y) of the synthetic zeolite. The Y type of synthetic zeolite is the most commonly applied type in preparation of heterogeneous catalysts [79]. One way of producing heterogeneous catalyst from synthetic zeolite is by impregnation of ferric ions followed by calcinations [80]. Another process is by the ion exchange, for example, where the sodium in zeolite containing high sodium content is replaced with ferric ions [79].

The synthetic zeolites have been used as heterogeneous catalytic materials for CWPO treatment of wastewater, where high effectiveness was reported. For instance, Arimi M. [39] studied the treatment of recalcitrants in industrial effluent using a modified natural zeolite as heterogeneous Fenton catalyst. In this case the effects of pH and temperature on heterogeneous Fenton were studied using the modified catalysts. The catalysts showed the highest affectivity which achieved removal of 90% of color and 60% of total organic carbon at 150 g/L pellet catalyst dosage, 2,000 mg/L H_2O_2, and 25 °C. The catalyst was also applied to pretreat the raw molasses distillery wastewater and increased its biodegradability by 4%. Probably even more remarkable was that heterogeneous Fenton with the same catalyst improved biodegradability (BOD_5/COD) of the anaerobic effluent from 0.07 to 0.55, making more feasible the reuse of the treated stream at least as dilution fresh water for the input raw wastewater before conventional microbiological treatment. The color of the resultant anaerobic effluent was also reduced. The kinetics of total TOC removal was found to depend on operation temperature [39].

3.3.2 Fe-Functionalized Activated Carbon

Definitely one type of functional materials more investigated along the past decade as active solids of the CWPO reaction has been that of the carbon-based ones. Several motivations have prompted the scientific community toward the application of hybrid magnetic carbon nanocomposites as catalytic materials for this still emerging technology. The most relevant literature on this topic has been recently reviewed by Ribeiro et al. [37], with a special focus on the synergies that can arise from the combination of highly active and magnetically separable iron species with the easily tuned properties of carbon-based materials. These are mainly ascribed to increased adsorptive interactions, together with good structural stability and low leaching levels of the metal species, as well as to increased regeneration and dispersion of the active sites, which are promoted by the presence of the carbon-based materials in the composites. According to the authors, the presence of stable metal impurities, basic active sites, and sulfur-containing functionalities, as well as high specific surface area, adequate porous texture, adsorptive interactions, and structural defects, was shown to increase the CWPO activity of carbon materials, while the presence of acidic oxygen-containing functionalities had the opposite effect.

Some efforts in the purpose of giving added value to waste raw materials in preparation of this kind of carbonaceous catalyst must be remarked. Sewage sludge was in recent times used as precursor to prepare catalysts for CWPO by both simple drying and pyrolysis [81]. Iron-functionalized biosolids (Fe/ABS) were obtained from the dried biosolids, upon contact with aqueous solution of $FeCl_3$ at several concentrations followed by thermal treatment at 750 °C for 30 min. The materials displayed narrow developed microporosity, with total iron contents up to 9.3% w/w. The catalysts showed a relatively high activity in the CWPO oxidation (80 °C, pH = 3.0) of three target pollutants (phenol, bromophenol blue, and dimethoate), allowing a high mineralization (65% TOC reduction for phenol and dimethoate). A fairly good stability was observed in long-term continuous experiments where the Fe leaching remained below 11% of the initial loading after 170 h on stream. More recently, highly stable iron on activated carbon catalysts was prepared by chemical activation of dry sewage sludge with iron chloride at different mass ratios [82]. The iron content of the resulting catalysts varied between 5.2 and 7.3%, and there was developed porosity (BET specific surfaces above 800 m^2/g). The catalysts were tested batchwise in the CWPO of antipyrine (50 °C), a recognized emergent pollutant. Working with 500 mg/L of the best catalyst prepared, complete conversion of antipyrine (20 mg/L) and almost 70% of TOC reduction were achieved in 1 h with the stoichiometric theoretical amount of H_2O_2. In spite that very soundly results have been claimed in this area so far, metal stability remains being one of the main drawbacks. In addition, the best results were reported under high temperatures, pH or reaction very low and close to the Fenton's optimal value (3.0), and conditions easily reproduced at lab scale but still far for application on a real-scale process where environmental conditions of reaction are mandatory (RT; natural pH, being in general a circumneutral one).

3.3.3 Biologically Activated Carbon

The BAC treatment is one of the most promising, environmentally friendly, and economically feasible processes. The BAC can overcome several limitations of AC treatment and other conventional water treatment processes. The BAC uses the highly porous AC as a medium to immobilize microorganisms and remove organic matters present in water.

It is possible to substantially reduce the GAC replacement costs by implementing a biofiltration process, where microbial activity on activated carbon possibly extends GAC adsorption capacity via in situ regeneration of adsorption sites on the external surface or in inner pores through the biodegradation of previously adsorbed organic matter [83]. The BAC process combines both biosorption/sorption and biodegradation functions, providing many benefits for the water treatment. In this work [84], DOC content in the influent and effluent ranged between 1.1–5.5 and 0.5–2.5 mg/L, respectively, showing a good DOC removal (see Fig. 3).

3.3.4 Carbon Nanotubes

NA possesses a series of unique physical and chemical properties. A very important one is that most of the atoms that have high chemical activity and adsorption capacity are on the surface of the nanomaterials. CNTs have been at the center of nanoscience and nanotechnology research for a variety of applications such as adsorbents. The adsorption capacity and strength strongly depend on the type of NOM and the type of CNT. Factors affecting adsorption have been reported to include (1) size and chemical characteristics of NOM and (2) pore structure and surface chemistry of CNT. Due to the polydisperse nature of NOM, various fractions of NOM tend to have a different degree of adsorptive interactions with the adsorbent. This preferential adsorption is reflected by the occurrence of dose-

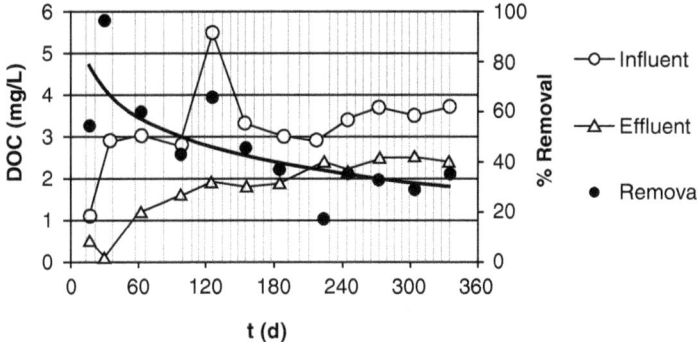

Fig. 3 Evolution of DOC concentration (mg/L) in the influent and effluent of the column and removal percentage over the whole experiment. *Vertical lines* indicate dates of backwashing events (reprinted from Gibert et al. [84], with permission of Elsevier)

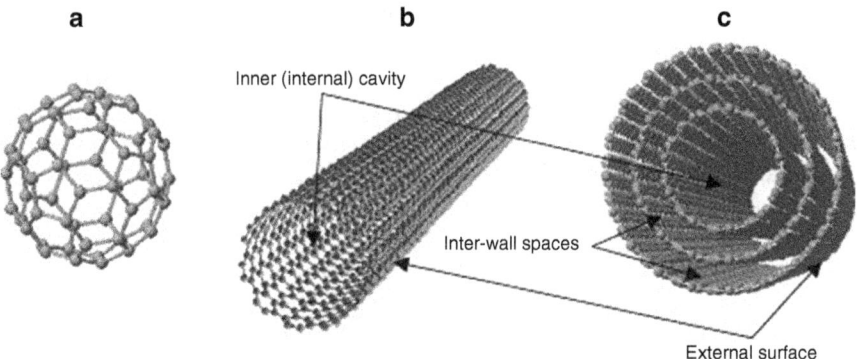

Fig. 4 Schematic structures of fullerene (**a**), single-walled carbon nanotubes (**b**), and multiwalled carbon nanotubes (**c**), showing inner cavity, inter-wall spaces, and external surface. Fullerene (C60) has only external surface (reprinted from Yang et al. [87], with permission of Elsevier)

dependent isotherm relationship. For example, the strongly adsorbable fraction of NOM exhibits a more favorable adsorption at lower CNT dose.

Finally, NOM adsorption is affected by water quality parameters such as ionic strength and pH which influence the charge and configuration of NOM. Specifically, the adsorption of negatively charged NOM to the activated carbon surface generally increases as ionic strength increases and pH decreases [85].

NOM generally carries a negative charge in the natural environment, due to the carboxylic and phenolic moieties distributed throughout the entire molecule [86]. These physical and chemical characteristics of NOM are likely to be closely related to the mechanism of NOM interaction with CNT. Compared to NOM adsorption onto CNT, the mechanism of NOM adsorption onto activated carbon is relatively well-known. A few characteristics of NOM interaction with activated carbon are noteworthy and might be helpful for interpretation of CNT-NOM interaction.

Differences between activated carbon and CNT also need to be recognized for the proper interpretation of CNT-NOM adsorption phenomena. The activated carbon consists of micropores with different sizes which provide sites for NOM adsorption. CNT in contrast provides adsorption sites only along the surface of a cylindrical structure; see Fig. 4 [87].

4 Conclusions

Removal of NOM and other synthetic surrogates by several AOPs, focused in the Fenton-like, heterogeneous CWPO, has been critically reviewed. Main advances achieved along the past few years by these still emerging technologies have mainly focused on increasing the removal efficiency, not only in terms of color and TOC but also in the aromatic content, under the lower possible temperature. Moreover, several efforts are still to be made in order to unify the information about the amount of H_2O_2 as oxidizing agent employed per unit of concentration of NOM, preferably in terms of stoichiometric loading against some widely accepted reference; a standard model synthetic solution is proposed for. In addition, recording of molecular weight distribution is also strongly advised in order to get better correlation between mineralization and changes in polar and size of NOM intermediaries and by-products.

Several types of solid materials have been studied in order to more efficiently catalyze the heterogeneous activation of the oxidizing agent (higher removal of organics through as low as possible either H_2O_2 consumption or temperature of reaction, under real-water pH values, typically circumneutral): being prepared from low-cost, widely available clay minerals, Al/Cu- but mainly Al/Fe-pillared clays have shown to gather in a great extent such a challenging set of features, but their preparation process must be first seriously intensified before getting proper applications in real-scale water treatment. Some naturally occurring minerals have lately shown good rates of reaction but under still high temperatures (e.g., >50 °C). Regarding this type of materials, Fe-functionalized allophane showed interesting activity in phenol degradation at 40 °C with outstanding stability of the active metal against leaching in comparison to many other supported catalysts reported so far in the literature for this reaction.

Some zeolites and carbon-based materials are extremely versatile and have been used as heterogeneous catalytic materials for CWPO treatment of wastewater. In particular, functional materials, biologically activated carbon, and carbon nanotubes were used as active solids of the CWPO reaction with a special focus on NOM removal showing higher efficiencies of reaction.

Finally, in the future several integrated technologies will be used to remove NOM from supply water, including nanofiltration membranes, coagulation with subsequent floc separation, and CWPO followed by biofiltration and sorption processes (chemisorption and physical adsorption). The evolution of water-related normativity and progressively more restrictive standards for drinking water, however, will seek the research of advanced, more efficient, and cost-effective water treatment processes.

Acknowledgment Financial support from CWPO Project for enhanced drinking water in Nariño (BPIN 2014000100020), CT&I Fund of SGR, Colombia, is kindly acknowledged.

References

1. Sadiq R, Rodriguez MJ (2004) Disinfection by-products (DBPs) in drinking water and predictive models for their occurrence: a review. Sci Total Environ 321(1–3):21–46. https://doi.org/10.1016/j.scitotenv.2003.05.001
2. Meylan S, Hammes F, Traber J, Salhi E, von Gunten U, Pronk W (2007) Permeability of low molecular weight organics through nanofiltration membranes. Water Res 41(17):3968–3976. https://doi.org/10.1016/j.watres.2007.05.031
3. de Ridder DJ, Verliefde AR, Heijman SG, Verberk JQ, Rietveld LC, van der Aa LT, Amy GL, van Dijk JC (2011) Influence of natural organic matter on equilibrium adsorption of neutral and charged pharmaceuticals onto activated carbon. Water Sci Technol 63(3):416–423. https://doi.org/10.2166/wst.2011.237
4. Matilainen A, Sillanpaa M (2010) Removal of natural organic matter from drinking water by advanced oxidation processes. Chemosphere 80(4):351–365. https://doi.org/10.1016/j.chemosphere.2010.04.067
5. Werner S (1992) Chemistry of the solid-water interface. Wiley, New York
6. Bhatnagar A, Sillanpaa M (2017) Removal of natural organic matter (NOM) and its constituents from water by adsorption – a review. Chemosphere 166:497–510. https://doi.org/10.1016/j.chemosphere.2016.09.098
7. Li Q, Snoeyink VL, Mariãas BJ, Campos C (2003) Elucidating competitive adsorption mechanisms of atrazine and NOM using model compounds. Water Res 37(4):773–784. https://doi.org/10.1016/S0043-1354(02)00390-1
8. Gauden PA, Szmechtig-Gauden E, Rychlicki G, Duber S, Garbacz JK, Buczkowski R (2006) Changes of the porous structure of activated carbons applied in a filter bed pilot operation. J Colloid Interface Sci 295(2):327–347. https://doi.org/10.1016/j.jcis.2005.08.039
9. Kim J, Kang B (2008) DBPs removal in GAC filter-adsorber. Water Res 42(1–2):145–152. https://doi.org/10.1016/j.watres.2007.07.040
10. Sorlini S, Biasibetti M, Collivignarelli MC, Crotti BM (2015) Reducing the chlorine dioxide demand in final disinfection of drinking water treatment plants using activated carbon. Environ Technol 36(9-12):1499–1509. https://doi.org/10.1080/09593330.2014.994043
11. Särkkä H, Vepsäläinen M, Sillanpää M (2015) Natural organic matter (NOM) removal by electrochemical methods – a review. J Electroanal Chem 755:100–108. https://doi.org/10.1016/j.jelechem.2015.07.029
12. Kimura M, Matsui Y, Kondo K, Ishikawa TB, Matsushita T, Shirasaki N (2013) Minimizing residual aluminum concentration in treated water by tailoring properties of polyaluminum coagulants. Water Res 47(6):2075–2084. https://doi.org/10.1016/j.watres.2013.01.037
13. Jiang J-Q, Wang H-Y (2009) Comparative coagulant demand of polyferric chloride and ferric chloride for the removal of humic acid. Sep Sci Technol 44(2):386–397. https://doi.org/10.1080/01496390802590020
14. Cheng Y-L, Wong R-J, Lin JC-T, Huang C, Lee D-J, Mujumdar AS (2010) Water coagulation using electrostatic patch coagulation (EPC) mechanism. Drying Technol 28(7):850–857. https://doi.org/10.1080/07373937.2010.490492

15. Hai FI, Yamamoto K, Fukushi K (2007) Hybrid treatment systems for dye wastewater. Crit Rev Environ Sci Technol 37(4):315–377. https://doi.org/10.1080/10643380601174723
16. Edzwald JK (1993) Coagulation in drinking water treatment: particles, organics and coagulants. Water Sci Technol 27(11):21–35
17. Jiang JQ, Graham NJD (1996) Enhanced coagulation using Al/Fe(III) coagulants: effect of coagulant chemistry on the removal of colour-causing NOM. Environ Technol 17(9):937–950. https://doi.org/10.1080/09593330.1996.9618422
18. Ghernaout D, Ghernaout B, Kellil A (2012) Natural organic matter removal and enhanced coagulation as a link between coagulation and electrocoagulation. Desalin Water Treat 2 (1-3):203–222. https://doi.org/10.5004/dwt.2009.116
19. Ghernaout D (2014) The hydrophilic/hydrophobic ratio vs. dissolved organics removal by coagulation – a review. J King Saud Univ Sci 26(3):169–180. https://doi.org/10.1016/j.jksus.2013.09.005
20. Bukhari AA (2008) Investigation of the electro-coagulation treatment process for the removal of total suspended solids and turbidity from municipal wastewater. Bioresour Technol 99 (5):914–921. https://doi.org/10.1016/j.biortech.2007.03.015
21. Zodi S, Merzouk B, Potier O, Lapicque F, Leclerc J-P (2013) Direct red 81 dye removal by a continuous flow electrocoagulation/flotation reactor. Sep Purif Technol 108:215–222. https://doi.org/10.1016/j.seppur.2013.01.052
22. El-Naas MH, Al-Zuhair S, Al-Lobaney A, Makhlouf S (2009) Assessment of electrocoagulation for the treatment of petroleum refinery wastewater. J Environ Manage 91 (1):180–185. https://doi.org/10.1016/j.jenvman.2009.08.003
23. Glaze WH, Kang J-W, Chapin DH (1987) The chemistry of water treatment processes involving ozone, hydrogen peroxide and ultraviolet radiation. Ozone Sci Eng 9(4):335–352. https://doi.org/10.1080/01919518708552148
24. Andreozzi R, Caprio V, Insola A, Marotta R (1999) Advanced oxidation processes (AOP) for water purification and recovery. Catal Today 53(1):51–59
25. Benítez FJ, Beltrán-Heredia J, Acero JL, Pinilla ML (1997) Ozonation kinetics of phenolic acids present in wastewaters from olive oil mills. Ind Eng Chem Res 36(3):638–644
26. Casero I, Sicilia D, Rubio S, Pérez-Bendito D (1997) Chemical degradation of aromatic amines by Fenton's reagent. Water Res 31(8):1985–1995. https://doi.org/10.1016/S0043-1354(96)00344-2
27. Bigda RJ (1995) Consider Fenton's chemistry for wastewater treatment. Chem Eng Prog 62–66
28. Safarzadeh-Amiri A, Bolton JR, Cater SR (1996) The use of iron in advanced oxidation processes. J Adv Oxid Technol 1(1). https://doi.org/10.1515/jaots-1996-0105
29. Pignatello JJ, Liu D, Huston P (1999) Evidence for an additional oxidant in the photoassisted Fenton reaction. Environ Sci Technol 33(11):1832–1839. https://doi.org/10.1021/es980969b
30. Schoonen MAA, Xu Y, Strongin DR (1998) An introduction to geocatalysis. J Geochem Explor 62(1–3):201–215
31. Sillanpää M (2014) Natural organic matter in water: characterization and treatment methods. doi: https://doi.org/10.1016/C2013-0-19213-6
32. Cehovin M, Medic A, Scheideler J, Mielcke J, Ried A, Kompare B, Zgajnar Gotvajn A (2017) Hydrodynamic cavitation in combination with the ozone, hydrogen peroxide and the UV-based advanced oxidation processes for the removal of natural organic matter from drinking water. Ultrason Sonochem 37:394–404. https://doi.org/10.1016/j.ultsonch.2017.01.036
33. Gregory J, Dhond RV (1972) Anion exchange equilibria involving phosphate, sulphate and chloride. Water Res 6(6):695–702. https://doi.org/10.1016/0043-1354(72)90184-4
34. Clarizia L, Russo D, Di Somma I, Marotta R, Andreozzi R (2017) Homogeneous photo-Fenton processes at near neutral pH: a review. Appl Catal Environ 209:358–371. https://doi.org/10.1016/j.apcatb.2017.03.011

35. Tatibouët J-M, Guélou E, Fournier J (2005) Catalytic oxidation of phenol by hydrogen peroxide over a pillared clay containing iron. Active species and pH effect. Top Catal 33 (1–4):225–232. https://doi.org/10.1007/s11244-005-2531-3
36. Debellefontaine H, Chakchouk M, Foussard JN, Tissot D, Striolo P (1996) Treatment of organic aqueous wastes: wet air oxidation and wet peroxide oxidation. Environ Pollut 92 (2):155–164. https://doi.org/10.1016/0269-7491(95)00100-X
37. Ribeiro RS, Silva AMT, Figueiredo JL, Faria JL, Gomes HT (2016) Catalytic wet peroxide oxidation: a route towards the application of hybrid magnetic carbon nanocomposites for the degradation of organic pollutants. A review. Appl Catal Environ 187:428–460. https://doi.org/10.1016/j.apcatb.2016.01.033
38. Kasprzyk-Hordern B, Raczyk-Stanisławiak U, Świetlik J, Nawrocki J (2006) Catalytic ozonation of natural organic matter on alumina. Appl Catal Environ 62(3–4):345–358. https://doi.org/10.1016/j.apcatb.2005.09.002
39. Arimi MM (2017) Modified natural zeolite as heterogeneous Fenton catalyst in treatment of recalcitrants in industrial effluent. Prog Nat Sci Mater Int 27(2):275–282. https://doi.org/10.1016/j.pnsc.2017.02.001
40. de Ridder DJ, Verberk JQJC, Heijman SGJ, Amy GL, van Dijk JC (2012) Zeolites for nitrosamine and pharmaceutical removal from demineralised and surface water: Mechanisms and efficacy. Sep Purif Technol 89:71–77. https://doi.org/10.1016/j.seppur.2012.01.025
41. Chen C, Apul OG, Karanfil T (2017) Removal of bromide from surface waters using silver impregnated activated carbon. Water Res 113:223–230. https://doi.org/10.1016/j.watres.2017.01.019
42. Galeano L-A, Vicente MÁ, Gil A (2014) Catalytic degradation of organic pollutants in aqueous streams by mixed Al/M-pillared clays (M = Fe, Cu, Mn). Catal Rev 56 (3):239–287. https://doi.org/10.1080/01614940.2014.904182
43. Galeano LA, Bravo PF, Luna CD, Vicente MÁ, Gil A (2012) Removal of natural organic matter for drinking water production by Al/Fe-PILC-catalyzed wet peroxide oxidation: effect of the catalyst preparation from concentrated precursors. Appl Catal Environ 111–112:527–535. https://doi.org/10.1016/j.apcatb.2011.11.004
44. Munoz M, de Pedro ZM, Menendez N, Casas JA, Rodriguez JJ (2013) A ferromagnetic γ-alumina-supported iron catalyst for CWPO. Application to chlorophenols. Appl Catal Environ 136–137:218–224. https://doi.org/10.1016/j.apcatb.2013.02.002
45. Garcia-Costa AL, Zazo JA, Rodriguez JJ, Casas JA (2017) Microwave-assisted catalytic wet peroxide oxidation. Comparison of Fe catalysts supported on activated carbon and γ-alumina. Appl Catal Environ 218:637–642. https://doi.org/10.1016/j.apcatb.2017.06.058
46. Xiang L, Royer S, Zhang H, Tatibouet JM, Barrault J, Valange S (2009) Properties of iron-based mesoporous silica for the CWPO of phenol: a comparison between impregnation and co-condensation routes. J Hazard Mater 172(2–3):1175–1184. https://doi.org/10.1016/j.jhazmat.2009.07.121
47. Zhong X, Barbier J, Duprez D, Zhang H, Royer S (2012) Modulating the copper oxide morphology and accessibility by using micro-/mesoporous SBA-15 structures as host support: effect on the activity for the CWPO of phenol reaction. Appl Catal Environ 121–122:123–134. https://doi.org/10.1016/j.apcatb.2012.04.002
48. Barrault J, Bouchoule C, Echachoui K, Frini-Srasra N, Trabelsi M, Bergaya F (1998) Catalytic wet peroxide oxidation (CWPO) of phenol over mixed (Al-Cu)-pillared clays. Appl Catal Environ 15(3–4):269–274. https://doi.org/10.1016/S0926-3373(97)00054-4
49. Barrault J, Tatibouët JM, Papayannakos N (2000) Catalytic wet peroxide oxidation of phenol over pillared clays containing iron or copper species. C R Acad Sci Ser IIc Chem 3 (10):777–783
50. Carriazo J, Guelou E, Barrault J, Tatibouet JM, Molina R, Moreno S (2005) Catalytic wet peroxide oxidation of phenol by pillared clays containing Al-Ce-Fe. Water Res 39 (16):3891–3899. https://doi.org/10.1016/j.watres.2005.06.034

51. Timofeeva MN, Khankhasaeva ST, Talsi EP, Panchenko VN, Golovin AV, Dashinamzhilova ET, Tsybulya SV (2009) The effect of Fe/Cu ratio in the synthesis of mixed Fe, Cu, Al-clays used as catalysts in phenol peroxide oxidation. Appl Catal Environ 90(3–4):618–627. https://doi.org/10.1016/j.apcatb.2009.04.024
52. Galeano LA, Gil A, Vicente MA (2010) Effect of the atomic active metal ratio in Al/Fe-, Al/Cu- and Al/(Fe–Cu)-intercalating solutions on the physicochemical properties and catalytic activity of pillared clays in the CWPO of methyl orange. Appl Catal Environ 100 (1–2):271–281. https://doi.org/10.1016/j.apcatb.2010.08.003
53. Khankhasaeva S, Dambueva DV, Dashinamzhilova E, Gil A, Vicente MA, Timofeeva MN (2015) Fenton degradation of sulfanilamide in the presence of Al, Fe-pillared clay: catalytic behavior and identification of the intermediates. J Hazard Mater 293:21–29. https://doi.org/10.1016/j.jhazmat.2015.03.038
54. Guélou E, Barrault J, Fournier J, Tatibouët JM (2003) Active iron species in the catalytic wet peroxide oxidation of phenol over pillared clays containing iron. Appl Catal Environ 44 (1):1–8. https://doi.org/10.1016/S0926-3373(03)00003-1
55. Banković P, Milutinović-Nikolić A, Mojović Z, Jović-Jovičić N, Žunić M, Dondur V, Jovanović D (2012) Al, Fe-pillared clays in catalytic decolorization of aqueous tartrazine solutions. Appl Clay Sci 58:73–78. https://doi.org/10.1016/j.clay.2012.01.015
56. Banković P, Milutinović-Nikolić A, Mojović Z, Jović-Jovičić N, Perović M, Spasojević V, Jovanović D (2013) Synthesis and characterization of bentonites rich in beidellite with incorporated Al or Al–Fe oxide pillars. Microporous Mesoporous Mater 165:247–256. https://doi.org/10.1016/j.micromeso.2012.08.029
57. Zhou S, Zhang C, Hu X, Wang Y, Xu R, Xia C, Zhang H, Song Z (2014) Catalytic wet peroxide oxidation of 4-chlorophenol over Al-Fe-, Al-Cu-, and Al-Fe-Cu-pillared clays: sensitivity, kinetics and mechanism. Appl Clay Sci 95:275–283. https://doi.org/10.1016/j.clay.2014.04.024
58. Carriazo J, Guélou E, Barrault J, Tatibouët JM, Molina R, Moreno S (2005) Synthesis of pillared clays containing Al, Al-Fe or Al-Ce-Fe from a bentonite: characterization and catalytic activity. Catal Today 107–108:126–132. https://doi.org/10.1016/j.cattod.2005.07.157
59. Sanabria NR, Centeno MA, Molina R, Moreno S (2009) Pillared clays with Al–Fe and Al–Ce–Fe in concentrated medium: synthesis and catalytic activity. Appl Catal A Gen 356 (2):243–249. https://doi.org/10.1016/j.apcata.2009.01.013
60. Gao H, Zhao B-X, Luo J-C, Wu D, Ye W, Wang Q, Zhang X-L (2014) Fe–Ni–Al pillared montmorillonite as a heterogeneous catalyst for the catalytic wet peroxide oxidation degradation of orange acid II: preparation condition and properties study. Microporous Mesoporous Mater 196:208–215. https://doi.org/10.1016/j.micromeso.2014.05.014
61. Pizarro AH, Molina CB, Munoz M, de Pedro ZM, Menendez N, Rodriguez JJ (2017) Combining HDC and CWPO for the removal of p-chloro-m-cresol from water under ambient-like conditions. Appl Catal Environ 216:20–29. https://doi.org/10.1016/j.apcatb.2017.05.052
62. Fetter G, Hernández V, Rodríguez V, Valenzuela MA, Lara VH, Bosch P (2003) Effect of microwave irradiation time on the synthesis of zirconia-pillared clays. Mater Lett 57 (5–6):1220–1223
63. Sanabria NR, Ávila P, Yates M, Rasmussen SB, Molina R, Moreno S (2010) Mechanical and textural properties of extruded materials manufactured with AlFe and AlCeFe pillared bentonites. Appl Clay Sci 47(3-4):283–289. https://doi.org/10.1016/j.clay.2009.11.029
64. Kooli F (2013) Pillared montmorillontes from unusual antiperspirant aqueous solutions: characterization and catalytic tests. Microporous Mesoporous Mater 167:228–236. https://doi.org/10.1016/j.micromeso.2012.09.012
65. Salerno P, Mendioroz S (2002) Preparation of Al-pillared montmorillonite from concentrated dispersions. Appl Clay Sci 22(3):115–123. https://doi.org/10.1016/S0169-1317(02)00133-3

66. Yan Y, Jiang S, Zhang H (2014) Efficient catalytic wet peroxide oxidation of phenol over Fe-ZSM-5 catalyst in a fixed bed reactor. Sep Purif Technol 133:365–374. https://doi.org/10.1016/j.seppur.2014.07.014
67. Munoz M, Domínguez P, de Pedro ZM, Casas JA, Rodriguez JJ (2017) Naturally-occurring iron minerals as inexpensive catalysts for CWPO. Appl Catal Environ 203:166–173. https://doi.org/10.1016/j.apcatb.2016.10.015
68. Garrido-Ramirez EG, Sivaiah MV, Barrault J, Valange S, Theng BKG, Ureta-Zañartu MS, Mora ML (2012) Catalytic wet peroxide oxidation of phenol over iron or copper oxide-supported allophane clay materials: influence of catalyst SiO2/Al2O3 ratio. Microporous Mesoporous Mater 162:189–198. https://doi.org/10.1016/j.micromeso.2012.06.038
69. Garrido-Ramírez EG, Theng BKG, Mora ML (2010) Clays and oxide minerals as catalysts and nanocatalysts in Fenton-like reactions – a review. Appl Clay Sci 47(3–4):182–192. https://doi.org/10.1016/j.clay.2009.11.044
70. Ayekoe CYP, Robert D, Lanciné DG (2017) Combination of coagulation-flocculation and heterogeneous photocatalysis for improving the removal of humic substances in real treated water from Agbô River (Ivory-Coast). Catal Today:281:2–28113. https://doi.org/10.1016/j.cattod.2016.09.024
71. González O, Justo A, Bacardit J, Ferrero E, Malfeito JJ, Sans C (2013) Characterization and fate of effluent organic matter treated with UV/H2O2 and ozonation. Chem Eng J 226:402–408. https://doi.org/10.1016/j.cej.2013.04.066
72. Bazri MM, Barbeau B, Mohseni M (2012) Impact of UV/H(2)O(2) advanced oxidation treatment on molecular weight distribution of NOM and biostability of water. Water Res 46(16):5297–5304. https://doi.org/10.1016/j.watres.2012.07.017
73. Black KE, Berube PR (2014) Rate and extent NOM removal during oxidation and biofiltration. Water Res 52:40–50. https://doi.org/10.1016/j.watres.2013.12.017
74. Moncayo-Lasso A, Rincon A-G, Pulgarin C, Benitez N (2012) Significant decrease of THMs generated during chlorination of river water by previous photo-Fenton treatment at near neutral pH. J Photochem Photobiol A Chem 229(1):46–52. https://doi.org/10.1016/j.jphotochem.2011.12.001
75. Deng Y, Englehardt JD (2006) Treatment of landfill leachate by the Fenton process. Water Res 40(20):3683–3694. https://doi.org/10.1016/j.watres.2006.08.009
76. García AM, Torres R, Galeano LA (2018) RSM optimization of the Al/Fe-pillared clay activated CWPO degradation of natural organic matter (NOM). Environ Sci Pollut Res. In preparation
77. Brame J, Long M, Li Q, Alvarez P (2015) Inhibitory effect of natural organic matter or other background constituents on photocatalytic advanced oxidation processes: mechanistic model development and validation. Water Res 84:362–371. https://doi.org/10.1016/j.watres.2015.07.044
78. Sen Kavurmaci S, Bekbolet M (2013) Photocatalytic degradation of humic acid in the presence of montmorillonite. Appl Clay Sci 75–76:60–66. https://doi.org/10.1016/j.clay.2013.03.006
79. Rache ML, García AR, Zea HR, Silva AMT, Madeira LM, Ramírez JH (2014) Azo-dye orange II degradation by the heterogeneous Fenton-like process using a zeolite Y-Fe catalyst – kinetics with a model based on the Fermi's equation. Appl Catal Environ 146:192–200. https://doi.org/10.1016/j.apcatb.2013.04.028
80. Noorjahan M, Durga Kumari V, Subrahmanyam M, Panda L (2005) Immobilized Fe(III)-HY: an efficient and stable photo-Fenton catalyst. Appl Catal Environ 57(4):291–298. https://doi.org/10.1016/j.apcatb.2004.11.006
81. Mohedano AF, Monsalvo VM, Bedia J, Lopez J, Rodriguez JJ (2014) Highly stable iron catalysts from sewage sludge for CWPO. J Environ Chem Eng 2(4):2359–2364. https://doi.org/10.1016/j.jece.2014.01.021
82. Bedia J, Monsalvo VM, Rodriguez JJ, Mohedano AF (2017) Iron catalysts by chemical activation of sewage sludge with FeCl3 for CWPO. Chem Eng J 318:224–230. https://doi.org/10.1016/j.cej.2016.06.096

83. Kim WH, Nishijima W, Baes AU, Okada M (1997) Micropollutant removal with saturated biological activated carbon (BAC) in ozonation-BAC process. Water Sci Technol 36 (12):283–298. https://doi.org/10.1016/S0273-1223(97)00722-1
84. Gibert O, Lefevre B, Fernandez M, Bernat X, Paraira M, Calderer M, Martinez-Llado X (2013) Characterising biofilm development on granular activated carbon used for drinking water production. Water Res 47(3):1101–1110. https://doi.org/10.1016/j.watres.2012.11.026
85. Hyung H, Kim JH (2008) Natural organic matter (NOM) adsorption to multi-walled carbon nanotubes: effect of NOM characteristics and water quality parameters. Environ Sci Technol 42(12):4416–4421. https://doi.org/10.1021/es702916h
86. Summers RSRP (1988) Activated carbon adsorption of humic substances. 2. Size exclusion and electrostatic interactions. J Colloid Interface Sci 122:382–397
87. Yang K, Xing B (2007) Desorption of polycyclic aromatic hydrocarbons from carbon nanomaterials in water. Environ Pollut 145(2):529–537. https://doi.org/10.1016/j.envpol.2006.04.020

Separation and Characterization of NOM Intermediates Along AOP Oxidation

Ana-María García, Ricardo A. Torres-Palma, Luis Alejandro Galeano, Miguel Ángel Vicente, and Antonio Gil

Abstract Removal of natural organic matter (NOM) in drinking water treatment systems has been a matter of thorough study in recent years. NOM affects organoleptic properties of water and causes membrane fouling; it may act as energy source for microorganisms in distribution systems and leads to the formation of undesired disinfection by-products through its interaction with chlorine. Currently the role played by advanced oxidation processes in the removal of NOM has gained

---

A.-M. García
Grupo de Investigación en Materiales Funcionales y Catálisis (GIMFC), Departamento de Química, Facultad de Ciencias Exactas y Naturales, Universidad de Nariño, Pasto, Colombia

Grupo de Investigación en Remediación Ambiental y Biocatálisis (GIRAB), Instituto de Química, Facultad de Ciencias Exactas y Naturales, Universidad de Antioquia – UdeA, Medellín, Colombia
e-mail: ana.garcia11@udea.edu.co; anamariagarcia@udenar.edu.co

R.A. Torres-Palma (✉)
Grupo de Investigación en Remediación Ambiental y Biocatálisis (GIRAB), Instituto de Química, Facultad de Ciencias Exactas y Naturales, Universidad de Antioquia – UdeA, Medellín, Colombia
e-mail: ricardo.torres@udea.edu.co

L.A. Galeano
Grupo de Investigación en Materiales Funcionales y Catálisis (GIMFC), Departamento de Química, Facultad de Ciencias Exactas y Naturales, Universidad de Nariño, Pasto, Colombia
e-mail: alejandrogaleano@udenar.edu.co

M.Á. Vicente
GIR-QUESCAT, Departamento de Química Inorgánica, Universidad de Salamanca, Salamanca, Spain
e-mail: mavicente@usal.es

A. Gil
Departamento de Química Aplicada, Edificio de los Acebos, Universidad Pública de Navarra, Pamplona, Spain
e-mail: andoni@unavarra.es

great interest; understanding the composition and behaviour of NOM throughout such a kind of processes may allow to get significant insight in order to improve efficiency. In this chapter the main techniques useful for characterization are described, and their use to investigate the changes undergone by NOM throughout several AOPs has been reviewed.

Keywords AOP, NOM characterization, NOM intermediates, NOM oxidation, NOM separation

Contents

1 Introduction 102
2 Separation of NOM and NOM Intermediates 104
   2.1 Fractionation by Resins 105
   2.2 Reversed-Phase Chromatography (RP-LC) 107
   2.3 Size-Exclusion Chromatography (LC-SEC) 107
3 Characterization of NOM Intermediates 108
   3.1 Ultraviolet Spectroscopy (UV-Vis) and SUVA 108
   3.2 Total Organic Carbon (TOC) 109
   3.3 Fourier Transform Infrared (FT-IR) Spectroscopy 110
   3.4 Fluorescence Excitation/Emission Matrix (FEEM) Spectroscopy 110
   3.5 Electrospray Ionization Fourier Transform Ion Cyclotron Resonance Mass Spectrometry (ESI-FT-ICR-MS) 111
4 Oxidation of NOM Through AOPs 111
   4.1 Ozone-Based Applications: O_3/H_2O_2 and O_3/UV 112
   4.2 UV Light-Based Applications: UV/H_2O_2 and UV/Cl_2 116
   4.3 Fenton, Photo-Fenton and Fenton-Like Catalysed Processes 122
   4.4 Heterogeneous Photocatalysis: TiO_2/UV 124
   4.5 Ultrasound-Based Applications 126
5 Conclusions 127
References 129

Abbreviations

AMW	Apparent molecular weight distribution
AOPs	Advanced oxidation processes
BA	Benzoic acid
CAS	Conventional activated sludge
C-DBPs	Carbonaceous disinfection by-products
CDOM	Coloured dissolved organic matter
COD	Chemical oxygen demand
DOC	Dissolved organic carbon (mg C/L)
DOM	Dissolved organic matter
DPBs	Disinfection by-products
ESI-FT-ICR-MS	Electrospray ionization Fourier transform ion cyclotron resonance mass spectrometry
FA	Fulvic acids
FEEM	Fluorescence excitation/emission matrix

FFA	Furfuryl alcohol
FT-IR	Fourier transform infrared spectroscopy
GC-MS	Gas chromatography-mass spectrometry
HA	Humic acids
HAAs	Haloacetic acids
HMW	High molecular weight
HPI	Hydrophilic fraction
HPI-A	Hydrophilic acids
HPI-B	Hydrophilic bases
HPI-N	Hydrophilic neutrals
HPO	Hydrophobic fraction
HPO-A	Hydrophobic acids
HPO-B	Hydrophobic bases
HPO-N	Hydrophobic neutrals
HS	Humic substances
LC-OCD	Liquid chromatography-organic carbon detection
LC-SEC	Size-exclusion liquid chromatography
LC-UVD	Liquid chromatography-ultraviolet detection
LMW	Low molecular weight
MBR	Membrane biological reactor
Mt	Montmorillonite
MTBE	Methyl tert-butyl ether
MW	Molecular weight
N-DBPs	Nitrogenous disinfection by-products
NMR	Nuclear magnetic resonance spectroscopy
NOM	Natural organic matter
NPOC	Non-purgeable organic carbon
OCD	Organic carbon detector
PCU	Platinum-cobalt units
PFBHA	Pentafluorobenzyl hydroxylamine hydrochloride
PS	Persulphate
PSS	Polystyrene sulphonate
RCSs	Reactive chlorine species
RID	Refractive index detector
ROS	Reactive oxygen species
RP	Reversed-phase liquid chromatography
RT	Room temperature
SEC	Size-exclusion chromatography
SHA	Slightly hydrophobic acid fraction
SUVA	Specific UV absorbance (L/mg m)
THMs	Trihalomethanes
TOC	Total organic carbon (mg C/L)
TPI	Transphilic fraction
US	Ultrasound
UV	Ultraviolet
UV_{254}	UV absorbance at 254 nm (m^{-1})
UVD	Ultraviolet detector
UV-Vis	Ultraviolet visible
VHA	Very hydrophobic acid fraction

1 Introduction

Natural organic matter (NOM) is a complex mixture of particulate components, about 10% of the carbon present in water [1], and soluble fractions varying significantly from one source to another [2]. The origin of NOM in both fresh and salt waters comes from natural processes in the environment, including the run-off leading to larger export of organic material from the terrestrial system, soil organic matter decomposition (allochthonous) and algal metabolic reactions (autochthonous) [3, 4]. As a consequence, NOM is a heterogeneous mixture of substances with wide ranges of molecular sizes, reactivity and chemical functionalities [5, 6] oscillating from large aliphatic chains (mainly hydroxylated, carbonyl and carboxylic acids) to highly coloured aromatics [7]. NOM can be divided into humic and non-humic substances: the humic substances, humic and fulvic acids, represent residual degradation products, and the non-humic ones include lignins and derivatives, tannins, carbohydrates, peptides and proteins, amino acids, aromatic acids and phenols, carboxylic acids and miscellaneous compounds [8].

When NOM is filtered through 0.45 μm porous membranes, dissolved organic matter (DOM) is obtained [7], which is commonly represented by the amount of dissolved organic carbon (DOC) in solution. The humic substances being hydrophobic compounds are the major constituents of NOM, in general reaching approximately 50% of DOC, represented mainly in humic acids usually including a large number of aromatic carbons, phenolic structures and conjugated double bonds [9]. The hydrophilic fraction is about 25–40% of DOC composed by non-humic substances [3] that include a high proportion of aliphatic carbons and nitrogen compounds, such as carbohydrates, proteins, sugars and amino acids (Fig. 1) [11]. Finally, the transphilic fraction represents approximately 25% of the DOC. This distribution, however, may vary far and wide from one water source to another.

NOM has been lately found to increase in many surface waters, and particularly it has been evidenced for its coloured fraction [7]. As a result, nowadays NOM can seriously affect the organoleptic properties of water (colour, taste and odour) [12],

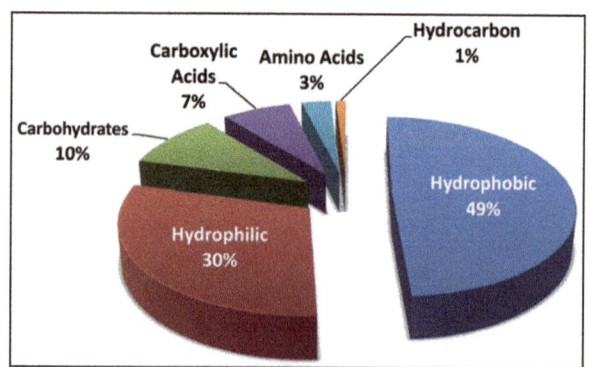

Fig. 1 Average distribution of NOM fractions present in surface waters based on dissolved organic carbon. Reproduced with permission of the Korean Society of Environmental Engineers [10]

increase the required doses of coagulant and disinfectant agents in drinking water plants, promote bacterial growth in distribution systems and increase levels of heavy metals and adsorbed organic pollutants [13, 14]. In addition, NOM blocks porosity and strives for sites during adsorption processes [15], imposing either continuous regeneration of filters based on activated carbon or cleaning of membrane surfaces [7]. Furthermore, during the chlorine disinfection process, it may react with natural organic matter to form carcinogenic disinfection by-products (DPBs) [16]. Two prevalent groups of DPBs are regulated in Canada and the United States: trihalomethanes (THMs) and haloacetic acids (HAAs). The US Environmental Protection Agency (US EPA) has defined maximum acceptable levels of THMs as 80 and 60 µg/L for the five most common HAAs (HAA_5) [17]. Therefore, it is very important to control and limit NOM content in water supplies for drinking water production in order to decrease the potential formation of DBPs.

Vulnerability of drinking water distribution systems is very important because of the more recent strict regulations in public health [18]. Oxidation strategies that could be used for NOM removal in the drinking water industry include ozonization combined with either filtration [19], biological process [20] or slow sand filtration [21] and more precisely advanced oxidation processes (AOPs) [22]. AOPs have gained great interest due to its high oxidation power over almost any organic compound. Currently several studies are being conducted to show the potential role that AOPs can play in the transformation of NOM from natural sources of water. AOPs are mainly based on the formation and use of hydroxyl (HO$^\bullet$) radicals, which because of its high oxidation potential [23] are capable to remove a wide range of substances of difficult degradation; since this type of radicals are highly reactive and then non-selective, they may react very quickly improving several parameters of the output stream [4]. Several applications of these processes have been studied involving combinations of oxidizing agents, radiation and catalysts in order to remove NOM and organic pollutants [9]. Examples of such processes have included UV-based (UV/H_2O_2) and ozone-based (O_3/H_2O_2, O_3/UV, O_3/H_2O_2/UV and O_3/H_2O_2/TiO_2) applications, heterogeneous photocatalysis (TiO_2/UV), ultrasound, electrochemical-based processes (anodic oxidation with BDD electrodes, electro-Fenton and photo-electro-Fenton), homogeneous Fenton, heterogeneous Fenton and the Fenton-like processes like the so-called catalytic wet peroxide oxidation (CWPO) [14].

Given the very high oxidizing power of HO$^\bullet$ radicals, it is expected that use of AOPs leads to deep mineralization of the organic matter, i.e. final reaction products corresponding to CO_2, water and inorganic ions – Eq. (1).

$$NOM + H_2O_2 \xrightarrow{mineralization} CO_2 + H_2O + inorganic\ ions \qquad (1)$$

However, depending on the complexity of the targeted organic molecule and the efficiency of the oxidation process, various by-products can be obtained which in general are expected to be of lower toxicity than the starting molecule. Typically, AOPs in drinking water treatment would be useful to degrade taste and odour-causing chemical compounds, as well as to destroy any residual toxicity in water resulting from these types of contaminants [22]. In many model molecules, it has been determined as an oxidative route through typical attack pathways, where HO˙ reacts mainly by abstracting H atoms or adding it to unsaturated bonds [24]. Due to the widely distributed chemical functionalities and molecular sizes present in NOM, the tracking of NOM and its by-products [4] through the degradation has become a challenging issue. Furthermore, other reactive oxygen species (ROS) such as the peroxyl (ROO˙) radical, hydroperoxyl $\left(HO_2^{\bullet}\right)$ radical, superoxide anion $\left(O_2^{-}\right)$ and singlet oxygen ($^1O_2$) can also participate together with the NOM itself in the degradation process. Thus, understanding the molecular and structural properties of the targeted NOM and its by-products is extremely important in order to elucidate the degradation as well as to understand the behaviour of several NOM fractions throughout the process.

Various analytical techniques have been employed to characterize DOM composition including (1) ultraviolet spectroscopy (UV-Vis), (2) Fourier transform infrared (FT-IR) spectroscopy, (3) nuclear magnetic resonance (NMR) spectroscopy, (4) fluorescence excitation/emission matrix (FEEM) spectroscopy, (5) mass spectrometric methods such as liquid/gas chromatography-mass spectrometry (LC-MS or GC-MS), (6) ultrahigh-pressure liquid chromatography (UPLC) coupled to quadrupole time-of-flight mass spectrometer (QTOF-MS) and (7) electrospray ionization (ESI) coupled to ultrahigh-resolution Fourier transform ion cyclotron resonance mass spectrometry (FT-ICR-MS) [25]. The fractionation of NOM into broad chemical classes is the first step to examine its structure. Fractionation by resins is the method most frequently used for isolation/fractionation of NOM based on its polar moieties (e.g. hydrophobic, hydrophilic and transphilic). This method currently uses non-ionic macroporous copolymers, such as XAD resin analogues, followed by ion exchange resins [26]. Therefore, the first part of this chapter revises the main methods of separation and characterization of natural organic matter. The first recommended step is fractionation, mostly based in two methods: (1) resin fractionation and (2) reverse-phase high-performance liquid chromatography. Although these have been also used as characterization methods, in this chapter they are separated into several sections for more useful approach. Size-exclusion liquid chromatography (LC-SEC) has been also employed to separate NOM based on its molecular weight. In the second part, the main characterization methods of NOM are briefly described: *UV-Vis, SUVA, TOC, FT-IR, FEEM* and *ECI-FT-ICR-MS*. Finally, the NOM intermediates found in several AOPs are compared, including UV light-based (UV/H_2O_2 and UV/Cl_2), ozone-based (O_3/H_2O_2 and O_3/UV), differences in reaction pathways of homogeneous and heterogeneous processes, photocatalytic (TiO_2/UV) and ultrasound-based applications.

2 Separation of NOM and NOM Intermediates

Complex mixtures of NOM in water supplies commonly affect operation of drinking water treatment plants [15]. Thus, elucidation of the chemical properties of the NOM present in a particular water source may greatly help choosing the more suitable treatment technology. For this purpose, the first step is to separate NOM in several fractions. The most frequently used methods to separate NOM are fractionation by resins through column-liquid chromatography or reversed-phase liquid chromatography (RP-LC). Afterwards, each fraction can be further characterized by size-exclusion liquid chromatography (LC-SEC) in order to obtain molecular weight distribution. These three methods for separation of NOM will be discussed in forthcoming sections.

2.1 Fractionation by Resins

Fractionation by adsorption in resins is an effective way of elucidating the chemical properties of DOM according to its hydrophobic, hydrophilic and transphilic nature. Often in this process, non-ionic macroporous resins (DAX-8 acrylic esters and XAD-4 styrene divinylbenzene) are used, which can split DOM into three classes: (1) hydrophobic fraction (HPO), mainly constituted of humic substances (HS) including humic acids (HA) and fulvic acids (FA), (2) transphilic fraction (TPI) including hydrophilic acids (HPI-A) and (3) hydrophilic fraction (HPI), involving hydrophilic bases (HPI-B) and neutrals (HPI-N). Subsequently, by employing cation-exchange resins and anion-exchange resins, six classes of DOM could be further obtained as follows: hydrophobic acids (HPO-A), humic and fulvic acids ranging from 450 to 1,000 Da; hydrophobic bases (HPO-B), proteins and amino acids ranging from 250 to 850 Da; hydrophobic neutrals (HPO-N), hydrocarbons ranging from 100 to 70,000 Da; hydrophilic acids, fatty acids ranging from 250 to 850 Da; hydrophilic bases (HPI-B), proteins and amino acids ranging from 100 to 1,000 Da; and hydrophilic neutrals (HPI-N), polysaccharides ranging from 120 to 900 Da [4, 27].

As an adaptation of the methods proposed by Leenheer et al. [11] and Fabris et al. [28], the fractionation method starts conditioning the chromatographic columns, glass tubes with varied diameters from 5 to 50 mm and heights from 5.0 cm to 1.0 m, which involves the following steps:

1. Cleaning with ultrapure type I water (MilliQ or similar purified water), which is passed through the resin packed into the column in order to remove any remaining methanol (the resins are preserved in methanol when not in use).
2. Cleaning with 0.1 M NaOH, which is used for conditioning resins.
3. Repetition of steps (1) and (2).
4. Keeping of the resins in 0.1 M NaOH for approximately 12 h.
5. Monitoring by UV-Vis and total organic carbon (TOC), which are used to detect interferences. If impurities are still present, water and methanol ($\cong$100 mL) can be added to each column to complete cleaning. Then, the conditioning process (NaOH/water wash) is repeated until the TOC of the effluent from each column becomes pretty close to zero and the absorbance (λ_{254}) < 0.0001 m^{-1}. In some cases a control of the conductivity (<10 μS/cm) is also recommended [4].

6. Conditioning with H_3PO_4 0.1 N and finally ultrapure water again to obtain a similar pH to surface water samples.
7. Set the pH of the sample to around 2.0 [29].

Once the columns are conditioned, 100 mL of the water containing the NOM is added first through a DAX-8 column using a flow of around 3 mL/min. From the total eluted volume, samples are taken and recorded by means of TOC measurements. The remaining volume is fed in the column of the XAD-4 resin and the TOC of the eluted solution also measured. In order to enrich the fraction retained in the XAD-4 resin, the eluted sample could be recirculated; it also serves as preconditioning of the resin, which improves the adsorption of the fraction of interest in the column (Fig. 2). Finally, the hydrophobic fraction of the NOM water retained in the DAX-8 resin and the transphilic fraction retained in the XAD-4 resin are eluted using 0.01 mol/L NaOH in each case. These solutions can also be stored for further TOC analysis.

When analysing a real water source, some limitations can be presented to apply the fractionation method to samples containing a low concentration of organic matter. When the NOM concentration is very low, let's say below 5.0 mg TOC/L, rotary evaporation could be advised not exceeding 50°C and if necessary followed by drying in vacuum oven.

Although this technique is recognized for its easy implementation, it has been also recognized as a time-consuming one [15]. In addition, other limitations have been reported and should be taken into account such as either the need of a relative large volume of sample (~100–300 mL), chemical alterations that may occur in the sample due to the use of extreme pH levels along resin's conditioning, contamination of the sample by resin's bleeding or irreversible binding of some DOM components on the resins, among others [7].

Fig. 2 Columns packed with DAX-8 and XAD-4 resins for NOM fractionation

2.2 Reversed-Phase Chromatography (RP-LC)

Methods by reversed-phase liquid chromatography (RP-LC) have been established as fast techniques to distinguish between hydrophilic and hydrophobic fractions of NOM, taking advantage of their differences in polarity [15]. RP-LC has been used to compare the hydrophilic and hydrophobic contents of the NOM fractions from different water sources. It employs a polar mobile phase, commonly mixtures methanol-water, and a non-polar stationary phase, typically C_{18} column (octadecyl carbon chain C_{18}-bonded silica); thus, NOM molecules get eluted later from the column as their polarity decreases [30].

2.3 Size-Exclusion Chromatography (LC-SEC)

Size-exclusion chromatography (LC-SEC) has been widely used to determine the molecular weight distribution of humic substances [15]. It should be noted that in SEC, the components of a sample are separated according to the hydrodynamic size of the molecules. Since the peak distribution is established according to the molecular size of the analyte, in the ideal case when the molecules are larger than the pore diameters, they more easily pass through the column (first peaks). Smaller molecules entering the pores of the stationary phase must diffuse in and out of them until they are able to leave the column (final peaks) [28].

LC-SEC is often used to obtain the apparent molecular weight distribution (AMW) of several compounds. The relationship between AMW and the resulting elution time obtained by LC-SEC must be determined using compounds of known molecular weight [31]. Therefore, the previous preparation of a calibration curve on the basis of commercial standards with known molecular weight is necessary.

The calibration curve can be raised from a range of polystyrene sulphonate (PSS) macromolecule standards (1, 5, 13 and 20 kDa) used in concentrations of 0.1–1.0 g/L; the PSS standards and NOM compounds can be traced at 254 nm. It is important to preadjust the ionic strength and pH of the standard solutions for the properties in solution to be very similar to those in the mobile phase. It must be done to suppress charge effects in order to ensure that separation takes place mainly by differences in the molecular size and not because of charge interactions [3, 31]. The mobile phase is usually phosphate buffer of pH 7.0 prepared in ultrapure water, together with sodium chloride to obtain an ionic strength equivalent to 0.1 mol/L NaCl. One of the drawbacks of LC-SEC with UV detection is the low response obtained for NOM structures with low molecular UV absorptivities such as proteins, sugars, amino acids and aliphatic acids. In this sense, LC-SEC has been improved in recent years by coupling to dissolved organic carbon/nitrogen [3] or refractive index detectors (RID).

3 Characterization of NOM Intermediates

The properties and amount of NOM can significantly affect the efficiency of the degradation process. It is also important to be able to understand and predict the reactivity of NOM and its fractions at different treatment steps [6]. Ultraviolet spectroscopy (UV-Vis) and total organic carbon (TOC) are the most common parameters employed to follow the overall composition of NOM surrogates. In addition to adsorption in resins, size-exclusion chromatography (LC-SEC) and reversed-phase chromatography (RP-LC) are used for both separation and characterization of NOM. Techniques such as nuclear magnetic resonance (NMR) spectroscopy, Fourier transform infrared (FT-IR) spectroscopy and fluorescence excitation/emission matrix (FEEM) spectroscopy can also be used to characterize these types of substances. FT-IR allows to elucidate the main chemical functionalities in the molecule, whereas the fluorescence spectroscopy is a relatively low-cost and easily handled analysis [32]. FEEM spectroscopy provides a unique perspective of the NOM profile, usually not available from other modes of detection [17]. Lately, new more sophisticated methods have been developed whereby NOM structures can be determined more precisely, among which pyrolysis coupled to gas chromatography-mass spectrometry (Py-GC-MS), multidimensional NMR techniques and electrospray ionization Fourier transform ion cyclotron resonance mass spectrometry (ESI-FT-ICR-MS) are worth mentioning [6]. The latter allows identification of thousands of mass peaks from single given isolated sample, usually in the range of 200–1,000 Da [7]. Then, hereafter the main characteristics of selected techniques (UV-Vis, TOC, FT-IR, FEEM and ESI-FT-ICR-MS) most frequently employed for the characterization of NOM intermediates are displayed.

3.1 Ultraviolet Spectroscopy (UV-Vis) and SUVA

The amount of natural organic matter in water has been determined by means of several parameters including UV-Vis and specific ultraviolet absorbance (SUVA). SUVA has recently become a useful surrogate's parameter for NOM characterization as a function of molecular weight, aromatic content and hydrophobic/hydrophilic nature [32]. Absorption at 254 nm by π electrons has been proposed as a key parameter to follow degradation of the aromatic content in arenes, phenols, benzoic acids, aniline derivatives, polyenes and polycyclic aromatic hydrocarbons with one, two or more aromatic rings [33]. SUVA corresponds to the ratio of absorbance in a water sample at determined wavelength within the ultraviolet range (usually at 254 nm), normalized to concentration of dissolved organic carbon (DOC) (UV_{254}/DOC) [6]. In general, a high value of SUVA indicates the presence of high proportion of aromatic compounds and π-conjugated functionalities in the sample. SUVA can be also interpreted qualitatively describing hydrophobic and hydrophilic contents of organics in water. Although no clear limit values have been defined,

SUVA > 4 indicates mainly hydrophobic and especially aromatic compounds, whereas SUVA < 3 can be used to indicate predominantly hydrophilic character [12]. In the same way, the quinoid structures and keto-enol-systems are well-known to absorb mainly in the visible range [32]. In addition to SUVA, colour and TOC measurements are also useful. Rodríguez and Núñez [31] reported a SUVA ratio of 0.029 for natural fulvic acids, 0.040 for natural humic acids and 0.050 for commercial humic acids, whereas 5.87, 23.33 and 35.53 were experimental values for the same set of samples in terms of the ratio (colour – Pt-Co units)/TOC in mg C/L), respectively.

Some other ratios of absorbance have been reported for the spectral differentiation of humic substances and expressed as energies [34]. The E_{254}/E_{456} (A_{254nm}/A_{456nm}) gives an indication about the intensity of the UV-absorbing functional groups compared to the coloured ones, whose values are in the range 4–11 as a consequence of higher content of organic matter due to the presence of tannin-like or humic-like substances derived from plants and soil organic matter. Meanwhile, the E_{465}/E_{665} is commonly used to indicate the degree of condensation of the aromatic carbon network and is characteristic for different NOM fractions: it is usually <5.0 for HAs and in the range 6.0–8.5 for fulvic acids.

3.2 Total Organic Carbon (TOC)

The content of organic carbon is widely used as a parameter to represent NOM concentration in water [35]. In drinking water systems, where the TOC concentration is too low, the work with high-sensitivity TOC apparatus is recommended. It may involve a special platinum wool catalyst, whose higher surface imposes larger injection volumes.

TOC is a measurement of non-purgeable carbon in organic compounds present in a water sample, of course including all NOM species. Dissolved organic carbon (DOC) can be measured when the sample is passed through a 0.45 μm filter. This parameter is very useful to complement other characterization methods, for example, in the recording along resin fractionation [4].

When examining content of natural organic matter in drinking water, it is expectable that the content of inorganic carbons such as carbonates and hydrogen carbonates could be much higher than the organic fraction. Usually the organic fraction is only around 1.0% of the total carbon. Therefore, a TOC determination via the difference method (TOC = TC – IC) will not be appropriate in this case due to large statistical errors that could get propagated. The non-purgeable organic carbon (NPOC) parameter is then more advised in this case. The drinking water sample is first acidified to a pH value of 2.0 to transform the carbonates and hydrogen carbonates into carbon dioxide; CO_2 is then removed by sparging with pure air as the carrier gas. What remains in the solution can be oxidized to CO_2, detected via NDIR and corresponds to non-volatile organic carbonaceous compounds.

3.3 Fourier Transform Infrared (FT-IR) Spectroscopy

The FT-IR allows to realize the chemical functionalities present in the NOM molecules, but it has been more rather scarcely reported. Main characteristics documented [31] for humic substances are O-H stretching (alcohols, phenols and carboxylic groups, υ: 3,400 cm^{-1}), C-H stretching (CH$_3$ and CH$_2$, υ: 2,850–2,960 cm^{-1}), O-H stretching (hydrogen-bonded carboxylic groups, υ: 2,620 cm^{-1}), C = O stretching (carboxylic groups, υ: 1,720 cm^{-1}), C = C stretching (alkenes and aromatic rings, υ: 1,630 cm^{-1}), N-H bending (N-H structures, υ: 1,540 cm^{-1}), C-H bending (CH$_3$ and CH$_2$, υ: 1,455 cm^{-1}), O-H bending (carboxylic groups, υ: 1,410 cm^{-1}), C-H bending (CH$_3$, υ: 1,375 cm^{-1}), C-O stretching (alcohols, aliphatic ethers, υ: 1,095 and 1,030 cm^{-1}) and C-H bending (tri- and tetra-substituted aromatic rings, υ: 805 cm^{-1}).

3.4 Fluorescence Excitation/Emission Matrix (FEEM) Spectroscopy

Fluorescence spectroscopy has become a very useful tool for the analysis of NOM in water. Fluorometers are capable of generating high-dimensional fluorescence excitation/emission matrices efficiently and without extensive sample preparation. In natural or treated waters, humic-like substances typically represent the majority of fluorophores in both lake and river waters [17]. Fluorophores can be categorized according to their tendency to fluoresce in five distinct regions of the FEEM (see Table 1), through the fluorescent regional integration (FRI) procedure; it is used to integrate fluorescence intensity within each region to make easier interpretation of FEEM and also to quantify region-specific changes in fluorescence.

Table 1 Excitation and emission wavelength ranges for fluorescent regions I–V

Region	Characteristics	Excitation wavelength (nm)	Emission wavelength (nm)
I	Aromatic protein I	200–250	200–330[a]
II	Aromatic protein II	200–250	330–380
III	Associated with fulvic acids	200–250	380–550
IV	Soluble microbial products	250–340	200–380[a]
V	Associated with humic acids	250–400	380–550

Taken from [17]
[a]Lower limit was extended from 280 to 200 nm to match the detector range

3.5 Electrospray Ionization Fourier Transform Ion Cyclotron Resonance Mass Spectrometry (ESI-FT-ICR-MS)

Electrospray ionization (ESI) is called a "soft" ionization technique that ionizes polar compounds from aqueous solutions prior to injection into a mass spectrometer. ESI has a large mass range, ionizing compounds $10 < m/z < 3{,}000$ as quasi-molecular ions, $(M + nH)^{n+}$ or $(M - nH)^{n-}$ eliminating the need of deep fragmentation. ESI has been coupled with Fourier transform ion cyclotron resonance mass spectrometry (FT-ICR-MS), an ultrahigh-resolution mass spectrometer that allows to obtain the highest resolution and mass accuracy in the characterization. Thus, individual molecules within a variety of natural organic mixtures can be detected, its elemental composition determined, and changes at molecular level examined [36]. Due to a high accuracy (<1 ppm), this technique allows differentiation between NOM components having small differences in molecular mass and unambiguous assignment of molecular formulas up to approximately 600 Da. However, structural information cannot be obtained, due to high number of possible isomers, but it can be coupled with other components in order to obtain optical and structural information of NOM [7].

4 Oxidation of NOM Through AOPs

The implementation of advanced oxidation processes and the determination of their actual effectiveness on natural waters are often difficult, considering that they largely depend on the particular water-contaminated matrix. It is well-known that AOP treatment of organic compounds at relatively high concentrations (>50 ppm) in complex matrices may be much energy and oxidant consuming [22]. High concentrations of NOM may lead to formation of recalcitrant oxidation by-products that negatively further impact the quality of water, interfere with the elimination of the targeted compounds and reduce the effectiveness of the selected AOP [23]. In recent times a significant part of the research has been focused in realizing the role played by AOPs in the catalytic degradation of NOM. Throughout the oxidation process, a series of intermediates can be formed (Eq. 2) [37]:

$$RH + HO^{\bullet} \rightarrow \text{Intermediates} \rightarrow CO_2 + H_2O \qquad (2)$$

Action of the AOPs can be divided in two stages: (1) formation of hydroxyl radicals and (2) reaction of such oxidizing species with organic pollutants in water. However, throughout AOPs in situ formation of other reactive oxygen species (ROS) [38] may also occur: hydroperoxyl radical ($HO_2^{\bullet}$), singlet oxygen ($^1O_2$) and superoxide radical ($O_2^{\bullet-}$). The kinetic constant reported for the reaction of NOM with hydroxyl radicals has ranged from 1.9×10^4 to 1.3×10^5 $(mg/L)^{-1} \cdot s^{-1}$, which according to the authors is comparable with those observed for other organic contaminants [23]. Moreover, NOM itself must be considered as an important scavenger of HO˙ radicals, e.g. high concentrations of NOM may result in significant reduction of MTBE destruction potential. The reaction of HO˙ radicals in aqueous systems has been discussed in detail in several reports [39, 40]. It can be summarized in three types of well-known pathways:

1. Addition to aromatic rings and double bonds between C-C, C-N and S-O (in sulphoxides), but not to C-O, double bonds; this addition is typically very fast, close to diffusion-controlled [24].

$$HO^{\bullet} + R - H \rightarrow H_2O + R^{\bullet} \tag{3}$$
$$HO^{\bullet} + C = C \rightarrow HO - C - C^{\bullet} \tag{4}$$

2. H-abstraction reactions from C-H, N-H or O-H bonds (related to the R-H bond-dissociation energy), reactions leading to formation of carbon-centred radicals, which in the presence of O_2 (Eqs. 5 and 6) are converted into the corresponding peroxyl radicals ($HO_2^{\bullet}$ (or $O_2^{\bullet-}$) and $R - O^{\bullet}$) [24]:

$$R^{\bullet} + O_2 \rightarrow \rightarrow R(-H^+) + HO_2^{\bullet} \tag{5}$$
$$R^{\bullet} + O_2 \rightarrow \rightarrow R - OO^{\bullet} \rightarrow \rightarrow R - O^{\bullet} \tag{6}$$

3. Electron transfer reactions [41]

The organic intermediates formed in the first stage of the oxidation may further react with HO˙ and oxygen. Ideally, the overall process eventually leads to full mineralization towards CO_2, H_2O and, if the contaminant contains heteroatoms such as N and O, inorganic acids [24]. ROS usually do not achieve complete mineralization in the oxidation of natural organic matter. Instead, a series of intermediates such as aldehydes, keto acids and carboxylic acids, among others, are formed, which affect the potential formation of disinfection by-products [12], since they may act as precursors of haloacetaldehydes, haloketones and haloacetonitriles [42].

The fundamental aspects of AOPs and the most important findings reported in the degradation of NOM along the past few years are discussed in the next sections. The general conditions of operation for several AOPs are summarized in Table 2.

4.1 Ozone-Based Applications: O_3/H_2O_2 and O_3/UV

Ozonation is a common technology applied in treatment of drinking water either for enhancing the subsequent conventional processes or for improving the quality and disinfection of the treated product [32]. In a conventional ozonation process, ozone reacts with NOM by an electrophilic addition to double bonds in a very selective

Table 2 General conditions of operation reported for several AOPs

AOPs	Reaction conditions	Oxidizing agent	Catalytic species	General mechanism
Ozone-based applications: O_3/H_2O_2 and O_3/UV	T = variable, P = atmospheric, irradiation: optional, UV	O_3	Some combinations with photocatalyst have been reported synergistic	$O_3 + H_2O \xrightarrow{h\nu} O_2 + H_2O_2$ $H_2O_2 + h\nu \rightarrow 2HO^\bullet$ $H_2O_2 \rightleftharpoons HO_2^- + H^+$ $H_2O_2 + HO^\bullet \rightarrow HO_2^\bullet + H_2O$ $O_3 + H_2O_2 \rightarrow HO_2^\bullet + HO^\bullet$
UV light-based applications: UV/H_2O_2 and UV/Cl_2	T = variable, P = atmospheric, irradiation: UV, solar	H_2O_2, H_2O, O_2	UV light	$H_2O_2/HO_2^- \xrightarrow{h\nu} 2HO^\bullet$ ($\lambda < 300$ nm) $H_2O_2 + HO^\bullet \rightarrow HO_2^\bullet + H_2O$ $HO_2^- + HO^\bullet \rightarrow HO_2^\bullet + HO^-$ $2HO^\bullet \rightarrow H_2O_2 + O_2$ $HO_2^\bullet + HO^\bullet \rightarrow H_2O + O_2$
Fenton and photo-Fenton	T = RT – 0°C, P = atmospheric, irradiation: optional, UV	H_2O_2	Mainly Fe^{2+}, Cu^{2+} in homogeneous phase	$Fe^{2+} + H_2O_2 \rightarrow HO^\bullet + Fe^{3+} + OH^-$ $Fe^{III}(OH)^{2+} + h\nu \rightarrow HO^\bullet + Fe^{2+}$
Fenton-like catalysed processes: catalytic wet peroxide oxidation (CWPO)	T = RT – 70°C, P = atmospheric, irradiation: not mandatory	H_2O_2	Mainly Fe^{3+} in heterogeneous phase	$Fe^{3+} + H_2O_2 \rightarrow HO_2^\bullet + Fe^{2+} + H^+$
Heterogeneous photocatalysis: (TiO_2/UV)	T = variable, P = atmospheric, irradiation: UV, solar	H_2O_2, H_2O, O_2	Mainly TiO_2, heterogeneous phase	$TiO_2 \xrightarrow{h\nu} e^- + h^+$ $h^+ + HO^- \rightarrow HO^\bullet$ $h^+ + H_2O \rightarrow H^+ + HO^\bullet$ $e^- + O_2 \rightarrow O_2^{\bullet-}$ $O_2^{\bullet-} + H^+ \rightarrow HO_2^\bullet$ $O_2^{\bullet-} + HO_2^\bullet \rightarrow O_2 + H_2O_2 + HO^\bullet$ $2O_2^{\bullet-} + 2H_2O \rightarrow 2HO^\bullet + O_2 + H_2O_2$ $H_2O_2 + O_2^{\bullet-} \rightarrow O_2 + HO^- + HO^\bullet$ $H_2O_2 + e^- \rightarrow HO^- + HO^\bullet$ $e^- + h^+ \rightarrow$ heat/light

(continued)

Table 2 (continued)

AOPs	Reaction conditions	Oxidizing agent	Catalytic species	General mechanism
Ultrasound-based applications (US, US/H_2O_2)	T = variable, P = atmospheric	H_2O, O_2	High-frequency sound waves (>16 kHz). Several combinations suitable with other AOPs	$H_2O \xrightarrow{)))} H^{\bullet} + HO^{\bullet}$ $O_2 \xrightarrow{)))} 2O$ $O + H_2O \rightarrow 2HO^{\bullet}$ $H^{\bullet} + O_2 \rightarrow O + HO^{\bullet}$ $H^{\bullet} + O_2 \rightarrow HO_2^{\bullet}$

Adapted from [37] and supplemented with [23, 43]
[a]*RT* room temperature

way. Simultaneous use of H_2O_2 or UV radiation has been found as a useful strategy to enhance degradation rates and to promote production of hydroxyl radicals, translating the conventional ozonation into an AOP [14, 43]. Lamsal et al. [44] compared the performance displayed by three advanced oxidation processes including O_3/UV, H_2O_2/O_3 and H_2O_2/UV in the NOM removal. They found that combination of O_3 or UV with H_2O_2 resulted in higher TOC and UV_{254} depletion in comparison to every individual process. Upon treatment with ozone alone, NOM oxidation occurred with the removal of conjugated double bonds, due to the high electrophilic character of ozone but rather minimal mineralization. O_3/UV displayed the most efficient removal (TOC, 31%; UV_{254}, 88%) followed by H_2O_2/UV and H_2O_2/O_3. Among the three assessed processes, H_2O_2/UV was found to be the most effective treatment for the reduction of THM and HAA formation potential. However, Stylianou et al. [32] recently found that although O_3/H_2O_2 increased the NOM mineralization degree, 9–17% and 8–15%, for Aliakmonas and Axios Rivers – northern Greece, respectively, it showed negligible impact in reduction of the UV_{254}. The application of single ozonation resulted in high reduction of humic-like peak fluorescence intensities (50–85%), whereas the co-addition of H_2O_2 did not present the expected reduced fluorescence intensity. It was argued that hydroxyl radicals might also get scavenged by dissolved organic carbon (DOC), carbonates and other inorganic compounds, normally found in natural waters; thus, the relatively high scavenging rate calculated based on DOC and alkalinity values for water of Aliakmonas River was $7.2 \times 10^4 \text{ s}^{-1}$. The emission comparative spectrum of raw and treated water is displayed in Fig. 3 (left); the ozonation of Aliakmonas River caused the formation of one new discrete peak with maximum absorbance at 315–335 nm (tryptophan-like), also observed in the emission spectra of O_3/H_2O_2-treated samples although with much lower intensity. It suggested that oxidation of humic-like components was the first step of the O_3–AOP treatment, while the produced protein-like intermediates were subsequently oxidized towards both nonfluorescent and probably also smaller molecular weight moieties.

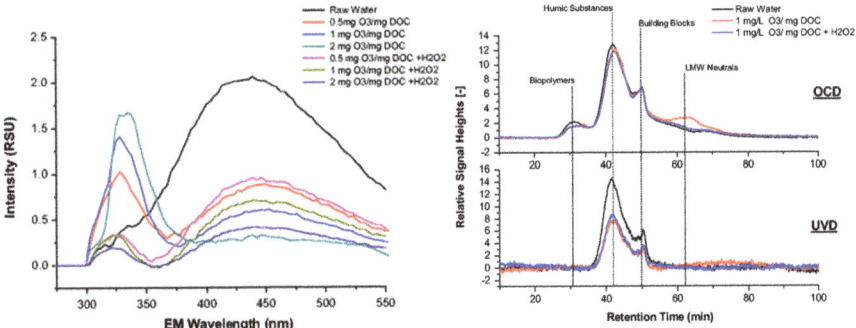

Fig. 3 (Left) Emission spectra of Aliakmonas River at $\lambda = 290$ nm (humic-like substances close to 325–340 nm) and (right) OCD and UVD chromatograms of Aliakmonas River before and after bubble-less treatment with applied dosage of 1.0 mg O_3/mg DOC. Both reproduced with permission from Springer Publishing Company [32]

Size-exclusion chromatography has been usually coupled to ultraviolet detector (LC-UVD) but recently also to organic carbon detector (LC-OCD). The latter is probably the most sensitive and reliable technique for the detailed characterization of NOM. For instance, a comparison between LC-OCD and fluorescence spectroscopy can be seen at Fig. 3; in the fluorescence measurements, the removal of humic-like components was almost completed with O_3 or O_3/H_2O_2 treatment, whereas the LC-OCD removal related with humic substances did not exceed 20% in any case (UVD and OCD).

Finally, the application of O_3 or simultaneous O_3–AOP decreased building blocks and low molecular weight neutral concentrations, indicating that hydroxyl radicals can further react with the intermediates towards CO_2 formation or other smaller (not detectable) by-products. This agreed with Lamsal et al. [44], who found by LC-SEC that during degradation of NOM, the reduction of larger molecular weight NOM (first peaks) was usually higher than that corresponding to lower molecular weight NOM (last peaks), probably as a result of the higher rate constants of reaction between $HO^{\bullet}$ and the fraction of larger MW NOM. It probably obeys to more extended intrinsic conjugated aromaticity, offering larger number of target reaction sites as also confirmed by Tubić et al. [45]. Then, NOM was partially oxidized, and higher molecular weights were transformed into smaller and more biodegradable compounds such as aldehydes and carboxylic acids [44].

Recently Zhong et al. [46], exploring the pathways of degradation of aromatic carboxylic acids in ozone solutions as the main by-products in the degradation of NOM by AOPs, found that the reaction mechanism in the ozonation of benzoic acid (BA) would involve three steps: (1) BA hydroxylation, (2) hydroxylated products of BA getting oxidized to generate ring-opened compounds together with unsaturated carbonyl compounds and (3) short-chain aldehydes (formaldehyde, glyoxal, methyl glyoxal) and carboxylic acids (formic and acetic acids) finally transformed into CO_2 and H_2O, instead due to the stability of carboxylic acids in O_3 solution, would be accumulated in the solution.

4.2 UV Light-Based Applications: UV/H_2O_2 and UV/Cl_2

The presence of UV light is essential in several advanced oxidation process as it can participate both directly and indirectly in the degradation of NOM. AOPs based on UV include UV/H_2O_2, UV/O_3, UV/chlorine, UV/persulphate (UV/PS) and heterogeneous and homogeneous photocatalysis, all of them very interesting methods producing reactive species [42]. UV radiation can degrade NOM, splitting large molecules into organic acids of lower molecular weight; changes of NOM resulting from the application of UV light may also subsequently affect the formation of disinfection by-products (DBPs) when a sequential disinfecting step with chlorine is used [16]. It happens since the generated intermediates through the oxidation of the humic and fulvic acids may react with chlorine, leading to DBPs [47].

Upon irradiation, the amount of HO• produced in the UV/H_2O_2 process strongly depends on the H_2O_2 concentration, and in turn, H_2O_2 dosage depends on intrinsic characteristics and concentration of the organic targets; there is an optimum dosage necessary to achieve the best oxidizing performance. The H_2O_2 can react with HO•, behaving itself as a HO•-inhibiting agent under certain conditions, but it also absorbs UV energy. Indeed, Wang et al. [48] in a previous study found that the HO•-scavenging effect became significant when the H_2O_2 concentration was higher than 0.1% (32.6 mM), with optimal conditions between 0.01 and 0.05% (3.25–16.3 mM) of H_2O_2. The UV-absorbing compounds in the humic acids were degraded almost completely under a 450 W high-pressure mercury vapour lamp used as light source (25°C), UV_{254} decreased from 0.433 to 0.006 cm^{-1}, and 90% of mineralization was achieved. The FT-IR spectra after the UV/H_2O_2 treatment displayed that most of the –OH stretching corresponding to –COOH and –COH (3,400–3,200 cm^{-1}) got removed from original structure. González et al. [49] treated DOM present in two secondary effluents from either a conventional activated sludge (CAS) or a membrane biological reactor (MBR) by means of UV/H_2O_2. The monitoring of the organic matter fraction by LC-OCD demonstrated that the reduction of the aromaticity in the effluent (decreasing SUVA) was not strictly correlated with the complete depletion of humic substances in the effluents (Fig. 4a) unlike Wang et al. [48] who found a clear decrease in the UV absorption together with almost full mineralization. During the first 30 min of oxidation, certain reduction of biopolymers and an important increase of low molecular weight (LMW) compounds (building blocks, neutrals and LMW acids) were achieved [49].

Xie et al. [42] employed the sulphate radical anion ($SO_4^{•-}$, 2.5–3.1 V) which features high redox potential but being more selective than HO• to react very fast with organic pollutants. The impact of UV/PS and UV/H_2O_2 pretreatments on the formation of both C-DBPs (carbonaceous disinfection by-products) and N-DBPs (nitrogenous disinfection by-products) was assessed. UV/H_2O_2 with an initial dosage of 30 μM in H_2O_2 led to significant increased formation of both C-DBPs and N-DBPs in comparison with UV/PS. In this treatment some C-DBPs such as chloroform and haloacetic acids only increased marginally, while N-DBPs such as haloacetonitriles and trichloronitromethane decreased slightly under low dosages of PS (10 μM). Recently several reports have described UV/Cl_2 as an alternative to traditional AOPs taking into account that this process is expected to provide substantial cost savings over conventional AOPs [47]. Aqueous chlorine solutions include two species, hypochlorous acid (HOCl) and hypochlorite ion (ClO^-), related in Eq. (7):

$$HOCl \leftrightarrow H^+ + ClO^- \quad pKa = 7.58 \text{ at } 20°C \quad (7)$$

The photolysis of both species with different photophysical properties (HOCl absorbs UV at 227 nm and ClO^- at 292 nm) leads to production of hydroxyl radicals in several reaction pathways (Eqs. 8–10):

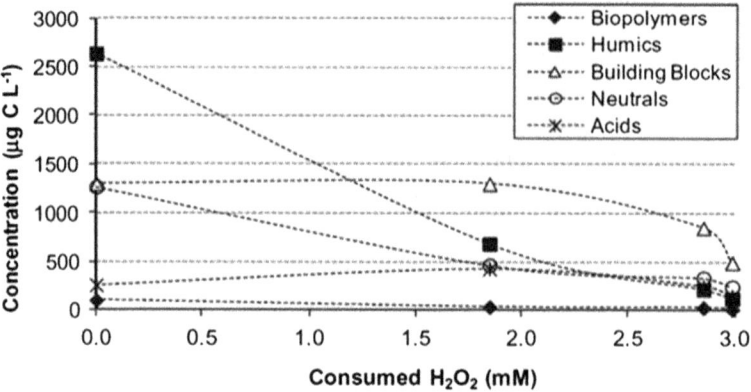
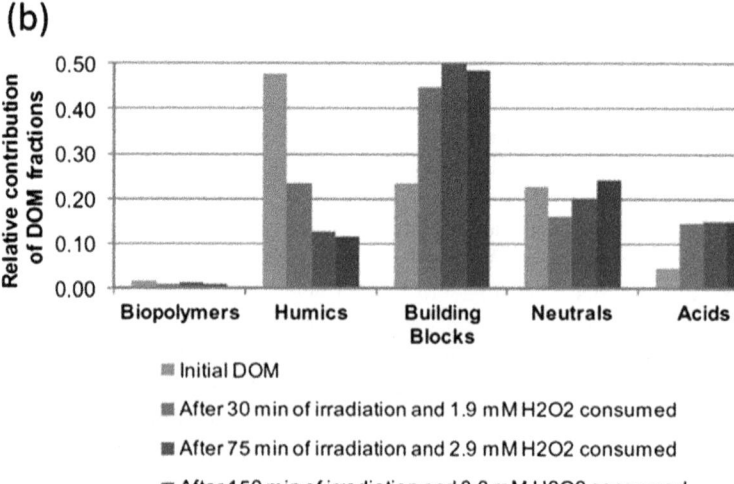

Fig. 4 (a) Evolution of DOM fractions as a function of the consumed H_2O_2 and (b) relative contribution of every fraction to total DOM for the MBR effluent after UV/H_2O_2 treatment. Reproduced with permission of Elsevier from [49]

$$HOCl + h\nu \rightarrow HO^{\bullet} + Cl^{\bullet} \quad (8)$$
$$ClO^- + h\nu \rightarrow O^{\bullet -} + Cl^{\bullet} \quad (9)$$
$$O^{\bullet -} + H_2O \rightarrow HO^{\bullet} + OH^- \quad (10)$$

Then, the UV/chlorine system can generate both non-selective $HO^{\bullet}$ and reactive chlorine species (RCSs) such as $Cl^{\bullet}$, $Cl_2^{\bullet -}$ and $ClO^{\bullet}$ [50].

Pisarenko et al. [47] before investigated the use of UV/chlorine in oxidation of NOM in surface water and the impact of the treatment on formation of disinfection by-products and the structure of NOM. The results showed the destruction of chromophoric components of the NOM at doses ranging 2–10 mg/L Cl_2. Wang et al. [50] indeed confirmed these results finding that the UV/Cl_2 system degraded ~80% of chromophores and 76.4–80.8% of fluorophores, including electron donor groups like double bonds, aromatic and phenolic functionalities getting preferentially degraded by the UV/chlorine process. Finally, the DOC removal was 15.1–18.6%, since the oxidation degraded high MW fractions into low MWs without appreciable decrease in the carbon loading by mineralization. According to Fang et al. [51], $HO^{\bullet}$ reacts with NOM under a second-order rate constant of 2.5×10^4 $(mg/L)^{-1}$ s^{-1}, while $Cl^{\bullet}$ reacts with NOM at 1.3×10^4 $(mg/L)^{-1}$ s^{-1}; therefore, the degradation through the $HO^{\bullet}$ pathway would be about 1.4 times faster than that through the $Cl^{\bullet}$ pathway. However, although there is a synergistic mechanism involved with the addition of chlorine [24], NOM treatment by the UV/chlorine system is similar to that of $HO^{\bullet}$-based AOPs, particularly as a function of the molecular weight; namely, high MW fractions are decomposed at a higher rate (~4.5 times) than medium MW fractions, generating low fractions. Finally, these results supported the conclusion that the chlorine-based AOP was also effective reducing the aromatic content in the NOM.

4.2.1 NOM as a Photosensitizing Agent

During direct photolysis electrons may migrate from basal to excited states of NOM, from where they can be transferred to oxygen either to form $^1O_2$ or to provoke homolytic breaks in NOM producing organic radicals that further react with oxygen. In this case, the dissolved organic matter (DOM) can be considered a photosensitizing agent [52] that by different pathways may generate ROS and provoke its self-degradation [53]. Thus, the chromophoric natural organic matter is one of the main sources promoting formation of ROS species through its interaction with light in a cascade of photochemical reactions, in the earlier stages of AOP reactions (see Fig. 5).

Birben et al. [55] recently reported that even photosensitization via light absorption leading to the formation of reactive oxygen species could also initiate self-degradation of HAs in a photocatalytic process. The hydroxyl radical is the most reactive and less selective of the ROS, and the formation of singlet oxygen can occur through the transfer of energy from excited triplet states of coloured dissolved organic matter (CDOM) to O_2:

$$^3CDOM^* + {}^3O_2 \rightarrow CDOM + {}^1O_2 \tag{11}$$

In the case of superoxide, although CDOM is the main source, the precise reactions forming this species remain unclear [54].

The triplet excited states of natural organic matter ($^3NOM^*$) were found by Li et al. [56] to play a dominant role in the photodegradation (1,700 W Xenon lamp filtered light for $\lambda > 290$ nm) of acetaminophen by photolysis (see Fig. 6). Similarly, Porras et al. [53] established, based in kinetic and analytical studies, an accelerating effect on the rate of ciprofloxacin decomposition caused by humic substances.

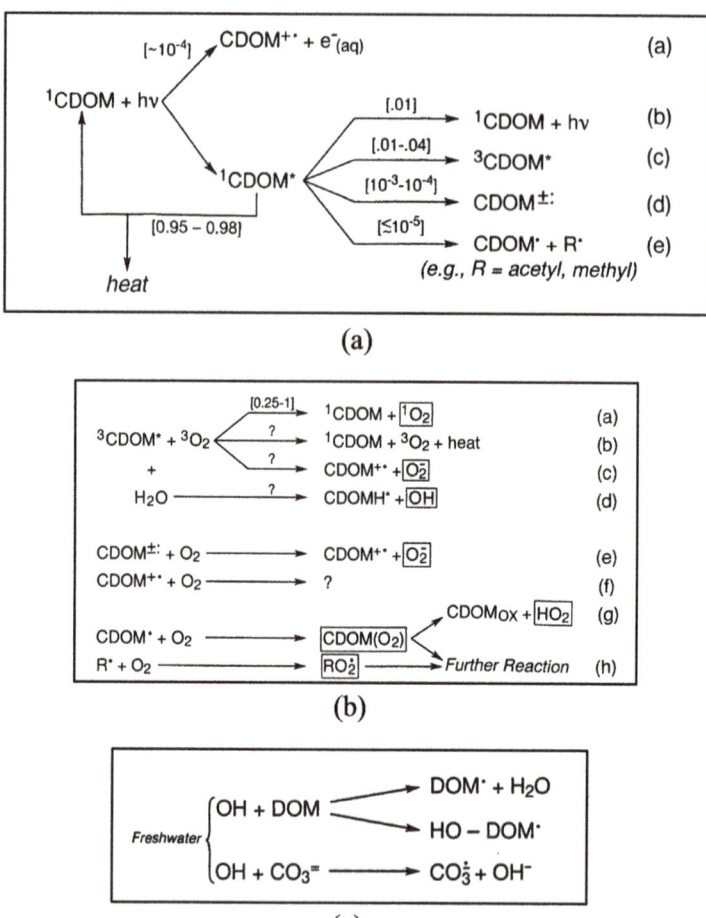

Fig. 5 Photophysical and photochemical reactions of CDOM species: (**a**) primary, (**b**) secondary and (**c**) secondary in fresh water. Reproduced with permission of Springer from [54]

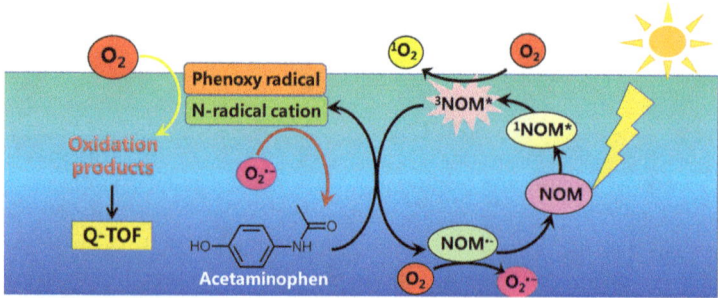

Fig. 6 Schematic representation of initial step in the indirect photodegradation of acetaminophen in NOM-enriched solutions. Reproduced with permission of Elsevier from [56]

In general, NOM reaches triplet excited states after irradiation, and dissolved oxygen acts as a quencher for these triplet excited states through an energy transfer process to generate singlet oxygen ($^1O_2$) and superoxide anion $O_2^{\cdot-}$. It was found that under increasing concentration of oxygen, the steady-state concentration of ^{3}NOM* would be expected to drop and the decay rate of the targeted process decreases too. However, the results showed that the degradation remained stable even in excess of oxygen in the case of acetaminophen, suggesting that the contribution of $^1O_2$ gradually stands out and offsets the decreased ^{3}NOM*. Porras et al. [53] employed furfuryl alcohol (FFA) scavenger, highly selective to $^1O_2$, and demonstrated that singlet oxygen also participates in the reaction.

Due to the strong complexability of iron with the humic substances, the leaching and stability of the iron species were investigated by Birben et al. [55] through photocatalytic experiments (Fe-doped Ti catalyst) in the presence or absence of humic acids in deionized water. This procedure was employed as a strategy to demonstrate the efficiency of the metal-ion doping. Thus, the catalyst was prepared to improve trapping of the photoexcited electrons of the conduction band towards the catalyst's surface while minimizing charge carrier recombination. Fe^{3+} was the chosen ion due to its similar ionic radius (0.69 Å) to that of Ti^{4+} (0.75 Å) as well as energy level pairs Fe^{2+}/Fe^{3+}–Ti^{3+}/Ti^{4+} favouring the separation of the photogenerated electron-hole pairs. It was found that in the presence of the Fe-doped TiO_2 (Evonik P-25), the concentration of the Fe species dissolved in the medium was 0.013 mg/L, whereas in the presence of catalyst and humic acids (HAs 50 mg/L, average molecular weight < 100 kDa), the concentration of iron species increased almost fivefold in the medium reaching a value of 0.070 mg/L. After 60 min of irradiation ($\lambda = 300$–800 nm; light intensity 250 W/m^2), a significant concentration of Fe leached (0.116 mg/L) was found in the presence of HAs. It was attributed to strong chelating effect of the humic sub-fractions, resulting in the release of iron into the aqueous medium. The reactions that can be triggered by the formation of the Fe(III)–HA complex under solar irradiation are shown in Eqs. (12)–(20) [55]:

$$HA + h\nu \rightarrow HA^* \tag{12}$$
$$HA + O_2 \rightarrow Products + O_2^{\cdot-}/HO_2^{\cdot} \tag{13}$$
$$O_2^{\cdot-}/HO_2^{\cdot} \rightarrow H_2O_2 \tag{14}$$
$$Fe(III) + HA \rightarrow Fe(III) - HA \tag{15}$$
$$Fe(III) + h\nu \rightarrow HA^{\cdot+} + Fe(II) \tag{16}$$
$$Fe(III) + h^+ \rightarrow Fe(IV) - HA \rightarrow HA^{*+} + Fe(III) \tag{17}$$
$$Fe(II) + O_2 \rightarrow Fe(III) + O_2^{\cdot-}/HO_2^{\cdot} \tag{18}$$
$$Fe(II) + H_2O_2 \rightarrow Fe(III) + {}^{\cdot}OH + OH^- \tag{19}$$
$$HA + OH \rightarrow Photocatalytic\ degradation\ products \tag{20}$$

Finally, low achieved efficiencies of the photocatalytic removal in terms of UV-Vis absorbance (254, 280, 365 and 456 nm) and DOC mineralization rate (0.044 mg/L min in the presence of HAs-doped catalyst; 0.188 mg/L min in the presence of the free catalyst) were attributed to reactions of complexation between Fe^{3+} and HA molecular fractions on the catalyst's surface.

4.3 Fenton, Photo-Fenton and Fenton-Like Catalysed Processes

The Fenton reaction is a non-expensive and environmental-friendly oxidation method, widely studied for wastewater treatment. The mechanism of the process has been extensively studied, and it is here summarized in Table 2. The hydroxyl radicals as usual attack the organic matter present in water, but some parallel reactions occur, and then the hydroxyl radicals also produce other radicals with less oxidizing power, the so-called scavenging effect of HO$^{\bullet}$ [18]. Many interferences in the water source may occur, e.g. other dissolved organic compounds present, alkalinity, etc., which may compete with NOM by radicals. For instance, both carbonate and bicarbonate anions scavenge hydroxyl radicals to form carbonate radicals; nitrates and nitrites absorb UV light around 230 and 300 nm, respectively, and then concentrations exceeding 1.0 mg/L may strongly limit the effectiveness of UV-based technologies; phosphates, sulphates, chloride, bromide and fluoride ions also may act as scavengers at concentration over 100 mg/L for phosphates and sulphates [23, 24].

Fenton and photo-Fenton processes have been established as alternative to coagulation in treatment of drinking water, usually when the water sources contain natural organic matter at levels of up to 15 mg/L [57]. A drawback in the application of conventional homogeneous Fenton in drinking water facilities is the influence of pH, given that optimum pH for this application is 2.54 when Fe^{3+} and $FeOH^{2+}$ species are in the same abundance. As pH increases, the precipitation of amorphous ferric oxyhydroxides occurs (Eq. 21), which do not redissolve readily and are considerably less Fenton-active in comparison to the free metal ions (Eqs. 18 and 19). The presence of coordinating ligands may affect the pH dependence considerably; lowering pH not only keeps Fe(III) soluble but also reduces parasite decomposition of H_2O_2 [24]:

$$Fe^{3+} \rightleftharpoons FeOH^{2+} \rightleftharpoons Fe(OH)_2^+ \rightleftharpoons Fe_2(OH)_2^{4+} \rightleftharpoons \text{other polynuclear species} \rightleftharpoons Fe_2O_3 \cdot nH_2O_{(s)} \qquad (21)$$

Molnar et al. [58] treated a groundwater rich in natural organic matter (10.6 ± 0.37 mg C/L) employing the Fenton process; they evaluated the influence of pH at 5.5 and 6.0, iron concentrations between 0.10 and 0.50 mM Fe(II) and molar ratios Fe(II):H_2O_2 of 1:5–1:20. High NOM removal was found at pH 5.5 (55% DOC removal) with a dose of 0.25 mM Fe(II) and under molar ratio 1:5. It was also confirmed that the Fenton process was much more effective in removing NOM than conventional coagulation with similar dose of $FeCl_3$. The distribution of NOM fractions upon treatment (HPI-N = 75%, HPI-A = 4%, FA = 21%) changed in comparison to the raw water (FA = 68%, HA = 14%). The fraction of humic acids was completely removed, whereas the fraction of fulvic acids decreased and the contribution of the hydrophilic fraction in the final effluent was 79% (HPI-N + HPI-A).

When the Fenton degradation is carried out in the dark, low molecular weight organic acids such as glyoxylic, maleic, oxalic, acetic and formic are accumulated because of their high stability in the reaction medium. Under light, however, these acids can be mineralized via Fe photocatalysed reactions [24]. The photo-Fenton process has been employed to enhance efficiency in the generation of hydroxyl radicals and also in disinfection units employing UV light sources [59]. A 55% of NOM mineralization was achieved by Moncayo-Lasso et al. [60] by the photo-Fenton process on a river surface water (5.3 mg C/L) at natural pH (~6.5) employing 0.010 mM (0.6 mg/L) of Fe^{3+} (almost 24-fold less than that employed by Molnar et al. [58]) and 10 mg/L of H_2O_2. In this case, the experiments were carried out by using a solar compound parabolic collector. In 2012, Moncayo-Lasso et al. [33] dramatically reduced THM formation potential, thanks to photo-degradation of the MON fraction more related to formation of THMs; the experiments were carried by chlorination of river surface water with 7.1 mg C/L at pH near 7.0, initial $[H_2O_2]$ = 60 mg/L and initial $[Fe^{3+}]$ = 1.0 mg/L. The mineralization reached 55% at 3 h of treatment (25–30°C). Low molecular weight products were obtained at the end of the process; according to the authors, higher mineralization rates were not achieved since the addition of HO$^{\bullet}$ on the aromatic rings generates radicals resembling hydroxy-cyclohexadienyl (HCHD$^{\bullet}$), whose subsequent oxidation leads to breaking of the aromatic rings towards less oxidizable, open-chain products.

Galeano et al. [13] studied the removal of humic acids using an Al/Fe-pillared clay catalyst, where it was established that the fraction of Fe inserted in true "mixed pillars" within the clay layers was responsible for initiating the Fenton-like, catalytic cycle. The results showed that once the first 15 min of the process passed (equilibrium period, without addition of hydrogen peroxide), it established a kind of induction period, explained by the competition between H_2O_2 and NOM molecules for active iron. It happened in the early stages of peroxide addition, taking into account that natural organic matter has an important impact complexing metal ions, and then an interaction mainly between the aromatic moieties of the HAs and the metal inserted in the mineral took place in advance. Afterwards, once the peroxide molecules achieved minimal interaction with the metal, the concentration of NOM started to get decreased by the attack of the formed oxidizing radicals. The results showed almost full depletion of the starting chemical oxygen demand (96.3% COD removal in 4 h of reaction) and complete colour removal (in less than 1 h of reaction) under the following conditions: $[colour_{455}]_0$ = 42 PCU, catalyst loading = 5.0 g/L, $[H_2O_2]_{added}$ = 0.047 mol/L, H_2O_2 addition flow rate = 6.0 cm^3/h, final stoichiometric ratio $[H_2O_2]/[COD]_0$ = 1.0, pH of reaction = 3.7, room temperature (291 ± 2.0 K) and pressure (72 kPa). The highly performing $[Fe]_{active}/[H_2O_2]$ ratio employed was 0.119; this ratio is very useful in order to realize the best set of operating parameters in Fenton and Fenton-like processes, since it may guarantee the most efficient use of the hydrogen peroxide by the catalyst, improving cost operation of the technology (see Fig. 7a).

As a result of the above study, two general statements can be raised: Natural organic matter could be not only complexed by iron but also adsorbed on the

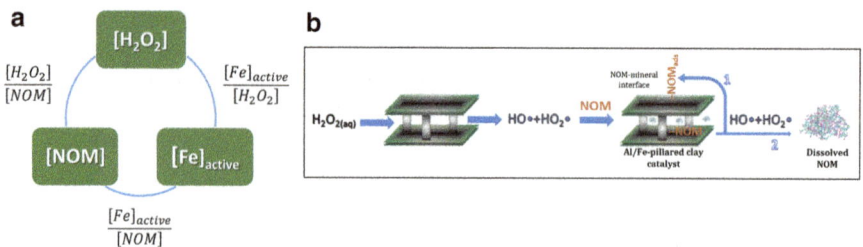

Fig. 7 Schematic representation of (**a**) relationship between three main factors governing CWPO degradation of NOM and (**b**) possible initial pathways of radical attack on NOM substrates in the heterogeneous Fenton-like CWPO system as activated by Al/Fe-PILCs: (1) attack of radicals on NOM adsorbed on the catalyst's surface and (2) radical attack on NOM dissolved in the reaction medium

catalyst's surface [61] so that the radicals can attack it, while it is inside the pores or on the surface of the solid (see Fig. 7b). A second pathway of attack can be established by an adsorption-desorption equilibrium where the dissolved NOM is in the fluid phase and gets attacked by the oxidizing radicals diffused from the catalyst surface.

4.4 Heterogeneous Photocatalysis: TiO_2/UV

The TiO_2 photocatalytic treatment has been considered to be effective in the destruction of NOM. TiO_2-catalysed system is attractive due to its potential to degrade organic macromolecules [62]. The photocatalysis offers a potentially cost-effective avenue for contaminant removal through extensive material reuse, use of solar illumination energy and reutilization of existing UV disinfection facilities to achieve more efficient treatment in drinking water treatment plants [63]. The TiO_2 photocatalytic mechanism initiates with the absorption of UV light with energy greater than +3.2 eV (TiO_2 band gap, energy corresponding to wavelengths below 370 nm). It results in generation of conduction band electrons (e^-) and valence band hole (h^+) pairs involved in the production of $HO^{\bullet}$, $O_2^{\bullet-}$ and $HO_2^{\bullet}$. H_2O_2 can also be formed in situ, what improves the production of $HO^{\bullet}$ and slows down the recombination of the charges [43].

Brame et al. [63] proposed a model to explain various inhibition ways in catalytic AOPs, including role played by NOM as scavenger itself during the removal of a target pollutant. The model is schematized at Fig. 8 and assumes that the organic solute is adsorbed on the photocatalyst surface and the degradation may occur both on catalyst surface and in the bulk solution.

Sen Kavurmaci and Bekbolet [64] investigated the influence of a montmorillonite (Mt) on the TiO_2 photocatalytic removal of HAs as the model compounds of

Fig. 8 Mass transport pathways during photocatalytic degradation. Reproduced with permission of Elsevier from [63]

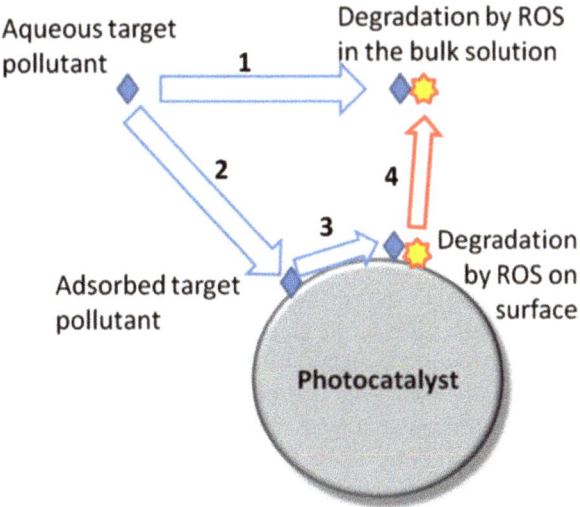

natural organic matter. The study about the interaction of Mt with organics dissolved in water is very important in order to better figure out the NOM degradation in the presence of the Al-/Fe-clay catalysts analysed above. Experiments were done in the absence or presence of TiO_2 under dark or light conditions and presence or absence of Mt. The adsorptive removal of colour$_{436}$ was higher than the removal of UV_{254} or UV_{365}, which was interpreted in terms of the coloured moieties in the HAs being the main responsible of the interactions with the TiO_2 surface. In fact, the oxidative removal displayed the following order: colour$_{436}$ > UV_{365} > UV_{254} > DOC. Similar results had been before reported for the interaction of HAs with the Al/Fe-PILC catalysed CWPO treatment [13], but the very fast colour removal was there explained instead by higher susceptibility of the chromophores present in HAs against the oxidizing species.

The presence of Mt slightly altered the photocatalytic reactivity of HA, predominantly the colour moieties, probably because of the increased turbidity in the colloidal medium. In addition, the presence of Mt and absence of TiO_2 did not change the removal efficiency of DOC under irradiation. The mechanism in absence of Mt is summarized in Eq. (22):

$$\begin{aligned} &TiO_2 + h\nu(\lambda < 388 \text{ nm}) \rightarrow TiO_2\left(h^+_{VB} + e^-_{CB}\right) \rightarrow \cdots \rightarrow ROS \\ &\left(HO^\bullet/HO^\bullet_2/O^-_2\right) \rightarrow \cdots \rightarrow HO^\bullet + HA_{ads} \rightarrow A^\bullet + H_2O \rightarrow \\ &\text{via radical reactions} \rightarrow \cdots \rightarrow HA_{ox} \rightarrow \cdots \\ &\text{Lower molecular weight degradation products} \rightarrow\rightarrow CO_2 + H_2O \end{aligned} \quad (22)$$

Short-chain aldehydes and ketones have been identified as key degradation products in this type of photocatalytic degradation of NOM by Liu et al. [62]. The addition of hydrogen peroxide did not change the reaction pathway producing similar degradation products. The GC-MS analyses of aldehydes and ketones in raw and treated waters were carried out by derivatization of the carbonyl compounds with O-2,3,4,5,6-(pentafluorobenzyl)hydroxylamine hydrochloride (PFBHA), and the oxime derivatives were subsequently recovered by extraction with hexane. The photocatalytic NOM oxidation (365 nm; 0.1 g catalyst/L; pH 7.0–7.6) reached around 80% of mineralization and 100% of UV_{254} removal after 4 h of reaction. The higher elimination of aromaticity in comparison with the DOC implied that loss of aromaticity and conjugation is easier to achieve than mineralization of the NOM. According to LC-SEC results, TiO_2-photocatalytic treatment preferentially degraded the high molecular weight fractions together with a considerable decrease in the fraction of hydrophobic acids, in pretty similar fashion above-mentioned for other AOPs. This fraction has shown to be more prone against degradation, unlike the hydrophilic charged and neutral fractions, which tend to increase throughout anyone of the processes, preventing the complete mineralization to be achieved. These fractions consist mostly of aldehydes, ketones, alcohols and small carbohydrates. The analyses of a raw water, Myponga Reservoir, Adelaide, Australia, revealed five carbonyl compounds being present: formaldehyde, acetaldehyde, acetone, n-propanal and n-butanal. In general, after 15 min of irradiation, formaldehyde and acetone notably increased, while acetaldehyde, propanal and butanal got easily degraded.

4.5 Ultrasound-Based Applications

Some applications have employed ultrasound waves to produce an oxidative environment due to cavitation bubbles generated during the rarefaction phase of sound waves. The cavitation bubble violently collapses during the compression cycle, and localized hot spots are formed, which may reach temperatures and pressures in excess of 5,000 K and 1,000 atm, respectively. The high temperatures result in the splitting and decomposition of chemical compounds present inside the bubbles, including water, which leads to formation of $HO^{\bullet}$ radicals and hydrogen peroxide [65].

The monitored parameters in an experimental ultrasound study are mainly sonication time, irradiation power, NOM concentration, temperature, pH, conductivity, redox potential and turbidity [35]. An acid pretreatment can be used to eliminate carbonates and bicarbonates in the sample and to enhance the TOC removal efficiency [66]. NOM removal is strongly influenced in this type of technologies by the power, initial NOM concentration and sonication time [67].

Sonochemical destruction of contaminants is particularly effective on volatile substrates; alternative decomposition mechanisms have been postulated to account for the destruction of semi-volatile and non-volatile solutes [66, 68, 69]. The easy application of the technology as well as no production of toxic by-products like THMs formed through chlorination make the system very attractive for NOM removal [35]. However, NOM may significantly influence the effectiveness of ultrasound, since it often interferes with the treatment process by binding organic and inorganic contaminants and scavenging reactive species [65], as explained before. Olson et al. [66] evaluated, more than two decades ago, the potential of an advanced oxidation process involving ozone and ultrasound for catalytic degradation of humic acids; they suggested that any volatile organic compound could potentially be oxidized directly by pyrolysis inside the cavitation bubbles. However, NOM does not have a volatile nature. In spite of that, volatile intermediates, which undergone pyrolysis, were produced through reaction between hydroxyl radicals with dissolved fulvic acids. The same authors showed that ultrasound combined with ozone showed better results in comparison with ozone alone (40% removal TOC) or low-frequency ultrasound (55 W, 20 kHz; no degradation). In fact, after 60 min the combination of both methods reached 90% of TOC removal under ultrasonic power of 27 W and 3.2 mg/min of ozone [66]. Interestingly, the pH increased during the process due to volatilization of small molecular weight carboxylic acids and carbon dioxide, a feature clearly different in comparison of most of the rest of AOPs. The same trend in pH evolution was later observed by Chen et al. [65]. Such changes could be not so evident in all cases due to relatively high pH-buffering capacity displayed by humic acids [35]. Interestingly, they also reported that the use of high-frequency ultrasound (354 kHz, energy density of 450 W/L) acting alone was able to remove NOM. The TOC removals reported over two samples treated by ultrasound, (1) a commercial Aldrich humic acid and (2) DOM extracted from Pahokee peat (purchased from the International Humic Substances Society Sonochemical reactions of dissolved organic matter, IHSS), were 33.3% and 19.1%, respectively. At 354 kHz under intensity of 120 W/L, no significant depletion was observed of none of the targeted substrates.

As a result of ultrasonication, the NOM structure changes on the contents of chromophores, −COOH, −OH-substituted benzene rings, intramolecular electron donor-acceptor complexes and complex unsaturated chromophores; it also changes in dissociation or protonation of carboxyl and phenolic groups in humic acids, as observed by Naddeo et al. [35].

Al-Juboori et al. [70] evidenced that pulse treatment at high power and long treatment time achieved the highest reduction in very hydrophobic acid fraction (VHA). For charged hydrophilic acids and neutral hydrophilic acids, the highest increment was attained under continuous treatment for a long time together with low and high powers, respectively. Ultrasound in combination with hydrogen peroxide (US/H_2O_2) has displayed better results in the efficiency removal of HA (91.5%) than US alone (69.3%) or H_2O_2 alone (20%) [67].

5 Conclusions

The selection of one particular AOP for degradation of NOM strongly depends on the physicochemical properties of the target water. For instance, the conventional Fenton reaction catalysed in homogeneous regime requires operating pH values below 4.0, whereas waters containing strongly UV-absorbing substrates may be difficult to be treated by UV/H_2O_2 and other UV-based technologies. Several NOM fractions used to have high aromatic content, whose UV adsorption may increase energy consumption. Moreover, the application of AOPs for NOM removal remains in general quite interesting since the reactions of $HO^{\bullet}$ radicals on this type of substrates have shown similar rate constants in comparison to other more studied organic pollutants. Like in other contaminated systems, the presence of carbonates, bicarbonates, sulphates, chlorides, bromides and fluorides may act as scavengers of the oxidizing radicals decreasing overall performance of the process. In addition, NOM may also get involved in metal complexing that could affect the Fenton processes. However, especially the heterogeneous Fenton variants remain being the most interesting ones to be further investigated, since they could operate efficiently even under circum-neutral pH values, typical of most supply sources feeding drinking water plants. Furthermore, it has been recently evidenced that coloured NOM triplet excited states can also promote side pathways increasing formation of ROS in photoactivated processes that could be useful in photocatalytic degradation of NOM. Finally, attention must be paid to the selective efficiency displayed by the AOPs as a function of the polar character. There is probably enough evidence demonstrating that the hydrophilic fraction is the most refractory in NOM, whereas the hydrophobic one is rapidly degraded and even mineralized to CO_2. Thus, it is recommendable to find a process that allows for the more efficient elimination of the hydrophilic fraction at the lowest possible cost of reagents and energy.

Since the natural organic matter is a complex mixture mainly of organic contents, its molecular tracing through AOP treatments is still challenging from the analytical point of view. It explains in part the more rather delayed assessment of intermediates and by-products occurring in the NOM degradation by AOPs that is evidenced in literature in comparison with other widely studied contaminants as, for instance, phenols and azo-dyes. However, a set of analytical tools is now available, making possible to efficiently differentiate both the polar character (resin fractionation; RP-LC) and the molecular size distribution (LC-SEC); besides, some spectroscopic measurements are very useful in order to infer several features of dissolved NOM like the aromatic character, the predominant polar nature and even the degree of condensation of the carbonaceous networks. Within these analytical issues, the role still to be played by special variants of fluorescence spectroscopy (FEEM) and mass spectrometry (ESI-FT-ICR-MS) is very interesting. It is noteworthy that TOC equipments remain being probably the most useful technique in the field of the overall characterization of NOM, its intermediates and by-products.

Acknowledgements Financial support from CWPO Project for Enhanced Drinking Water in Nariño (BPIN 2014000100020), CT&I Fund of SGR, Colombia, is kindly acknowledged. MAV and AG thank the support from the Spanish Ministry of Economy and Competitiveness (MINECO) and the European Regional Development Fund (FEDER) (projects MAT2013-47811-C2-R and MAT2016-78863-C2-R). AMG also gratefully thanks PhD scholarship granted by Nariño Department (BPIN 2013000100092) and Managed by CEIBA Foundation, Colombia.

References

1. Camargo Valero M, Cruz Torres LE (1999) Sustancias húmicas en aguas para abastecimiento. Ing Invest 44:63–72
2. Chiou CT, Malcolm RL, Brinton TI, Kile DE (1986) Water solubility enhancement of some organic pollutants and pesticides by dissolved humic and fulvic acids. Environ Sci Technol 20(5):502–508
3. Leenheer JA, Croué J–P (2003) Aquatic organic matter: understanding the unknown structures is key to better treatment of drinking water. Environ Sci Technol 37(1):18A–26A
4. CBCL Limited, Newfoundland and Labrador, Department of Environment and Conservation, Water Management Division (2011) Study on characteristics and removal of natural organic matter in drinking water systems in newfoundland and labrador – final report. Department of Environment and Conservation, Water Management Division, St. John's
5. Frimmel FH, Abbt-Braun G, Heumann KG, Hock B, Lüdemann H-D, Spiteller M (2002) Refractory organic substances in the environment. Wiley–VCH Verlag GmbH, Weinheim
6. Matilainen A, Gjessing ET, Lahtinen T, Hed L, Bhatnagar A, Sillanpaa M (2011) An overview of the methods used in the characterisation of natural organic matter (NOM) in relation to drinking water treatment. Chemosphere 83(11):1431–1442
7. Lavonen E (2015) Tracking changes in dissolved natural organic matter composition. PhD thesis, Swedish University of Agricultural Sciences, Uppsala
8. Navalón S (2010) Parámetros de calidad del agua relacionados con la presencia de materia orgánica y microorganismos. PhD thesis, Universidad Politécnica de Valencia, Valencia
9. Matilainen A, Vepsalainen M, Sillanpaa M (2010) Natural organic matter removal by coagulation during drinking water treatment: a review. Adv Colloid Interf Sci 159(2):189–197
10. Cui X, Choo K-H (2014) Natural organic matter removal and fouling control in low–pressure membrane filtration for water treatment. Environ Eng Res 19(1):1–8
11. Leenheer JA (1981) Comprehensive approach to preparative isolation and fractionation of dissolved organic carbon from natural waters and wastewaters. Environ Sci Technol 15(5):578–587
12. Cehovin M, Medic A, Scheideler J, Mielcke J, Ried A, Kompare B, Zgajnar Gotvajn A (2017) Hydrodynamic cavitation in combination with the ozone, hydrogen peroxide and the UV–based advanced oxidation processes for the removal of natural organic matter from drinking water. Ultrason Sonochem 37:394–404
13. Galeano LA, Bravo PF, Luna CD, Vicente MÁ, Gil A (2012) Removal of natural organic matter for drinking water production by Al/Fe–PILC–catalyzed wet peroxide oxidation: effect of the catalyst preparation from concentrated precursors. Appl Catal B 111–112:527–535
14. Matilainen A, Sillanpaa M (2010) Removal of natural organic matter from drinking water by advanced oxidation processes. Chemosphere 80(4):351–365
15. Xing L, Murshed MF, Lo T, Fabris R, Chow CWK, van Leeuwen J, Drikas M, Wang D (2012) Characterization of organic matter in alum treated drinking water using high performance liquid chromatography and resin fractionation. Chem Eng J 192:186–191
16. Wang H, Zhu Y, Hu C (2017) Impacts of bacteria and corrosion on removal of natural organic matter and disinfection byproducts in different drinking water distribution systems. Int Biodeterior Biodegrad 117:52–59
17. Trueman BF, MacIsaac SA, Stoddart AK, Gagnon GA (2016) Prediction of disinfection by–product formation in drinking water via fluorescence spectroscopy. Environ Sci Water Res Technol 2(2):383–389
18. Giannino M (2014) Drinking water and water management: new research. Nova Science Publishers, New York
19. Fan X, Tao Y, Wang L, Zhang X, Lei Y, Wang Z, Noguchi H (2014) Performance of an integrated process combining ozonation with ceramic membrane ultra–filtration for advanced treatment of drinking water. Desalination 335(1):47–54

20. Abdul Hamid KI, Sanciolo P, Gray S, Duke M, Muthukumaran S (2017) Impact of ozonation and biological activated carbon filtration on ceramic membrane fouling. Water Res 126:308–318. https://doi.org/10.1016/j.watres.2017.09.012
21. Papageorgiou A, Voutsa D, Papadakis N (2014) Occurrence and fate of ozonation by-products at a full-scale drinking water treatment plant. Sci Total Environ 481:392–400
22. Linden KG, Mohseni M (2014) Advanced oxidation processes: applications in drinking water treatment. In: Ahuja S (ed) Comprehensive water quality and purification. Elsevier, Amsterdam
23. Kommineni S, Zoeckler J, Stocking A, Liang S, Flores A, Kavanaugh M, Rodriguez R, Browne T, Roberts R, Brown A, Stocking A (2008) 3.0 advanced oxidation processes. National Water Research Institute, Fountain Valley
24. Pignatello JJ, Oliveros E, MacKay A (2006) Advanced oxidation processes for organic contaminant destruction based on the Fenton reaction and related chemistry. Crit Rev Env Sci Technol 36(1):1–84
25. Li Y, Fang Z, He C, Zhang Y, Xu C, Chung KH, Shi Q (2015) Molecular characterization and transformation of dissolved organic matter in refinery wastewater from water treatment processes: characterization by fourier transform ion cyclotron resonance mass spectrometry. Energy Fuel 29(11):6956–6963
26. Świetlik JS, Sikorska E (2005) Characterization of natural organic matter fractions by high pressure size–exclusion chromatography, specific UV absorbance and total luminescence spectroscopy. Pol J Environ Stud 15(1):145–153
27. Hu HY, Du Y, Wu QY, Zhao X, Tang X, Chen Z (2016) Differences in dissolved organic matter between reclaimed water source and drinking water source. Sci Total Environ 551–552:133–142
28. Fabris R, Chow C, Tran T, Gray S, Drikas M (2008) Development of combined treatment processes for the removal of recalcitrant organic matter. CMSE Internal Technical Report 2008–316, CSIRO, Canberra
29. Pavlik JW, Perdue EM (2015) Number–average molecular weights of natural organic matter, hydrophobic acids, and transphilic acids from the suwannee river, georgia, as determined using vapor pressure osmometry. Environ Eng Sci 32(1):23–30
30. Pan Y, Li H, Zhang X, Li A (2016) Characterization of natural organic matter in drinking water: sample preparation and analytical approaches. Trends Environ Anal Chem 12:23–30
31. Rodríguez FJ, Núñez LA (2011) Characterization of aquatic humic substances. Water Environ J 25(2):163–170
32. Stylianou SK, Katsoyiannis IA, Ernst M, Zouboulis AI (2017) Impact of O_3 or O_3/H_2O_2 treatment via a membrane contacting system on the composition and characteristics of the natural organic matter of surface waters. Environ Sci Pollut Res Int. https://doi.org/10.1007/s11356-017-9554-8
33. Moncayo-Lasso A, Rincon A-G, Pulgarin C, Benitez N (2012) Significant decrease of THMs generated during chlorination of river water by previous photo–Fenton treatment at near neutral pH. J Photochem Photobiol A 229(1):46–52
34. Uyguner CS, Bekbolet M (2005) Implementation of spectroscopic parameters for practical monitoring of natural organic matter. Desalination 176(1–3):47–55
35. Naddeo V, Belgiorno V, Napoli RMA (2007) Behaviour of natural organic matter during ultrasonic irradiation. Desalination 210(1–3):175–182
36. Kujawinski EB, Del Vecchio R, Blough NV, Klein GC, Marshall AG (2004) Probing molecular–level transformations of dissolved organic matter: insights on photochemical degradation and protozoan modification of DOM from electrospray ionization Fourier transform ion cyclotron resonance mass spectrometry. Mar Chem 92(1–4):23–37
37. Galeano LA, Vicente MA, Gil A (2014) Catalytic degradation of organic pollutants in aqueous streams by mixed Al/M–pillared clays (M = Fe, Cu, Mn). Catal Rev 56(3):239–287

38. Tsydenova O, Batoev V, Batoeva A (2015) Solar–enhanced advanced oxidation processes for water treatment: simultaneous removal of pathogens and chemical pollutants. Int J Environ Res Public Health 12(8):9542–9561
39. Von Sonntag CS, Schuchmann HP (1991) The elucidation of peroxyl radical reactions in aqueous solution with the help of radiation–chemical methods. Angew Chem Int Ed Engl 30(10):1229–1253
40. Von Sonntag C, Schuchmann HP (1997) Peroxyl radicals in aqueous solutions. Peroxyl radicals. Wiley, New York
41. Von Sonntag C (2008) Advanced oxidation processes: mechanistic aspects. Water Sci Technol 58(5):1015–1021
42. Xie P, Ma J, Liu W, Zou J, Yue S (2015) Impact of UV/persulfate pretreatment on the formation of disinfection byproducts during subsequent chlorination of natural organic matter. Chem Eng J 269:203–211
43. Rubio-Clemente A, Torres-Palma RA, Peñuela GA (2014) Removal of polycyclic aromatic hydrocarbons in aqueous environment by chemical treatments: a review. Sci Total Environ 478:201–225
44. Lamsal R, Walsh ME, Gagnon GA (2011) Comparison of advanced oxidation processes for the removal of natural organic matter. Water Res 45(10):3263–3269
45. Tubić A, Agbaba J, Dalmacija B, Perović SU, Klašnja M, Rončević S, Ivančev-Tumbas I (2011) Removal of natural organic matter from groundwater using advanced oxidation processes at a pilot scale drinking water treatment plant in the central banat region (Serbia). Ozone Sci Eng 33(4):267–278
46. Zhong X, Cui C, Yu S (2017) Exploring the pathways of aromatic carboxylic acids in ozone solutions. RSC Adv 7(55):34339–34347
47. Pisarenko AN, Stanford BD, Snyder SA, Rivera SB, Boal AK (2013) Investigation of the use of chlorine based advanced oxidation in surface water: oxidation of natural organic matter and formation of disinfection byproducts. J Adv Oxid Technol 16(1):137–150
48. Wang GS, Liao CH, Chen HW, Yang HC (2006) Characteristics of natural organic matter degradation in water by UV/H_2O_2 treatment. Environ Technol 27(3):277–287
49. González O, Justo A, Bacardit J, Ferrero E, Malfeito JJ, Sans C (2013) Characterization and fate of effluent organic matter treated with UV/H_2O_2 and ozonation. Chem Eng J 226:402–408
50. Wang WL, Zhang X, Wu QY, Du Y, Hu HY (2017) Degradation of natural organic matter by UV/chlorine oxidation: molecular decomposition, formation of oxidation byproducts and cytotoxicity. Water Res 124:251–258
51. Fang JY, Fu Y, Shang C (2014) The roles of reactive species in micropollutant degradation in the UV/free chlorine system. Environ Sci Technol 48(3):1859–1868
52. Giannakis S, Gamarra Vives FA, Grandjean D, Magnet A, De Alencastro LF, Pulgarin C (2015) Effect of advanced oxidation processes on the micropollutants and the effluent organic matter contained in municipal wastewater previously treated by three different secondary methods. Water Res 84:295–306
53. Porras J, Bedoya C, Silva-Agredo J, Santamaria A, Fernandez JJ, Torres-Palma RA (2016) Role of humic substances in the degradation pathways and residual antibacterial activity during the photodecomposition of the antibiotic ciprofloxacin in water. Water Res 94:1–9
54. Blough NV, Zepp RG (1995) Active oxygen in chemistry. In: Foote CS, Valentine JS, Greenberg A, Liebman JF (eds) Structure energetics and reactivity in chemistry series1st edn. Chapman & Hall, London, pp 280–333
55. Birben NC, Uyguner-Demirel CS, Kavurmaci SS, Gürkan YY, Turkten N, Cinar Z, Bekbolet M (2017) Application of Fe–doped TiO_2 specimens for the solar photocatalytic degradation of humic acid. Catal Today 281:78–84
56. Li Y, Pan Y, Lian L, Yan S, Song W, Yang X (2017) Photosensitized degradation of acetaminophen in natural organic matter solutions: the role of triplet states and oxygen. Water Res 109:266–273

57. Murray CA, Parsons SA (2004) Removal of NOM from drinking water: Fenton's and photo–Fenton's processes. Chemosphere 54(7):1017–1023
58. Molnar J, Agbaba J, Watson M, Tubić A, Kragulj M, Maletić S, Dalmacija B (2015) Groundwater treatment using the Fenton process: changes in natural organic matter characteristics and arsenic removal. Int J Environ Res 9(2):467–474
59. Giannakis S, Polo Lopez MI, Spuhler D, Sanchez Perez JA, Fernandez Ibanez P, Pulgarin C (2016) Solar disinfection is an augmentable, in situ–generated photo–Fenton reaction part 2: a review of the applications for drinking water and wastewater disinfection. Appl Catal B 198:431–446
60. Moncayo-Lasso A, Sanabria J, Pulgarin C, Benitez N (2009) Simultaneous E. coli inactivation and NOM degradation in river water via photo–Fenton process at natural pH in solar CPC reactor. A new way for enhancing solar disinfection of natural water. Chemosphere 77 (2):296–300
61. Greathouse J, Johnson K, Greenwell H (2014) Interaction of natural organic matter with layered minerals: recent developments in computational methods at the nanoscale. Minerals 4(2):519–540
62. Liu S, May L, Fabris R, Chow C, Drikas M, Amal R (2008) TiO_2 photocatalysis of natural organic matter in surface water: impact on trihalomethane and haloacetic acid formation potential. Environ Sci Technol 42:6218–6223
63. Brame J, Long M, Li Q, Alvarez P (2015) Inhibitory effect of natural organic matter or other background constituents on photocatalytic advanced oxidation processes: mechanistic model development and validation. Water Res 84:362–371
64. Sen Kavurmaci S, Bekbolet M (2013) Photocatalytic degradation of humic acid in the presence of montmorillonite. Appl Clay Sci 75–76:60–66
65. Chen D, Ziqi H, Weavers LK, Chin Y-P, Walker HW, Hatcher PG (2004) Sonochemical reactions of dissolved organic matter. Res Chem Intermed 30(7–8):735–753
66. Olson TM, Barbier PF (1994) Oxidation kinetics of natural organic matter by sonolysis and ozone. Water Res 28(6):1383–1391
67. Pourzamani H, Majd AMS, Attar HM, Bina B (2015) Natural organic matter degradation using combined process of ultrasonic and hydrogen peroxide treatment. Anu Inst Geocienc 38 (1):63–72
68. Serna-Galvis EA, Silva-Agredo J, Giraldo-Aguirre AL, Flórez-Acosta OA, Torres-Palma RA (2016) High frequency ultrasound as a selective advanced oxidation process to remove penicillinic antibiotics and eliminate its antimicrobial activity from water. Ultrason Sonochem 31:276–283
69. Guzman-Duque FL, Pétrier C, Pulgarin C, Peñuela G, Herrera-Calderón E, Torres-Palma RA (2016) Synergistic coupling between electrochemical and ultrasound treatments for organic pollutant degradation as a function of the electrode material (IrO_2 and BDD) and the ultrasonic frequency (20 and 800 kHz). Int J Electrochem Sci 11(9):7380–7394
70. Al-Juboori RA, Yusaf T, Aravinthan V, Bowtell L (2016) Investigating natural organic carbon removal and structural alteration induced by pulsed ultrasound. Sci Total Environ 541:1019–1030

Photo(Catalytic) Oxidation Processes for the Removal of Natural Organic Matter and Contaminants of Emerging Concern from Water

Monica Brienza, Can Burak Özkal, and Gianluca Li Puma

Abstract Natural organic matter (NOM) is a heterogeneous complex of organic materials and is ubiquitous in natural aquatic systems. The amount of NOM in the environment is continuously increasing because of global warming and/or changes in precipitation patterns and has negative impact on drinking water as it produces an undesirable colour and as a vector for the introduction of contaminants. For these reasons, several technologies have been proposed to address the impact of NOM in aqueous systems. Among these, advanced oxidation processes (AOPs) refer to oxidation processes that result in the formation of highly reactive radical species. This chapter presents an overview of recent research studies dealing with photon-activated AOPs for the removal of NOM and emerging contaminants in water.

Keywords Advanced oxidation processes, Degradation, Energy efficiency, NOM, Water pollutants

Contents

1 Introduction 134
2 Advanced Oxidation Processes (AOPs) 136
3 Photocatalysis 137
   3.1 Heterogeneous Photocatalysis 137
   3.2 Homogenous Photocatalysis 138
4 Photocatalysis Treatment for Contaminant Removal 141
   4.1 Photocatalytic NOM Removal 141
   4.2 Pesticides 142

M. Brienza (✉), C.B. Özkal, and G. Li Puma
Faculty of Engineering, Namik Kemal University, Tekirdağ, Turkey
e-mail: monica.brienza@ird.fr

4.3 Pharmaceuticals and Personal Care Products (PPCPs) 143
5 Efficiency of AOPs 143
  5.1 AOP Efficiency Based on Toxicity 143
  5.2 AOPs Efficiency Based on Energy Consumption 147
6 Conclusions 149
References 150

Abbreviations

AOPs	Advanced oxidation processes
COD	Chemical oxygen demand
DBPs	Disinfection by-products
DOC	Dissolved organic carbon
DOM	Dissolved organic matter
ECs	Emerging contaminants
EDCs	Endocrine-disrupting compounds
E_{EO}	Electrical energy per order
FA	Fulvic acid
Fe^{2+}/H_2O_2	Fenton
GAC	Granular activated carbon
H_2O_2	Hydrogen peroxide
HA	Humic acid
IC	Inorganic carbon
NOM	Normal organic matter
O_3/H_2O_2	Peroxone
O_3/UV	Ozonation
PPCPs	Pharmaceuticals and personal care products
ROS	Radical oxygen species
TCE	Trichloroethylene
THMFP	Trihalomethanes formation potential
TOC	Total organic carbon
US	Ultrasound
UV254	Ultraviolet absorbance at 254 nm
UWW	Urban wastewater
VCOCs	Volatile chlorinated organic carbons
VUV	Vacuum UV

1 Introduction

The treatment, reuse and disposal of polluted water originating from many different sources, including agriculture, industrial processes and rapid urban development, require the strengthening of advanced treatment systems that prevents the introduction of contaminants to surface and groundwaters. The development of such systems is best addressed, from a technical and economical standpoint, through the development of treatment systems that are capable of handling relatively small volumes of effluents containing toxic, persistent organics (xenobiotics, etc.), rather than systems that focus on the decontamination of large volumes of water [1].

A large body of research and development in recent years has demonstrated that advanced oxidation processes (AOPs) can effectively degrade a large variety of micropollutants and microcontaminants via the production of hydroxyl and other radical oxygen species (ROS). These species unselectively and through multistep oxidative pathways may lead to the removal of water contaminants and to the reduction of its toxicity [2].

AOPs include a range of advanced oxidative processes such as ozone (O_3), Fenton (Fe^{2+}/H_2O_2), electrolysis (electrodes with current), sonolysis (ultrasounds), photolysis, photocatalysis (light with catalyst), photo-Fenton (Fenton reaction with light).

The selection of the most effective AOP is highly dependent on the characteristics of the wastewater such as the environmental matrix, the organic load, pH, dissolved oxygen, dissolved organic matter, temperature and the target pollutants. Wastewaters can be defined as complex environmental matrices often containing dissolved organic matter (DOM), carbonate/bicarbonate, anions and many other species, which collectively act as hydroxyl radical scavengers limiting the efficiency of AOPs [3].

Natural organic matter (NOM) results from the decomposition of organic matter, primarily from vegetation and plants. Its chemical composition contains a large range of molecular size and is ubiquitous to all surface, ground and soil waters [4]. It is an important component that interacts with metal ions and minerals in water and soil forming complexes of widely differing chemical and biological nature. One common approach for the characterization of NOM includes the fractionation of the mixture into its hydrophobic and hydrophilic fractions. The hydrophobic fraction includes humic substances such as fulvic acids (FA) and humic acid (HA) that account for 50% of the total organic carbon (TOC) in water [5]. The hydrophilic fraction contains high amount of aliphatic carbons and nitrogenous compounds, such as carbohydrates, sugars and amino acids [4]. The amount, character and properties of NOM vary considerably according to the origins of the water. In the last decade, the amount of NOM in the environment has been continuously increasing, as a result of global warming and/or changes in precipitation patterns [6]. In consequence, a negative impact on drinking water has occurred, such as undesirable colour, taste and odour. NOM also favours bacterial regrowth in water distribution systems which represent a significant problem [7]. NOM is also a precursor for the formation of disinfection by-products (DBPs) in water, such as trihalomethanes (THMFP) which have been linked to an increase of the incidence of cancer in humans, low birth weight and bird defects [8]. For these reasons, several treatment processes for the removal of NOM have been considered and installed in water treatment plants, such as ozonation [9], adsorption [10], coagulation [11], electrochemical treatment [12] and AOPs.

This chapter reviews AOPs activated by photons used for the treatment of water in the presence of NOM and contaminants of emerging concern. The fundamental aspects of each AOPs and the efficiency of the radicals as oxidative species for the destruction of common pesticides, pharmaceuticals and emerging contaminants will be presented. The impact of the water matrix components that have a determining role on the AOPs oxidative mechanism will be discussed in detail.

2 Advanced Oxidation Processes (AOPs)

AOPs are based on the generation of very powerful, non-selective radical oxidative species (ROS) such as hydroxyl (•OH), peroxyl (•OHH), superoxide (•O_2^-) radicals and singlet oxygen (•O) that rapidly attack and oxidize a wide spectrum of organic matter in water, unselectively [13]. Glaze and colleagues [13] were the first to investigate the effectiveness of AOPs on the treatment of chlorinated organics in contaminated groundwater. AOPs include heterogeneous and homogenous photocatalysis, Fenton and Fenton-like processes, ozonation, ultrasound, microwave and γ-irradiation and electrochemical and wet oxidation processes.

Among chemical-driven AOPs, Fenton and Fenton-like processes involving the use of Fe^{2+} or other metal species in combination with H_2O_2 are in general pH dependent and function within a narrow operating range, usually at acidic pH (<4.0). These processes have found numerous applications for organic degradation such as the removal of pharmaceuticals, pesticides and endocrine-disrupting chemicals. Fenton processes present several advantages including the operation in the absence of light and a very fast initial rate of reaction. However, the accumulation of iron sludge at the end of treatment requires post-treatment and the cost of pH adjustment on large volumes of water are major drawback. Fenton processes have been coupled with a range of conventional water treatment processes such as coagulation, biological oxidation and membrane filtration to achieve a higher degree of contaminant degradation and mineralization [14, 15].

Peroxone, a mixture of O_3 and H_2O_2, is mostly applied for the removal of toxic contaminants (hydrocarbons, pesticides) and micropollutants, frequently as a pretreatment step before adsorption in granular activated carbon (GAC) beds. Peroxone is able to reduce the concentration of micropollutants in the influent stream to the beds and prolongs its lifetime. The energy requirement and the necessity of on-site production of O_3 are the main limitations, although peroxone is known to provide an outstanding bactericidal performance [16].

Sonochemical oxidation processes involve the use of ultrasound waves propagating through the aqueous medium. They produce intense vibrations to the water molecules that cause the production of hydroxyl radicals. In general, the production of •OH radicals is localized near the sound emission source, decreasing exponentially with distance, and this presents a significant limitation for the treatment of large volumes of water. Thereby, it is more often applied in combination with other oxidants such as H_2O_2 and dioxygen or combined with UV radiation and other AOPs (including Fenton reagent and Fenton-type reactions). Sonochemical processes have found a variety of applications including the degradation of pesticides, aromatic compounds, endocrine disrupters and pharmaceuticals and disinfection by-products [17–20]. Despite its technical potential for decontamination purposes, most studies have been executed at laboratory scale, and industrial applications are limited [21].

Electrochemical oxidation using direct electrochemistry (anodic oxidation) and indirect electrochemistry (electro-Fenton) and their combination with sonochemical, physicochemical and photochemical treatment methods have fulfilled many applications in water detoxification/decontamination including industrial, agricultural and pharmaceutical treatment [21, 22]. Low-energy requirements, using small amounts of chemical reagents and the capability of producing efficient mineralization, make electrochemical oxidation a promising process for full-scale industrial implementation in comparison to other AOPs. A detailed evaluation and a comparison with other AOPs have been presented in a recent review paper [1].

Alongside chemical-driven AOPs, photon-driven AOPs have the advantage of providing a simple, relatively inexpensive, clean and efficient treatment of contaminated water in terms of degradation, mineralization and total detoxification. Coupling UV irradiation with various oxidants such as O_3 and H_2O_2 and photocatalysis such as Fe^{3+} and semiconductor nanomaterials (e.g. TiO_2) results in efficient oxidative processes. Among various photon-driven AOPs, the photolysis of H_2O_2, (H_2O_2/UV) and O_3 (O_3/UV), the photo-Fenton process (H_2O_2/Fe^{2+}/UV) and heterogeneous photocatalysis (TiO_2/UV) have attracted the utmost attention in the last decades [1, 23]. The main mechanisms by which photon-driven AOPs degrade and mineralize water contaminants also exhibiting a bactericidal effect include ROS attack and the direct photochemical action of UV irradiation [24]. Among all of the oxidative treatment, photocatalytic processes have received significant attention for the treatment of water.

3 Photocatalysis

The word photocatalysis contains a prefix and a word, "photo" and "catalysis". Generally speaking, photocatalysis involves the activation of a photocatalytic material or substance by light photons, which in turn increases the rate of a chemical reaction without being consuming. The photocatalyst can be species dissolved in the liquid phase (homogeneous photocatalysis) or can be solids suspended in the liquid or immobilized on surfaces (heterogeneous photocatalysis).

3.1 Heterogeneous Photocatalysis

Heterogeneous photocatalysis makes use of semiconductor solid materials (e.g. ZnO, TiO_2, ZrO_2, CdS, ZnS) which catalyse the production of ROS and the removal of organic and inorganic contaminants when photoactivated by light photons. Among many semiconductors, titanium dioxide (TiO_2) has received significant attention in fundamental research and in practical applications, due to its favourable characteristics, such as abundance and low cost, chemically inert and stable properties, low toxicity and relatively high photoactivity compared to other semiconductor materials [25].

The energy gap between the valence and the conduction bands of TiO_2 (anatase) is 3.2 eV, which implies that the absorption of photons of wavelength of less than 384 nm may photoactivate electrons (e^-) from the valence band towards the conduction band, leaving holes (h^+) in the valence band (Reaction 1) [26]. The electrons and the holes can in turn drive reduction and oxidation reactions, respectively. For example, in the presence of an oxygenated water solution, (e^-) can reduce adsorbed O_2 to form the superoxide radical ($\bullet O_2^-$) (Reaction 3), and the (h^+) can oxidize an electron acceptor such as water to form the hydroxyl radical ($\bullet OH$) (Reaction 3).

$$TiO_2 + h\nu \rightarrow e_{cb}^- + h_{vb}^+ \quad (1)$$
$$O_2 + e_{cb}^- \rightarrow O_2^{\bullet -} \quad (2)$$
$$H_2O_{ads} + h_{vb}^+ \rightarrow OH_{ads}^{\bullet} + H^+ \quad (3)$$
$$O_2^{\bullet -} + H^+ \rightarrow HO_2^{\bullet} \quad (4)$$
$$HO_2^{\bullet} + HO_2^{\bullet} \rightarrow H_2O_2 + O_2 \quad (5)$$
$$H_2O_2 + e_{cb}^- \rightarrow HO^{\bullet} + HO^- \quad (6)$$
$$D + h_{vb}^+ \rightarrow D^{\bullet +} \quad (7)$$
$$A + e_{cb}^- \rightarrow A^{\bullet -} \quad (8)$$
$$e_{cb}^- + h_{vb}^+ \rightarrow TiO_2 + \text{heat} \quad (9)$$

Reactions 5 and 6 in the sequence above show the formation of hydrogen peroxide, which is known to split into two hydroxyl radicals through aqueous photolysis or to accept an electron (Reaction 6). Reactions 7 and 8 show how the hole (h^+_{vb}) can react directly with an adsorbed organic electron donor (D) and how a conduction band hole (e^-_{cb}) can reduce an adsorbed electron acceptor. The competing recombination of the two charge carriers release heat (Reaction 9).

3.2 Homogenous Photocatalysis

3.2.1 Photo-Fenton Treatment

The Fenton process was discovered by Fenton in 1894 [27] during the oxidation of maleic acid. The Fenton reaction involved the reaction between dissolved Fe^{2+} and hydrogen peroxide (H_2O_2) which results in the formation of hydroxyl and peroxyl radicals (Reactions 10 and 11).

$$Fe^{2+} + H_2O_2 \rightarrow Fe^{3+} + \bullet OH + OH^- \quad (10)$$
$$Fe^{3+} + H_2O_2 \rightarrow Fe^{2+} + \bullet OOH + H^+ \quad (11)$$

The efficiency of the above reactions is improved by irradiation with near UV and visible light [28], which drives the photoreduction of Fe^{3+} to Fe^{2+}, regenerating the catalyst and simultaneously producing further •OH (Reaction 12).

$$Fe^{3+} + H_2O + hv \rightarrow Fe^{2+} + H^+ + \bullet OH \qquad (12)$$

In addition, Fe^{3+} hydroxyl complexes under acidic condition (e.g. $Fe(OH)^{2+}$) undergo photoreduction as reported in Reaction 13 [28].

$$Fe(OH)^{2+} + hv \rightarrow Fe^{2+} + \bullet OH \qquad (13)$$

The Fenton reaction carried out in the presence of irradiation is known as photo-Fenton. This reaction, of course, is wavelength dependent, and the quantum yields of •OH and Fe^{2+} ion formation decrease as the wavelength increases. The quantum yield of •OH formation is 0.14 at 313 nm and 0.017 at 360 nm [29]. The basic photo-Fenton process has found many applications for the degradation and mineralization of pesticides, dyes, chlorophenols and chlorinated compounds [28, 30].

3.2.2 Photocatalytic Ozonation (O_3/UV)

Ozone is a powerful oxidant with an oxidation potential of 2.07 eV that increases to 2.8 eV in the presence of irradiation. The AOP with ozone and UV irradiation is initiated by the photolysis of ozone. The photodecomposition of ozone leads to two hydroxyl radicals, which do not act if they recombine producing hydrogen peroxide [31].

$$O_3 + H_2O + hv \rightarrow H_2O_2 + O_2 \qquad (14)$$

In this system, there are three components that are able to produce hydroxyl radicals: UV radiation, ozone and hydrogen peroxide. Therefore, the reaction mechanisms involving O_3/H_2O_2, as well the combination UV/H_2O_2, are of great importance. O_3/UV has attracted great attention for the elimination of volatile chlorinated organic carbons (VCOCs) such as $CHCl_3$, CCl_4, trichloroethylene (TCE), tetrachloroethylene and 1,1,2-trichloroethane [32]. Besides effluents containing pesticides, endocrine disrupters, pharmaceutical compounds, antibiotics, surfactants, dyes and nitrobenzene have been set as targets in different O_3/UV applications.

3.2.3 Photolysis of Hydrogen Peroxide (UV/H_2O_2)

In this process, H_2O_2 is split into two hydroxyl radicals by adsorption of UVC light. The most accepted mechanism for the H_2O_2 photolysis is the rupture of the O–O bond by activation of ultraviolet light.

$$H_2O_2 + h\nu \rightarrow 2\,^\bullet OH \tag{15}$$

This reaction is pH dependent and becomes more efficient under alkaline conditions, probably because at 253.7 nm the anion peroxide $HO_2^{\circ-}$ has a high molar adsorption coefficient (240 vs 18.6 M^{-1} cm^{-1}) [13]. The hydroxyl radical formed can further attack hydrogen peroxide leading to the reactions sequence (Reactions 16, 17, and 18):

$$H_2O_2 + \bullet OH \rightarrow HO_2^\circ + H_2O \tag{16}$$
$$\bullet OH + HO_2^- \rightarrow HO_2^\circ + OH^- \tag{17}$$
$$H_2O_2 + HO_2^- \rightarrow \bullet OH + O_2 + H_2O + \bullet OH \tag{18}$$

One of the pioneering studies in the field of UV/H$_2$O$_2$ was the study by Eckenfelder et al., which set the target on groundwater decontamination [33]. The H$_2$O$_2$/UV process has been found to be efficient for the elimination of cyanides and other organic pollutants, such as benzene, trichloroethylene and tetrachloroethylene [34].

3.2.4 Peroxone Process (O$_3$/H$_2$O$_2$)

The peroxone process was studied by Staehelin and Hoigné [35]. The decomposition of O$_3$ into •OH radicals is accelerated by the presence of hydrogen peroxide according to Reaction 19 [36].

$$H_2O_2 + 2O_3 \rightarrow 2\bullet OH + 3\,O_2 \tag{19}$$

The main advantage of peroxone treatment is the short reaction time; in fact, due the quick reaction, time is not necessary to use high doses of ozone to obtain faster reactions with pollutants. Von Gunten [37, 38] reported a good review including ozone-AOPs as drinking water treatment.

3.2.5 Vacuum Ultraviolet (VUV)

The application of ultraviolet irradiation in water treatment can be distinguished as UV-A (380–315 nm), UV-B (315–280 nm), UV-C (280–200 nm), V-UV (200–100 nm) and extreme UV (100-1) [39]. Photolysis of water at 185 nm is a highly efficient process for the generation of hydroxyl radicals. At this wavelength, the homolysis and photochemical ionization of water lead to the formation of •OH with quantum yields of 0.33 and 0.045, respectively [40].

$$H_2O \xrightarrow{h\nu\ (185\ nm)} {}^\circ OH + H^+ \quad \Phi = 0.33 \tag{20}$$
$$H_2O \xrightarrow{h\nu\ (185\ nm)} {}^\circ OH + H^+ \quad \Phi = 0.045 \tag{21}$$

The problem connected with the VUV treatment is the formation of by-products, especially nitrite, which occurs from the photolysis of nitrate in the water.

Among the photochemical reactions described above, the photo-Fenton process and heterogeneous photocatalysis have been reported to provide superior oxidation conditions compared to H_2O_2 and O_3 photolysis processes, although the latter have been commercialized. Solar-driven photo-Fenton and heterogeneous photocatalysis have drawn significant interest due to the potential use of solar radiation (solar photocatalytic processes) and have been investigated in numerous studies, from bench to pilot and to industrial scale.

4 Photocatalysis Treatment for Contaminant Removal

In the last decade, various AOPs have been evaluated for their efficacy and economical feasibility for the removal of water and wastewater contaminants. The literature data presented in this chapter focuses primarily on the photocatalytic treatment and its effectiveness in degrading emerging contaminants such as pesticides, pharmaceuticals and endocrine-disrupting compounds (EDCs) and for the control of taste and odour.

4.1 Photocatalytic NOM Removal

The characterization of NOM in the contaminated water is essential prior to the design and optimization of an AOP treatment, to determine the potential of the water for the formation of DBPs and other toxic by-products. Further characterization of the water effluent is also required after AOPs treatment, to determine the possible formation of DBPs and toxic degradation products, which may determine the suitability of an AOP treatment for the reduction of NOM. The reactions of •OH radicals with NOM proceed through three different ways: (1) by the addition of •OH radicals to double bonds; (2) by H-atom abstraction, which yields carbon-centred radicals; and (3) by a reaction mechanism in which the OH radical accept electron from an organic substituent [2].

The degradation of NOM is much more rapid compared to total DOC mineralization; thereby, the reduction in the absorbance at 254 (UV254) nm may indicate the degradation of organics, while the reduction in DOC may take much longer time. A rapid reduction in UV254 could indicate the presence of NOM with high molecular mass, such as aromatic rings and their breakdown into lower molecular mass by-products presenting negligible UV absorbance. These by-products may be less prone to oxidation by •OH radical and may act as a barrier against total mineralization. As the degradation of NOM fractions continue upon •OH radical attack, the by-products formed may display several hydrophilic/hydrophobic properties compared to its parent compound. In consequence, fractionation studies represent good manuals for unravelling the possible reactions and prevailing mechanisms occurring during the photocatalytic oxidation of NOM species [41, 42].

The shift of the molecular weight distribution towards lower values during oxidation has been reported in numerous literature findings but most frequently with the UV/H_2O_2 process [42–45]. Further studies are now focusing on the identification of DBPs and other by-products during the photocatalytic oxidation of NOM [4]. Current investigations in the field of AOPs have already extended their vision by setting simultaneous objectives on the removal of DBPs, NOM, mineralization and toxicity in the water. This multidisciplinary approach is essential for determining the most effective AOP capable of producing a safe effluent.

Traditional treatment processes, such as filtration, physical or chemical disinfection processes, biological processes alone or combined with other processes, have shown several limitations in terms of removing complex organic matter, the bacterial content and/or other recalcitrant contaminants in the aqueous matrix [46]. The determination of the desired treatment goals are essential factors for the proper AOP choice and design. These goals could include biodegradability enhancement, the total mineralization of organic compounds or the desired level of water detoxification. Any of these scenarios necessitate the optimization of the process conditions considering the type, fractional distribution and amount of organic matter present in the water. The alkalinity and pH of the water also play a major role on the photocatalytic oxidation of NOM. NOM is less efficiently adsorbed onto the surface of photocatalysts at elevated pH with a decisive decline between pH 5–10 [47]. Concurrently at alkaline pH, carbonate or bicarbonate species act as efficient ROS scavengers that actively decrease the rates of degradation of the target pollutants, as shown for the degradation of pharmaceuticals in alkaline secondary treated urban wastewater (UWW) effluents [47].

Biologically treated urban wastewater contains trace and bulk organics such as humic acids and other low molecular organic acids with a wide range of molecular size distribution, which would act as a strict barrier against the oxidative conditions of a photocatalytic process. In most of the scenarios, these defined factors act as unavoidable radical scavengers, which favour surface charge recombination in TiO_2 photocatalysis.

In order to make a proper comparison of the efficiency of AOPs and most adequate choice of a pretreatment, it is necessary to consider simultaneously the degradation/mineralization performance of the process and the energetic requirements of the treatment processes.

4.2 Pesticides

The European Water Framework Directive 2000/60/EC [48] established that 33 priority substances present high toxicity, high environmental persistence, endocrine-disrupting capabilities and bioaccumulation potential. Among these hazardous chemicals, there are also pesticides, such as atrazine, alachlor, isoproturon, pentachlorophenol and chlorfenvinphos. These recalcitrant compounds and their metabolites (e.g. glyphosate AMPA) have been detected in groundwater and drinking wells of various EU countries, for example, Italy [49], Hungary [50], Spain [51] and the UK [52]. Table 1 summarizes the representative AOP treatment of common pesticides.

4.3 Pharmaceuticals and Personal Care Products (PPCPs)

Pharmaceuticals and personal care products (PPCPs) are a wide group of emerging environmental contaminants. Increasing number of studies have confirmed the presence of various PPCPs in several environmental matrices such as groundwater [65], surface water and within the water distribution system (e.g. at the tap) [66]. Nowadays, it is known that some antibiotics may cause long-term and irreversible change to the microorganism genome even at trace concentrations. Some PPCPs have also been demonstrated to disrupt the human endocrine system, and hence, their presence in aquatic systems has been a source of concern [67]. In addition to their potential negative effect on human and wildlife, pharmaceuticals and personal care products are often resistant to biological degradation processes [68]. The class and range of molecules studied are vast. Table 2 reports selected representative AOP studies for the degradation of three commonly investigated pharmaceuticals detected in water.

5 Efficiency of AOPs

This section evaluates the efficiency of AOPs on the basis of effluent toxicity and the energy consumption. Comparison is not trivial due to the different processes involved in the production of radicals. Furthermore, for the case of AOPs and oxidative treatment, water matrix constituents which may become target for non-selective oxidants (most pronounced the •OH) has to be paid the utmost attention.

5.1 AOP Efficiency Based on Toxicity

Several bioassays have been used to measure the response of organisms exposed to target compounds and complex water matrices such as surface water, groundwater, wastewater or seawater. Standard bioassays are best developed through the use of a battery of organisms such as plants and algae, invertebrates, microorganisms and fishes. The most common toxicity tests used to evaluate the effectiveness of AOP treatment are *Daphnia magna* [69], *Vibrio fischeri* [70], *lettuce* seeds [71] and *zebrafish* [72]. Wastewater from several sources (industrial, hospital or domestic) necessitates choice of proper methodology for toxicity evaluation.

Table 1 Representative AOP treatment of common pesticides

Pesticides	AOPs	Experimental conditions	Efficiency	Reference
Atrazine [ATZ]	Coupled solar photo-Fenton and heterogeneous TiO_2 photocatalysis processes	Gesaprim solution with [atrazine] = 35 mg/l (19 mg/l of TOC); $[H_2O_2]_0/COD_0 = 3$ and 10 ppm Fe^{2+} at pH 2.8; $[TiO_2] = 200$ mg/l, UV = solar irradiation; volume treated = 60 l	TOC removal (%): 46 (1 h) and 30 (4 h) after solar photo-Fenton and TiO_2 photocatalysis, respectively; COD removal (%): 90 (2 h) of photo-Fenton reaction; Atrazine: 68% (6 h) of photoreaction, the treated effluent was nontoxic	[53]
	UV/H_2O_2	Natural, drinking water and wastewater mixed with organic micropollutants; [ATZ] = 2.6(±0.2) mg/l; pH = 7.4 (±0.5); $[H_2O_2] = 10$ mg/l	ATZ = 80% degradation was achieved at 990 mJ/cm^2	[54]
	O_3/UV	Distilled-deionized water solution with non-ionic surfactant, Brij35; treated volume, 203 ml; [ATZ] = 0.01 mM; total light intensity = 300×10^{-6} E/l/S (35 W)	Improve the efficiency of the treatment by adding the surfactant. The enhancement (%) was 42 and 52 at pH 2.5 and 7, respectively	[55]
	VUV photolysis; UV/photocatalysis	Xe lamp (150 W) at 172 nm; T = 25 °C; [ATZ] = 1×10^{-4} M	VUV produce OH• and H•; in absence of oxygen, H• are more prone to participate in the mineralization reaction; total mineralization of ATZ	[56]
Alachlor [ALC]	Photo-Fenton	Alachlor = 10 mg/l; volume = 20 ml; T = 25 ± 1 °C; UV radiation = <300 nm; ligand (citrate, EDTA and pyrophosphate) = 1×10^{-4} M; $[H_2O_2] = 4 \times 10^{-3}$ M; [Fe(II)] = 1×10^{-4} M	Degradation (%) of alachlor = 60 (60 min) without ligand; Degradation (%) of alachlor = 100 in 15 min and in 30 min with citrate and EDTA, respectively; Total TOC abatement under photo-Fenton with citrate	[57]
	Ozonation processes; UV photolysis; UV/H_2O_2	Distilled (6 mg/l IC) and surface water from Genova River (2 mg/l IC); gas flow rate = 20 l/h; [ALC] = 3.7×10^{-5} M; $[O_3] = 1.8 \times 10^{-4}$ M; pH = 7 ± 0.2; $[H_2O_2] = 10^{-3}$ M; $I_0 = 1.6 \times 10^{-6}$ E/L s	Ozonation processes depend on pH and water quality; Ozonation rate of alachlor is improved when ozone is combined with hydrogen peroxide or UV radiation	[58]
	TiO_2 photocatalysis	UV light = 400 W metal-halide lamp with major peak of spectrum; $[TiO_2]$ =0.1 g/l; [ALC] = 5 mg/l	Efficiency depend on the thickness of a TiO_2 film; Removal rate of alachlor with Fe^{3+} and UV (without TiO_2) was 0.28 mg/l/h in 10 h; Removal rate of alachlor with Fe^{3+} and UV (with TiO_2) was 0.32 mg/l/h in 10 h; TOC decrease (%) under TiO_2/UV from 100 to 81, 51, 44 in 4, 8 and 10 h, respectively; TOC decrease (%) under Fe^{3+}/UV (without TiO_2) from 100 to 70 in 10 h	[59]
	$FeCl_3$ photocatalysis, TiO_2 photocatalysis and photolysis	UV lamp = Xenon arc lamp; volume = 25 ml; [ALC] = 30-60 µg/l; $[TiO_2]$ = 150 mg/l; $[FeCl_3]$ = 15 mg/l; $[H_2O_2]$ = 0.05% (v/v)	The most effective method was to use 15 mg/l of $FeCl_3$ that in 15 min was able to destroy alachlor; $FeCl_3$ is cheaper and offers shorter reaction half-live for ALC	[60]

Compound	Process	Conditions	Results	Ref
Isoproturon	TiO$_2$ photocatalysis	Volume = 250 ml double-distilled water; [Isoproturon] = 0.5 mM; UV irradiation = 125 W medium pressure mercury lamp; λ = 238 nm; [TiO$_2$] = 1 g/l (P25; Hombikat UV100; PC500 and TTP)	94.8% degradation and 78.45% mineralization in 80 min	[61]
			Degussa P25 was more efficient as compared to other photocatalysts	
			The degradation rate is not influenced from pH but improves adding at 3 g/l of catalyst	
			Addition of electron acceptors (hydrogen peroxide, potassium bromate, potassium persulphate) gives beneficial effect on the degradation of isoproturon	
	UV/H$_2$O$_2$; UV/TiO$_2$; UV/H$_2$O$_2$/Fe(II); UV/H$_2$O$_2$/TiO$_2$	V = 0.5 l; UV intensity = 5 × 10^{-6} E/s; [Isoproturon] = 5 × 10^{-5} mol/l; [H$_2$O$_2$] = 5 × 10^{-4} mol/l; [Fe(II)] = 5 × 10^{-5} mol/l; [TiO$_2$] = 1 g/l	The best performance was obtained by photo-Fenton: Isoproturon was completely removed in 15 min, while TOC is reduced of 89% after 60 min of irradiation	[62]
Glyphosate [glyph]	UVC/H$_2$O$_2$	Λ = 253.7 nm; [glyph] = 0.30 mM, [H$_2$O$_2$] = 2.20 mM; pH = 3, 5, 7 and 10	TOC conversion after 5 h was 29%	[63]
			63.5% glyph conversion was obtained under irradiation of two lamps of 40 W	
			Best condition for degradation was at the highest value of pH (pH 7 and pH 10)	
	Ozonation, photolysis and heterogeneous photocatalysis	[Glyph] = 42.275 mg/l; [O$_3$] = 14 mg/l; pH = 7 and 10; [TiO$_2$] = 0.1 g/l; V = 200 ml; λ > 290 nm	Ozonation at pH 6.5 = 20% TOC reduction while at pH 10 97.5% TOC removal	[64]
			TiO$_2$ photocatalysis at pH 6.5 = 92% and 99.9% TOC and glyph removal, respectively	
			pH 10 the degradation rate showed to be highest in both processes	
			The half-lives obtained for glyphosate degradation were 1.8 and 6.2 min for O$_3$/pH 10 and TiO$_2$/UV, respectively	

Table 2 Representative AOP treatment for widely used pharmaceutical compounds

PCPPs	AOPs	Experimental conditions	Reference
Carbamazepine (CBZ)	UV/H_2O_2	Surface river water; [CBZ] = 240–710 µg/l; UV = 254 nm; [H_2O_2] = 10 mg/l	[74, 75]
	UV/H_2O_2	Water; solar photodegradation; volume = 500 ml, [CBZ] = 0.5 mmol, [H_2O_2] = 100 mM	
	TiO_2 P25 photocatalysis; TiO_2 Hombikat UV 100 photocatalysis	Surface lake water; irradiation = simulation solar light; [catalyst] = various; [CBZ] = various	[76]
	UV-AOP *(UV/PS, UV/H_2O_2, UV/PMS)	UV lamp = 75 W (253.7 nm); [CBZ] = 21.16 µM, [oxidant] = 0.5–5 mM, pH = 3–11	[77]
	UVC and VUV	Pure water, Lamp1 = 254 nm; Lamp2 = 254 and 185 nm	[78]
Diclofenac (DCF)	Sunlight	Surface lake water; [DCF] = 1 µg/l, volume = 30 ml, irradiation = full sunlight for 2 and 4 h	[79]
	TiO_2/solar light solar photo-Fenton	Freshwater; solar irradiation; [DCF] = 50 mg/l; volume = 35 l (22 l irradiated); [H_2O_2] = 200–400 mg/l; [TiO_2] = 0.2 g/l; [Fe(II)] = various	[80]
	Ozonation, photolysis, photolytic ozonation, photocatalysis and photocatalytic ozonation	Water and wastewater; volume = 250 ml; [DCF] = 0.1 mM; [O_3] = 50 g/nm^3; UV = mercury vapour lamp; [TiO_2] = 0.5 g/l	[81]
Sulfamethoxazole [SMX]	TiO_2/UV-A	UVA: $324 \leq \lambda \leq 400$ nm; [SMX] = 100 µM; [TiO_2] = 0.1 g/l; pH = various; [NOM] = 0–20 mg/l; [$NaHCO_3$] = 0–100 mM	[82]
	Hydrolysis, photolysis, solar photo-Fenton	Distillate and seawater; [SMZ] = 50 mg/l; intensity = 250 W/m^2; [Fe(II)] = 2.6, 5.2 and 10.4 mg/l; [H_2O_2] = 30 to 210 mg/l; volume = 39 l; pH = 2.5–2.8	[83]
	O_3; O_3/H_2O_2; UV/H_2O_2	Milli-Q and surface lake waters; T = 5–25°C; pH various; [SMX] = 1 µM; LP-UV; [O_3] = 20–200 µM	[84]

Brienza and co-workers [73] reported that for a specific domestic wastewater effluent, among the four standard bioassays (*Vibrio fischeri, Daphnia magna, Pseudokirchneriella subcapitata* and *Brachionus calyciflorus*), the most sensitive to the variation of contaminant concentrations was *Pseudokirchneriella subcapitata*. The oxidative processes evaluated through specific estrogenic assays [endocrine-disrupting substances were measured with cell lines expressing oestrogen receptor alpha (ERα) based on human cell] were able to remove emerging contaminants and the detected oestrogenic activity.

Heringa and co-workers [85] demonstrated through a UV/H$_2$O$_2$ treatment process that water matrices' effects must be considered and that hydroxyl radicals were not solely responsible for the increase in toxicity.

The EU Water Framework Directive requires the achievement of a "good chemical and biological status" of all water bodies by 2015 [48]. In order to obtain water bodies that meet both standards/characteristics, assessment of tertiary treatment efficiency should be supplemented by proper toxicity evaluation. In Table 3, the effect of selected AOPs on final toxicity of water solution is summarized.

5.2 AOPs Efficiency Based on Energy Consumption

There is increasing attention on developing cost and energy-efficient AOPs especially in the municipal water sector for the treatment of the so-called emerging contaminants (ECs) [90]. In addition to AOPs, other technologies that can be applied for the removal of ECs include ozonation, activated carbon or membrane technologies (reverse osmosis) which have relatively high capital and operating costs.

The major operating cost in AOPs is electrical energy consumption. The most common way to compare AOPs based on their electrical energy efficiency is to use the parameter of electrical energy per order (E_{EO}). The E_{EO} is the electric energy in kilowatt hours (kWh) required to reduce the concentration of a target contaminant by one order of magnitude (90% removal) in 1 m^3 of contaminated water [91]. E_{EO} can be calculated using the following equations:

$$E_{EO} = \frac{1,000 P_t}{V \log\left(\frac{C_i}{C_f}\right)} \quad \text{Batch operation} \quad (22)$$

$$E_{EO} = \frac{P}{F \log\left(\frac{C_i}{C_f}\right)} \quad \text{Continuous operation} \quad (23)$$

where P is the electrical power input in the processes [kW], V is the volume [l] of water treated, c_i and c_f correspond to the initial and final the concentration after one order of magnitude reduction (mol l^{-1}) and F is the flow rate [m^3/h] in continuous flow system [91].

E_{EO} can be also used to estimate if the treatment is economically acceptable for commercial applications as suggested by Arslan-Altan [92]; in fact, in the last

Table 3 Toxic evaluation of AOP treatment

Emerging contaminants	AOP	Toxicity test	Comment	Reference
Cyanide	TiO$_2$/UV-A	V. fischeri	• EC$_{50}$ values was significantly different for V. fischeri commercial kit, V. fischeri lab culture and Polystichum setiferum fern spore • In all test, the toxicity decrease in correspondence of cyanide or phenol degradation	[86]
Phenol		Fern spore bioassay		
Mepanipyrim	O$_3$/Hg-UV	V. fischeri	• Toxic unit (TU) increase from 2.22 to 16.66 at 120 min • At 120 min, formation of main photoproducts was observed • TU decrease rapidly from 120 to 180 min	[87]
Five pesticides (alachlor, atrazine, chlorfenvinphos, diuron, isoproturon)	Photo-Fenton	V. fischeri	• Decrease in the beginning of the treatment • Followed stable toxic level during the rest of the photo-Fenton treatment	[88]
4-Methylaminoantipyrine 4-Formylaminoantipyrine 4-Acetylaminoantipyrine	Solar irradiation	D. magna	• Increasing of acute and chronic toxicity from 27–55% to 60–85%, respectively • Target compounds were totally removed; the toxicity on D. magna was made up by formation of intermediate product	[89]

decades, several researches have used EEO to evaluate their investigation of AOPs for contaminant degradation.

Mehrjouei and co-workers [93] determined the E_{EO} for economic comparative studies of three heterogeneous advanced oxidation treatments for removing oxalic acid and dichloroacetic acid. The treatments involved in this study were catalytic ozonation (TiO$_2$/O$_3$), photocatalytic oxidation (TiO$_2$/UVA) and photocatalytic ozonation (TiO$_2$/O$_3$/UVA). The catalytic ozonation has been the most energy efficient with EEO values of 0.017 and 0.050 kWh/mM per order for oxalic acid and dichloroacetic acid, respectively. The equivalent values for the TiO$_2$/UVA/O$_3$ and the TiO$_2$/UVA/O$_2$ processes were 0.017 and 0.063 kWh/mM for oxalic acid degradation and 0.050 and 0.35 kWh/mM for dichloroacetic acid degradation [93].

In another study, a comparison of AOPs was made among ozone and Fenton processes combined with UV and H_2O_2 for the treatment of water [94]. The parameters considered were COD and colour removal rates and electrical energy per order. This interesting work reported that the most efficient treatment able to remove 100% colour and COD with less energy consumption of 0.01 kWh/m^3 order[1] was O_3/UV/Fe^{2+}/H_2O_2. Hence, this process can be suitable for the treatment of industrial or highly organic effluents.

The treatment costs with AOPs and their combination with ultrasound (US) are relatively higher for hydrophilic pollutants such as reactive azo dyes and phenol than that of hydrophobic contaminants such as trichloroethylene [95]. It has been also pointed out that high costs of ultrasonic waste treatment reported in the literature so far is the expected result of very high densities used in treatability studies with single ultrasonic irradiation. To get to the energy efficiency standpoint, the mechanism of degradation plays the key role in this kind of situation.

In the case of phenol degradation, among UV (254 nm), US (300 kHz) and O_3 (2 mg/l) and TiO_2 photocatalysis and their combination, US was found to be the least energy-efficient process alternative. Energy efficiency can be improved by proper combination of AOPs; significant reduction in process duration can be established by the use of ultrasonic irradiation combined with AOP technologies [95].

For the case of azo-dye degradation, US-coupled H_2O_2 process was found to be ten times more energy efficient compared to US/UV alternative, while UV/O_3 was found to be ten times more energy efficient compared to US/UV and US/O_3 alternatives. Adequate use of H_2O_2 with ultrasound and/or UV cavitation may promote oxidative capacity of a well-known reaction. From the reduced total process time and energy requirement standpoint, sono-photocatalysis was defined as an adequate option for oxidative degradation of pollutants that have less complex organic structure [95].

6 Conclusions

Advanced oxidation processes are effective treatments for the purification of water and reuse. Heterogeneous and homogenous processes are able to destroy emerging contaminants including pesticides, pharmaceuticals and personal care products, hormones, etc. On one hand, the complex mechanisms that are involved during oxidative processes affects the range of their applications, and the major research is focused to determine by-products and to understand the potential toxicity. In addition, the requirement of energy for the pre- and/or post-treatment of water is the main limit of AOP applications. On the other hand, the concern about water scarcity and its contamination brings a growing demand for AOPs. So, there is a need to focus more research to the development of these technologies and their application in treating water.

References

1. Oturan MA, Aaron J (2014) Advanced oxidation processes in water/wastewater treatment: principles and applications. A review. Crit Rev Environ Sci 44:2577–2641
2. Kleiser G, Frimmel FH (2000) Removal of precursors for disinfection by-products (DBPs) – differences between ozone-and OH-radical-induced oxidation. Sci Total Environ 256:1–9
3. Robert D, Malato S (2002) Solar photocatalysis: a clean process for water detoxification. Sci Total Environ 291:85–97
4. Matilainen A, Sillanpää M (2010) Removal of natural organic matter from drinking water by advanced oxidation processes. Chemosphere 80:351–365
5. Thurman EM (1985) Classification of dissolved organic carbon. In: Organic geochemistry of natural waters. Martinus Nijhoff/Dr. W. Junk Publishers, The Hague
6. Eikebrokk B, Vogt RD, Liltved H (2004) NOM increase in northern European source waters: discussion of possible causes and impacts on coagulation/contact filtration processes. Water Sci Tech-W Sup 4:47–54
7. Volk C, Bell K, Ibrahim E, Verges D, Amy G, LeChevallier M (2000) Impact of enhanced and optimized coagulation on removal of organic matter and its biodegradable fraction in drinking water. Water Res 34:3247–3257
8. Richardson SD, Plewa MJ, Wagner ED, Schoeny R, DeMarini DM (2007) Occurrence, genotoxicity, and carcinogenicity of regulated and emerging disinfection by-products in drinking water: a review and roadmap for research. Mutat Res 636:178–242
9. Siddiqui MS, Amy GL, Murphy BD (1997) Ozone enhanced removal of natural organic matter from drinking water sources. Water Res 31:3098–3106
10. Newcombe G, Drikas M (1997) Adsorption of NOM onto activated carbon: electrostatic and non-electrostatic effects. Carbon 35:1239–1250
11. Matilainen A, Vepsäläinen M, Sillanpää M (2010) Natural organic matter removal by coagulation during drinking water treatment: a review. Adv Colloid Interf Sci 159:189–197
12. Särkkä H, Vepsäläinen M, Sillanpää M (2015) Natural organic matter (NOM) removal by electrochemical methods – a review. J Electroanal Chem 755:100–108
13. Glaze WH, Kang JW, Chapin DH (1987) The chemistry of water treatment processes involving ozone, hydrogen peroxide and ultraviolet radiation. Ozone Sci Eng 9:335–352
14. Lucas MS, Dias AA, Sampaio A, Amaral C, Peres JA (2007) Degradation of a textile reactive Azo dye by a combined chemical–biological process: Fenton's reagent-yeast. Water Res 41:1103–1109
15. Mantzavinos D, Psillakis E (2004) Enhancement of biodegradability of industrial wastewaters by chemical oxidation pre-treatment. J Chem Technol Biot 79:431–454
16. Tarr MA (2003) Chemical degradation methods for wastes and pollutants: environmental and industrial applications. CRC Press, New York
17. Hua I, Hoffmann MR (1997) Optimization of ultrasonic irradiation as an advanced oxidation technology. Environ Sci Technol 31:2237–2243
18. Liang J, Komarov S, Hayashi N, Kasai E (2007) Improvement in sonochemical degradation of 4-chlorophenol by combined use of Fenton-like reagents. Ultrason Sonochem 14:201–207
19. Ma Y-S, Sung C-F (2010) Investigation of carbofuran decomposition by a combination of ultrasound and Fenton process. J Environ Eng Manag 20:213–219
20. Namkung K-C, Burgess AE, Bremner DH, Staines H (2008) Advanced Fenton processing of aqueous phenol solutions: a continuous system study including sonication effects. Ultrason Sonochem 15:171–176
21. Ma Y-S (2012) Short review: current trends and future challenges in the application of sono-Fenton oxidation for wastewater treatment. Sustain Environ Res 22:271–278
22. Martínez-Huitle CA, Brillas E (2009) Decontamination of wastewaters containing synthetic organic dyes by electrochemical methods: a general review. Appl Catal B-Environ 87:105–145
23. Zaviska F, Drogui P, Mercier G, Blais J-F (2009) Procédés d'oxydation avancée dans le traitement des eaux et des effluents industriels: application à la dégradation des polluants réfractaires. Rev Sci Eau 22:535–564

24. Gumy D, Rincon AG, Hajdu R, Pulgarin C (2006) Solar photocatalysis for detoxification and disinfection of water: different types of suspended and fixed TiO_2 catalysts study. Sol Energy 80:1376–1381
25. Daghrir R, Drogui P, Robert D (2013) Modified TiO_2 for environmental photocatalytic applications: a review. Ind Eng Chem Res 52:3581–3599
26. Fujishima A, Zhang X, Tryk DA (2008) TiO_2 photocatalysis and related surface phenomena. Surf Sci Rep 63:515–582
27. Fenton HJH (1894) LXXIII. Oxidation of tartaric acid in presence of iron. J Chem Soc Trans 65:899–910
28. Pignatello JJ, Oliveros E, MacKay A (2006) Advanced oxidation processes for organic contaminant destruction based on the Fenton reaction and related chemistry. Crit Rev Environ Sci Technol 36:1–84
29. Faust BC, Hoigné J (1990) Photolysis of Fe (III)-hydroxy complexes as sources of OH radicals in clouds, fog and rain. Atmos Environ 24:79–89
30. Kaichouh G, Oturan N, Oturan MA, El Kacemi K, El Hourch A (2004) Degradation of the herbicide imazapyr by Fenton reactions. Environ Chem Lett 2:31–33
31. Peyton GR, Glaze WH (1988) Destruction of pollutants in water with ozone in combination with ultraviolet radiation. 3. Photolysis of aqueous ozone. Environ Sci Technol 22:761–767
32. Bhowmick M, Semmens MJ (1994) Ultraviolet photooxidation for the destruction of VOCs in air. Water Res 28:2407–2415
33. Eckenfelder WW, Bowers AR, Roth JA (1993) Chemical oxidation: technology for the nineties, vol 2. CRC Press, New York
34. Doré M (1989) Chimie des Oxydants et Traitement des Eux. Tec & Doc
35. Staehelin J, Hoigné J (1982) Decomposition of ozone in water in the presence of organic solutes acting as promoters and inhibitors of radical reactions. Environ Sci Technol 19:1206–1213
36. Glaze WH, Kang JW (1989) Advanced oxidation processes – description of a kinetic model for the oxidation of hazardous materials in aqueous-media with ozone and hydrogen-peroxide in a semibatch reactor. Ind Eng Chem Res 28(11):1573–1580
37. von Gunten U (2003) Ozonation of drinking water: part I. Oxidation kinetics and product formation. Water Res 37:1443–1467
38. von Gunten U (2003) Ozonation of drinking water: part II. Disinfection and by product formation in presence of bromide, iodide or chlorine. Water Res 37:1469–1487
39. Oppenlander T (2007) Photochemical purification of water and air: advanced oxidation processes (AOPs) – principles, reaction mechanisms, reactor concepts. Wiley, Weinheim
40. Gonzalez MG, Oliveros E, Wörner M, Braun AM (2004) Vacuum-ultraviolet photolysis of aqueous reaction systems. J Photoch Photobio C 5:225–246
41. Buchanan W, Roddick F, Porter N (2006) Formation of hazardous by-products resulting from the irradiation of natural organic matter: comparison between UV and VUV irradiation. Chemosphere 63:1130–1141
42. Sanly Lim M, Chiang K, Amal R, Fabris R, Cho C, Drikas M (2007) A study on the removal of humic acid using advanced oxidation processes. Sep Sci Technol 42:1391–1404
43. Espinoza LAT, Frimmel FH (2008) Formation of brominated products in irradiated titanium dioxide suspensions containing bromide and dissolved organic carbon. Water Res 42:1778–1784
44. Goslan EH, Gurses F, Banks J, Parsons SA (2006) An investigation into reservoir NOM reduction by UV photolysis and advanced oxidation processes. Chemosphere 65:1113–1119
45. Katsumata H, Sada M, Kaneco S, Suzuki T, Ohta K, Yobiko Y (2008) Humic acid degradation in aqueous solution by the photo-Fenton process. Chem Eng J 137:225–230
46. Birben NC, Uyguner-Demirel CS, Bekbolet M (2016) Photocatalytic removal of microbiological consortium and organic matter in greywater. Catalysts 6(6):91
47. Valencia S, Marín J, Velásquez J, Restrepo G, Frimmel FH (2012) Study of pH effects on the evolution of properties of brown-water natural organic matter as revealed by size-exclusion chromatography during photocatalytic degradation. Water Res 46:1198–1206

48. Directive 2000/60/EC of the European Parliament and of the Council of 23 October 2000 establishing a framework for community action in the field of water policy. Off J Eur Commun 22.12.2000, L327/1-L327/69
49. Meffe R, de Bustamante I (2014) Emerging organic contaminants in surface water and groundwater: a first overview of the situation in Italy. Sci Total Environ 481:280–295
50. Székács A, Mörtl M, Darvas B (2015) Monitoring pesticide residues in surface and ground water in Hungary: surveys in 1990-2015. J Chem. Article ID717948
51. Jurado A, Vàzquez-Suñé E, Carrera J, López de Alda M, Pujades E, Barceló D (2012) Emerging organic contaminants in groundwater in Spain: a review of sources, recent occurrence and fate in a European context. Sci Total Environ 440:82–94
52. Stuart M, Lapworth D, Crane E, Hart A (2012) Review of risk from potential emerging contaminants in UK groundwater. Sci Total Environ 416:1–21
53. Pineda Arellano CA, González AJ, Martínez SS, Salgado-Tránsito I, Franco CP (2013) Enhanced mineralization of atrazine by means of photodegradation processes using solar energy at pilot plant scale. J Photoch Photobio A 272:21–27
54. Rozas O, Vidal C, Baeza C, Jardim WF, Rossner A, Mansilla HD (2016) Organic micropollutants (OMPs) in natural waters: oxidation by UV/H_2O_2 treatment and toxicity assessment. Water Res 98:109–118
55. Chu W, Chan KH, Graham NJD (2006) Enhancement of ozone oxidation and its associated processes in the presence of surfactant: degradation of atrazine. Chemosphere 64:931–936
56. Gonzalez MC, Braun AM, Prevot AB, Pelizzetti E (1994) Vacuum-ultraviolet (VUV) photolysis of water: mineralization of atrazine. Chemosphere 28:2121–2127
57. Katsumata H, Kaneco S, Suzuki T, Ohta K, Yobiko Y (2006) Photo-Fenton degradation of alachlor in the presence of citrate solution. J Photoch Photobio A 180:38–45
58. Beltrán FJ, González M, Rivas FJ, Acedo B (2000) Determination of kinetic parameters of ozone during oxidations of alachlor in water. Water Environ Res 72:689–697
59. Ryu CS, Kim M-S, Kim B-W (2003) Photodegradation of alachlor with the TiO_2 film immobilised on the glass tube in aqueous solution. Chemosphere 53:765–771
60. Peñuela GA, Barceló D (1996) Comparative degradation kinetics of alachlor in water by photocatalysis with $FeCl_3$, TiO_2 and photolysis, studied by solid-phase disk extraction followed by gas chromatographic techniques. J Chromatogr A 754:187–195
61. Haque MM, Muneer M (2003) Heterogeneous photocatalysed degradation of a herbicide derivative, isoproturon in aqueous suspension of titanium dioxide. J Environ Manag 69:169–176
62. Bobu MM, Siminiceanu I, Lundanes E (2005) Photodegradation of Isoproturon in water by several advanced oxidation processes. Chem Bull "Politehnica" Univ 50(64):45–48
63. Manassero A, Passalia C, Negro AC, Cassano AE, Zalazar CS (2010) Glyphosate degradation in water employing the H2O2/UVC process. Water Res 44:3875–3882
64. Assalin MR, De Moraes SG, Queiroz SCN, Ferracini VL, Duran N (2009) Studies on degradation of glyphosate by several oxidative chemical processes: ozonation, photolysis and heterogeneous photocatalysis. J Environ Sci Health B 45:89–94
65. Sui Q, Cao X, Lu S, Zhao W, Qiu Z, Yu G (2015) Occurrence, sources and fate of pharmaceuticals and personal care products in the groundwater: a review. Emerg Contam 1:14–24
66. Benotti MJ, Trenholm RA, Vanderford BJ, Holady JC, Stanford BD, Snyder SA (2009) Pharmaceuticals and endocrine disrupting compounds in U.S. drinking water. Environ Sci Technol 43:597–603
67. Bredhult C, Bäcklin B-M, Olovsson M (2007) Effects of some endocrine disruptors on the proliferation and viability of human endometrial endothelial cells in vitro. Reprod Toxicol 23:550–559
68. Verlicchi P, Al Aukidy M, Zambello E (2012) Occurrence of pharmaceutical compounds in urban wastewater: removal, mass load and environmental risk after a secondary treatment: a review. Sci Total Environ 429:123–155
69. ISO 6341:2012 (2012) Water quality. Determination of the inhibition of the mobility of *Daphnia Magna* straus (*Cladocera, Crustacea*) – acute toxicity test. International Organization for Standardization

70. ISO 1134-3:2007 (2007) Water quality. Determination of the inhibitory effect of water samples on the light emission of *Vibrio fischeri* (luminescent bacteria test) – part 3: method using freeze-dried bacteria. International Organization for Standardization
71. Greene JC, Bartels CL, Warren-Hicks WJ, Parkhurst BR, Linder GL, Peterson SA, Miller WE (1988) EPA 600/3-88/029. Protocol for short term toxicity screening of hazardous waste sites. US Environmental Protection Agency
72. ISO 7346-1:1996 (1996) Water quality. Determination of the acute lethal toxicity of substances to a freshwater fish [*Brachydanio rerio* Hamilton-Buchanan (Teleostei, Cyprinidae)] – part 1: static method. International Organization for Standardization
73. Brienza M, Mahdi Ahmed M, Escande A, Plantard G, Scrano L, Chiron S, Bufo SA, Goetz V (2016) Use of solar advanced oxidation processes for wastewater treatment: follow-up on degradation products, acute toxicity, genotoxicity and estrogenicity. Chemosphere 148:473–480
74. Pereira VJ, Weinberg HS, Linden KG, Singer PC (2007) UV degradation kinetics and modeling of pharmaceutical compounds in laboratory grade and surface water via direct and indirect photolysis at 254 nm. Environ Sci Technol 41:1682–1688
75. Vogna D, Marotta R, Andreozzi R, Napolitano A, D'Ischia M (2004) Kinetic and chemical assessment of the UV/H_2O_2 treatment of antiepileptic drug carbamazepine. Chemosphere 54:497–505
76. Doll TE, Frimmel FH (2005) Photocatalytic degradation of carbamazepine, clofibric acid and iomeprol with P25 and Hombikat UV100 in the presence of natural organic matter (NOM) and other organic water constituents. Water Res 39:403–411
77. Kim I, Yamashita N, Tanaka H (2009) Performance of UV and UV/H_2O_2 processes for the removal of pharmaceuticals detected in secondary effluent of a sewage treatment plant in Japan. J Hazard Mater 166:1134–1140
78. Kim I, Tanaka H (2009) Photodegradation characteristics of PPCPs in water with UV treatment. Environ Int 35:793–802
79. Buser HR, Poiger T, Müller MD (1998) Occurrence and fate of the pharmaceutical drug diclofenac in surface waters: rapid photodegradation in a lake. Environ Sci Technol 32:3449–3456
80. Pérez-Estrada LA, Maldonado MI, Gernjak W, Agüera A, Fernández-Alba AR, Ballesteros MM, Malato S (2005) Decomposition of diclofenac by solar driven photocatalysis at pilot plant scale. Catal Today 101:219–226
81. Moreira NFF, Orge CA, Ribeiro AR, Faria JL, Nunes OC, Pereira MFR, Silva AMT (2015) Fast mineralization and detoxification of amoxicillin and diclofenac by photocatalytic ozonation and application to an urban wastewater. Water Res 87:87–96
82. Hu L, Flanders PM, Miller PL, Strathmann TJ (2007) Oxidation of sulfamethoxazole and related antimicrobial agents by TiO_2 photocatalysis. Water Res 41:2612–2626
83. Trovó AG, Nogueira RFP, Agüera A, Fernandez-Alba AR, Sirtori C, Malato S (2009) Degradation of sulfamethoxazole in water by solar photo-Fenton. Chemical and toxicological evaluation. Water Res 43:3922–3931
84. Huber MM, Canonica S, Park GY, Von Gunten U (2003) Oxidation of pharmaceuticals during ozonation and advanced oxidation processes. Environ Sci Technol 37:1016–1024
85. Heringa MB, Harmsen DJH, Beerendonck EF, Reus AA, Krul CAM, Metz DH, Ijpelaar GF (2011) Formation and removal of genotoxic activity during UV/H_2O_2-GAC treatment of drinking water. Water Res 45:366–374
86. Marugán J, Bru D, Pablos C, Catalá M (2012) Comparative evaluation of acute toxicity by Vibrio fischeri and fern spore based bioassays in the follow-up of toxic chemicals degradation by photocatalysis. J Hazard Mater 213–214:117–122
87. Lelario F, Brienza M, Bufo SA, Scrano L (2016) Effectiveness of different advanced oxidation processes (AOPs) on the abatement of the model compound mepanipyrim in water. J Photoch Photobio A 321:187–201
88. Lapertot M, Ebrahimi S, Dazio S, Rubinelli A, Pulgarin C (2007) Photo-Fenton and biological integrated process for degradation of a mixture of pesticides. J Photoch Photobio A 186:34–40

89. Gómez MJ, Sirtori C, Mezcua M, Fernández-Alba AR, Agüera A (2008) Photodegradation study of three dipyrone metabolites in various water systems: identification and toxicity of their photodegradation products. Water Res 42:2698–2706
90. Sichel C, Garcia C, Andre K (2011) Feasibility studies: UV/chlorine advanced oxidation treatment for the removal of emerging contaminants. Water Res 45:6371–6380
91. Bolton JR, Bircher KG, Tumas W, Tolman C (2001) Figures-of-merit for the technical development and application of advanced oxidation technologies for both electric-and solar-driven systems. Pure Appl Chem 73:627–637
92. Arslan-Altan L (2004) Advanced oxidation of textile industry dyes. In: Parsons S (ed) Advanced oxidation processes for water and wastewater treatment. IWA Publishing, London, pp 302–328
93. Mehrjouei M, Mülle S, Möller D (2014) Energy consumption of three different advanced oxidation methods for water treatment: a cost-effectiveness study. J Clean Prod 65:178–183
94. Asaithambi P, Saravanathamizhan R, Matheswaran M (2015) Comparison of treatment and energy efficiency of advanced oxidation processes for the distillery wastewater. Int J Environ Sci Technol 12:2213–2220
95. Mahamuni NN, Adewuyi YG (2010) Ultrasonics sonochemistry advanced oxidation processes (AOPs) involving ultrasound for waste water treatment: a review with emphasis on cost estimation. Ultrason Sonochem 17:990–1003

Homogeneous Fenton and Photo-Fenton Disinfection of Surface and Groundwater

María Inmaculada Polo-López, Samira Nahim-Granados, and Pilar Fernández-Ibáñez

Abstract Polluted surface water and groundwater represent a significant human health risk as it is a vehicle for a number of diseases derived from the exposition to untreated drinking water. Historically, chlorination became a great advance on reducing the impact of many pathogens associated with polluted drinking water in developed countries, with the consequent benefits to societies growing and welfare. Nevertheless, other treatments have been investigated during the last decades with the aim of increasing their capability for treating water and overcoming the limitations of chlorination and other conventional technologies including UVC radiation and ozone. Fenton and photo-Fenton process have been demonstrated to be a good option as alternative water disinfection process during the last years. The aim of this chapter is to briefly describe the fundamentals of this process with special focus on particular aspects related to pathogens inactivation in water. Moreover, the most recent scientific contributions on the application of Fenton and photo-Fenton for water disinfection are discussed.

Keywords Compound parabolic collector, Fenton, Photo-Fenton, Solar treatment, Water disinfection

Contents

1 Background of Fenton and Photo-Fenton Reaction 156
2 Fundamental Aspects and Parameters 157
   2.1 Effect of Water pH 157

M.I. Polo-López (✉) and S. Nahim-Granados
Plataforma Solar de Almería – CIEMAT, Almería, Spain
e-mail: mpolo@psa.es; snahim@psa.es

P. Fernández-Ibáñez
Nanotechnology and Integrated BioEngineering Centre, School of Engineering, University of Ulster, Newtownabbey, Northern Ireland
e-mail: p.fernandez@ulster.ac.uk

2.2	Effect of Inorganic Compounds	158
2.3	Effect of Natural Organic Matter (NOM)	159
3	Microbial Contamination of Groundwater and Surface Water	159
4	Inactivation Mechanisms of Fenton and Photo-Fenton Process	161
5	Fenton Process: Examples of Application	163
6	Solar Driven Photo-Fenton Process: Examples of Application	165
7	Concluding Remarks	173
References		174

Abbreviations

AOP	Advanced oxidation process
ARB	Antibiotic-resistant bacteria
ARG	Antibiotic-resistant gene
CPC	Compound parabolic collector
DOM	Dissolved organic matter
MWWE	Municipal wastewater effluent
NOM	Natural organic matter
PS	Persulfate
ROS	Reactive oxygen species
SE	Secondary effluents

1 Background of Fenton and Photo-Fenton Reaction

Currently, photo-Fenton process constitutes one of the most powerful water treatments belonging to the so-called advanced oxidation processes (AOPs). It takes the advantage of using solar light as source of photons and generates a high rate of hydroxyl radicals (HO$^{\bullet}$). Historically, the first report on Fenton reactions was done by Henry J. Fenton, where iron(II) salt activation by H_2O_2 could oxidize tartaric acid [1]. Later, in 1934, the hydroxyl radical, the second most powerful oxidant ($E° = 2.73$ V) after fluoride, was proposed as the main responsible for the oxidative capacity of the Fenton reaction [2]. During the 1950s, a number of reactions that today still describe the classical thermal Fenton reactions were reported [3–5]. The main reactions involving the decomposition of H_2O_2 in dark and pure acid solutions are:

$$Fe^{2+} + H_2O_2 \rightarrow Fe^{3+} + OH^- + HO^{\bullet} \ (K \approx 70 \ M^{-1}s^{-1}) \tag{1}$$

$$Fe^{3+} + H_2O_2 \rightarrow Fe^{2+} + HO_2^{\bullet} + H^+ \ (K = 1 - 2 \times 10^{-2} \ M^{-1}s^{-1}) \tag{2}$$

$$HO^{\bullet} + H_2O_2 \rightarrow HO_2^{\bullet} + H_2O \ (K = 1.7 - 2.5 \times 10^{-7} \ M^{-1}s^{-1}) \tag{3}$$

$$HO^{\bullet} + Fe^{2+} \rightarrow Fe^{3+} + OH^- \ (K = 3.2 \times 10^8 \ M^{-1}s^{-1}) \tag{4}$$

$$\text{Fe}^{3+} + \text{HO}_2^{\bullet} \rightarrow \text{Fe}^{2+} + \text{O}_2 + \text{H}^+ \ (K = 1.2 \times 10^6 \ \text{M}^{-1}\text{s}^{-1}) \ \text{pH 3} \quad (5)$$
$$\text{Fe}^{2+} + \text{HO}_2^{\bullet} + \text{H}^+ \rightarrow \text{Fe}^{3+} + \text{H}_2\text{O}_2 \ (K = 1.3 \times 10^6 \ \text{M}^{-1}\text{s}^{-1}) \ \text{pH 3} \quad (6)$$
$$\text{HO}_2^{\bullet} + \text{HO}_2^{\bullet} \rightarrow \text{H}_2\text{O}_2 \quad (7)$$

Reaction 2 is extremely slow as compared with Reaction 1, which limits the whole process of HO$^{\bullet}$ generation. Later, in the 1990s, the strong acceleration of this reaction in the presence of radiation started to be investigated as novel process for water treatment [6]. This photo-assisted Fenton reaction, so-called photo-Fenton, leads to much faster HO$^{\bullet}$ generation rates and therefore a higher degree of oxidation than the Fenton (dark) reaction. Briefly, the efficiency of the process increases when the system is irradiated at wavelengths below 580 nm due to the light absorption of the Fe^{3+} aquo complexes generated by Reaction 2 to be reduced to Fe^{2+} complexes (Reaction 8), generating an extra HO$^{\bullet}$ and enabling the iron cycle to restart [7].

$$\text{Fe}^{3+}(\text{L})_n + h\nu \rightarrow \text{Fe}^{2+}(\text{L})_{n-1} + \text{L}_{\text{ox}}^{\bullet} \quad (8)$$

Since the first application of the Fenton reaction as oxidative process to degrade toxic organic compounds in water in the middle of the 1960s [8], the interest of researchers on the fundamental knowledge and practical application of Fenton and photo-Fenton process to treat polluted water has increased tremendously. Nevertheless, the application of this process to disinfect water, i.e., to inactivate water-borne pathogens, was initiated in the last decade. The first demonstration of the capacity of photo-Fenton to disinfect water was reported by Rincón and Pulgarín in 2006. These authors demonstrated that the use of low reagent concentrations (0.3 mg L^{-1} Fe and 10 mg L^{-1} H$_2$O$_2$) enhanced markedly the inactivation kinetics of *E. coli* cells in water [9]. Since then, other pathogens, chemical and biological parameters related to the iron chemistry, and the efficiency of the treatment in several types of surface waters have been investigated.

2 Fundamental Aspects and Parameters

2.1 Effect of Water pH

It is very well known that pH affects strongly the speciation and therefore the solubility of iron in water, establishing the optimum for Fenton or photo-Fenton reaction a pH $\leq$ 3. The three fractions of iron(II) species found in water as a function of the water pH are Fe(H$_2$O)$_6^{2+}$ or Fe^{2+}, which appears as predominant at acid conditions; Fe(H$_2$O)$_4$(OH)$_2$ or Fe(OH)$_2$ at alkaline conditions; and Fe(H$_2$O)$_5$(OH)$^+$ or FeOH$^+$ coexisting at pH lower than 5 [7].

However, the speciation of iron(III) differs from the iron(II), and in strongly acidic solution, the main specie is the hexaaqua ion, Fe(H$_2$O)$_6^{3+}$ [7]. When pH increases, the hydrolysis of Fe(H$_2$O)$_6^{3+}$ occurs depending on several water parameters including the ionic strength and the total iron concentration. Consequently, this ion precipitates as amorphous ferric oxyhydroxides [10], which are considerably less active. Therefore, the reaction of ferrous ion and H_2O_2 will lead to the precipitation of iron when Fe(III) oxyhydroxides are formed (Reaction 1) when pH is higher than 3.

Recently, the research trend is to investigate the naturally occurring or artificially prepared iron oxides (like magnetite, goethite, hematite, etc.) formed after the oxidation of ferric oxyhydroxides at near-neutral pH; quelates of iron such as citrate, EDDS, etc. are also under investigation. These Fe complexes are investigated from the point of view of their capability for promoting efficient Fenton or photo-Fenton reactions for water and wastewater treatment at near-neutral pH.

2.2 Effect of Inorganic Compounds

Iron chemistry and Fenton reactivity are negatively affected by the presence of several inorganic chemical compounds in water. They mainly produce precipitation of iron, scavenging of radicals, and/or less reactive Fe complex formation. One of the most negative effects is produced by the presence of carbonates (CO_3^{2-}) and bicarbonates (HCO_3^-). These anions may react with $HO^\bullet$, scavenging it and ending in a reduction of their availability to oxidize the target pollutants [7, 11, 12]. The reactions involved include:

$$HO^\bullet + HCO_3^- \rightarrow CO_3^{-\bullet} + H_2O \qquad (9)$$
$$HO^\bullet + CO_3^{2-} \rightarrow CO_3^{-\bullet} + H_2O \qquad (10)$$

Therefore, for the treatment of any kind of water by Fenton or photo-Fenton, the oxidation of pollutants in presence of CO_3^{2-} will be in direct competition with the oxidation by $HO^\bullet$, and if $CO_3^{-\bullet}$ formation rate exceeds the hydroxyl radical formation, the balance of the process will be negative from the point of view of capacity to treat contaminated waters [12]. On the other hand, $CO_3^{-\bullet}$ is also a radical with a wide range of reactivity with organic molecules, but with slower reactions than those of $HO^\bullet$, while the main known scavengers of both ($CO_3^{-\bullet}$ and $HO^\bullet$) are the dissolved natural organic matter and, if present, H_2O_2. In addition and regarding water disinfection, it has been also described by some authors that HCO_3^- may protect microorganisms from sunlight as it is photo absorptive and therefore acts as a screen of light [13, 14].

On the other hand, phosphate ions have been shown to have an important detrimental role during Fenton reactions because they react with iron and coprecipitate as stable complex at neutral and acidic conditions as well as they scavenge hydroxyl radicals [7, 15]. Other anions present in natural waters, including sulfates, nitrates, and chlorides, may also react with iron, H_2O_2, or other ROS, limiting the capability of Fenton or photo-Fenton to oxidize organic chemical and biological pollutants due to the reduction of the $HO^\bullet$ generated [11, 14–16].

Table 1 Main oxidative reactions occurring during photo-Fenton in the presence of NOM

Pathway	Reactions	Reference (equation)
Classical photo-Fenton reactions (main reactions)	$Fe^{2+} + H_2O_2 \rightarrow Fe^{3+} + OH^- + HO^{\bullet}$	(1)
	$Fe^{3+}(L)_n + h\nu \rightarrow Fe^{2+}(L)_{n-1} + L^{\bullet}_{ox}$	(8)
Direct oxidation of DOM by sunlight and $^3DOM^*$ energy transfer	$DOM + h\nu ^1DOM^3DOM^*$	(11)
	$^3DOM^* + O_2 \rightarrow DOM + {}^1O_2$	(12)
Complexation of iron with DOM	$Fe^{3+} - (DOM)_n + h\nu Fe^{2+} - (DOM)_{(n-1)} + DOM^+_{ox}$	(13)
DOM reaction with O_2	$DOM + O_2 + h\nu \rightarrow DOM^+_{ox} + O_2^{\bullet -}$	(14)

2.3 Effect of Natural Organic Matter (NOM)

In contrast to inorganic compounds, the presence of NOM in water has been reported to have both a positive and negative effect during Fenton or photo-Fenton reaction. The natural organic compounds present in any water source include humic and non-humic substances. In general, humic composition is the fraction including aromatic and aliphatic compounds, carboxylic, phenolic-OH, amino and quinone groups, etc. However, the humic composition that may differ significantly from one to another water sources due to their composition is the result of many random environmental transformations (microbiological, chemical, and photochemical reactions) of biological rests [14, 15, 17]. Therefore, and due to the high variability of NOM in water sources, the presence of humic substances may be a limiting factor [18] or not [19–21] for the Fenton processes. Briefly, the most significant oxidative reactions of NOM during photo-Fenton are summarized in Table 1.

On the other hand, NOM may react with free radicals, including $HO^{\bullet}$ (Reaction 15). This reaction was reported as a second-order rate kinetics with a kinetic constant value of $\sim 2 \times 10^4 \text{ s}^{-1} \text{ (mgC/L)}^{-1}$ [22].

$$DOM + HO^{\bullet} \rightarrow DOM^+_{ox} + OH^- \quad (15)$$

In summary, the oxidative capability of photo-Fenton in the presence of NOM will depend on several reasons including the iron complexation generated in the specific water source, the iron (Fe^{2+}/Fe^{3+}) transformation, and the types of oxidative radicals generated during the process [23].

3 Microbial Contamination of Groundwater and Surface Water

Traditionally, groundwater has been considered a microbe-free water source, especially in deep and confined aquifers. Nevertheless, the presence of indicator bacteria or human pathogens has been widely detected in the last years in many groundwater sources. Their pollution has been attributed to a nearby connection with contaminated surface waters and environment, including a leaking from a waste lagoon, septic tanks, sewer line, old or improperly designed landfill, etc. Obviously, surface water shows higher microbial contamination load than groundwater, like rivers, lagoon, dams, and bacterial "hot spots" like municipal wastewater treatment plants.

From a realistic point of view, it is impossible to monitor every possible pathogen present in a contaminated surface water source due to each type of pathogen requires a specific test, which is time-consuming and would be prohibitive in terms of cost. Therefore, microbial indicators of water quality have been established as model of pathogens for water guidelines and regulations, and therefore they have been also studied for research purposes. They include *E. coli* (the most widely investigated bacterium in water) as an indicator of fecal contamination or some specific phages (MS2) as indicator of the presence of viral contamination. However, the wide spectrum of naturally occurring microorganisms in surface waters is not well represented by these strains, as many other groups of pathogenic species may be present, including bacteria, viruses, fungi, protozoa, and parasites. Moreover, *E. coli* shows the lowest resistance to be inactivated by any disinfection treatment [24], including photo-Fenton as it is clearly observed in Fig. 1.

Today the advanced analytical methods in molecular biology include genetic tools as quantitative PCR (Polymerase Chain Reaction) and metagenomics analysis are transforming this area of research. However, culture-based methodologies are still being the most used technique for monitoring and quantification of

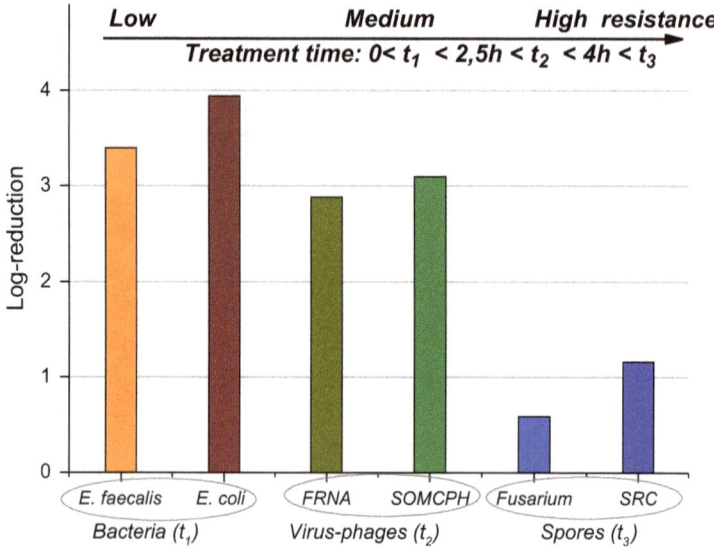

Fig. 1 Resistance of several microbial groups of bacteria, viruses, and spores of fungi and bacteria using solar photo-Fenton (natural pH, 10–20 mg L^{-1} of Fe^{2+} and H$_2$O$_2$) in secondary effluents (Reprinted from [15], with permission from Elsevier)

Table 2 Summary of priority pathogen list reported by the WHO (Publication date: 27 February 2017) (http://www.who.int/medicines/publications/global-priority-list-antibiotic-resistant-bacteria/en/)

Microorganism	Antibiotic resistance	Priority
Acinetobacter baumannii	Carbapenem	Critical[a]
Pseudomonas aeruginosa	Carbapenem	Critical[a]
Enterobacteriaceae[b]	Carbapenem and third-generation cephalosporin	Critical[a]
Enterococcus faecium	Vancomycin	High
Staphylococcus aureus	Methicillin, vancomycin intermediate and resistant	High
Helicobacter pylori	Clarithromycin	High
Campylobacter	Fluoroquinolone	High
Salmonella spp.	Fluoroquinolone	High
Neisseria gonorrhoeae	Third-generation cephalosporin, fluoroquinolone	High
Streptococcus pneumoniae	Penicillin-non-susceptible	Medium
Haemophilus influenzae	Ampicillin	Medium
Shigella spp.	Fluoroquinolone	Medium

[a]Mycobacteria (including *Mycobacterium tuberculosis*) was not included as it is already a globally established priority for innovative new treatments
[b]*Klebsiella pneumonia*, *Escherichia coli*, *Enterobacter* spp., *Serratia* spp., *Proteus* spp., *Providencia* spp., and *Morganella* spp.

microbiological contamination of water sources [25]. The use of qPCR for microbial monitoring in wastewater and groundwater sources has made possible to detect the presence of the so-called antibiotic-resistant bacteria (ARB), which currently is becoming one of the most important concerns for public health. In relation to water issues, ARB are an environmental water health issue associated with municipal wastewater treatment plants and hospital wastewater [26]. Their spread in the aquatic environment favors also the spread of genetic resistance material that can be introduced in food chains resulting in a reduction of the antibiotics' efficacy against animal and human diseases. To date, methicillin-resistant *Staphylococcus aureus* and vancomycin-resistant *Enterococcus* spp. are the most investigated ARB [27]. Very recently (Feb. 2017), the World Health Organization (WHO) reported a list of "priority pathogens list for R&D of new antibiotics" highlighting the importance of this type of pathogens on research and the implementation of new water treatments (Table 2).

4 Inactivation Mechanisms of Fenton and Photo-Fenton Process

The inactivation of any microbial agent means the loss of their growth capability, and the objective of any water disinfection process is to effectively inactivate waterborne pathogens. To obtain a successful inactivation, the chemical or physical agent applied must generate injuries in key biological structures (mainly proteins, lipids or DNA) that leads to functionality losses ending in the cells' inactivation or death. In Fenton and photo-Fenton process, as it has been explained before, the generation of reactive oxygen species (ROS), and mainly hydroxyl radicals, is considered the main responsible inactivation agents. ROS may react with proteins and lipids at cell membrane site altering their functionality directly as well as initiating the lipid peroxidation chains inside the cells increasing cell permeability and leading to final inactivation [28, 29]. On the other hand, ROS may also react with internal cells' components, including DNA. In this case, the attack occurs at the DNA bases or sugars leading to strand breakage and base release, generating mutations and other genetic alterations [30].

The proposed mechanisms to explain the inactivation of microorganisms by ROS and HO• generation during Fenton and photo-Fenton process are shown in Fig. 2. It is accepted that generated HO• (via Reactions 1–8) attacks mainly the external cell membrane leading its degradation and resulting in microorganism death. This degradation will depend on the reaction rate of HO• generation, and considering the short living of this radical, the attacks will occur only locally, at cell membrane level, and therefore, diffusion to inside cells cannot be considered. However, other reactions may take place at internal cell level which may also be considered as source of oxidative damage, especially in the case of photo-Fenton driven at near-neutral pH, where the reaction rate of external HO• generation is reduced compared with those occurring at acid solutions [14]. Briefly, the main reactions that may occur inside cells are summarized as follows:

1. Near UVA light may inactivate enzymes, including catalase (CAT) or super oxygen dismutase (SOD). Their inactivation will lead to increased intracellular

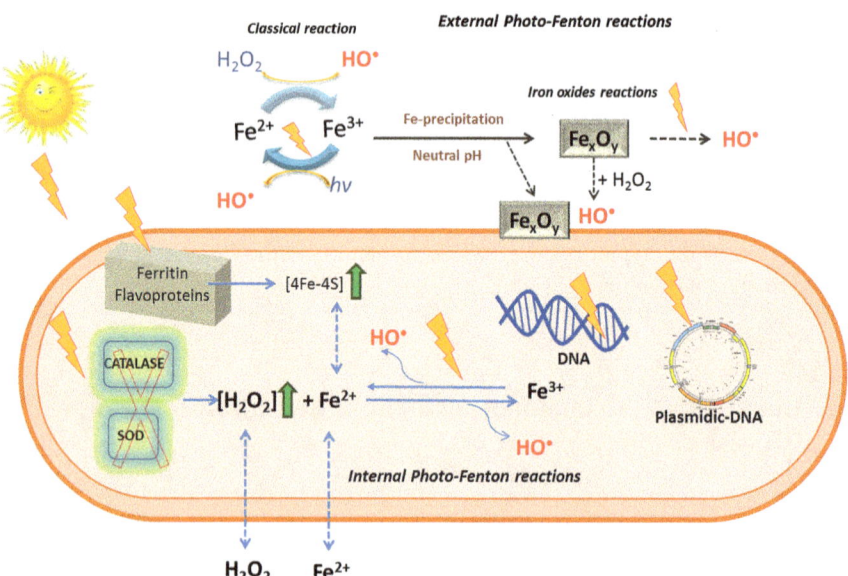

Fig. 2 Summary of the main reactions pathways involved during bacterial inactivation by photo-Fenton

concentrations of H_2O_2 and $O_2^{\bullet-}$, as they are responsible for the elimination of these metabolic generated ROS [31]. Moreover, other protein like ferritin may be also damaged, which will determine an increase release of Fe^{2+} [31].

2. Some intracellular photosensitizers may absorb photons in the UVA and visible spectrum and attack biomolecules or react with O_2 generating ROS such as $HO^{\bullet}$, H_2O_2, $^1O_2$, and $O_2^{\bullet-}$. Another source of intracellular ROS increment can be generated by the oxidation of flavoproteins ($FADH_2$). As a consequence, intracellular iron-sulfur clusters [4Fe-4S] may be oxidized leading other proteins, release of Fe^{2+} and generation of H_2O_2.

3. Fenton and photo-Fenton intracellular reactions by diffusion of added H_2O_2 and Fe^{2+}. The diffusion may occur as H_2O_2 is relatively stable and uncharged (unlike $HO^{\bullet}$ and $O_2^{\bullet-}$) and Fe^{2+} may freely diffuse favored by osmotic forces into the cells.

4. Fe^{3+} (with a higher charge density than Fe^{2+}) can be adsorbed on specific membrane-binding proteins, leading the deposition of iron on the external cell membrane and the formation of Fe^{3+}-bacteria exciplexes. Therefore, this means a possible way of direct oxidation over the cell membrane (proteins and lipids) by $HO^{\bullet}$ generation, and the subsequent regeneration of Fe^{2+}.

5. Iron oxides including magnetite, goethite, lepidocrocite, or feroxyhyte can be generated after precipitation of iron at near-neutral pH as oxyhydroxides and further oxidation [32]. Iron oxides can be adsorbed onto bacterial surfaces, and if H_2O_2 and light are present, they can act as semiconductors and/or heterogeneous catalysts.

5 Fenton Process: Examples of Application

The application of "classical" Fenton process for water disinfection has not been widely extended or intended for real applications due to several practical reasons:

1. The limitations of $HO^{\bullet}$ generation by the low kinetic rate of Reaction 2; meanwhile, the opposite trend occurs in the case of classical photo-Fenton, as the irradiation of the system strongly increase the oxidative capacity of the process. This is clearly observed in many works in the literature where the efficiency of Fenton and photo-Fenton applications for water treatment is compared, for example, the inactivation of *F. solani* spores using solar photo-Fenton (Fig. 3).

2. The higher amount of iron is needed to obtain good disinfection rates, considering that reagents are one of the most important costs for real or industrial application. In line with this, the development of modified Fenton reagents to increase the oxidative capacity of the system will be also a disadvantage as advanced and costly techniques are required to obtain doped or modified iron oxides. Nevertheless, several recent contributions on this area have been reported in literature. A recent study on modified Fenton system including

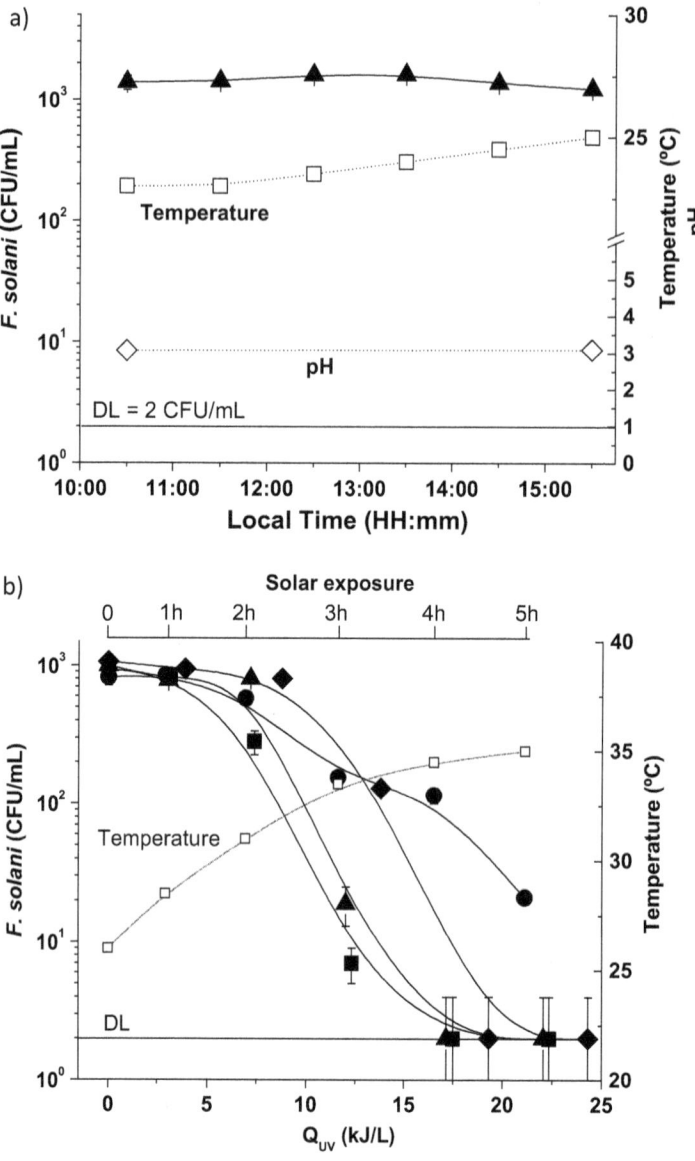

Fig. 3 *F. solani* microconidia concentrations in simulated municipal wastewater effluents (SMWWE) for 5 h dark Fenton (at pH 3) (**a**) and solar photo-Fenton (at pH 3 (filled triangle), pH 4 (filled square), pH 5 (filled diamond), and pH 8 (filled circle) as function of solar cumulative irradiance (Q_{UV})) (**b**) with 5 mg L^{-1} Fe^{2+} and 10 mg L^{-1} H$_2$O$_2$. Temperature and pH are shown with dashed lines (Reprinted from [33], with permission from Elsevier)

ascorbic acid and prooxidants for the inactivation of *Cryptosporidium*-contaminated drinking water samples has been published. In this work, the higher efficiency on *Cryptosporidium* viability decrease (0% after 5 min of exposure time) was obtained with 80 mg L^{-1} of iron(II), 30 mg L^{-1} of ascorbic acid, 30 mmol of H_2O_2, and 25 mg L^{-1} of prooxidants [34]. Castro-Rios et al. investigated the capability of heterogeneous Fenton, using pillared clay prepared by ultrasound and microwave in a semi-batch reactor for the inactivation of total coliforms and *E. coli* in natural water from Pasto River [35]. Relative good inactivation efficiencies were obtained in this work, between 42 and 75% in 180–240 min of treatment and where best operational conditions found were pH 3.7, 0.5 g L^{-1} of catalyst load, and 0.12–0.18 mg L^{-1} of H_2O_2 concentration [35]. The inactivation of *Ascaris lumbricoides* eggs through a Fenton reaction has been also investigated recently. In this study, maximum degradation (91.2%) was achieved with 500 mg L^{-1} dose of H_2O_2 and a 3:1 (H_2O_2/Fe^{2+}) molar ratio at pH 6 [36]. The inactivation of *Ascaris* eggs with FeOx/C nanocatalyst and H_2O_2 and further modification procedure using iron oxide nanocatalyst supported on activated carbon has been also investigated [37, 38]. In all these works, in spite of the efficiency obtained, the amount of iron or modified iron needed was very high which, from the economic point of view, is a clear disadvantage.

6 Solar Driven Photo-Fenton Process: Examples of Application

One of the driving forces on the progress of solar photo-Fenton process for water disinfection has been their application using the solar Compound Parabolic Collector (CPC) reactors (Fig. 4) [39]. This static system consists on a cylindrical photo-

Fig. 4 View of several solar CPC pilot plants located at Plataforma Solar de Almeria, The Solar Research Centre in Spain

reactor placed on the linear focus of a parabolic reflective surface. This configuration enhanced the collection of photons due to the reflector (mirror) geometry that reflects indirect light onto the photo-reactor (receiver tube). Briefly, the main advantages of this technology are collecting both direct and diffuse solar radiation, the use of non-imaging concentration optics with diffuse focus, complete and homogeneous distribution of photons at the absorber wall, and constant concentration factor (CF: 1) for all values of sun zenith angle within the acceptance angle limit of 90° [39]. The CPC mirror made of anodized aluminum has the advantage of a high reflectivity in the UVA range (87–90%) and 90% for the visible and infrared fraction of the solar spectrum, with high resistance to the environmental conditions. Typically, this reactor is operated by water recirculation; therefore centrifugal pumps are used to permit a flow rate enough to guarantee a turbulent regime that favors the homogenization of the sample.

The application of solar photo-Fenton for wastewater and surface water disinfection using CPC has been deeply investigated during the last years. A summary of those works reporting experimental results of this technology to date are shown in Table 3.

Since the first study reported by Moncayo-Lasso et al. [40] that demonstrated the efficiency of solar photo-Fenton for surface water disinfection, a number of contributions have investigated other aspects related to this process. A great part of the research carried out by Pulgarin's group has been focused on the use of this process at very low levels of iron, H_2O_2, and near-neutral pH with the aim of disinfecting drinking water. The treatment for drinking purposes requires obviously very mild reaction conditions. In line with this, there are a number of articles investigating the lethal effectiveness of low amounts of H_2O_2 under sunlight for inactivating a variety of pathogens for drinking also at pilot scale [45, 47, 53]. The effectiveness of this process is also attributed to photo-Fenton mechanisms at intracellular level [57, 58]. On the other hand, disinfection of wastewater involves a number of chemical and biological complex reactions which determine the use of photo-Fenton under strongest conditions, i.e., higher concentrations of Fe and H_2O_2, to guarantee the quality of the treated water for disposal or reuse [52].

Recently, significant research has been done to show the most important findings related to the effect of fundamental parameters as described below:

1. *Type of pathogen*: different microbial or waterborne pathogens have been evaluated, ranging from the very well-known *Escherichia coli* to the high-resistant spores of *Clostridium* sp. and human viruses [59]. In 2013, Agulló-Barceló et al. [43] reported an experimental work investigating, among other solar processes, the efficiency of solar photo-Fenton for secondary effluents disinfection focusing on the inactivation of several groups of naturally occurring pathogens. Figure 5 shows the influence of the type of microorganism on the inactivation efficiency at acidic (optimal for photo-Fenton) and neutral pH. These results clearly show that monitoring a single pathogen may not be sufficient to assure a low microbial risk of the water. Moreover, the influence of the pH on the processes was investigated, demonstrating the benefits of acidic

Table 3 Chronology experimental contributions on surface and wastewater disinfection by solar photo-Fenton at acid and near-neutral pH in CPC reactors

Water source	Microbial target	Water volume (L)	$[Fe^{2+}]:[H_2O_2]^a$ (mg L^{-1})	Reference
River	E. coli	20	0.6:10	[40]
SE	Total coliforms	80	5.58-$Fe_2(SO_4)_3^{3+}$ + 71.5-EDDS:50	[41]
SE	Antibiotic resistant (ofloxacin and tri-methoprim) Enterococci sp.	250	5:75	[42]
SE	E. coli, Spores of sulfite reducing clostridia, somatic coliphages, F-specific RNA bacteriophages	10	10:20	[43]
Synthetic SE	E. faecalis	7	20:50	[44]
Well	Total coliforms, E. coli, Salmonella spp.	25	0.01-natural Fe:10; 0.23-Fe oxides: 10	[45]
Real and synthetic SE	E. coli and E. faecalis	20	10:50	[46]
Well	Total coliforms, E. coli, Salmonella spp.	25	0.01-natural Fe:10	[47]
SE	Wild enteric E. coli and total coliform	7	20:50	[48]
SE	Antibiotic resistant (clarithromycin and sulfamethoxazole) Enterococci sp.	60	5.02-Fe^{3+}:160 (pH 4)	[49]
SE and synthetic SE	Fusarium sp. No inactivation at natural pH	60	5:60	[50]
SE	Clostridium sp.	60	10:50	[51]
SE	Multidrug resistant E. coli to ampicillin, ciprofloxacin, and tetracycline	8.5	5.02:10	[52]
Alkaline surface	Wild total coliforms, E. coli, Salmonella spp.	25	0.10-total Fe: 10	[53]
SE	Bacillus sp. and Clostridium sp.	60	10:80 (pH 3)	[54]
SE	Curvularia sp.	60	10:20 (pH 3)	[55]
Groundwater	E. coli and Klebsiella pneumoniae		Natural iron-0.3:10	[56]

Modified from [15] and updated
SE secondary effluents from municipal wastewater treatment plant
[a]Reagent concentrations leading faster inactivation kinetics

conditions against neutral pH. This work also confirmed that solar photo-Fenton at pilot plant scale is capable to strongly reduce the microbial load on secondary effluents achieving the required levels established in different national and international guidelines for wastewater reuse [43]. Recently, the presence of antibiotic-resistant bacteria (ARB) has attracted the interest of research, and the efficiency of Fenton and photo-Fenton for reduction of both ARB and ARG are

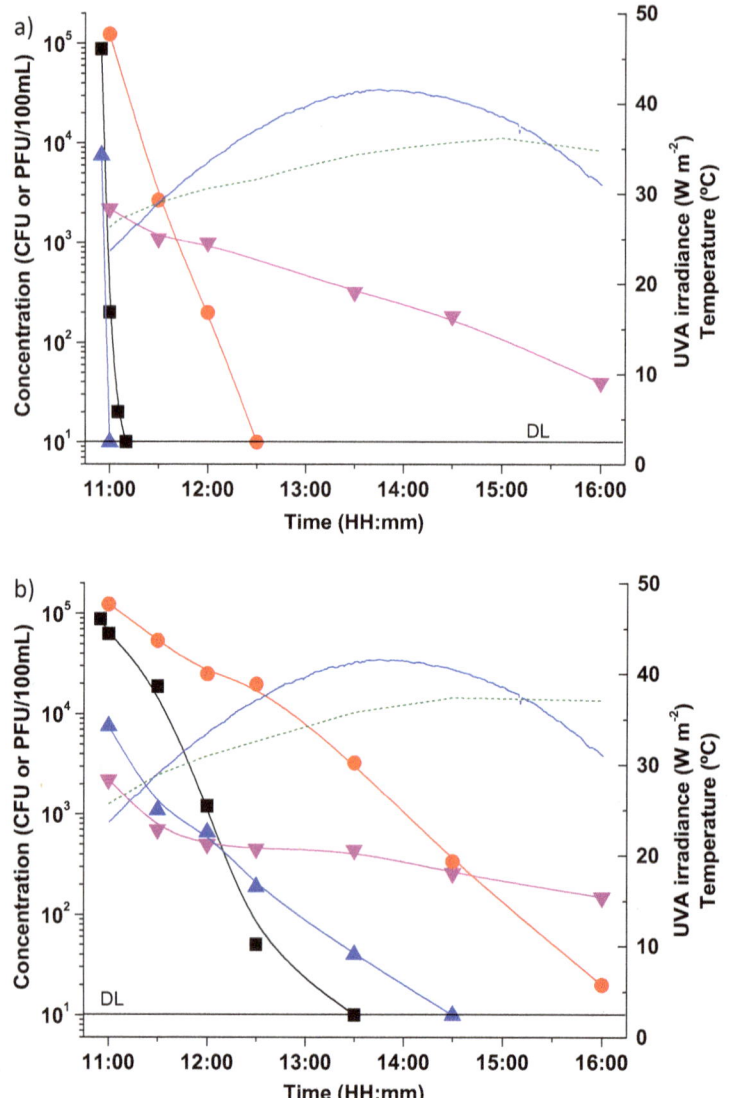

Fig. 5 Inactivation of naturally occurring *E. coli* (filled square); SRC, sulfite reducing clostridia (filled inverted triangle); SOMCPH, somatic coliphages (filled circle); and FRNA, F-specific RNA bacteriophages (filled triangle) in secondary effluents by solar photo-Fenton with 10 mg L^{-1} of Fe^{+2} and 20 mg L^{-1} of H$_2$O$_2$ at pH 3 (**a**) and natural pH (**b**). UVA irradiance (dashed line) and temperature (dashed line) (Reprinted from [43], with permission from Elsevier)

also found in literature [49, 60]. Ferro et al. [52] investigated at pilot scale the capability of solar photo-Fenton on the inactivation of a multidrug resistant *E. coli* to ampicillin, ciprofloxacin, and tetracycline. In this work, performed at pH 4, best inactivation rate was attained with 0.090/0.294 mM of Fe^{2+}/H$_2$O$_2$. In addition, the antibiotic resistance of the survived colonies was analyzed by Kirby-Bauer disk diffusion method. These results demonstrated that the antibiotic resistance did not decrease the ARB quality by this solar process, therefore other processes with a higher oxidative capacity may be found to combat this issue in water [52].

2. *Type of water matrix and influence of natural organic matter*: this is a critical parameter, where the majority of the studies conducted at pilot scale show a negative effect or a reduced effect produced by the presence of NOM on the photo-Fenton efficiency. Ortega-Gómez et al. [61] demonstrated the competitive role of NOM (using resorcinol as a model of organic matter) in clean water for the oxidant species (HO•) against bacteria (*Enterococcus faecalis*) when a photo-oxidation and photo-disinfection processes are occurring simultaneously. These authors confirmed this negative effect at high resorcinol concentrations (20, 30, and 40 mg L^{-1}) and for different photo-Fenton reagent concentrations (Fe^{2+}/H_2O_2: 2.5/5, 5/10 and 10/20 mg L^{-1}). For high concentrations of NOM, the photo-Fenton might be worked at stronger oxidative conditions.

As previously mentioned, secondary effluents from a wastewater treatment plant represent a challenging condition, due to it contains not only a high complexity of ions and organic matter but also the presence of a rich consortium of microorganisms which makes this water harder to purify. Recent contributions have demonstrated the important influence of complex wastewater on the efficiency of photo-Fenton for disinfection using real secondary effluents in a solar pilot reactor [46, 50]. As an example, Fig. 6 shows a comparison between the inactivation of spores of *Curvularia* spp. (a phytopathogenic fungi) by solar photo-Fenton at pilot scale and near-neutral pH in distilled water and in real secondary effluents [55]. The high difference of both graphs clearly demonstrated that under similar operational conditions, the significant effect of the water matrix plays an important role on the efficiency of the process.

In addition, the very well-known effect of hydroxyl radical scavenging by inorganic compounds such as carbonate/bicarbonate has been also investigated in secondary effluents. Ortega-Gómez et al. [48] reported the influence of reducing the amount of bicarbonate from secondary effluents in the inactivation rate of naturally occurring total coliforms and *E. coli* by solar photo-Fenton in CPC reactors (Fig. 7). They concluded that although reducing the concentration of bicarbonate in water may enhance the inactivation rates of bacteria, when an excess of hydroxyl radical is not generated, the negative effect of bicarbonate is attenuated.

3. *Optimization of operational conditions*: process optimization can be done via reagents concentrations and pH, which is very important and desirable to remain close to neutral for either drinking of wastewater reuse. Most of contributions on this demonstrate that the addition of low concentrations of iron and H_2O_2 lead to high log removal of water pathogens (4–6 log reduction value, LRV), although the treatment times to achieve these values are still longer (30 min to few hours) than conventional disinfection methods, chlorination, UVC, and ozone. When the process is run without the interference of other pathogens, NOM, or ions, the

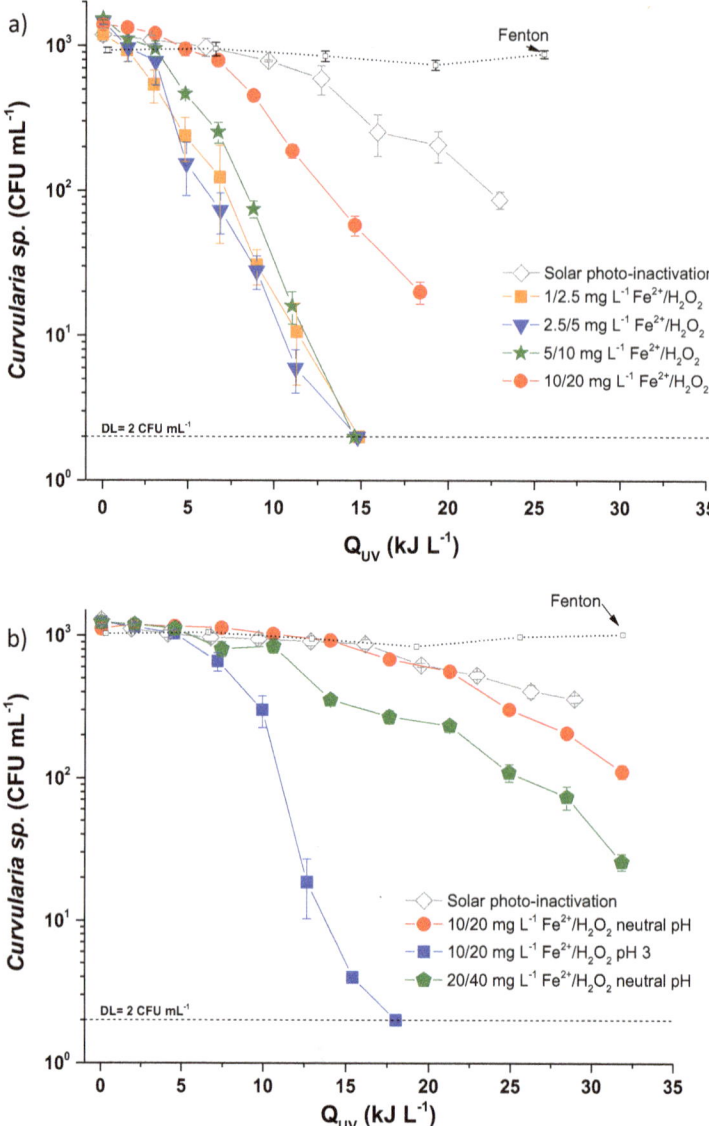

Fig. 6 Inactivation of *Curvularia* sp. by solar photo-Fenton at pilot scale reactor using several reagents' concentration in (**a**) distilled water at near-neutral pH and (**b**) real secondary effluents at acid and near-neutral pH (Reprinted from [55], with permission from Elsevier)

behavior of disinfection at higher reagents concentrations did not show a clear enhancement, as observed in Fig. 6a, where *Curvularia* spp. is inactivated in distilled water [55]. Similar results have been reported for other pathogens [33, 50, 62].

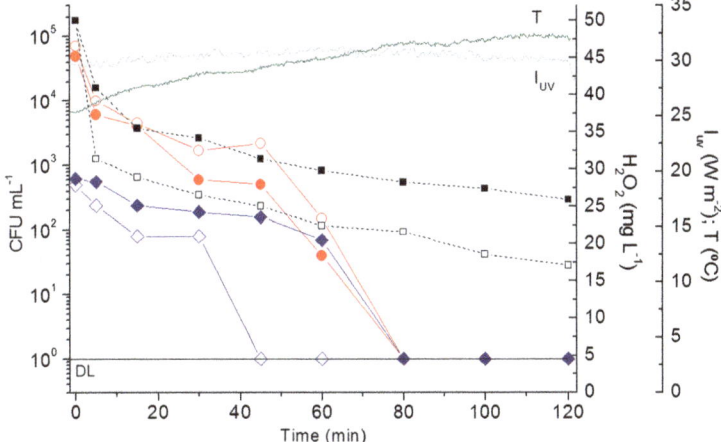

Fig. 7 Influence of bicarbonate concentration on the inactivation of total coliform (open circle) and *E. coli* (open diamond) in secondary effluent by solar photo-Fenton (20/50 mg L^{-1} of Fe^{2+}/H_2O_2). H_2O_2 (open square). Empty symbols, $HCO_3^- = 100 \pm 5$ mg L^{-1}. Full symbols, $HCO_3^- = 250 \pm 12$ mg L^{-1} (Reprinted from [48], with permission from Elsevier)

On the contrary, it has been also reported that when the concentration of iron is high enough (>10 mg L^{-1}), the negative effect of iron precipitates due to light screening makes the disinfection process slower. Rodríguez-Chueca et al. [46] investigated the influence of the presence of precipitated iron in the efficiency of solar photo-Fenton under natural sunlight for the inactivation of *E. coli* and *E. faecalis*. The precipitated formed at pH 5 with 10 mg L^{-1} of iron added to the sample negatively affected the inactivation results as compared with the samples without iron precipitated and only in the presence of the little solved iron (below 0.1 mg L^{-1}); this was especially clear in the case of *E. faecalis* (Fig. 8). In these experiments, the lower efficiency in the presence of iron precipitates was attributed to light screening of these suspended particles which induced higher turbidity in the sample, reducing therefore the overall capability of the oxidative process. Meanwhile, this effect was not so significant in the case of *E. coli*, probably because *E. coli* is a more sensitive bacterium as compared to *E. faecalis*, and the limited oxidative action produced by this process in the presence of precipitated iron still produced enough injuries in this higher sensitive bacteria, leading to complete inactivation without differences between the presence or not of precipitated iron (Fig. 8).

The modification of the classical photo-Fenton reactions is also actively investigated for water disinfection with the aim of increasing the microbial inactivation rate by performing the process at natural water pH. In this sense, the

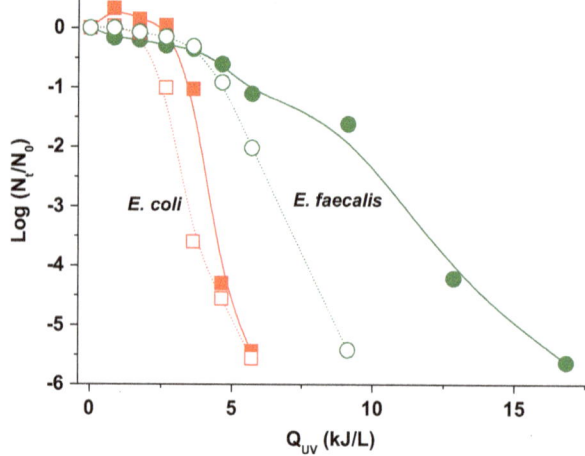

Fig. 8 Comparison of *E. coli* and *E. faecalis* inactivation by solar photo-Fenton with 10/20 mg L^{-1} of Fe^{2+}/H$_2$O$_2$ in the presence (full symbol) and absence (empty symbols) of precipitated iron (Reprinted from [46], with permission from Elsevier)

main modifications on water disinfection by solar photo-Fenton are summarized as follows:

1. *Heterogeneous Fenton-like processes catalyzed by iron (hydr)oxide particles.* Nieto-Juarez and Kohn reported the removal by absorption and inactivation of MS2 coliphage at near-neutral pH, with 200 mg L^{-1} of commercial hematite (α-Fe$_2$O$_3$), goethite (α-FeOOH), magnetite (Fe$_3$O$_4$), and amorphous iron(iii) hydroxide (Fe(OH)$_3$). In general, all particles exhibited a similar virus adsorption capacity and inactivation, although their exposition to sunlight and H$_2$O$_2$ drastically enhanced the inactivation rate [63]. Recently, hematite (α-Fe$_2$O$_3$), goethite (α-FeOOH), wüstite (FeO), and magnetite (Fe$_3$O$_4$) at concentration of 0.6 mg L^{-1} Fe^{3+} with and without addition of 10 mg L^{-1} H$_2$O$_2$ under solar radiation were also investigated for bacterial inactivation at neutral pH. It was demonstrated that all the semiconductor iron (hydr)oxides showed photoactivity under sunlight with the exception of magnetite which was the only one that needs H$_2$O$_2$ as electron acceptor [64].

2. *H$_2$O$_2$ versus persulfate (PS).* To date, sulfate radical (SO$_4^{\bullet-}$) has been increasingly applied as an efficient oxidant for water treatment. A recent contribution has demonstrated an *E. coli* O157:H7 inactivation five times faster via persulfate (S$_2$O$_8^{2-}$) activation using Fe^{2+} than HO$^{\bullet}$. This capacity was attributed to the highly selective reactivity of SO$_4^{\bullet-}$ with electron-rich moieties present on the cell membrane of *E. coli* [65]. Garkusheva et al. [66] also have reported about the feasibility of the solar/PS/Fe^{2+} system for removal of herbicide atrazine (4 mg L^{-1}) and inactivation of *E. coli* (10^5 CFU mL^{-1}) in several water matrixes (milliQ, lake, and diluted wastewater) [66]. Another interesting application of this process was performed for the treatment of ballast water using PS activated with zerovalent iron (Fe^0) as a chemical biocide against two species of marine phytoplankton, *Pseudonitzshia delicatissima* (a diatom) and *Dunaliella tertiolecta* (green alga) [67].

3. *Iron chelates.* The complexation of iron with several chelates to make it soluble for longer periods has been widely investigated during the last years. Many examples of the investigation of complexes with polycarboxylate ligands like ethylenediaminetetraacetic acid (EDTA), citric acid and oxalic acid, nitrilotriacetic acid, tartaric acid, and humic acids are found in literature [15]. Recently the use of natural and/or artificial but biodegradable substances like (*S,S*)-ethylenediamine-*N,N'*-disuccinic acid (EDDS), an structural isomer of EDTA. Regarding water disinfection, the first approach of the use of EDDS was performed by Klamert et al. in 2012. These authors reported a slight increased efficiency on the reduction of total coliforms (>3-log in 120 min) with 0.1 mM of iron, 0.2 mM of EDDS, and 50 mg L^{-1} of H_2O_2 in secondary effluents at neutral pH [41]. Recently, Bianco et al. [68] reported on Fe/EDDS complex activation by H_2O_2 and persulfate ions for water disinfection. This work concluded that HO$^•$ seems to be more effective on the *E. faecalis* inactivation than sulfate radical. Moreover, it is also discussed the dual role that EDDS concentration plays on the final efficiency, as it kept iron in dissolution at neutral pH but also may act trapping generated radicals. And, the latter effect may overpass the benefits of using EDDS and affect negatively the efficiency of the process if the concentration of EDDS is not well selected [68].

7 Concluding Remarks

Fenton and photo-Fenton reactions have widely demonstrated their capability to treat and disinfect different types of water sources. In the last years, especially solar photo-Fenton driven by natural sunlight has gained great attention due to its high efficiency for disinfection of secondary effluents (wastewater).

The chemical composition of secondary effluents plays a very important role on the efficiency of photo-Fenton for water disinfection. The presence of organic matter is considered as the main competitor for hydroxyl radicals against bacteria; nevertheless, other parameters like inorganic compounds and natural occurring pathogens also play a competition for HO$^•$ and, therefore reducing the efficiency of the process.

As for other disinfection techniques, the biological nature of the waterborne pathogens is also a key parameter on the photo-Fenton efficiency, with the following gradient of susceptibilities: non-forming spore bacteria < fungi < viruses < forming spore bacteria < helminths < protozoa. When using photo-Fenton for disinfection of secondary wastewater effluents, the presence of different groups of pathogens and the composition in terms of organic matter make the scenario quite complex, which may require additional investigation to assure the final microbial quality of the treated effluent and low health microbial risk when considering reclaimed wastewater.

Up to date, there is no commercial application of this process for water disinfection yet; nevertheless, the scaling up of the process from laboratory to pilot plant scale (dozens of liters) has been widely demonstrated. In line with this, the use of solar CPC reactors has been demonstrated to be a good technological option for solar photo-Fenton application as a tertiary treatment for secondary wastewater effluents disinfection.

The current trend on photo-Fenton research is to increase the process efficiency at near-neutral pH. Nevertheless, many aspects related to microbial inactivation in water by iron chelates and iron oxides under solar radiation are unknown, and new findings are expected to increase the knowledge in this area.

Acknowledgments The authors thank the financial supports given by the Spanish Ministry of Economy and Competitiveness under the WATER4CROP project (CTQ2014-54563-C3-3) and the European project WATERSPOUTT H2020-Water-5c-2015 (GA 688928).

References

1. Fenton HJH (1984) Oxidation of tartaric acid in presence of iron. J Chem Soc 65:899–910
2. Haber F, Weiss J (1934) The catalytic decomposition of hydrogen peroxide by iron salts. Proc R Soc A 134:332–351
3. Barb WG, Baxendale JH, George P, Hargrave KR (1949) Reactions of ferrous and ferric ions with hydrogen peroxide. Nature 163:692–694
4. Barb WG, Baxendale JH, George P, Hargrave KR (1951) Reactions of ferrous and ferric ions with hydrogen peroxide. Part I. The ferrous ion reaction. Trans Faraday Soc 47:462–500
5. Barb WG, Baxendale JH, George P, Hargrave KR (1951) Reactions of ferrous and ferric ions with hydrogen peroxide. Part II. The ferric ion reaction. Trans Faraday Soc 47:591–616
6. Bauer R (1990) Applicability of solar irradiation for photochemical wastewater treatment. Cancer Lett 49:1225–1233
7. Pignatello JJ, Oliveros E, MacKay A (2006) Advanced oxidation processes for organic contaminant destruction based on the Fenton reaction and related chemistry. Crit Rev Environ Sci Technol 36:1–84
8. Brown RF, Jamison SE, Pandit UK, Pinkus J, White GR, Braendlin HP (1964) The reaction of Fenton's reagent with phenoxyacetic acid and some halogen-substituted phenoxyacetic acids. J Org Chem 29(1):146–153
9. Rincón AG, Pulgarín C (2006) Comparative evaluation of Fe^{3+} and TiO_2 photoassisted processes in solar photocatalytic disinfection of water. Appl Catal B 63:222–231
10. Sylva RN (1972) The hydrolysis of iron(III). Rev Pure Appl Chem 22:115–130
11. Pignatello JJ (1992) Dark and photoassisted iron(3+)-catalyzed degradation of chlorophenoxy herbicides by hydrogen peroxide. Environ Sci Technol 26:944–951
12. Canonica S, Kohn T, Mac M, Real FJ, Wirz J, von Gunten U (2005) Photosensitizer method to determine rate constants for the reaction of carbonate radical with organic compounds. Environ Sci Technol 39:9182–9188
13. Fernández-Ibáñez P, Sichel C, Polo-López MI, de Cara-García M, Tello J (2009) Photocatalytic disinfection of natural well water contaminated by *Fusarium solani* using TiO_2 slurry in solar CPC photo-reactors. Catal Today 144:62–68
14. Giannakis S, Polo López MI, Spuhler D, Sánchez Pérez JA, Fernández-Ibáñez P, Pulgarin C (2016) Solar disinfection is an augmentable, in situ-generated photo-Fenton reaction – Part 1: A review of the mechanisms and the fundamental aspects of the process. Appl Catal B 199:199–223
15. Giannakis S, Polo López MI, Spuhler D, Sánchez Pérez JA, Fernández-Ibáñez P, Pulgarin C (2016) Solar disinfection is an augmentable, in situ-generated photo-Fenton reaction – Part 2:

A review of the applications for drinking water and wastewater disinfection. Appl Catal B 198:431–446
16. Kiwi J, Lopez A, Nadtochenko V (2000) Mechanism and kinetics of the OH-radical intervention during Fenton oxidation in the presence of a significant amount of radical scavenger (Cl-). Environ Sci Technol 34:2162–2168
17. Rose AL, Waite TD (2003) Kinetics of iron complexation by dissolved natural organic matter in coastal water. Mar Chem 84:85–103
18. Bogan BW, Trbovic V (2003) Effect of sequestration on PAH degradability with Fenton's reagent: roles of total organic carbon, humin, and soil porosity. J Hazard Mater 100:285–300
19. Bissey LL, Smith JL, Watts RJ (2006) Soil organic matter-hydrogen peroxide dynamics in the treatment of contaminated soils and groundwater using catalyzed H_2O_2 propagations (modified Fenton's reagent). Water Res 40:2477–2484
20. Li ZM, Shea PJ, Comfort SD (1998) Nitrotoluene destruction by UV-Catalyzed Fenton oxidation. Chemosphere 36:1849–1865
21. Tyre BW, Watts RJ, Miller GC (1991) Treatment of four biorefractory contaminants in soils using catalyzed hydrogen peroxide. J Environ Qual 20:832–838
22. Goldstone JV, Pullin MJ, Bertilsson S, Voelker BM (2002) Reactions of hydroxyl radical with humic substances: bleaching, mineralization, and production of bioavailable carbon substrates. Environ Sci Technol 36:364–372
23. Ng TW, Chow AT, Wong PK (2014) Dual roles of dissolved organic matter in photo-irradiated Fe(III)-contained waters. J Photochem Photobiol A 290:116–124
24. Rutala WA, Weber DJ (2004) Registration of disinfectants based on relative microbicidal activity. Infect Control Hosp Epidemiol 25:333–341
25. Garrido-Cardenas JA, Polo-López MI, Oller-Alberola I (2017) Advanced microbial analysis for wastewater quality monitoring: metagenomics trend. Appl Microbiol Biotechnol. https://doi.org/10.1007/s00253-017-8490-3
26. Rizzo L, Manaia C, Merlin C, Schwartz T, Dagot C, Ploy MC, Michael I, Fatta-Kassinos D (2013) Urban wastewater treatment plants as hotspots for antibiotic resistant bacteria and genes spread into the environment: a review. Sci Total Environ 447:345–360
27. Ferro G, Polo-Lopez MI, Fernández-Ibáñez P (2016) Conventional and new processes for urban wastewater disinfection: effect on emerging and resistant microorganisms. In: Fatta-Kassinos D, Dionysiou DD, Kümmerer K (eds) Advanced treatment technologies for urban wastewater reuse. Handbook of Environmental Chemistry, vol 45. Springer, Cham, pp 107–128
28. Cabiscol E, Tamarit J, Ros J (2000) Oxidative stress in bacteria and protein damage by reactive oxygen species. Int Microbiol 3:3–8
29. Reed RH (2004) The inactivation of microbes by sunlight: solar disinfection as a water treatment process. Adv Appl Microbiol 54:333–365
30. Halliwell B, Gutteridge JM (1992) Biologically relevant metal ion-dependent hydroxyl radical generation. FEBS Lett 307:108–112
31. Castro-Alférez M, Polo-López MI, Fernández-Ibáñez P (2016) Intracellular mechanisms of solar water disinfection. Sci Rep 6:38145
32. Jolivet JP, Chanéac C, Tronc E (2004) Iron oxide chemistry. From molecular clusters to extended solid networks. Chem Commun 10:481–487
33. Polo-López MI, García-Fernández I, Velegraki T, Katsoni A, Oller I, Mantzavinos D, Fernández-Ibáñez P (2012) Mild solar photo-Fenton: an effective tool for the removal of *Fusarium* from simulated municipal effluents. Appl Catal B 111–112:545–554
34. Matavos-Aramyan S, Moussavi M, Matavos-Aramyan H, Roozkhosh S (2017) Cryptosporidium-contaminated water disinfection by a novel Fenton process. Free Radic Biol Med 106:158–167
35. Castro-Ríos K, Corpas EJ, Cárdenas V, Taborda G (2017) Inactivation efficiency of total coliforms and *Escherichia coli* in doped natural water by heterogeneous Fenton: effect of process factors. J Mater Environ Sci 8:364–369

36. Escobar-Megchún SI, Nájera-Aguilar HA, González-Hilerio M, Gutiérrez-Jiménez J, Gutiérrez-Hernández RF, Rojas-Valencia MN (2014) Application of the Fenton process in the elimination of Helminth eggs. J Water Health 12:722–726
37. Morales AA, Ramírez-Zamora RM, Schouwenaars R, Pfeiffer H (2013) Inactivation of *Ascaris* eggs in water using hydrogen peroxide and a Fenton type nanocatalyst (FeOx/C) synthesized by a novel hybrid production process. J Water Health 11:419–429
38. Morales-Pérez AA, Maravilla P, Solís-López M, Schouwenaars R, Durán-Moreno A, Ramírez-Zamora RM (2016) Optimization of the synthesis process of an iron oxide nanocatalyst supported on activated carbon for the inactivation of *Ascaris* eggs in water using the heterogeneous Fenton-like reaction. Water Sci Technol 73(5):1000–1009
39. Malato S, Fernández-Ibáñez P, Maldonado MI, Blanco J, Gernjak W (2009) Decontamination and disinfection of water by solar photocatalysis: recent overview and trends. Catal Today 147:1–59
40. Moncayo-Lasso A, Sanabria J, Pulgarin C, Benitez N (2009) Simultaneous *E. coli* inactivation and NOM degradation in river water via photo-Fenton process at natural pH in solar CPC reactor. A new way for enhancing solar disinfection of natural water. Chemosphere 77:296–300
41. Klamerth N, Malato S, Agüera A, Fernández-Alba A, Mailhot G (2012) Treatment of municipal wastewater treatment plant effluents with modified photo-Fenton as a tertiary treatment for the degradation of micro pollutants and disinfection. Environ Sci Technol 46:2885–2892
42. Michael I, Hapeshi E, Michael C, Varela A, Kyriakou S, Manaia C, Fatta-Kassinos D (2012) Solar photo-Fenton process on the abatement of antibiotics at a pilot scale: degradation kinetics, ecotoxicity and phytotoxicity assessment and removal of antibiotic resistant enterococci. Water Res 46:5621–5634
43. Agulló-Barceló M, Polo-López MI, Lucena F, Jofre J, Fernández-Ibáñez P (2013) Solar advanced oxidation processes as disinfection tertiary treatments for real wastewater: implications for water reclamation. Appl Catal B 136:341–350
44. Ortega-Gómez E, Esteban García B, Ballesteros Martín MM, Fernández-Ibáñez P, Sánchez Pérez JA (2013) Inactivation of *Enterococcus faecalis* in simulated wastewater treatment plant effluent by solar photo-Fenton at initial neutral pH. Catal Today 209:195–200
45. Ndounla J, Pulgarin C (2014) Evaluation of the efficiency of the photo Fenton disinfection of natural drinking water source during the rainy season in the Sahelian region. Sci Total Environ 493:229–238
46. Rodríguez-Chueca J, Polo-López MI, Mosteo R, Ormad M, Fernández-Ibáñez P (2014) Disinfection of real and simulated urban wastewater effluents using a mild solar photo-Fenton. Appl Catal B 150:619–629
47. Ndounla J, Kenfack S, Wéthé J, Pulgarin C (2014) Relevant impact of irradiance (vs. dose) and evolution of pH and mineral nitrogen compounds during natural water disinfection by photo-Fenton in a solar CPC reactor. Appl Catal B 148:144–153
48. Ortega-Gómez E, García BE, Martín MB, Ibáñez PF, Pérez JS (2014) Inactivation of natural enteric bacteria in real municipal wastewater by solar photo-Fenton at neutral pH. Water Res 63:316–324
49. Karaolia P, Michael I, García-Fernández I, Agüera A, Malato S, Fernández-Ibáñez P, Fatta-Kassinos D (2014) Reduction of clarithromycin and sulfamethoxazole-resistant *Enterococcus* by pilot-scale solar-driven Fenton oxidation. Sci Total Environ 468:19–27
50. Polo-López MI, Castro-Alférez M, Oller I, Fernández-Ibáñez P (2014) Assessment of solar photo-Fenton, photocatalysis, and H_2O_2 for removal of phytopathogen fungi spores in synthetic and real effluents of urban wastewater. Chem Eng J 257:122–130
51. Ruiz-Aguirre A, Polo-López MI, Fernández-Ibáñez P, Zaragoza G (2015) Assessing the validity of solar membrane distillation for disinfection of contaminated water. Desalin Water Treat 55:2792–2799

52. Ferro G, Fiorentino A, Alférez MC, Polo-López MI, Rizzo L, Fernández-Ibáñez P (2015) Urban wastewater disinfection for agricultural reuse: effect of solar driven AOPs in the inactivation of a multidrug resistant *E. coli* strain. Appl Catal B 178:65–73
53. Ndounla J, Pulgarin C (2015) Solar light (hv) and H_2O_2/hv photo-disinfection of natural alkaline water (pH 8.6) in a compound parabolic collector at different day periods in Sahelian region. Environ Sci Pollut Res 22:17082–17094
54. Ruiz-Aguirre A, Polo-López MI, Fernández-Ibáñez P, Zaragoza G (2017) Integration of membrane distillation with solar photo-Fenton for purification of water contaminated with *Bacillus* sp. and *Clostridium* sp. spores. Sci Total Environ 595:110–118
55. Aguas Y, Hincapie M, Fernández-Ibáñez P, Polo-López MI (2017) Solar photocatalytic disinfection of agricultural pathogenic fungi (*Curvularia* sp.) in real urban wastewater. Sci Total Environ 607–608:1213–1224
56. Gutiérrez-Zapata HM, Sanabria J, Rengifo-Herrera JA (2017) Addition of hydrogen peroxide enhances abiotic sunlight-induced processes to simultaneous emerging pollutants and bacteria abatement in simulated groundwater using CPC solar reactors. Sol Energy 148:110–116
57. Sichel C, Fernández-Ibañez P, De Cara M, Tello J (2009) Lethal synergy of solar UV-radiation and H_2O_2 on wild *Fusarium solani* spores in distilled and natural well water. Water Res 43:1841–1850
58. Spuhler D, Rengifo-Herrera JA, Pulgarín C (2010) The effect of Fe^{2+}, Fe^{3+}, H_2O_2 and the photo-Fenton reagent at near neutral pH on the solar disinfection (SODIS) at low temperatures of water containing *Escherichia coli* K12. Appl Catal B 96:126–141
59. Ortega-Gómez E, Ballesteros Martín MM, Carratalà A, Fernández Ibañez P, Sánchez Pérez JA, Pulgarín C (2015) Principal parameters affecting virus inactivation by the solar photo-Fenton process at neutral pH and µM concentrations of H_2O_2 and $Fe^{2+/3+}$. Appl Catal B 174–175:395–402
60. Macku'ak T, Nagyov K, Faberov M, Grabic R, Koba O, Gal M, Birosova L (2015) Utilization of Fenton-like reaction for antibiotics and resistant bacteria elimination in different parts of WWTP. Environ Toxicol Phar 40:492–497
61. Ortega-Gómez E, Ballesteros Martín MM, Esteban García B, Sánchez Pérez JA, Fernández Ibáñez P (2014) Solar photo-Fenton for water disinfection: an investigation of the competitive role of model organic matter for oxidative species. Appl Catal B 148–149:484–489
62. García-Fernández I, Polo-López MI, Oller I, Fernández-Ibáñez P (2012) Bacteria and fungi inactivation using Fe^{3+}/sunlight, H_2O_2/sunlight and near neutral photo-Fenton: a comparative study. Appl Catal B 121–122:20–29
63. Nieto-Juarez JI, Kohn T (2013) Virus removal and inactivation by iron (hydr)oxide-mediated Fenton-like processes under sunlight and in the dark. Photochem Photobiol Sci 12:1596–1605
64. Ruales-Lonfat C, Barona JF, Sienkiewicz A, Bensimon M, Vélez-Colmenares J, Benítez N, Pulgarín C (2015) Iron oxides semiconductors are efficient for solar water disinfection: a comparison with photo-Fenton processes at neutral pH. Appl Catal B 166–167:497–508
65. Wordof DN, Walker SL, Liu H (2017) Sulfate radical-induced disinfection of pathogenic *Escherichia coli* O157:H7 via iron-activated persulfate. Environ Sci Technol Lett 4:154–160
66. Garkusheva N, Matafonova G, Tsenter I, Beck S, Batoev V, Linden K (2017) Simultaneous atrazine degradation and *E. coli* inactivation by simulated solar photo-Fenton-like process using persulfate. J Environ Sci Health A 52:849–855
67. Ahn S, Peterson TD, Righter J, Miles DM, Tratnyek PG (2013) Disinfection of ballast water with iron activated persulfate. Environ Sci Technol 47:11717–11725
68. Bianco A, Polo-López MI, Fernández Ibañez P, Brigante M, Mailhot G (2017) Disinfection of water inoculated with *Enterococcus faecalis* using solar/Fe(III)EDDS-H_2O_2 or $S_2O_8^{2-}$ process. Water Res 118:249–260

AOPs Methods for the Removal of Taste and Odor Compounds

M. Antonopoulou and I. Konstantinou

Abstract The production of drinking water with good quality including the chemical, microbiological, and aesthetic characteristics remains one of the main contemporary challenges for drinking water industry. As the most predominant and problematic earthy-musty taste and odor (T&O) compounds are recalcitrant to conventional water treatment, advanced oxidation processes (AOPs) have been recently studied and employed in drinking water treatment for taste and odor control. In the light of recent developments, the present chapter reviews the effectiveness of various AOPs for T&O compounds removal from aqueous media. More specifically, an overview of the recent research studies dealing with AOPs for the removal of geosmin, 2-methylisoborneol, 2,4,6-trichloroanisole, 2-isopropyl-3-methoxypyrazine, and 2-isobutyl-3-methoxypyrazine from water reservoirs and drinking water, is presented. The fundamentals and experimental setup involved in relative technologies and the effectiveness of each process are further discussed. Special attention was also given to the degradation products and mechanisms that have been proposed for all the compounds in interest. Future research directions regarding the application of AOPs for T&O control and recommendations for further development are also highlighted.

Keywords AOPs, Drinking water, Odor, Removal, Taste

M. Antonopoulou
Laboratory of Industrial Chemistry, Department of Chemistry, University of Ioannina, Ioannina 45110, Greece

Department of Environmental and Natural Resources Management, University of Patras, Seferi 2, Agrinio 30100, Greece

I. Konstantinou (✉)
Laboratory of Industrial Chemistry, Department of Chemistry, University of Ioannina, Ioannina 45110, Greece
e-mail: iokonst@cc.uoi.gr

Contents

1 Introduction 180
2 Geosmin (GSM) and 2-Methylisoborneol (2-MIB) 186
3 2,4,6-Trichloroanisole (TCA) 197
4 2-Isopropyl-3-Methoxypyrazine (IPMP) and 2-Isobutyl-3-Methoxypyrazine (IBMP)... 200
5 Coupling AOPs with Other Treatment Processes 202
6 Conclusions 205
References 206

1 Introduction

Taste and odor (T&O) are two important aesthetic characteristics of the drinking water and problems associated with both of them are of growing interest. Various compounds causing taste and odor problems have been identified up to now in source water and in distribution systems. The majority of them can be perceived at very low concentrations in the ng L^{-1} range and can be assessed directly with human senses [1–3]. Although T&O compounds usually do not pose a health risk in commonly detected concentrations, their presence has been associated with its quality and safety, resulting in general decreased consumer acceptability and consequently in an increase of bottled-water consumption [4]. Changes in aesthetic properties of drinking water can also impart major costs to the drinking water industry, whereas important T&O episodes can affect treatment plant operation and even cause severe drinking water shortage, as well [3].

Numerous compounds have been identified during T&O events in drinking water including β-ionone, β-cyclocitral, cis-3-hexen-1-ol, 1-penten-3-one, various aldehydes [5–7], sulfur-containing compounds such as hydrogen sulfide, mercaptans and organic sulfides often emitted from sewage and wastewater treatment plants [8], benzothiazole and its derivatives [9], chlorophenols, bromophenols [10], and trihalomethanes [11]. The most reported compounds in literature that cause unpleasant taste and odor in ng L^{-1} levels (Table 1) are the geosmin (GSM), 2-methylisoborneol (2-MIB), 2,4,6-trichloroanisole (TCA), 2-isopropyl-3-methoxypyrazine (IPMP), and 2-isobutyl-3-methoxypyrazine (IBMP) [12–14]. The basic physicochemical properties of T&O of interest are summarized in Table 2 [15–20].

In view of the global concern about the aesthetic characteristics, a drinking water taste and odor wheel was developed by Suffet et al. [22] and identifies the main tastes and odors in water. According to Fig. 1, the eight major odor categories are earthy-musty, chlorinous, grassy, sulfurous/septic, fragrant/vegetable, fishy/rancid, medicinal, and chemical [22–24]. Specific descriptors for each odor category are also given.

Compounds giving rise to taste and odor in water can be derived by both natural and anthropogenic sources (Fig. 2), and the corresponding episodes may be developed from either one or the other or a combination of both. They are mainly produced by numerous benthic and pelagic aquatic microorganisms, such as cyanobacteria and

Table 1 Chemical structure and odor/taste threshold of the selected T&O compounds [1, 21]

Compound	Structure	Odor threshold (ng L^{-1})	Taste threshold (ng L^{-1})
Geosmin		4	7.5
2-Methylisoborneol (2-MIB)		15	2.5
2-Isopropyl-3-methoxypyrazine (IPMP)		0.2	9.9
2-Isobutyl-3-methoxypyrazine (IBMP)		1	0.4
2,4,6-Trichloroanisole (TCA)		0.03	25

actinomycetes that emerged in eutrophic surface waters. In addition, odorous compounds can be formed as by-products during oxidation and disinfection of drinking water as well as in the distribution systems through leaching from the polyethylene pipes [23].

Conventional technologies such as adsorption with the use of granular activated carbon (GAC), powered activated carbon (PAC), alum coagulation, and sand filtration [25, 26] have been used to remove T&O from the water. Although adsorption is widely applied for the control of taste and odor, extremely high dosages of the adsorbent are required many times, whereas the presence of natural

Table 2 Basic physicochemical properties of T&O compounds of interest [15–20]

Compound	Molecular formula	Molecular weight	Solubility mg L^{-1}	log K_{ow}	Polarizability 10^{-24} cm^3	Henry's law constant atm m^3 mol^{-1}	Boiling point °C
Geosmin	$C_{12}H_{22}O$	182.31	157.0	3.7	21.8	1.2×10^{-5}	165
2-Methylisoborneol (2-MIB)	$C_{11}H_{20}O$	168.28	194.5	3.13	20	8.9×10^{-5}	197
2-Isopropyl-3-methoxypyrazine (IPMP)	$C_8H_{12}N_2O$	152.19	2438.0	2.41	17.2	3.1×10^{-6}	120–125
2-Isobutyl-3-methoxypyrazine (IBMP)	$C_9H_{14}N_2O$	166.22	1034.0	2.72	19	4.1×10^{-6}	83–86
2,4,6-Trichloroanisole (TCA)	$C_7H_5Cl_3O$	211.47	10.0	3.91	18.9	1.3×10^{-4}	240

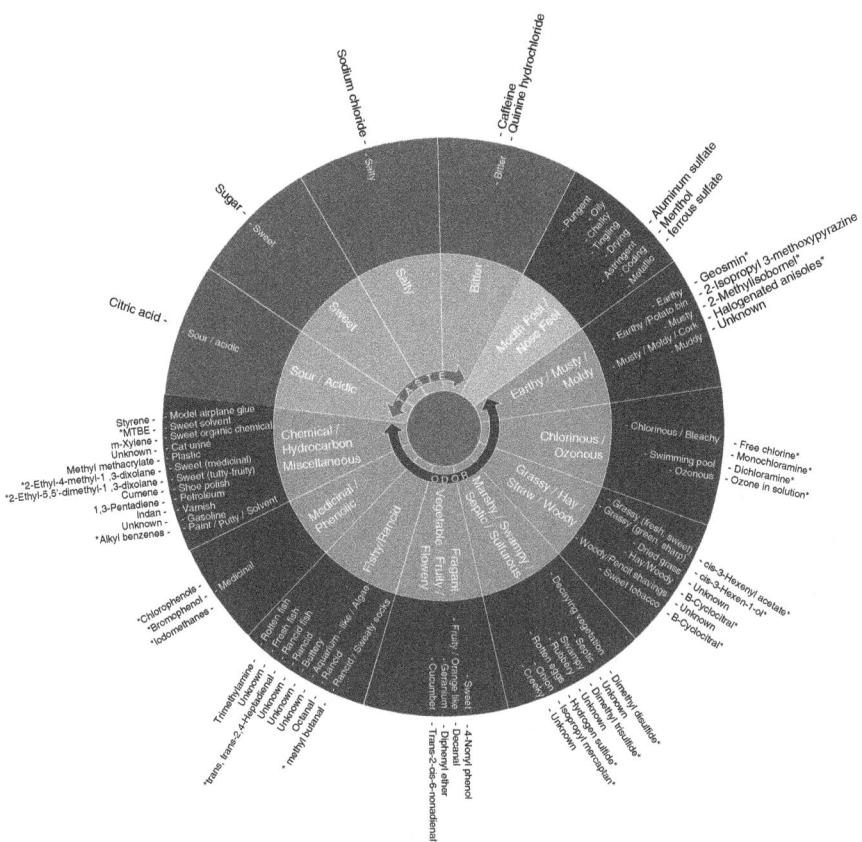

Fig. 1 The drinking water taste and odor wheel (reproduced from (Suffet et al. 1999) [22] with permission from the copyright holders, IWA Publishing)

organic matter (NOM) usually leads to reduced adsorption capacity and simultaneously requires additional steps for the regeneration of the used adsorbent [13, 27]. Common chemical oxidants such as chlorine, chlorine dioxide, or potassium permanganate have also been applied mainly for the removal of earthy and musty odor compounds, GSM and 2-MIB [1, 28]. Nonetheless, reduced removal efficiency was often exhibited due to the resistance of the tertiary alcohols like GSM and 2-MIB toward oxidation [29]. Rate constants for the reactions of both aforementioned compounds with different oxidants are summarized in Table 3. Chlorine dioxide, a selective oxidant which mainly reacts with activated aromatic systems and deprotonated amines, has shown relatively low rate constants. Similar low reactivity was exhibited for permanganate, which reacts predominantly with olefins and phenols.

Other conventional water treatment methods such as thermal oxidation and biofiltration involve high operating costs and could generate toxic secondary

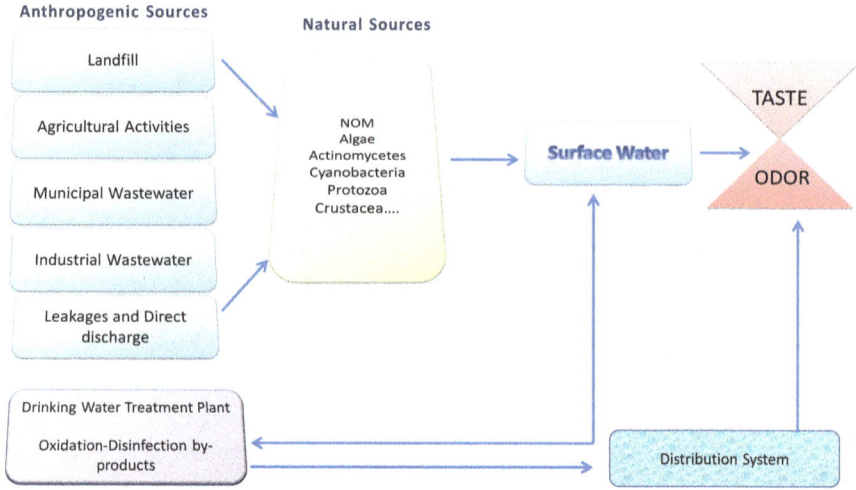

Fig. 2 Origins and sources of T&O compounds in the aquatic environment

Table 3 Second-order rate constants for the reaction of selected T&O compounds with common oxidants ($M^{-1} s^{-1}$) [5, 30, 31]

Compound	O_3	$HO^{\bullet}$	ClO_2	$KMnO_4$
Geosmin (GSM)	0.1	7.8×10^9	<1	<0.01
2-Isopropyl-3-methoxypyrazine (IPMP)	50	4.9×10^9	<10	<1
2-Methylisoborneol (2-MIB)	0.4	5.1×10^9	<1	<0.01
2,4,6-Trichloroanisole (TCA)	0.06	5.1×10^9	$<10^{-3}$	<1

pollutants [8]. As conventional water treatment processes presented significant weaknesses in order to act as a reliable barrier toward T&O control, the scientific community puts the focus on the development of more efficient technologies for the treatment and removal of T&O compounds [1, 32].

Among different alternative treatment options, AOPs have been proven as emerging technologies with significant importance in environmental remediation applications. In the last two decades, research activities have centered on AOPs for the destruction of various categories of organic contaminants and especially nonbiodegradable compounds [1, 14, 33]. More recently, AOPs have been applied increasingly for the removal of T&O compounds, and until now numerous studies have been devoted to the treatment of mainly five compounds, i.e., geosmin (GSM), 2-methylisoborneol (2-MIB), 2,4,6-trichloroanisole (TCA), 2-isopropyl-3-methoxypyrazine (IPMP), and 2-isobutyl-3-methoxypyrazine (IBMP) [1, 14].

AOPs can be broadly defined as aqueous-phase oxidation methods based on the formation of highly reactive species. Hydroxyl radicals ($HO^{\bullet}$), with oxidation potential equal to 2.8 V, are the primary oxidants in AOPs, while other radical and active oxygen species such as superoxide radical anions ($O_2^{\bullet-}$), hydroperoxyl radicals ($HOO^{\bullet}$), singlet oxygen ($^1O_2$), and holes (h^+) are also involved. A number of systems can be termed as AOPs, and typical AOP systems can be divided into chemical, photochemical, and electrochemical processes, as depicted in Fig. 3. Most of them use a combination of strong oxidizing agents (e.g., H_2O_2, O_3) with catalysts (e.g., transition metal ions) and irradiation (e.g., ultraviolet, visible) and have considerable similarities due to the participation of hydroxyl radicals which enhance the degradation process. In most cases, the complete mineralization of the target pollutant into CO_2, H_2O, and mineral acids is achieved [1, 33, 34]. Hydroxyl radicals are short-lived, simply produced, powerful electrophiles and unselective oxidants [34] which have shown high rate constants with T&O compounds (Table 3). Ozone has also been found to be a relatively good oxidant for T&O compounds removal [5].

Even though the study of T&O compounds removal by AOPs has attracted the scientific research quite recently, many scientific studies have already been published as presented in a recent review on the evaluation and comparison of various AOPs for the treatment of T&O compounds. The efficiency of the applied process, reaction kinetics, effect of operational parameters and water quality, identity of intermediate, and the possible transformation pathways have been presented [1].

This chapter aims to give an overview of the most recent progresses and achievements of various AOPs for the control of five T&O compounds in aqueous media together with possible future challenges that should be addressed. Geosmin, 2-methylisoborneol, 2,4,6-trichloroanisole, 2-isopropyl-3-methoxypyrazine, and 2-isobutyl-3-methoxypyrazine are reviewed herein as: (1) they were identified to

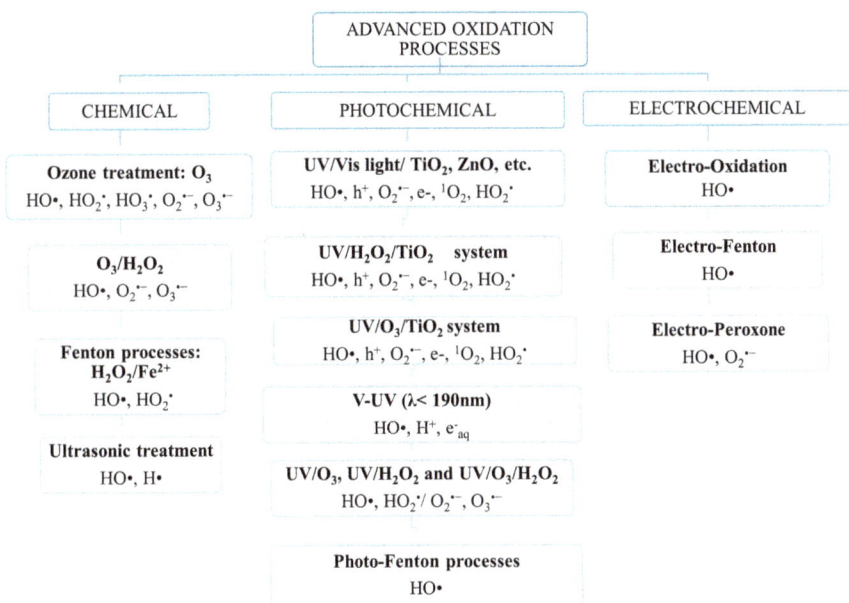

Fig. 3 Types and classifications of main advanced oxidation processes

be the major T&O compounds with extremely low taste and odor thresholds, frequently detected in aquatic systems; (2) adequate data concerning their removal with the application of AOPs are existing up to now in the literature. An extensive presentation of the recent literature published in the past 4 years concerning the removal of the selected T&O compounds using various AOPs is provided to gain up-to-date information on the degree of degradation, reaction kinetics, identification of transformation by-products, and possible degradation pathways. Furthermore, special attention is devoted to present important gaps that still need to be investigated by the scientific community, giving the most important research directions. In general, special emphasis is given to provide an updated integrated picture of the current status and trends prevailing in the specific field.

2 Geosmin (GSM) and 2-Methylisoborneol (2-MIB)

Geosmin (GSM) (trans-1,10-dimethyl-trans-9-decalol) and 2-methylisoborneol (2-MIB) are semi-volatile alicyclic alcohols and the most frequent and undesirable earthy-musty T&O compounds detected in drinking water [35, 36]. Both compounds are secondary metabolites produced by various microorganisms such as cyanobacteria, actinomycetes, protozoa, molds, and fungi, causing major concern for the drinking water industry [35]. GSM and 2-MIB are highly potent (detectable to humans at <10 ng L^{-1}) and extremely stable, resisting natural degradation and conventional treatment processes, as well [3].

Besides their frequent occurrence in water bodies, the concentration of these T&O compounds in drinking water is not regulated in many countries, as they have no known toxicological effects in environmental relevant concentrations toward humans and most aquatic organisms. However, for satisfactory quality of drinking water, a guideline level of 20 ng L^{-1} in South Korea and 10 ng L^{-1} in Japan and China was established [36].

Several chemical and photochemical AOPs in both homogeneous and heterogeneous phase have been studied for the removal of GSM and 2-MIB. Ozonation and associated oxidation processes (O_3/H_2O_2, UV/O_3, UV/O_3/H_2O_2), Fenton (Fe^{2+}/H_2O_2) and photo-Fenton process (UVC/Fe^{2+}/H_2O_2, UVA/Fe^{2+}/H_2O_2, solar/Fe^{2+}/H_2O_2), UV/H_2O_2 oxidation, heterogeneous photocatalysis using various catalysts, and irradiation sources as well as ultra-sonication have been applied with promising results [1]. The feasibility of a number of AOPs was tested at low initial concentrations relevant to those found in environmental matrices. The efficiency of the aforementioned processes was based on the formation of HO$^{\bullet}$ radicals, resulting in final concentrations lower than the human taste threshold. However, it was worth mentioning that undesired transformation products (TPs) have been formed in some cases, pointing out that simultaneous with the parent compound removal, the overall abatement of T&O is vitally important. The presence of organic matter [e.g., humic acids (HA) and/or fulvic acids (FA)] or inorganic ions (e.g., carbonate and bicarbonate) in real waters was acknowledged as a limitation factor because they can act as scavengers of HO• radicals leading to slower degradation kinetics. 2-MIB was found to be more resistant to the removal by different AOPs than GSM, due to the greater steric hindrance in its chemical structure [1].

Current research on the removal of GSM and 2-MIB by various AOPs in water has been compiled in Tables 4 and 5, considering only the works published in the last 4 years. Detailed experimental conditions, and water matrix, applied for each treatment process, as well as results expressed as the observed degradation efficiency, are presented.

UV/H_2O_2 process showed promising results for the removal of GSM and 2-MIB in both pilot and laboratory scales [37, 38, 45]. The efficiency of this process is based on the formation of HO• radicals produced via photolysis of H_2O_2 according to the reactions 1–5 [48, 49]:

$$H_2O_2 + h\nu \rightarrow 2HO^• \quad k = \varphi_{H2O2} I_a \tag{1}$$
$$H_2O_2 + HO^• \rightarrow HO_2^• + H_2O \quad k = 2.7 \times 10^7 \ M^{-1}s^{-1} \tag{2}$$
$$HO_2^• + HO_2^• \rightarrow H_2O_2 + O_2 \quad k = 8.3 \times 10^5 \ M^{-1}s^{-1} \tag{3}$$
$$H_2O_2 + HO_2^• \rightarrow HO^• + O_2 + H_2O \quad k = 3.0 \ M^{-1}s^{-1} \tag{4}$$
$$HO^• + HO^• \rightarrow H_2O_2 \quad k = 5.5 \times 10^9 \ M^{-1}s^{-1} \tag{5}$$

where M is molarity (mol dm^{-3}), φ_{H2O2} is the quantum yield for photodissociation of H_2O_2, I_a (M s^{-1}) is the intensity of absorbed light, and k (M^{-1} s^{-1}) is the rate constant.

Optimal conditions during the UV/H_2O_2 process were assessed using 300 ng L^{-1} and 275 ng L^{-1} of GSM and 2-MIB as initial concentrations, respectively, and an efficiency higher than 96% was achieved [37, 45]. The process is influenced by many factors (e.g., parent compound concentration, pH, H_2O_2 concentration, UV intensity), and the removal by UV/H_2O_2 was enhanced with an increase in H_2O_2 concentration and UV intensity. In contrast, removal efficiency decreased with increasing parent compound concentration [37] and pH values [38]. Besides the investigation of UV/H_2O_2 process, UV/chlorine AOP for the destruction of both compounds was also studied by Wang et al. [38]. This process is based on the formation of both HO• and chlorine radicals (•Cl) produced from photolysis of HOCl/OCl$^-$ at wavelengths less than 400 nm via Eqs. (6)–(8) [50]:

$$HOCl + h\nu \rightarrow HO^• + Cl^• \tag{6}$$
$$OCl^- + h\nu \rightarrow O^{•-} + Cl^• \tag{7}$$
$$O^{•-} + H_2O \rightarrow HO^• + HO^- \tag{8}$$

The UV/chlorine process exhibited comparable efficiency at pH 7.5 and 8.5 with the UV/hydrogen peroxide (UV/H_2O_2) process under parallel conditions but showed higher efficiency at pH 6.5 [38]. The higher performance of UV/chlorine process was attributed to the following factors: (a) at lower pH values, HOCl absorbs UV light about 2.3 times more efficiently than H_2O_2 producing more HO•; (b) HOCl reacts more slowly with HO• than H_2O_2 (rate constant with HO• for HOCl, 8.46 × 10^4 M^{-1} s^{-1} vs. 2.7 × 10^7 M^{-1} s^{-1} for H_2O_2) [38, 50–52]. A possible disadvantage of

Table 4 Summary of AOPs removal performance for GSM

Initial concentration	Matrix scale	AOP features	Removal efficiency	References
514 ng L^{-1}	Raw water (from drinking water treatment plant)	UV/H_2O_2 (laboratory scale) Low-pressure UV lamp (254 nm) [H_2O_2]: 1.93 mg L^{-1} UV intensity: 86 mJcm^{-2}	99.03%	[37]
300 ng L^{-1}	Raw water (from drinking water treatment plant)	UV/H_2O_2 (pilot scale) UV system (low-pressure UV lamps / 254 nm) [H_2O_2]: 7.5 mg L^{-1} UV intensity: 400 mJcm^{-2}	97.14%	[37]
~350 ng L^{-1}	Water (from water purification plant)	UV/chlorine MP UV reactors [Chlorine]: 10 mg L^{-1} UV dose: 2000 ± 150 mJ cm^{-2} pH: 6.5	>90%	[38]
~350 ng L^{-1}	Water (from water purification plant)	UV/H_2O_2 MP UV reactors [H_2O_2]: 4.8 mg L^{-1} UV dose: 2000 ± 150 mJ cm^{-2} pH: 6.5	~85%	[38]
100 ng L^{-1}	Distilled water	UV/H_2O_2 Low-pressure mercury UV-C lamp (254 nm) UV fluence = 3,348 mJ cm^{-2} [H_2O_2]: 20 mg/L pH: 4.1–5.0	38.28%	[36]
100 ng L^{-1}	Distilled water	Fenton [Fe(II)], 2 mg L^{-1}; [H_2O_2], 20 mg L^{-1} pH: 4.1–5.0	< 20%	[36]
100 ng L^{-1}	Distilled water	Photo-Fenton Low-pressure mercury UV-C lamp (254 nm) UV fluence: 5,022 mJ cm^{-2} [Fe(II)], 2 mg L^{-1}; [H_2O_2], 20 mg L^{-1} pH: 3	78.86%	[36]
50 ng L^{-1}	River water	Photo-Fenton Low-pressure mercury UV-C lamp (254 nm) UV fluence: 3,348 mJ cm^{-2} [Fe(II)], 2 mg L^{-1}; [H_2O_2], 20 mg L^{-1} pH: 7.2–7.4	Final concentration <20 ng/L	[36]
40 μg L^{-1}	Milli-Q water	UV/persulfate (PDS) 6 W low-pressure Hg UV lamp (254 nm/ 1.79 mW cm^{-2}) I_0/V: 1.26 μE s^{-1} L^{-1} 2 mM phosphate buffer [PDS]$_0$, 10 μM; pH, 7.0 600 s reaction time	94.5%	[32]
40 μg L^{-1}	Reservoir water	UV/persulfate (PDS) 6 W low-pressure Hg UV lamp (254 nm/ 1.79 mW cm^{-2}) I_0/V: 1.26 μE s^{-1} L^{-1} [PDS]$_0$ = 100 μM, 20 °C 1,200 s reaction time	>95%	[32]
40 μg L^{-1}	River water	UV/persulfate (PDS) 6 W low-pressure Hg UV lamp (254 nm/ 1.79 mW cm^{-2}) I_0/V: 1.26 μE s^{-1} L^{-1} [PDS]$_0$ = 100 μM, 20 °C 1,200 s reaction time	>95%	[32]

(continued)

Table 4 (continued)

Initial concentration	Matrix scale	AOP features	Removal efficiency	References
1 mg L^{-1}	Milli-Q water	UV-A (315–400 nm)/TiO_2	~100% (in 30 min)	[39]
		F15 W/T8 black light tubes (max. emission ~365 nm; I, 71.7 mW cm^{-2})		
		[TiO_2]: 200 mg L^{-1}		
1 mg L^{-1}	Milli-Q water	UV-A (315–400 nm)/$SiW_{12}O_{40}^{4-}$	~100% (in 120 min)	[39]
		F15 W/T8 black light tubes (max. emission ~365 nm; I, 71.7 mW cm^{-2})		
		[$SiW_{12}O_{40}^{4-}$]: 200 mg L^{-1}		
100 ng L^{-1}	Milli-Q water	UV-A/TiO_2–USY zeolite composite coatings	99% (in 120 min)	[40]
		UV-A lamps (Hitachi, FL8BL-B, 8 W wavelength, 320–4,000 nm)		
		TiO_2–USY zeolite composite coatings (6 mg of zeolite +4 mg of TiO_2)		
100 ng L^{-1}	Milli-Q water	UV/TiO_2 pellets	96%	[41]
		Bed reactor-UVB lamps (100 W; spectral output, 280–330 nm)		
		Hombikat K01/C TiO_2 pellets (22 kg)		
18.5 ng L^{-1}	Fish farm water	UV/TiO_2 pellets	91%	[41]
		Bed reactor-UVB lamps (100 W; spectral output, 280–330 nm)		
		Hombikat K01/C TiO_2 pellets (22 kg)		
50 ng L^{-1}	Distilled water	UV-A/immobilized TiO_2	~80%	[42]
		GE black light blue bulbs (8 W; spectral range, 350–400 nm)		
		Immobilized TiO_2 (0.25 mg/cm^2)		
50 ng L^{-1}	Tainted water (from a recirculating aquaculture system)	UV-A/immobilized TiO_2	~60%	[42]
		GE black light blue bulbs (8 W; spectral range, 350–400 nm)		
		Immobilized TiO_2 (0.25 mg cm^2)		
1 µg L^{-1}	Milli-Q water	Simulated solar light (SSL)/Pd/WO_3	>99% (in 20 min)	[43]
		Simulated solar lamp		
		Light intensity: 0.6 $mWcm^{-2}$/0.8 $mWcm^{-2}$		
		[Pd/WO_3]: 150 mg L^{-1}		
~10 µg L^{-1}	Surface water	Ozonation	45% (reaction time, 40 min)	[13]
		pH: 7.9–8.1		
		T: 23 ± 1 °C		
		O_3 dose: 2 mg O_3/mg DOC		
~10 µg L^{-1}	Surface water	Electro-peroxone process	54% (reaction time, 5 min)	[13]
		Electrodes: anode was a Pt plate/cathode was a carbon-polytetrafluorethylene (carbon-PTFE)		
		pH: 7.9–8.1		
		T: 23 ± 1 °C		
		O_3 dose: 2 mg O_3/mg DOC		
		Current: 40 mA		
4 ng L^{-1}	Ultrapure water	Electrochemical oxidation (EO)/persulfate	~100%	[44]
		Boron-doped diamond (BDD) electrode		
		Current density: 5.0 mA cm^{-2}		
		[Na_2SO_4], 30 mM; pH, 2.0		
		Reaction time: 15 min (undivided cell)		

Table 5 Summary of AOPs removal performance for 2-MIB

Initial concentration	Matrix scale	AOP features	Removal efficiency	References
764 ng L^{-1}	Raw water (from drinking water treatment plant)	UV/H_2O_2 (laboratory scale) Low-voltage light lamp (254 nm; 70 W; dose, 86 mJ cm^{-2}) [H_2O_2]: 3.93 mg L^{-1} UV dose: 86 mJ cm^{-2}	99.99%	[45]
275 ng L^{-1}	Raw water (from drinking water treatment plant)	UV/H_2O_2 (pilot scale) UV system (low-voltage light lamp, wavelength of 254 nm, 2000 W) [H_2O_2]: 6 mg L^{-1} UV dose: 350 mJ cm^{-2}	96.58%	[45]
100 ng L^{-1}	Distilled water	UV/H_2O_2 Low-pressure mercury UV-C lamp (254 nm) UV fluence = 3,348 mJ cm^{-2} [H_2O_2]: 20 mg/L pH: 4.1–5.0	52.10%	[36]
100 ng L^{-1}	Distilled water	Fenton [Fe(II)], 2 mg L^{-1}; [H_2O_2], 20 mg L^{-1} pH: 4.1–5.0	< 20%	[36]
100 ng L^{-1}	Distilled water	Photo-Fenton Low-pressure mercury UV-C lamp (254 nm) UV fluence: 5,022 mJ cm^{-2} [Fe(II)], 2 mg L^{-1}; [H_2O_2], 20 mg L^{-1} pH: 3	100%	[36]
50 ng L^{-1}	River water	Photo-Fenton Low-pressure mercury UV-C lamp (254 nm) UV fluence: 3,348 mJ cm^{-2} [Fe(II)], 2 mg L^{-1}; [H_2O_2], 20 mg L^{-1} pH: 7.2–7.4	Final concentration < 20 ng/L	[36]
1 mg L^{-1}	Milli-Q water	UV-A (315–400 nm)/TiO_2 F15 W/T8 black light tubes (max. emission ~365 nm; I, 71.7 mW cm^{-2}) [TiO_2]: 200 mg L^{-1}	~100% (in 25 min)	[39]

AOPs Methods for the Removal of Taste and Odor Compounds

Concentration	Water type	Method/Conditions	Removal	Ref
1 mg L^{-1}	Milli-Q water	UV-A (315–400 nm)/$SiW_{12}O_{40}^{4-}$ F15 W/T8 black light tubes (max. emission ~365 nm; I, 71.7 mW cm^{-2}) [$SiW_{12}O_{40}^{4-}$]: 200 mg L^{-1}	~100% (in 100 min)	[39]
4 ng L^{-1}	Ultrapure water	Electrochemical oxidation (EO)/persulfate Boron-doped diamond (BDD) electrode Current density: 5.0 mA cm^{-2} [Na_2SO_4], 30 mM; pH, 2.0 Reaction time: 15 min (undivided cell)	>95%	[44]
100 ng L^{-1}	Milli-Q water	UV/TiO_2 pellets Bed reactor-UVB lamps (100 W; spectral output, 280–330 nm) Hombikat K01/C TiO_2 pellets (22 kg)	96%	[41]
14 ng L^{-1}	Fish farm water	UV/TiO_2 pellets Bed reactor-UVB lamps (100 W; spectral output, 280–330 nm) Hombikat K01/C TiO_2 pellets (22 kg)	84%	[41]
50 ng L^{-1}	Distilled water	UV-A/immobilized TiO_2 GE black light blue bulbs (8 W; spectral range, 350–400 nm) Immobilized TiO_2 (0.25 mg/cm^2)	~67%	[42]
50 ng L^{-1}	Tainted water (from a recirculating aquaculture system)	UV-A/immobilized TiO_2 GE backlight blue bulbs (8 W; spectral range, 350–400 nm) Immobilized TiO_2 (0.25 mg cm^2)	~64%	[42]
~10 µg L^{-1}	Surface water	Ozonation pH: 7.9–8.1 T: 23 ± 1 °C O_3 dose: 2 mg O_3/mg DOC	~40% (reaction time, 40 min)	[13]
~10 µg L^{-1}	Surface water	Electro-peroxone process Electrodes: anode was a Pt plate/cathode was a carbon-polytetrafluoroethylene (carbon-PTFE) pH: 7.9–8.1 T: 23 ± 1 °C O_3 dose: 2 mg O_3/mg DOC Current: 40 mA	48% (reaction time, 5 min)	[13]
40 µg L^{-1}	Milli-Q water	UV/persulfate (PDS) 6 W low-pressure Hg UV lamp (254 nm/1.79 mW cm^{-2}) I_0/V: 1.26 µE s^{-1} L^{-1} 2 mM phosphate buffer [PDS]$_0$, 10 µM; pH, 7.0	86.0%	[32]

(continued)

Table 5 (continued)

Initial concentration	Matrix scale	AOP features	Removal efficiency	References
40 µg L^{-1}	Reservoir water	UV/persulfate (PDS) 6 W low-pressure Hg UV lamp (254 nm/1.79 mW cm^{-2}) I_0/V: 1.26 µE s^{-1} L^{-1} [PDS]$_0$ = 100 µM, 20 °C 1,200 s reaction time	~80%	[32]
40 µg L^{-1}	River water	UV/persulfate (PDS) 6 W low-pressure Hg UV lamp (254 nm/1.79 mW cm^{-2}) I_0/V: 1.26 µE s^{-1} L^{-1} [PDS]$_0$ = 100 µM, 20 °C 1,200 s reaction time	~80%	[32]
1 mg L^{-1}	Distilled water	UV/PDMS-coated SiO$_2$/N-TiO$_2$ UV lamp irradiation (365 nm) 5:5 w/w PDMS-coated SiO$_2$/N-TiO$_2$	>80%	[46]
1 mg L^{-1}	Distilled water	Vis/PDMS-coated SiO$_2$/N-TiO$_2$ Blue light-emitting diode (LED) ~455 nm 5:5 w/w PDMS-coated SiO$_2$/N-TiO$_2$	~80%	[46]
400 ng L^{-1}	Water (from water purification plant)	UV/chlorine MP UV reactors [chlorine]: 10 mg L^{-1} UV dose: 2000 ± 150 mJ cm^{-2} pH: 6.5	~90%	[38]
400 ng L^{-1}	Water (from water purification plant)	UV/H$_2$O$_2$ MP UV reactors [H$_2$O$_2$]: 4.8 mg L^{-1} UV dose: 2000 ± 150 mJ cm^{-2} pH: 6.5	~80%	[38]
23.2 µg L^{-1}	Milli-Q water	Ozonation [O$_3$] 0.5 mg L^{-1} pH: 6.7	29.1%	[47]
23.2 µg L^{-1}	Milli-Q water	Catalytic ozonation by γ-AlOOH [catalyst]: 200 mg L^{-1} [O$_3$]: 0.5 mg L^{-1} pH: 6.7	27.5%	[47]

UV/chlorine AOP which needs attention is related with the potential formation of chlorination by-products (DBPs), especially with the usage of high chlorine dose. However, this limitation can be overcome by the careful management of chlorate formation using low doses of free chlorine, promoting further research and potential applications of the UV/chlorine process for the T&O control [38].

Photo-Fenton process, a combination of UV–vis irradiation with the so-called Fenton's reagent, consisting of a mixture of H_2O_2 and Fe^{2+}, has also been used for T&O control. Photo-Fenton process was found to enhance the degradation of GSM and 2-MIB compared to Fenton and UV/H_2O_2 processes, by generating more HO$^{\bullet}$ radicals according to the following reactions (9–11) [36]:

$$Fe^{2+} + H_2O_2 \rightarrow Fe^{3+} + OH^- + HO^{\bullet} \quad (9)$$
$$Fe^{3+} + H_2O_2 + h\nu \rightarrow Fe^{2+} + HO^{\bullet} + H^+ \quad (10)$$
$$H_2O_2 + h\nu \rightarrow 2HO^{\bullet} \quad (11)$$

The comparative study shows that the degradation followed the decreasing order: photo-Fenton > UV/H_2O_2 > UV > Fenton. The degradation was enhanced by decreasing the pH, concentration of organic matter, and initial concentration of the parent compound. However, the presence of organic matter (NOM, HA, and FA) had only a slight influence on the degradation of T&O compounds, due to sufficient UV fluence during the photo-Fenton process. The rates were found to be dependent upon organic matter molecular size. The degradation efficiencies in the presence of NOM increased with decreasing molecular size and followed the order: NOM < HA < FA [36].

In real river water supplying a drinking water treatment plant (DWTP), a slightly reduced degradation efficiency of both compounds was observed. Nevertheless, with the application of photo-Fenton process, the guideline level in South Korea of both GSM and 2-MIB was reached using initial concentrations under 30–50 ng L^{-1}. The toxicity of the treated solutions was evaluated using *D. magna*, and the results revealed the formation of nontoxic intermediates. During the photo-Fenton process, dehydration and open-ring pathways were the main common degradation pathways. For GSM, six aliphatic compounds were identified, i.e., nonanoic acid, isobutyl isobutyrate, butyl butyrate, methyl isobutyrate, methyl methacrylate, and methyl propionate [36]. 2-Methylenebornane, 2-methyl-2-bornene, camphor, 2-ethyl-1-hexanol, nonanal, and isooctyl alcohol were identified during the treatment of 2-MIB by photo-Fenton process showing that 2-MIB degradation proceeds through C–C dissociation and subsequent skeleton rearrangement as well [36].

In recent years, UV/persulfate process has proved to be an effective method for organic compounds degradation, including GSM and 2-MIB [32]. This process is based on the formation of sulfate radical ($SO_4^{\bullet-}$), a strong one-electron oxidant with high-standard redox potential [$E = 2.60$ V vs. normal hydrogen electrode (NHE)] [12, 53, 54], according to the following Eq. 12:

$$S_2O_8^{2-} + h\nu \rightarrow 2SO_4^{\cdot-} \quad \varepsilon = 21.1 \text{ M}^{-1}\text{cm}^{-1}, \varphi = 0.7 \text{ mol Einstein}^{-1} \quad (12)$$

HO$^{\cdot}$ is simultaneously formed from the reaction of water and hydroxyl ions with $SO_4^{\cdot-}$ [32, 55–57]:

$$SO_4^{\cdot-} + S_2O_8^{2-} \leftrightarrow S_2O_8^{\cdot-} + SO_4^{2-} \quad k = 5.5 \times 10^5 \text{ M}^{-1}\text{s}^{-1} \quad (13)$$
$$HO^{\cdot} + S_2O_8^{2-} \leftrightarrow S_2O_8^{\cdot-} + HO^- \quad k = 1.4 \times 10^7 \text{ M}^{-1}\text{s}^{-1} \quad (14)$$
$$SO_4^{\cdot-} + H_2O \rightarrow HSO_4^- + HO^{\cdot} \quad k = 8.3 \text{ M}^{-1}\text{s}^{-1} \quad (15)$$
$$SO_4^{\cdot-} + HO^- \rightarrow SO_4^{2-} + HO^{\cdot} \quad k = 6.5 \times 10^7 \text{ M}^{-1}\text{s}^{-1} \quad (16)$$

Second-order rate constants for 2-MIB and GSM reacting with $SO_4^{\cdot-}$ were calculated to be $4.2 \pm 0.6 \times 10^8$ M^{-1} s^{-1} and $7.6 \pm 0.6 \times 10^8$ M^{-1} s^{-1}, respectively, at a pH of 7.0 [32]:

$$SO_4^{\cdot-} + GSM \rightarrow \text{Oxidation products} \quad k = 7.6 \pm 0.6 \times 10^8 \text{ M}^{-1}\text{s}^{-1} \quad (17)$$
$$HO^{\cdot} + GSM \rightarrow \text{Oxidation products} \quad k = 5.7 \pm 0.2 \times 10^9 \text{ M}^{-1}\text{s}^{-1} \quad (18)$$
$$SO_4^{\cdot-} + 2-MIB \rightarrow \text{Oxidation products} \quad k = 4.2 \pm 0.6 \times 10^8 \text{ M}^{-1}\text{s}^{-1} \quad (19)$$
$$HO^{\cdot} + 2-MIB \rightarrow \text{Oxidation products} \quad k = 4.3 \pm 0.6 \times 10^9 \text{ M}^{-1}\text{s}^{-1} \quad (20)$$

At neutral pH, HO$^{\cdot}$ contribution was found to be more significant than $SO_4^{\cdot-}$ for 2-MIB and GSM in ultrapure water. Critical operating parameter dictating process performance was found to be the dosage of persulfate. Although pH did not affect the degradation directly, faster degradations of both compounds were observed in acidic media using phosphate buffer. This trend was connected with the different scavenging effects of hydrogen phosphate and dihydrogen phosphate species in the aqueous media. The degradation rate was appreciably reduced in the presence of bicarbonate and natural organic matter (NOM) acting as HO$^{\cdot}$ and $SO_4^{\cdot-}$ radical scavengers and directly affecting process efficiency.

The process was also effective in removing 2-MIB and GSM in real waters, showing, however, slower degradation kinetics [32].

Heterogeneous photocatalysis, a process combining semiconducting materials with various activating irradiation sources, can be characterized as the most studied photochemical AOPs for the removal of GSM and 2-MIB. From a mechanistic point of view, the reactions that take place, generating various species that can oxidize water or organic compounds, are described below [35]:

$$TiO_2 + h\nu \rightarrow e_{CB}^- + h_{VB}^+ \quad (21)$$
$$h_{VB}^+ + H_2O \rightarrow HO^{\cdot} + H^+ \quad (22)$$
$$O_2 + e_{CB}^- \rightarrow O_2^{\cdot-} \quad (23)$$
$$O_2^{\cdot-} + H^+ \rightarrow HO_2^{\cdot} \quad (24)$$
$$HO_2^{\cdot} + HO_2^{\cdot} \rightarrow H_2O_2 + O_2 \quad (25)$$
$$H_2O_2 + e_{CB}^- \rightarrow HO^{\cdot} + OH^- \quad (26)$$

The most employed catalyst as clearly seen in Tables 4 and 5 is TiO_2 in different forms (powder, pellets, films) [39, 41, 42]. Except this semiconductor, Pd nanoparticle-modified WO_3 (Pd/WO_3) catalysts [43] and polyoxometalates [39] have also been used for the photocatalytic degradation of 2-MIB and GSM. The available studies showed that the degradation rates of both compounds are highly influenced by the catalyst concentration. Other parameters that affect performance are the light wavelength and intensity, the solution pH, the initial concentration of the parent compound, dissolved oxygen (DO), and the water matrix (i.e., the presence of humic substances, bicarbonates). Photocatalytic reactions usually obey to Langmuir–Hinshelwood kinetic model which is reduced mainly to pseudo first-order kinetics for $\mu g\ L^{-1}$ to $ng\ L^{-1}$ concentration ranges [39, 41, 42]. The photocatalytic degradation and mineralization of GSM and 2-MIB in water were studied and compared using $SiW_{12}O_{40}^{4-}$ (POM) and TiO_2. Both photocatalysts were found to be effective in degradation of target compounds. However, the degradation of both compounds was slightly slower in the presence of POM. Degradation pathway is proposed for target compounds during their photocatalytic treatment. Quite similar intermediates have been identified, using either POM or TiO_2 with $HO^\bullet$ radical attack to be mainly responsible for the mechanism [39].

In the case of GSM, the majority of the primary identified intermediates were cyclic ketones, whereas linear saturated and unsaturated products were produced as a consequence of ring opening. During 2-MIB photocatalytic degradation, alcohol-, ketone-, and diketone- derivatives of 2-MIB, oxygenated cyclic compounds, and open-chain aliphatic compounds were identified [39].

As the use of catalyst in slurry form requires an additional treatment step to remove them from the treated aqueous media, Hombikat K01/C titanium dioxide pellets, immobilized TiO_2, films consisting of $N-TiO_2/PDMS$-coated SiO_2, and stable TiO_2–USY zeolite composite coatings have also been used [40–42, 46]. All the materials showed significant potential for the removal of extremely low GSM and 2-MIB concentrations in ultrapure and real water.

Heterogeneous photocatalysis is likely to benefit from the use of renewable energy sources, thus the use of novel semiconductor materials with photo response to visible light is also necessary. Xue et al. [43] studied the photocatalytic removal of GSM in the presence of Pd/WO_3, and high removal percentages were achieved. This catalyst also exhibited a superior reuse performance. 8a-Hydroxy-4a-methyl-octahydro-naphthalen-2-one, 8,8-dimethyl-decahydro-naphthalen-1-ol, and 2-ethyl-cyclohexanone were identified as intermediate products [43].

In addition, E-peroxone process is an emerging technology which has been recently used for the treatment of GSM and 2-MIB and combines conventional ozonation with an electrolysis process according to the following reactions 27 and 28 [13]:

$$H_2O_2 + O_3 \rightarrow HO^\bullet + O_2^{\bullet-} + H^+ + O_2 \qquad (27)$$
$$O_2 + 2H^+ + 2e^- \rightarrow H_2O_2 \qquad (28)$$

E-peroxone process typically involves the electrochemical conversion of O_2 in the reactor (sparkled with O_2/O_3 gas mixture) to H_2O_2. The process presented great potential for the treatment of 2-MIB and GSM with significantly lower bromate formation compared to conventional ozonation, for a given specific ozone dose [13]. The efficiency of the process was attributed to the higher HO$^\bullet$ production than in conventional ozonation, thus enhancing the removal of the parent compounds during water treatment. One more advantage of the applied technology was the in situ formation of H_2O_2 that could eliminate the problems associated with the use and storage of concentrated H_2O_2 [13].

However, it should be pointed out that the E-peroxone process may lead to a reduced disinfection efficiency during water treatment, due to enhanced O_3 decomposition in the presence of H_2O_2 [58]. Moreover, a posttreatment step is required for the removal of residual H_2O_2 [13].

Electrochemical oxidation, which uses boron-doped diamond (BDD) electrodes in the presence of a suitable electrolyte, is a relatively new process in water treatment and, therefore, less studied than other AOPs, for the control of T&O in water [44].

Two mechanisms can be considered for electrochemical degradation of the organic contaminants, (a) oxidation by the adsorbed HO$^\bullet$ formed at the anode surface, and (b) indirect oxidation in the liquid bulk by the oxidants that are formed electrochemically in the presence of inorganic ions such as chloride, sulfate, or phosphate [44, 59].

The relatively high removal efficiency of 2-MIB and GSM at BDD electrodes in sulfate solution can be attributed to the formation of persulfate according the reactions 29–31 [44]:

$$BDD + H_2O \rightarrow BDD(HO^\bullet) + H^+ + e^- \qquad (29)$$
$$2HSO_4^- - 2e^- \rightarrow S_2O_8^{2-} + 2H^+ \qquad (30)$$
$$2SO_4^{2-} - 2e^- \rightarrow 2S_2O_8^{2-} \qquad (31)$$

Other reactive species (e.g., superoxide and singlet oxygen) can also be generated in the EO, participating in the degradation process [44]:

$$2HO^\bullet + S_2O_8^{2-}(BDD) \rightarrow 2SO_4^{2-} + {}^1O_2 + 2H^+ \qquad (32)$$
$$4HO^\bullet + S_2O_8^{2-}(BDD) \rightarrow 2SO_4^{2-} + 2O_2^{\bullet -} + 4H^+ \qquad (33)$$

High-current density and low-solution pH were found to be favorable for 2-MIB and GSM degradation. In contrast, NOM acting as a scavenger of the generated reactive species could reduce the removal efficiency. 2-MIB and GSM electrochemical degradation proceeded mainly through dehydration, b-scission ring opening, and oxidation reactions [44].

Ultrasonic irradiation, an attractive AOP, has been considered as alternative for 2-MIB and GSM removal [28]. Although the formation of reactive species (e.g., H$^\bullet$ and HO$^\bullet$) takes place through water sonolysis, according to the reactions 34–43 [60], the degradation of both compounds was found to proceed via pyrolysis, upon cavitational collapse:

$$H_2O +))))) \rightarrow H^{\cdot} + HO^{\cdot} \quad (34)$$
$$O_2 +))))) \rightarrow O^{\cdot} + O^{\cdot} \quad (35)$$
$$O^{\cdot} + H_2O \rightarrow 2HO^{\cdot} \quad (36)$$
$$O^{\cdot} + HO^{\cdot} \rightarrow HO_2^{\cdot} \quad (37)$$
$$H^{\cdot} + O_2 \rightarrow HO_2^{\cdot} \quad (38)$$
$$2HO^{\cdot} \rightarrow H_2O_2 \quad (39)$$
$$2HO^{\cdot} \rightarrow H_2O + O^{\cdot} \quad (40)$$
$$H^{\cdot} + HO^{\cdot} \rightarrow H_2O \quad (41)$$
$$2HO_2^{\cdot} \rightarrow H_2O_2 + O_2 \quad (42)$$
$$2H^{\cdot} \rightarrow H_2 \quad (43)$$

A summary of transformation pathways of GSM and 2-MIB degradation applying various AOPs is presented in Figs. 4 and 5, respectively.

3 2,4,6-Trichloroanisole (TCA)

2,4,6-Trichloroanisole (TCA) is a typical T&O compound, which can cause unpleasant taste and odor even at ng L^{-1} levels in drinking water [12]. Three main routes can be considered for its occurrence in aquatic systems: (a) biomethylation of the parent halophenols by actinomycetes [65] and fungi [66] during the biological treatment process of wastewater or in water distribution systems, (b) biotransformation of the parent phenolic compound by biofilm in waterworks [67, 68], and (c) the leaching from the polyethylene pipes in the distribution system [69]. In contrast to other T&O compounds, TCA has been reported to cause cancer and potential genetic damages in human cells [12]. Several methods have been investigated to remove TCA in different phases [70–72]. Based on literature data, TCA is extremely resistant to removal by the conventional treatment processes including coagulation and sand filtration [71]. Although, up to now limited publications have been devoted to the removal of TCA from water by advanced oxidation processes, the results are promising.

Ozonation and catalytic ozonation processes showed significant efficacy to destroying TCA. Catalytic ozonation by various solid catalysts exhibited significant removal efficiency for TCA in drinking water compared with the sole ozonation [71, 72]. It was proposed that HO$^{\bullet}$ radicals produced during decomposition of ozone contributed significantly to TCA degradation compared with molecular ozone [5, 71, 72]. A possible reaction mechanism for catalytic ozonation of TCA has been proposed by Qi and co-workers [71] and includes (1) chemisorption of ozone on the catalyst surface leading to the formation of active species (HO$^{\bullet}$) that could oxidize TCA, (2) chemisorption of TCA molecules on the catalyst surface and their further oxidation by gaseous or aqueous ozone, and (3) reaction between both chemisorbed species of ozone and TCA. The factors which may affect TCA removal are mainly the catalyst and ozone dosage. Ozonation-based processes

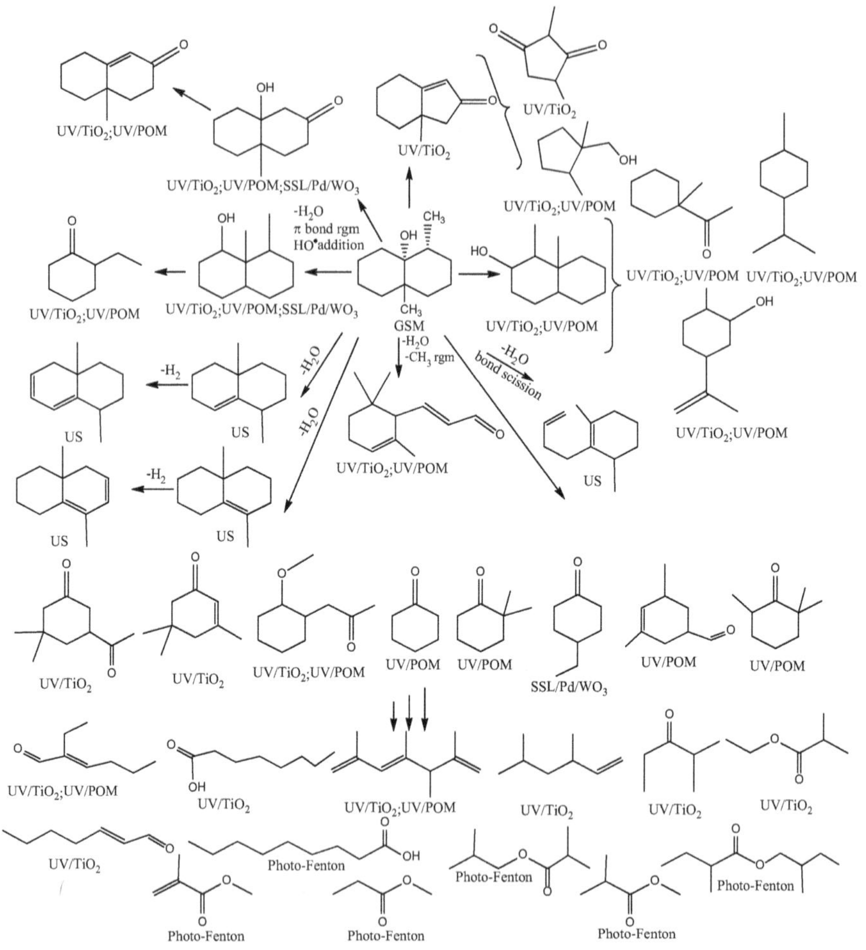

Fig. 4 Proposed transformation pathways of GSM under various AOPs (based on the studies [1, 28, 36, 39, 43, 61, 62])

may be also affected by the presence of natural organic matter and the pH of the aqueous media [71, 72].

UV/persulfate process was also an effective method for the removal of TCA [12]. The second-order rate constant of TCA reaction with sulfate radical ($SO_4^{\cdot-}$) was determined to be $3.72 \pm 0.1 \times 10^9$ M^{-1} s^{-1} [12].

$$SO_4^{\cdot-} + TCA \rightarrow \text{Oxidation products } k = 3.72 \pm 0.1 \times 10^9 \text{ M}^{-1}\text{s}^{-1} \text{ [12]} \quad (44)$$
$$HO^{\cdot} + TCA \rightarrow \text{Oxidation products } k = 5.1 \times 10^9 \text{ M}^{-1}\text{s}^{-1} \text{ [5]} \quad (45)$$

As expected, increasing dosage of persulfate increased the observed pseudo first-order rate constant for TCA degradation. The effects of water matrix [i.e., NOM, pH, carbonate/bicarbonate (HCO_3^-/CO_3^{2-}), and chloride ions (Cl^-)] were also

AOPs Methods for the Removal of Taste and Odor Compounds 199

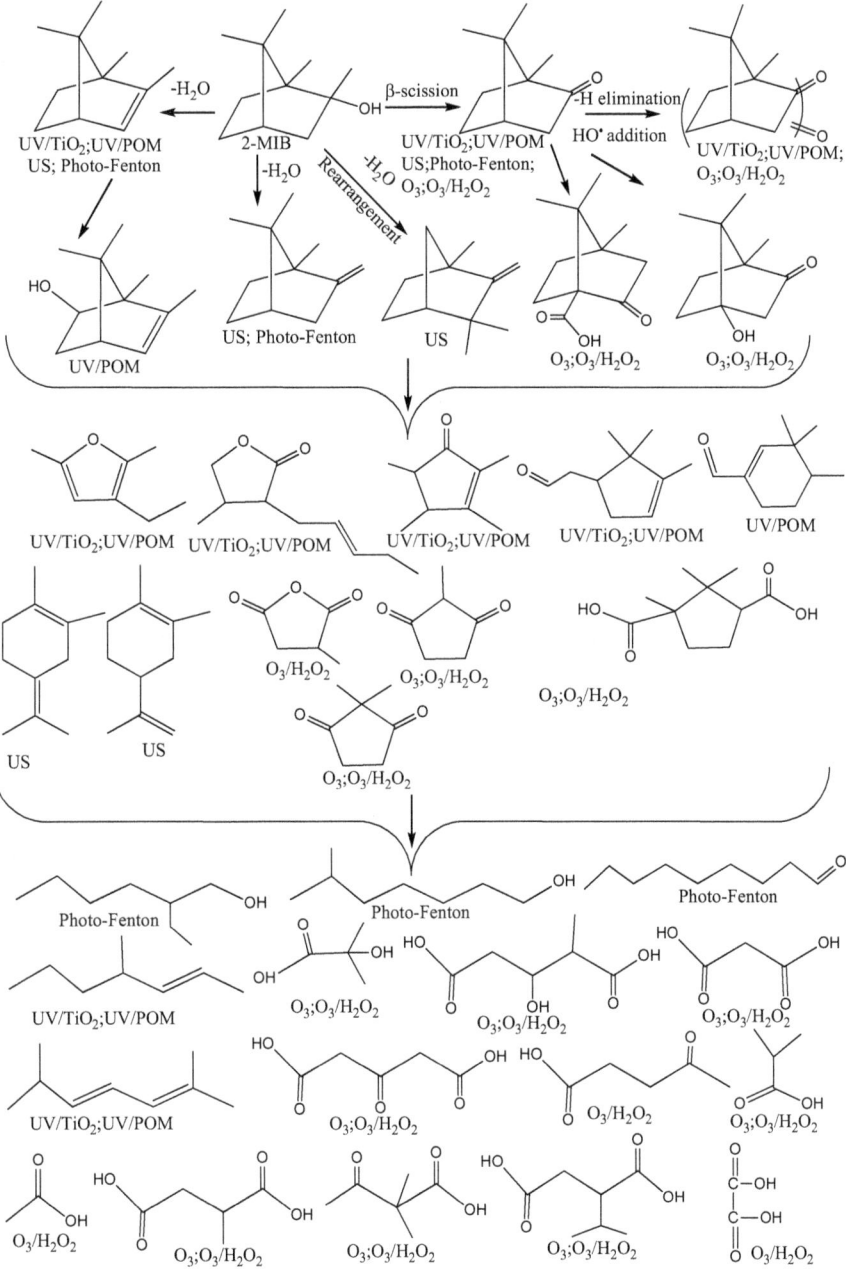

Fig. 5 Proposed transformation pathways of 2-MIB under various AOPs (based on the studies [1, 28, 36, 39, 43, 63, 64])

evaluated. Degradation rate of TCA decreased with pH increase from 4.0 to 9.0, which could be explained by the lower radical scavenging effect of dihydrogen phosphate than hydrogen phosphate in acidic condition (pH < 6). An inhibitory effect of NOM on TCA degradation was observed that could be attributed to the ability of NOM to act: (a) as an inner filter hindering the photolysis of persulfate and (b) as scavenger of the generated radicals [12]. The presence of CO_3^{2-}/HCO_3^- significantly decreased the efficiency, whereas a slight decrease of TCA degradation rate constant was observed in the presence of 10 mM Cl^-.

The degradation mechanism of TCA was also investigated, and four oxidation products [i.e., 2,4,6-trichlorophenol (TCP), 2,6-dichloro-1,4-benzoquinone, and two aromatic ring-opening products] were detected (Fig. 6). According to the literature data, TCA degradation is initiated by an electron abstraction of the aromatic ring by $SO_4^{\cdot-}$ radicals and the consequent formation of a cation radical (Fig. 6). Sequential reaction of the cation radical with H_2O and O_2 produces an organic peroxy radical, with subsequent elimination of methanol and superoxide radical-generating TCP. Further hydroxylation and Cl^- release from TCP formed the corresponding hydroquinone which is converted to 2,6-dichloro-1,4-benzoquinone by two-electron oxidation (Fig. 6). Subsequent oxidation of the identified aromatic intermediates leads to ring cleavage and formation of aliphatic products [12].

4 2-Isopropyl-3-Methoxypyrazine (IPMP) and 2-Isobutyl-3-Methoxypyrazine (IBMP)

2-Isopropyl-3-methoxypyrazine (IPMP) and 2-isobutyl-3-methoxypyrazine (IBMP) are two main taste and odor compounds found in aquatic systems and drinking water. They cause a rotten vegetable odor and have been frequently detected in water reservoir systems worldwide in concentrations ranging from 10 to 65 ng L^{-1} [14, 20, 73, 74]. Up to now ozonation, catalytic ozonation, and heterogeneous photocatalysis have been studied for the removal of IPMP [14, 75, 76]. In general ozonation was found to be efficient for the degradation of IPMP in aqueous solution with initial concentration in ng L^{-1} level [14]. The removal efficiency of IPMP increased in alkaline pH and decreased with the increase of initial concentrations of the parent compound. In comparative studies between ozonation and catalytic ozonation, the latter was reported to be more efficient [76]. However, in both processes the significant participation of $HO^\bullet$ radicals was proved [14, 76]. During ozonation, the TPs of IPMP were identified by GC/MS techniques and included 2-isopropyl-3-methylpyrazine, 2-hydroxy-3-methylbutyric acid methyl ester, 3-methyl-2-oxobutanoic acid methyl ester, 4-methyl-2-oxovaleric acid, 2-methylpropanoic acid, 3-methyl-2-oxobutanoic acid, and 3-methylbutanoic acid [14]. Qi et al. [76] found that the major intermediate by-products of IPMP in catalytic ozonation by γ-AlOOH were aliphatic

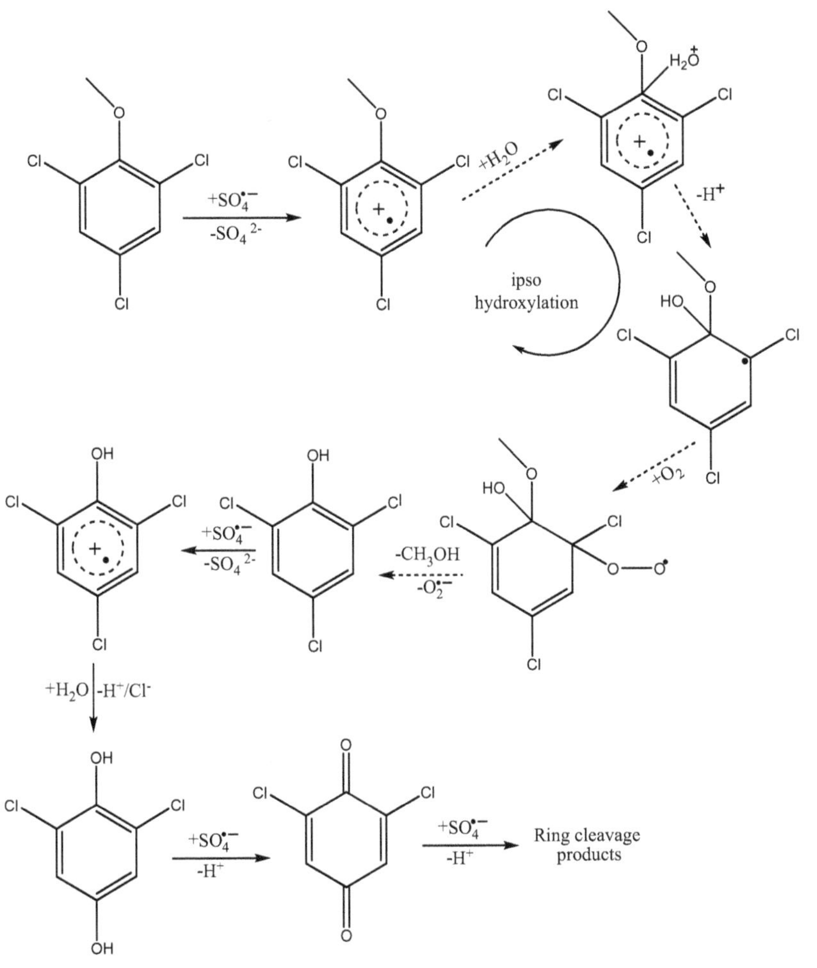

Fig. 6 Proposed transformation pathways and main TPs of 2,4,6-TCA using UV/persulfate (reproduced from Luo et al. [12] with permission of the copyright holders, Elsevier)

amines, i.e., methylamine, ethylamine, dimethylamine, n-propylamine, n-methylethylamine, and diethylamine.

The photocatalytic degradation of IPMP has been also investigated using TiO_2 Degussa P25 as catalyst in ultrapure water [75]. Under the studied conditions (C = 10 mg L^{-1}, C_{TiO2} = 100 mg L^{-1}, and I = 600 W m^{-2}), 95% of IPMP was removed within 20 min of irradiation. The major transformation products of TiO_2 photocatalysis of IPMP have been characterized by accurate mass measurements, and the transformation was found to proceed through hydroxylation, oxidation, and demethylation pathways. HO$^•$ radicals were found to be the most significant reactive species during the photocatalytic process. The contribution of $O_2^{•-}$ and h$^+$ was also confirmed. Based on the identified TPs during the applied AOPs (ozonation, catalytic ozonation, and heterogeneous photocatalysis) for the degradation of IPMP, the transformation pathways are presented in Fig. 7.

Finally, the oxidation of IBMP in aquatic samples by ozone has been recently reported [14]. At pH 6.8 the removal efficiency of IBMP (100 ng L^{-1}) reached over 90% within 15 min of ozonation. Five intermediates have been identified by GC-MS, i.e., 2-isobutyl-3-methylpyrazine, 1-(3-methoxy-2-pyrazinyl)-2-methyl-1-propanol, 4-methyl-2-oxovaleric acid, 1-(3-methoxy-2-pyrazinyl)-2-methyl-1-propanone, 5-methylcaproic acid, and 4-methyl-2-oxovaleric acid, and the possible degradation pathway of IBMP by ozonation is depicted in Fig. 8.

5 Coupling AOPs with Other Treatment Processes

Despite that AOPs are effective processes for the removal of numerous organic pollutants, their major limitation is related with their cost compared with other conventional treatments [77]. The coupling of AOPs with conventional processes has gained a lot of attention over the past several years for the treatment of water and wastewater. In particular, a combination of both processes is expected to lead to the highest efficiency and simultaneously reduce the overall treatment costs [78]. These hybrid methods are also becoming popular for T&O control, as they can generate complementary oxidation conditions and overcome the limitations of the individual conventional methods.

UV/H_2O_2 in pilot scale has been employed in combination with GAC adsorption for the removal of GSM and 2-MIB during the bloom season (September to November) [79]. The authors pointed out the benefits of the integration of processes for the removal of emerging cyanobacterial metabolites, as well as the need for further investigation of the process performance under extreme environmental conditions [79].

Conventional water treatment processes integrated with ozonation for the T&O control have also been recently studied at a drinking water treatment plant (DWTP). GSM and 2-MIB were not removed effectively during the drinking water treatment process and even seem to be increasing in the drinking water due to cell lyses and re-circulation of filter backwash water. However, optimization of the process can lead to the effective oxidation of T&O compounds [80].

Other configurations, like biologically active carbon (BAC) filters, have also been combined with ozonation achieving high removal efficiency for GSM and 2-MIB. Simultaneously, bromate formation was controlled by lowering the pH less than 7 [81]. A water treatment plant adopting O_3-BAC process was evaluated by Guo et al. [2], for the removal of multiple odorants causing different types of odor. The combined process has exhibited high efficiency in the simultaneous removal of GSM and 2-MIB, causing musty odor, and bis(2-chloroisopropyl) ether and the thioethers, diethyl disulfide and dimethyl disulfide, causing septic odor. However, for complete removal of the septic odorants, further treatment by sand filtration and chlorination was required. It is worth pointing out that careful management of

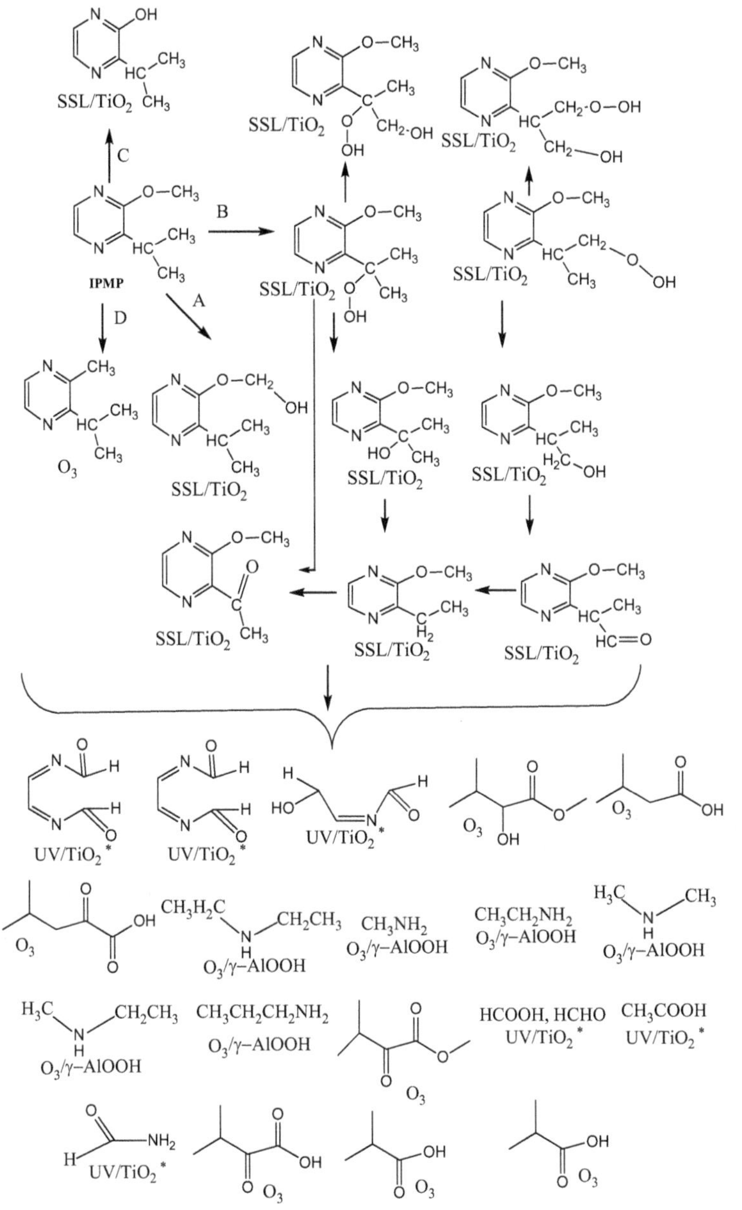

Fig. 7 Proposed transformation pathways and main TPs of IPMP under various AOPs (based on the studies [14, 75, 76])

sedimentation process should be considered in order to avoid the release of odorants in this treatment stage [2].

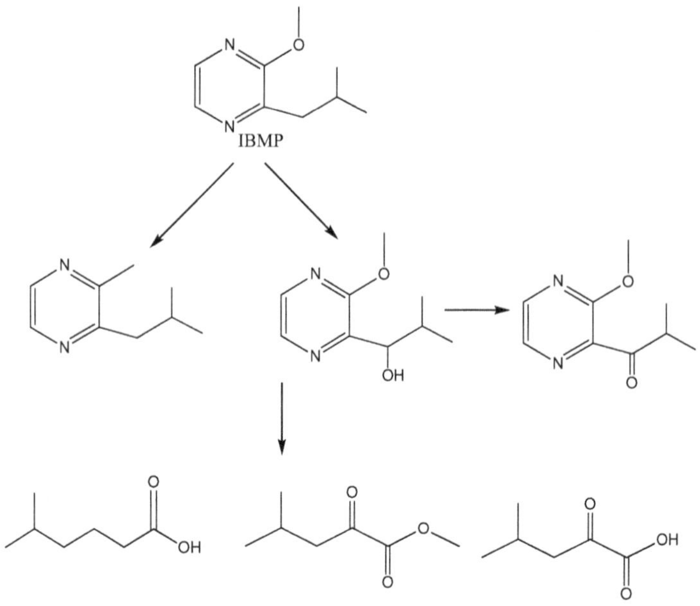

Fig. 8 Proposed transformation pathways and main TPs of IBMP using ozonation (based on the study [14])

Investigations in pilot scale for the removal of 2-MIB with a low- or medium-pressure ultraviolet-based advanced oxidation process (AOP) were performed in a water treatment plant (WTP) located in South Korea using flocculation/coagulation followed by sedimentation, sand filtration, and final chlorination as the final step. The comparison of the different experimental setup showed that low-pressure UV AOP can provide higher removal rates with lower electrical energy demand compared to medium-pressure systems [82].

An integrated process including coagulation, ozonation, ceramic membrane ultrafiltration, and activated carbon filtration was investigated for the removal of different pollutants including 2-MIB and GSM from micro-polluted source water. The removal efficiencies were enhanced in the combined system reaching the values of 96% for GSM and 88% for 2-MIB [83].

From a practical point of view, the integration of processes may be technically and economically feasible. The possibility to minimize the dose of the oxidant agent usually contributes significantly in the lower cost of the chemical process and followed also by a lower cost of conventional process that can reduce the overall cost while improving the overall removal.

6 Conclusions

The number of studies dedicated to T&O control with the application of AOPs follows an increasing trend during the recent years. Investigations on advanced oxidation methods for degradation of T&O compounds are mainly referred to ultrapure, river, and reservoir water in lab and pilot scale. With the application of various AOPs, high removal percentages were achieved for the T&O compounds of interest in ultrapure and real water samples. Percentages from ~60 to 100% of GSM and 2-MIB and 95% of IPMP were removed by heterogeneous photocatalysis using different catalysts and experimental conditions. UV/persulfate process led to the removal higher than about 95%, 80%, and 90% for GSM, 2-MIB, and 2,4,6-TCA, respectively. The removal of 2-MIB was approximately in the range of ~52–99.99% for UV/H_2O_2 and up to 100% for photo-Fenton in acidic pH. In case of GSM, ~38.28–99%, >90%, and 79% degradation were achieved with the application of UV/H_2O_2, UV/chlorine, and photo-Fenton (pH = 3), respectively. Large variations in compounds degradation from 27.5 to 90% were observed by O_3-based processes (ozonation, catalytic ozonation, and electro-peroxone process), depending on the studied compound and the applied experimental conditions.

The water quality parameters like pH, NOM, and inorganic species played a significant role in the whole processes. The presence of inorganic and organic water constituents often inhibited the degradation of target pollutants through radical scavenging mechanisms. The majority of the studies are focused mainly on the investigation of the operational parameters and kinetics, while the identification of the TPs formed during the processes is less frequently studied. Elucidation of chemical structures of by-products to evaluate the degradation pathways is an important demand. Data regarding the T&O potency of the TPs are limited up to now, as well. T&O assessment of TPs is of great importance and should be included in the various studies. Structure characterization of by-products and their simultaneous assessment of their possible taste and odor properties are closely related topics of interest that need to be addressed and will give useful insights for the applicability of a process on a practical level under real conditions. Additionally, the existing studies are generally performed for individual compounds at concentrations close to those found in the environment and limited studies addressing the degradation of mixtures of compounds.

The main conclusion arrived from the overall assessment of the literature is that despite the current extensive research in T&O removal, significant knowledge gaps remain. Overall, while there are promising results in removing T&O compounds using AOPs under laboratory conditions and pilot scale, more intensive research is needed before these methods can be considered as economically feasible and practically sustainable alternatives in water treatment facilities. In order to promote the feasibility of AOPs in the future, several aspects have to be more extensively addressed. Thus, more work needs to be done (1) on the elucidation of the structure of the products generated during the treatment, (2) on the study of the formation of potential T&O intermediate products, (3) on the performance of complete economic studies for the estimation of the overall costs, (4) on the application of integrated or hybrid systems for enhanced removal, and (5) on the assessment of the processes for the degradation of mixtures of compounds. Moreover, a lot more is needed from engineering design and modeling, related to their industrial development and process scaling-up.

References

1. Antonopoulou M, Evgenidou E, Lambropoulou D, Konstantinou I (2014) A review on advanced oxidation processes for the removal of taste and odor compounds from aqueous media. Water Res 53:215–234
2. Guo Q, Yang K, Yu J, Wang C, Wen X, Zhang L, Yang M, Xia P, Zhang D (2016) Simultaneous removal of multiple odorants from source water suffering from septic and musty odors: verification in a full-scale water treatment plant with ozonation. Water Res 100:1–6
3. Watson SB, Monis P, Baker P, Giglio S (2016) Biochemistry and genetics of taste- and odor-producing cyanobacteria. Harmful Algae 54:112–127
4. Watson SB (2013) Ecotoxicity of taste and odour compounds. Chapter 33 In: Férard JF, Blaise C (eds) Encyclopedia of aquatic ecotoxicology, Springer, pp 337–352
5. Peter A, von Gunten U (2007) Oxidation kinetics of selected taste and odor compounds during ozonation of drinking water. Environ Sci Technol 41:626–631
6. Cotsaris E, Bruchet A, Mallevialle J, Bursill DB (1995) The identification of odorous metabolites produced from algal monocultures. Water Sci Technol 31:251–258
7. Jo CH, Dietrich AM (2009) Removal and transformation of odorous aldehydes by UV/H_2O_2. J Water Supply Res Technol 58(8):580–586
8. Sun A, Xiong Z, Xu Y (2008) Removal of malodorous organic sulfides with molecular oxygen and visible light over metal phthalocyanine. J Hazard Mater 152:191–195
9. Andreozzi R, Caprio V, Marotta R (2001) Oxidation of benzothiazole, 2-mercaptobenzothiazole and 2-hydroxybenzothiazole in aqueous solution by means of H_2O_2/UV or photoassisted Fenton systems. J Chem Technol Biotechnol 76(2):196–202
10. Acero JL, von Gunten U, Piriou P (2005) Kinetics and mechanisms of formation of bromophenols during drinking water chlorination: assessment of taste and odor development. Water Res 39:2979–2993
11. Bichsel Y, von Gunten U (2000) Formation of iodo-trihalomethanes during disinfection and oxidation of iodide-containing waters. Environ Sci Technol 34(13):2784–2791
12. Luo C, Jiang J, Ma J, Pang S, Liu Y, Song Y, Guan C, Li J, Jin Y, Wu D (2016) Oxidation of the odorous compound 2,4,6-trichloroanisole by UV activated persulfate: kinetics, products, and pathways. Water Res 96:12–21
13. Yao W, Qu Q, von Gunten U, Chen C, Yu G, Wang Y (2017) Comparison of methylisoborneol and geosmin abatement in surface water by conventional ozonation and an electro-peroxone process. Water Res 108:373–382
14. Wang C, Li C, Li H, Lee H, Yang Z (2017) The removal efficiency and degradation pathway of IPMP and IBMP in aqueous solution during ozonization. Sep Purif Technol 179:297–303
15. Pirbazari M, Borow HS, Craig S, Ravindran V, McGuire MJ (1992) Physical chemical characterization of five earthy-musty-smelling compounds. Water Sci Technol 25(2):81–88
16. An N, Xie H, Gao N, Deng Y, Chu W, Jiang J (2012) Adsorption of two taste and odor compounds IPMP and IBMP by granular activated carbon in water. Clean Soil Air Water 40 (12):1349–1356
17. Ding Z, Peng S, Xi W, Zheng H, Chen X, Yin L (2014) Analysis of five earthy-musty odorants in environmental water by HS-SPME/GC-MS. Int J Anal Chem 2014:11 p. Article ID 697260

18. Li X, Lin P, Wang J, Liu Y, Li Y, Zhang X, Chen C (2016) Treatment technologies and mechanisms for three odorants at trace level: IPMP, IBMP, and TCA. Environ Technol 37(3):308–315
19. Meylan W, Howard PH (1991) Bond contribution method for estimating Henry's law constants. Environ Toxicol Chem 10:1283–1293
20. Deng X, Liang G, Chen J, Qi M, Xie P (2011) Simultaneous determination of eight common odors in natural water body using automatic purge and trap coupled to gas chromatography with mass spectrometry. J Chromatogr A 1218:3791–3798
21. Young WF, Horth H, Crane R, Ogden T, Arnott M (1996) Taste and odour threshold concentrations of potential potable water contaminants. Water Res 30:331–340
22. Suffet IH, Khiari D, Bruchet AE (1999) A drinking water taste and odor wheel for the millennium: beyond geosmin and methylisoborneol. Water Sci Technol 40(6):1–13
23. Khiari D, Barrett S, Chinn R, Bruchet A, Piriou P, Matia L, Ventura F, Suffet IH, Gittelman T, Leutweiler P (2002) Distribution generated taste-and-odor phenomena: AWWA Research Foundation and American Water Works Association, Denver, CO, USA
24. Suffet IH, Rosenfeld P (2007) The anatomy of odour wheels for odours of drinking water, wastewater, compost and the urban environment. Water Sci Technol 55(5):335–344
25. Jung S, Baek K, Yu M (2004) Treatment of taste and odor materials by oxidation adsorption. Water Sci Technol 49(9):289–295
26. Cook D, Newcombe G, Sztajnbok P (2001) The application of powdered activated carbon for MIB and geosmin removal: predicting PAC doses in four raw waters. Water Res 35:1325–1333
27. Chestnutt TE, Bach MT, Mazyck TW (2007) Improvement of thermal reactivated carbon for the removal of 2-methylisoborneol. Water Res 41(1):79–86
28. Song W, O'Shea KE (2007) Ultrasonically induced degradation of 2-methylisoborneol and geosmin. Water Res 41(12):2672–2678
29. Rosenfeldt EJ, Melcher B, Linden KG (2005) UV and UV/H_2O_2 treatment of methylisoborneol (MIB) and geosmin in water. J Water Supply Res Technol 54(7):423–434
30. Hoigne J, Bader H (1994) Kinetics of reactions of chlorine dioxide (Oclo) in water. 1. Rate constants for inorganic and organic-compounds. Water Res 28(1):45–55
31. Waldemer RH, Tratnyek PG (2006) Kinetics of contaminant degradation by permanganate. Environ Sci Technol 40(3):1055–1061
32. Xie P, Ma J, Liu W, Zou J, Yue S, Li X, Wiesner MR, Fang J (2015) Removal of 2- MIB and geosmin using UV/persulfate: contributions of hydroxyl and sulphate radicals. Water Res 69:223–232
33. Ribeiro A, Nunes OC, Pereira MFR, Silva AMT (2015) An overview on the advanced oxidation processes applied for the treatment of water pollutants defined in the recently launched Directive 2013/39/EU. Environ Int 75:33–51
34. Boczkaj G, Fernandes A (2017) Wastewater treatment by means of advanced oxidation processes at basic pH conditions: a review. Chem Eng J 320:608–633
35. Liato V, Aïder M (2016) Geosmin as a source of the earthy-musty smell in fruits, vegetables and water: origins, impact on foods and water, and review of the removing techniques. Chemosphere 181:9–18
36. Park JA, Nam HL, Choi JW, Ha J, Lee SH (2017) Oxidation of geosmin and 2-methylisoborneol by the photo-Fenton process: kinetics, degradation intermediates, and the removal of microcystin-LR and trihalomethane from Nak-Dong River water, South Korea. Chem Eng J 313:345–354
37. Tan F, Chen H, Wu D, Wang N, Gao Z, Wang L (2016) Optimization of geosmin removal from drinking water using UV/H_2O_2. J Resid Sci Tech 13(1):23–30
38. Wang D, Bolton JR, Andrews SA, Hofmann R (2015) UV/chlorine control of drinking water taste and odour at pilot and full-scale. Chemosphere 136:239–244

39. Fotiou T, Triantis TM, Kaloudis T, Papaconstantinou E, Hiskia A (2014) Photocatalytic degradation of water taste and odour compounds in the presence of polyoxometalates and TiO_2: intermediates and degradation pathways. J Photochem Photobiol A Chem 286:1–9
40. Wee LH, Janssens N, Vercammen J, Tamaraschi L, Thomassen LCJ, Martens JA (2015) Stable TiO_2 USY zeolite composite coatings for efficient adsorptive and photocatalytic elimination of geosmin from water. J Mater Chem A 3:2258–2264
41. Pestana CJ, Robertson PKJ, Edwards C, Wilhelm W, Mckenzie C, Lawton LA (2014) A continuous flow packed bed photocatalytic reactor for the destruction of 2-methylisoborneol and geosmin utilizing pelletised TiO_2. Chem Eng J 235:293–298
42. Pettit S, Rodriguez-Gonzalez L, Michaels J, Alcantar N, Ergas S, Kuhn J (2014) Parameters influencing the photocatalytic degradation of geosmin and 2- methylisoborneol utilizing immobilized TiO_2. Catal Lett 144:1460–1465
43. Xue Q, Liu Y, Zhou Q, Utsumi M, Zhang Z, Sugiura N (2016) Photocatalytic degradation of geosmin by Pd nanoparticle modified WO_3 catalyst under simulated solar light. Chem Eng J 283:614–621
44. Bu L, Zhou S, Shi Z, Deng L, Gao N (2017) Removal of 2-MIB and geosmin by electrogenerated persulfate: performance, mechanism and pathways. Chemosphere 168:1309–1316
45. Tan F, Chen H, Wu D, Lu N, Wang L, Gao Z (2016) Optimization of removal of 2-methylisoborneol from drinking water using UV/H_2O_2. J Adv Oxid Technol 19(1):98–104
46. Lee JH, Park EJ, Kim DH, Jeong MG, Kim YD (2016) Superhydrophobic surfaces with photocatalytic activity under UV and visible light irradiation. Catal Today 260:32–38
47. Xu B, Qi F (2016) Reaction mechanism of 2-methylisoborneol and 2,4,6-trichloroanisole in catalytic ozonation by γ-AlOOH: role of adsorption. Clean Soil Air Water 44(9):1099–1105
48. Buxton GV, Greenstock CL (1988) Critical review of rate constants of reactions of hydrated electrons, hydrogen atoms and hydroxyl radicals (·OH/O−) in aqueous solution. J Phys Chem Ref Data Monogr 17:513–886
49. Popov E, Mametkuliyev M, Santoro D, Liberti L, Eloranta J (2010) Kinetics of $UV-H_2O_2$ advanced oxidation in the presence of alcohols: the role of carbon centered radicals. Environ Sci Technol 44:7827–7832
50. Wang D, Bolton JR, Hofmann R (2012) Medium pressure UV combined with chlorine advanced oxidation for trichloroethylene destruction in a model water. Water Res 46 (15):4677–4686
51. Goldstein S, Aschengrau D, Diamant Y, Rabani J (2007) Photolysis of aqueous H_2O_2: quantum yield and applications for polychromatic UV actinometry in photoreactors. Environ Sci Technol 41(21):7486–7490
52. Watts MJ, Linden KG (2007) Chlorine photolysis and subsequent OH radical production during UV treatment of chlorinated water. Water Res 41(13):2871–2878
53. Avetta P, Pensato A, Minella M, Malandrino M, Maurino V, Minero C, Hanna K, Vione D (2015) Activation of persulfate by irradiated magnetite: implications for the degradation of phenol under heterogeneous photo-Fenton like conditions. Environ Sci Technol 49 (2):1043–1050
54. Lau TK, Chu W, Graham NJD (2007) The aqueous degradation of butylated hydroxyanisole by $UV/S_2O_8^{2-}$: study of reaction mechanisms via dimerization and mineralization. Environ Sci Technol 41(2):613–619
55. Yu XY, Bao ZC, Barker JR (2004) Free radical reactions involving $Cl^·$, $Cl_2^{·-}$, and $SO_4^{·-}$ in the 248 nm photolysis of aqueous solutions containing $S_2O_8^{2-}$ and Cl^-. J Phys Chem A 18 (2):295–308
56. Buxton GV, Salmon GA, Wood N (1990) A pulse radiolysis study of the chemistry of oxysulphur radicals in aqueous solution. In: Physico-chemical behaviour of atmospheric pollutants, Springer, The Netherlands, pp 245–250
57. Neta P, Huie RE, Ross AB (1988) Rate constants for reactions of inorganic radicals in aqueous solution. J Phys Chem Ref Data Monogr 17(3):1027–1284

58. Von Gunten U (2003) Ozonation of drinking water: part II. Disinfection and by-product formation in presence of bromide, iodide or chlorine. Water Res 37(7):1469–1487
59. Darsinou B, Frontistis Z, Antonopoulou M, Konstantinou I, Mantzavinos D (2015) Sono-activated persulfate oxidation of bisphenol A: kinetics, pathways and the controversial role of temperature. Chem Eng J 280:623–633
60. Sharma VK, Triantis TM, Antoniou MG, He X, Pelaez M, Han C, Song W, O'Shea KE, de la Cruz AA, Kaloudis T, Hiskia A, Dionysiou DD (2012) Destruction of microcystins by conventional and advanced oxidation processes: a review. Sep Purif Technol 91:3–17
61. Bamuza-Pemu EE, Chirwa EMN (2011) Photocatalytic degradation of geosmin: intermediates and degradation pathway analysis. Chem Eng Trans 24:91–96
62. Bamuza-Pemu EE, Chirwa EM (2012) Photocatalytic degradation of geosmin: reaction pathway analysis. Water SA 38(5):689–696
63. Qi F, Xu B, Chen Z, Ma J, Sun D, Zhang L (2009) Efficiency and products investigations on the ozonation of 2-methylisoborneol in drinking water. Water Environ Res 81(12):2411–2419
64. Li X, Huang Y, Wang D (2010) Efficiency and mechanism of degradation of 2-methylisoborneol (2-MIB) by O_3/H_2O_2 in water. 4th international conference on bioinformatics and biomedical engineering (iCBBE), Chengdu, China, pp 1–4
65. Zaitlin B, Watson SB (2006) Actinomycetes in relation to taste and odour in drinking water: myths, tenets and truths. Water Res 40(9):1741–1753
66. Urase T, Sasaki Y (2013) Occurrence of earthy and musty odor compounds (geosmin, 2-methylisoborneol and 2,4,6-trichloroanisole) in biologically treated wastewater. Water Sci Technol 68(9):1969–1975
67. Benanou D, Acobas F, de Roubin MR, Sandra FDP (2003) Analysis of off-flavors in the aquatic environment by stir bar sorptive extraction-thermal desorption capillary GC/MS/ olfactometry. Anal Bioanal Chem 376(1):69–77
68. Copete ML, Zalacain A, Lorenzo C, Carot JM, Esteve MD, Climent M, Salinas MR (2009) Haloanisole and halophenol contamination in Spanish aged red wines. Food Addit Contam A 26(1):32–38
69. Maillet L, Lenes D, Benanou D, Le Cloirec P, Correc O (2009) The impact of private networks on off-flavour episodes in tap water. J Water Supply Res Technol 58(8):571–579
70. Vlachos P, Stathatos E, Lyberatos G, Lianos P (2008) Gas-phase photocatalytic degradation of 2,4,6-trichloroanisole in the presence of a nanocrystalline titania film. Applications to the treatment of cork stoppers. Catal Comm 9(10):1987–1990
71. Qi F, Xu B, Zhao L, Chen Z, Zhang L, Sun D, Ma J (2012) Comparison of the efficiency and mechanism of catalytic ozonation of 2,4,6-trichloroanisole by iron and manganese modified bauxite. Appl Catal Environ 121–122:171–181
72. Qi F, Xu B, Chen Z, Ma J, Sun D, Zhang L, Wu F (2009) Ozonation catalyzed by the raw bauxite for the degradation of 2,4,6-trichloroanisole in drinking water. J Hazard Mater 168:246–252
73. Ma Z, Xie P, Chen J, Niu Y, Tao M, Qi M, Zhang W, Deng X (2013) Microcystis blooms influencing volatile organic compounds concentrations in lake Taihu. Fresen Environ Bull 22:95–102
74. Peter A, Koster O, Schildknecht A, von Gunten U (2009) Occurrence of dissolved and particle-bound taste and odor compounds in Swiss lake waters. Water Res 43:2191–2200
75. Antonopoulou M, Konstantinou I (2015) TiO_2 photocatalysis of 2-isopropyl-3-methoxypyrazine taste and odor compound in aqueous phase: kinetics, degradation pathways and toxicity evaluation. Catal Today 240:22–29
76. Qi F, Xu B, Chen Z, Feng L, Zhang L, Sun D (2013) Catalytic ozonation of 2-isopropyl-3-methoxypyrazine in water by gamma-AlOOH and gamma-Al_2O_3: comparison of removal efficiency and mechanism. Chem Eng J 219:527–536
77. Sarria V, Kenfack S, Guillod O, Pulgarin C (2003) An innovative coupled solar biological system at field pilot scale for the treatment of biorecalcitrant pollutants. J Photochem Photobiol A Chem 139(1):89–99

78. Klavarioti M, Mantzavinos D, Kassinos D (2009) Removal of residual pharmaceuticals from aqueous systems by advanced oxidation processes. Environ Int 35:402–417
79. Zamyadi A, Sawade E, Ho L, Newcombe G, Hofmann R (2015) Impact of UV–H_2O_2 advanced oxidation and aging processes on GAC capacity for the removal of cyanobacterial taste and odor compounds. Environ Health Insights 9(S3):1–10
80. Swanepoel A, Du Preez HH, Cloete N (2017) The occurrence and removal of algae (including cyanobacteria) and their related organic compounds from source water in Vaalkop Dam with conventional and advanced drinking water treatment processes. Water SA 43(1):67–80
81. Edwards-Brandt J, Shorney-Darby H, Neemann J, Hesby J, Conrad T (2007) Use of ozone for disinfection and taste and odor control at proposed membrane facility. Ozone Sci Eng 29:281–286
82. Scheideler J, Lee KH, Raichle P, Choi T, Dong HS (2015) UV-advanced oxidation process for taste and odor removal – comparing low pressure and medium pressure UV for a full-scale installation in Korea. Water Pract Technol 10(1):66–72
83. Fan X, Tao Y, Wang L, Zhang X, Lei Y, Wang Z, Noguchi H (2014) Performance of an integrated process combining ozonation with ceramic membrane ultra-filtration for advanced treatment of drinking water. Desalination 335:47–54

Wastewater Treatment by Heterogeneous Fenton-Like Processes in Continuous Reactors

Bruno M. Esteves, Carmen S.D. Rodrigues, and Luis M. Madeira

Abstract The treatment of several industrial effluents, such as textile, pharmaceutical, and phenol-containing wastewaters, often face limitations toward conventional treatment processes. Solutions for such problematic include the use of several advanced oxidation processes (AOPs) and particularly the Fenton one. However, most of the research has been focused in the homogeneous process, while recent trends point for the use of heterogeneous systems, with the catalyst immobilized in a solid support. In addition, process optimization and catalyst screening are commonly carried out in batch reactors, which often are not the best solution for continuous industrial units. In this chapter, a review is made regarding the application of heterogeneous Fenton-like advanced oxidation processes in continuous systems (fixed-bed, fluidized-bed, and continuous stirred-tank reactors). The application of this catalytic process for pollutant/wastewater treatment is summarized, giving emphasis to the effect of the key operational parameters (e.g., pH, feed dose of H_2O_2, catalyst load, and feed flow rate) affecting the oxidative performance of such systems. Moreover, the main physicochemical properties of heterogeneous catalysts (e.g., source of support and particle size) and preparation methods (e.g., type of precursor and metal ion) affecting the catalytic efficiency of Fenton's oxidation, and the stability of the catalyst itself, are also discussed. Finally, some operational issues of concern regarding solid catalysts operating in continuous-flow reactors are addressed.

Keywords Catalyst, Continuous reactor, Fenton, Heterogeneous, Wastewater treatment

---

B.M. Esteves, C.S.D. Rodrigues, and L.M. Madeira (✉)
LEPABE – Laboratório de Engenharia de Processos, Ambiente, Biotecnologia e Energia, Departamento de Engenharia Química, Faculdade de Engenharia da Universidade do Porto, R. Dr. Roberto Frias, 4200-465 Porto, Portugal
e-mail: mmadeira@fe.up.pt

Contents

1. Introduction 213
2. Chemical Reactors 215
   2.1 Batch Reactor 215
   2.2 Continuous Reactors 217
3. Heterogeneous Fenton/Fenton-Like Processes 222
   3.1 Fenton's Oxidation: Mechanism and Characteristics 222
   3.2 Operational Parameters of Influence 225
   3.3 Influence of Catalyst's Physicochemical Properties 241
   3.4 Operational Issues Regarding Heterogeneous Catalysts: Stability .. 244
   3.5 Performance of Continuous vs Batch Reactors 246
4. Conclusions 247
References 249

Abbreviations

AB	Alcian Blue-tetrakis dye
AC	Activated carbon
AOPs	Advanced oxidation processes
AY36	Acid yellow 36 dye
BB3	Basic blue 3 dye
BOD_5	Biochemical oxygen demand (5 days)
CNF	Carbon nanofibers
CNT	Carbon nanotubes
COD	Chemical oxygen demand
CSB	Chicago Sky Blue dye
CSTR	Continuous stirred-tank reactor
CWHPO	Catalytic wet hydrogen peroxide oxidation
CWPO	Catalytic wet peroxide oxidation
EDCs	Endocrine disrupting compounds
FBR	Fluidized-bed reactor
LHSV	Liquid hourly space velocity
NM	Natural magnetite
NP	Natural pyrite
OII	Orange II dye
PAN	Polyacrylonitrile
PBR	Packed-bed reactor
POPs	Persistent organic pollutants
PPCPs	Pharmaceutical and personal care products
RB19	Reactive blue 19
S_{BET}	Specific surface area
SW	Sewage
TDS	Total dissolved solids
TOC	Total organic carbon
UFBR	Upflow fixed-bed reactor

UV   Ultraviolet
WHPCO   Wet hydrogen peroxide catalytic oxidation
WW   Wastewater

1 Introduction

The increasing worldwide demand for clean water requires the development of cost-effective techniques for wastewater reclamation. The unceasing research and development of new products, technologies, and manufacturing processes lead to the exponential growth of many water-intensive industries, with new and potentially hazardous compounds being continuously introduced into the environment through different industrial wastewater streams.

Management of such effluents is a subject of major importance and has been under great scrutiny by the scientific community over the last decades [1–3]. Wastewater coming from major polluting industries, such as the chemical (e.g., fertilizer, petrochemical), pharmaceutical, dye and textile, pulp and paper, food, and many others, often contain persistent organic pollutants (POPs) that are biorefractory in nature and resistant to conventional wastewater treatments. Moreover, several organic compounds are potentially hazardous to public health and the environment, since they exhibit toxicity toward microorganisms and plants [4, 5], as well as carcinogenic [6, 7], mutagenic, and endocrine disruptor effects on the metabolic activity [8, 9].

With the fast industrial development, environmental regulation and health quality standards imposed by different governments also become more stringent. Consequently, for many industries, the selection and practice of any given treatment process relies on the characteristics of the wastewater produced, as to its pollutant's nature and quantity, and the level of reduction required for safe discharge, at the lowest operation and maintenance cost available.

Wastewater treatment technologies such as biological degradation [10, 11], precipitation [12, 13], membrane separation [14–16], coagulation/flocculation [17–19], adsorption [20–22], direct oxidation [23–25], as well as different combinations of the cited processes [26] have been extensively studied for the treatment of different pollutants and widely applied at industrial scale.

However, apart from biological and chemical processes, most technologies do not involve chemical transformations, hence only transfer the pollutant from one phase to another, causing secondary loading to the environment [27]. On the other hand, biological processes are often inefficient given the toxicity and low biodegradability of many molecules found in industrial wastewaters [28–30]. Direct chemical oxidation is widely used to degrade biorefractory substances, and high degradation efficiencies are possible with such techniques; nevertheless, the use of strong oxidants at slightly demanding operating conditions (namely, temperatures) and equipment necessary often make the process economically unviable [23].

In this context, the development of economically attractive and efficient wastewater treatment technologies has been an interesting topic of discussion. Among all treatment practices available, the advanced oxidation processes (AOPs) may constitute a useful technology for the mineralization of aromatics, pesticides, solvents, and volatile organics, just to give a few examples, into stable compounds – H_2O, CO_2, and inorganic salts. Even if complete mineralization is not possible/viable, the partial oxidation of nonbiodegradable organics often originates biodegradable intermediates, allowing the application of subsequent treatment steps (e.g., biological processes) [2].

AOPs have in common the fact that they form and take advantage of the great oxidative capacity of the hydroxyl radical ($HO^{\bullet}$) which, immediately after the fluorine, is the known species with greater oxidation potential (Table 1) [31, 32]. Also, hydroxyl radicals are nonselective and highly reactive species that attack most organic and many inorganic solutes, with high rate constants, typically in the range of 10^{-6}–10^{-9} L mol^{-1} s^{-1} [33–35].

The versatility of the AOPs is also enhanced by the fact that there are numerous ways for $HO^{\bullet}$ production, such as chemical, photochemical, and electrochemical, allowing the selection of the method that best fulfills specific treatment requirements. AOPs are commonly classified in respect to the reactive phase as homogeneous, if the catalyst is dissolved in solution (thus forming a single phase), or heterogeneous, if the metal ions are impregnated in the matrix of a solid phase (e.g., activated carbon, clay, zeolite, carbon nanotube). Table 2 shows some conventional and nonconventional AOPs based on the type of process and source used for hydroxyl radical generation.

Heterogeneous Fenton/Fenton-like processes, often called wet hydrogen peroxide catalytic oxidation (WHPCO) or catalytic wet peroxide oxidation (CWPO), were extensively studied over the last three decades and will be the subject of this chapter. Particular emphasis will be given to their application in continuous systems for the treatment of model compounds and industrial wastewaters.

Table 1 Oxidation potential of common oxidants in acidic medium oxidation potential of the highly reactive hydroxyl radical formed from the reaction between hydrogen peroxide and ferrous iron – Fenton's reagent

Oxidant	Oxidation potential (V)
Fluorine (F_2)	3.03
Hydroxyl radical ($HO^{\bullet}$)	**2.80**
Atomic oxygen (O)	2.42
Ozone (O_3)	2.07
Hydrogen peroxide (H_2O_2)	1.78
Hydroperoxyl radical ($HO_2^{\bullet}$)	1.70
Potassium permanganate ($KMnO_4$)	1.68
Hypochlorous acid (HClO)	1.49
Chlorine (Cl_2)	1.36
Oxygen (O_2)	1.23
Bromine (Br_2)	1.09

Adapted from Bigda [31]

Table 2 Examples of common advanced oxidation processes (AOPs)

Advanced oxidation processes			
Homogeneous		Heterogeneous	
Using energy	Without energy	Using energy	Without energy
O_3 + UV	O_3	$TiO_2 + O_2$ + UV	O_3 + solid catalyst
H_2O_2 + UV	$O_3 + H_2O_2$	$TiO_2 + H_2O_2$ + UV	$H_2O_2 + Fe^{2+}/Fe^{3+}/m^{n+}$ solid
$O_3 + H_2O_2$ + UV	$H_2O_2 + Fe^{2+}$ (Fenton)	H_2O_2 + solid catalyst + UV (photo-Fenton)	$H_2O_2 + Fe^0/Fe$ (nano-zerovalent iron)
$H_2O_2 + Fe^{2+}$ + UV (photo-Fenton)	$H_2O_2 + Fe^{3+}/m^{n+}$ (Fenton like)		
Electro-Fenton			
Photo-electro-Fenton			

Adapted from Poyatos et al. [36]

2 Chemical Reactors

2.1 Batch Reactor

Discontinuous reactors, also called batch reactors, are commonly used for liquid phase reactions, with the possibility of using solid catalysts in suspension (so-called slurry reactors). For both cases, agitation is required to homogenize the solution and equalize the temperature inside the reactor, to keep it at the desired level by efficient heat exchange through a recirculation jacket (or coil), which is connected to a cooling/heating system (see Fig. 1). In batch operation, the reactants and solution to be treated are fed to the reactor at the beginning of the process. Therefore, the operation proceeds at constant volume, and it is commonly admitted that there are no spatial variations of composition and temperature within the vessel; composition, however, varies over time [37].

Typically, batch reactors are used in the processing industry and medical products manufacturing, in laboratory applications (e.g., whenever a process is still in test phase), or in the study of reaction kinetics (essential stage for modeling and upscale). Batch reactors are also vastly used in wastewater treatment studies by AOPs, namely, by the Fenton process, to assess the effect of operating conditions, to optimize the process, or simply for catalysts screening.

Because Fenton's oxidation runs commonly at moderate conditions of temperature and pressure, a batch reactor consists essentially of a non-pressurized stirred device (Fig. 2a); for industrial operation it must be coupled with metering pumps for the addition of acid, base, a ferrous sulfate catalyst solution (in the case of homogeneous processes), and industrial strength hydrogen peroxide (35–50%) [38]. For heterogeneous Fenton process operation, the catalyst is usually kept in suspension (see Fig. 2b).

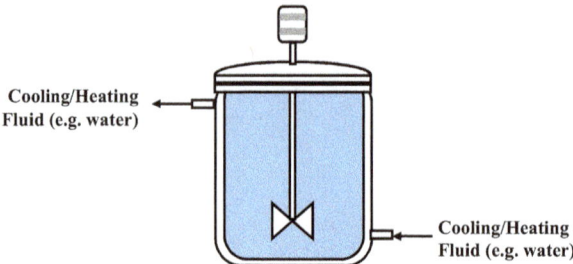

Fig. 1 Scheme of a typical batch reactor

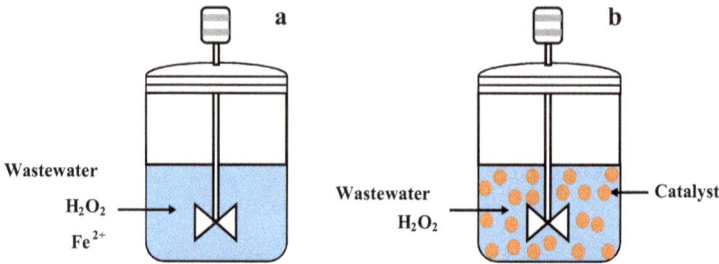

Fig. 2 Scheme of wastewater treatment by homogeneous (**a**) and heterogeneous (**b**) Fenton process in batch reactors

It is recommended the coating of the reactor vessel with an acid resistant material, as Fenton reagent is very aggressive and corrosion may become a serious operational issue. The addition of reactants is usually performed in the following sequence: wastewater (or solution to be treated), followed by diluted sulfuric acid (for maintaining acidic conditions), catalyst in acidic solutions, and base or acid for the adjustment of pH; reaction starts with the addition of the oxidant (hydrogen peroxide solution) which can be either instantaneous (batch reactor) or progressive over time (semicontinuous reactor). Industrially, reactants (and particularly the H_2O_2) must be added slowly with proper maintenance of temperature. Downstream Fenton's reactor, the treated solution/wastewater is discharged into a neutralizing tank for pH adjustment, followed by a flocculation tank and a solid liquid separation tank, where the adjustment of TDS (total dissolved solids) content is commonly performed. A schematic representation of the described Fenton oxidation treatment process in discontinuous mode is presented in Fig. 3.

The batch reactor has been widely used in lab-scale studies, where homogeneous and heterogeneous Fenton processes are applied for the treatment of model pollutants and different wastewaters [39, 40]. Also, catalytic efficiency and stability tests of solid catalysts are regularly performed in slurry reactors, due to the easiness in operation [41, 42]. However, the use of such reactors for these applications at industrial level has several disadvantages since effluents are normally generated continuously, so an equalization tank is required to store the effluent prior to the treatment. Other shortcomings such as the high concentration of intermediate

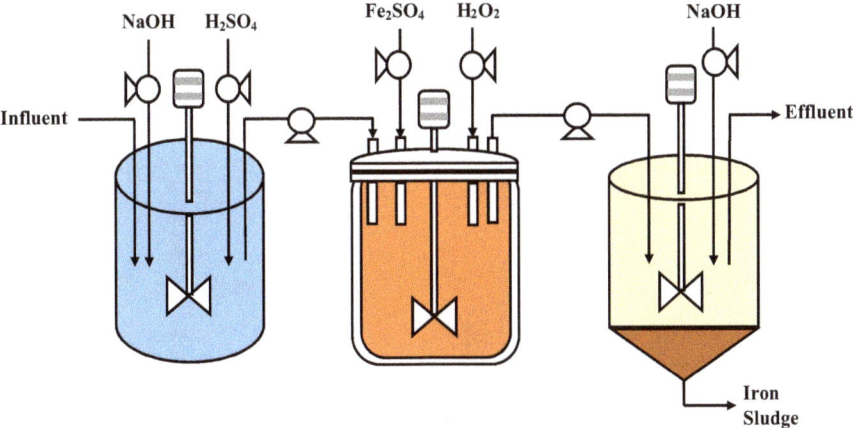

Fig. 3 Representative scheme of wastewater treatment by Fenton process

products formed and their long contact time with the catalyst, which may result in its loss of longevity due to deactivation mechanisms (further discussion regarding this subject is compiled in Sect. 3.4), are a concern regarding the slurry configuration in some applications. Moreover, this type of reactors usually has high cost of labor per unit of production as well as difficulty in maintaining large-scale production.

2.2 Continuous Reactors

In continuous reactors, or flow reactors, the reactants are continuously fed into the vessel at one point, reaction occurs, and then a continuous stream of end product emerges at another point of the reactor. There are two basic (and somehow ideal) configurations for continuous reactors: the plug flow reactor (Fig. 4a) and the perfectly mixed flow reactor or continuous stirred-tank reactor (CSTR) – see Fig. 4b. Plug flow reactors are commonly implemented in tubular units. Depending on different operational parameters, namely, the velocity of the fluid phase, heterogeneous reactions are implemented in fixed-bed or packed-bed reactors (PBR) or in fluidized-bed reactors (FBR); CSTRs with the catalyst in suspension (often within a basket) are another possible configuration. A brief description of the typical operation of heterogeneous Fenton processes in such reactors is provided below.

2.2.1 Fixed-Bed Reactor

In fixed-bed reactors, the solid catalysts are deposited along the length of the column (single-bed) (Fig. 5a) or in multi-bed column/multi-tubular reactor (Fig. 5b); the reactants are typically fed upflow into the column, so a common

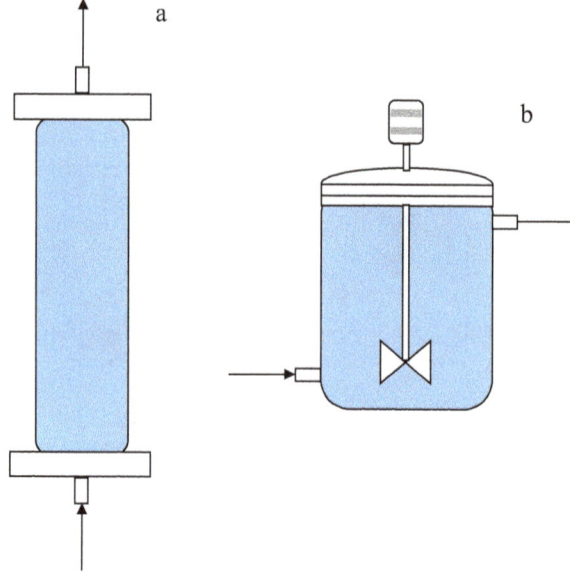

Fig. 4 Scheme of a plug flow reactor (**a**) and a CSTR (**b**)

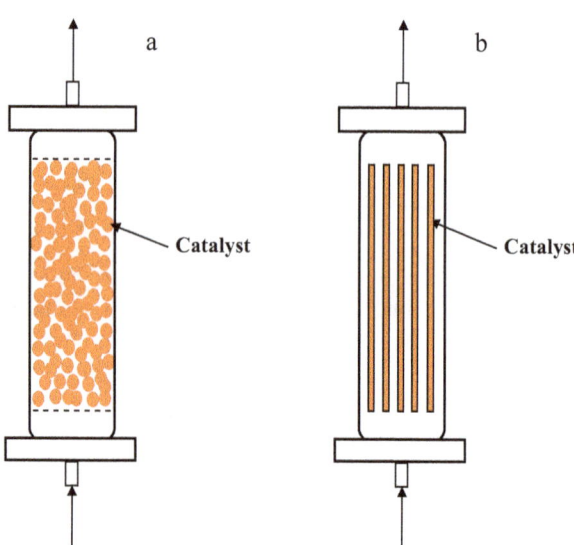

Fig. 5 Scheme of a single-bed (**a**) and multi-bed (**b**) fixed-bed reactor

designation for this type of reactors is UFBR (upflow fixed-bed reactor). Even if the flow pattern is not the ideal one (i.e., is far from plug flow due to, e.g., axial dispersion effects), oxidant is consumed and the solution/effluent is treated as they flow along the column. In this configuration, the composition (and temperature) has spatial variations in the axial direction and eventually also in the radial direction. For the hypothetical and idealized plug flow pattern, it is assumed that the transfer phenomenon is only due to convective effects, in which all molecules of the effluent and the oxidant move at the same velocity along parallel streamlines [37].

Single-bed reactors have been used in research studies for the treatment of pharmaceutical [43] and textile [44] wastewaters, in dye removal [44–46], and oxidation of phenolic compounds [47–49] by heterogeneous Fenton-like processes. In the literature it is also reported a reduced number of works dealing with the combination of this configuration with downstream biological degradation processes. Applications include, for example, the treatment of agrochemical wastewaters, where a fixed-bed reactor is used in series with rotating biological contactors [50], and the treatment of textile effluents, where in the first tubular reactor the classical homogeneous process takes place and, in a second reactor coupled in parallel, the heterogeneous Fenton-like oxidation occurs [39].

Industrially, this configuration is widely used in numerous chemical reactions using solid catalysts, namely, in the chemical industry (e.g., steam reforming of methane, synthesis of ammonia, sulfuric acid, and methanol), in the petrochemical industry (e.g., in the production of ethylene oxide, vinyl acetate, and butadiene), in petroleum refining (e.g., in isomerization, polymerization, and hydrocracking processes), and in fermentation processes. The vast industrial application of the fixed-bed reactor is linked with the high conversion per unit mass of catalyst attained and the relatively low operating cost.

To the best of our knowledge, fixed-bed reactors have never been implemented industrially for the treatment of wastewaters by the heterogeneous Fenton-like process. This is probably related to the fact that there are no catalysts available for such application that are commercially attractive (low cost, no leaching, stable). Other drawbacks associated to this technology are the temperature gradients in the reactor for exothermic reactions (e.g., oxidation of strongly polluted wastewater), making temperature control problematic, pressure drop, and the occurrence of mass transfer limitations – which result from the catalyst itself or from the gas bubbles, most likely CO_2 resulting from the oxidation of pollutants inside the column [44, 46]. Also, an additional operational problem regarding their application in continuous systems is the catalyst's deactivation, especially for effluents with complex composition; another issue is the fact that catalysts must be shaped as spheres, granules, and/or extrudates for the application in fixed-bed reactors, in order to decrease pressure drop along the column and to achieve good mechanical strength.

2.2.2 Fluidized-Bed Reactor

Fluidized-bed reactor (FBR) operation comprises a fluid phase (liquid or gas) and a solid phase with small particle size. In this reactor's configuration, the fluid passes through the catalyst bed and promotes the continuous suspension of the solid particles. This type of reactor involves a multiphase flow system, namely, two-phase (solid-gas or solid-liquid) or three-phase (solid-liquid-gas), with the inherent heat exchange and mass transfer [51]. When small particles are suspended in an upward-flowing stream of fluid, the solid presents a behavior similar to a fluid, thus improving the catalytic kinetics and performance of the system. Moreover, problems related with pressure drop are avoided in the FBR, while also improving heat and mass transfer.

The fluidization of the catalyst is achieved by the high velocity at which the fluid passes through the catalytic bed. The fluid velocity must be sufficient to suspend the particles but not too high to carry the catalyst out of the reactor. Depending on the fluid velocity and solid distribution in the FBR, three different stages may occur (see Fig. 6) [52]:

1. Fixed bed: at this stage the bed behaves similar to a packed-bed, where the fluid flow rate is low and the catalyst is compacted;
2. Incipient fluidization: as the velocity of the fluid passing through the catalyst bed increases, the catalyst starts dispersing in the column – this can be considered the inflection point that indicates the minimum velocity for fluidization;
3. Fluidized bed: this stage occurs when the minimal velocity for fluidization is surpassed and the catalyst particles expand and remain agitated.

It is possible to have several flow regimes, such as particulate fluidization, bubbling fluidization, slugging fluidization, turbulent fluidization, and pneumatic conveying regimes, that depend on the fluidization velocity [53].

Compared to other types of reactors (e.g., fixed-bed reactors), FBRs have a number of advantages, namely, high degradation efficiencies, shorter reaction times, better recirculation, and uniform temperature profiles even for highly exothermal reactions, along with high mixing and liquid/solid contacting [54]. Fenton-like FBR operation uses carriers that help in reducing the sludge formation that occurs in the homogeneous phase, since crystallization of the target pollutant happens in the carrier's surface [52].

FBRs are mostly applied in catalytic cracking in the petroleum industry, combustion, gasification, and catalytic processes. Also, fluidized-bed reactors have

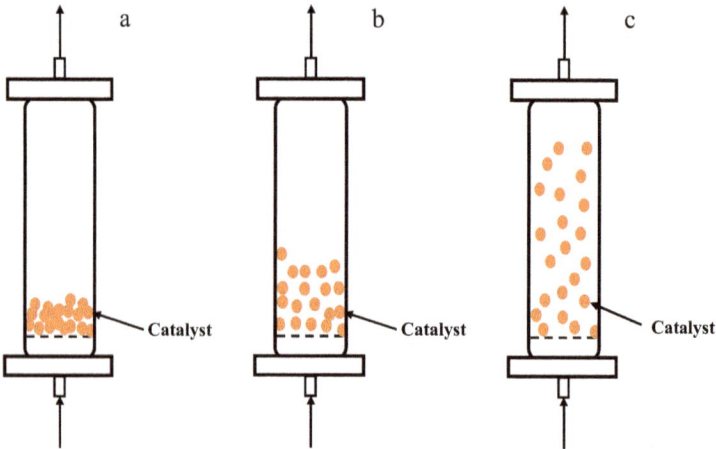

Fig. 6 Fixed-bed (**a**), incipient fluidization (**b**), and fluidized-bed (**c**) stages in a FBR

benefits such as the simplicity of construction and operation, low operating cost, and high flexibility for liquid and solid phase residence times. Recently, and regarding their use in pollutants removal or wastewater treatment, FBRs have been applied in:

1. Research studies by adsorption [55], biological degradation [56] and advanced oxidation processes, namely, electro-Fenton [57], photo-Fenton [58], and homogeneous [59] and heterogeneous Fenton-like processes [60], as well as combination of both [61] and integration of Fenton oxidation with adsorption [62] or ultrasound [63] processes;
2. Semi-pilot [60] and pilot-scale [64] operation for dye solutions oxidation by Fenton-like processes;
3. Industrial application for the treatment of effluents from the textile dyeing, pulp and paper mill, and petrochemical industries by the heterogeneous Fenton-like process, as mentioned in the review of Garcia-Segura et al. [52].

2.2.3 Continuous Stirred-Tank Reactor

In the CSTR configuration, mechanical or hydraulic agitation is required (see Fig. 7) in order to obtain a uniform composition and temperature inside the reactor. The choice of agitation method is strongly influenced by process considerations.

In this reactor typology, the composition of the output mixture is considered to be the same as the one that is inside the reactor (perfect mixing is assumed), so the reaction's driving force (usually the concentration of reactants) is necessarily low. Therefore, with few exceptions (e.g., zero-order reactions), CSTRs require larger volumes to achieve the desired conversions, when compared to plug flow reactors. When high conversions are required, several CSTRs can be used in series or, alternatively, a single reactor can be divided into stages while minimizing subsequent mixing and short circuits. The higher the number of CSTR stages, the closer the performance approaches a tubular flow reactor (with the same residence or contact time). CSTRs in series are (1) simpler and easier to design for isothermal operation (i.e., when operating temperature ranges are narrower) than tubular

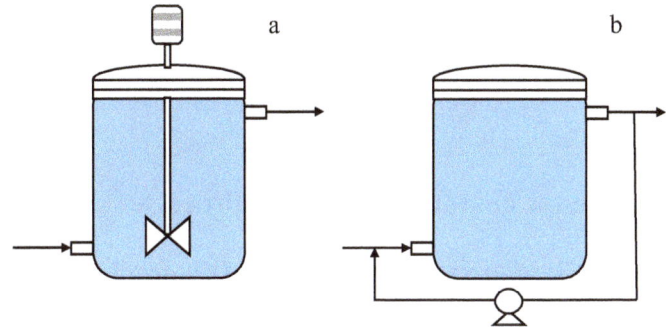

Fig. 7 Mechanical (**a**) and hydraulic (**b**) agitation in a CSTR

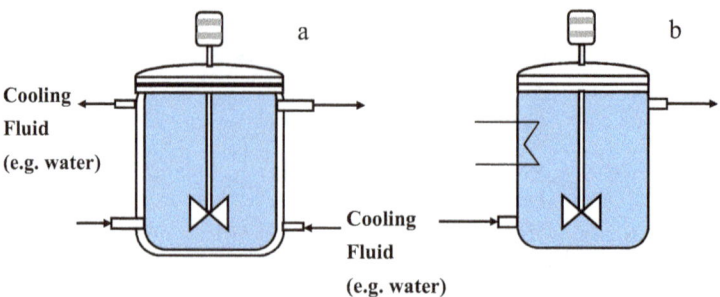

Fig. 8 Cooling/heating system out (**a**) and into (**b**) the CSTR

reactors and also (2) easier to operate in reactions where close control of reagent concentrations is required [65].

In such reactors, if severe heat transfer is necessary, the heating or cooling system may be incorporated into or out of the CSTR (see Fig. 8), for example, through single or multi-tube(s), where the heating/cooling fluid circulates. Likewise, the contents of the reactor can be recycled through external heat exchangers.

The CSTR configuration has advantages such as the good temperature control, cheap construction, large heat capacity, and easiness in the access to the interior of the reactor. Moreover, for heterogeneous systems, pressure-drop issues are minimized, as well as mass transfer resistances (see description above in Sect. 2.2.1). However, the disadvantage is that the conversion per unit of volume is relatively small when compared to other flow reactors.

CSTRs are widely used in chemical, pharmaceutical, and biofuel industrial applications and in wastewater treatment plants (i.e., activated sludge reactors). This reactor configuration has also been used in laboratorial scale for the treatment of industrial wastewater by anaerobic [66, 67] and aerobic [68, 69] processes. Its application in wastewater treatment by advanced oxidation processes has been performed mainly in lab scale. Some examples include the degradation of model compounds (phenol [70] and dyes [71]) and the treatment of textile effluents by heterogeneous Fenton-like processes [72], landfill leachate [73], orange II dye [74] and industrial wastewaters containing 2,4,6-trinitrotoluene [75] by the homogeneous Fenton process, as well as pilot-scale studies for the treatment of olive mill effluents [76]. Other works report the application of hybrid methods, such as electro-Fenton-like processes for landfill leachate treatment [77] and 4-nitrophenol removal [78].

3 Heterogeneous Fenton/Fenton-Like Processes

3.1 Fenton's Oxidation: Mechanism and Characteristics

Fenton process has emerged by observing that iron species have special oxygen transfer properties, influencing the oxidation performance of hydrogen peroxide.

This was first stated in 1984 by H. J. Fenton, when revising the oxidation of tartaric acid with hydrogen peroxide. H. J. Fenton found that the H_2O_2 oxidation potential was strongly elevated in the presence of a trace of ferrous salts (Fe^{2+}) [79]. Later, in 1934, Haber and Weiss [80] stated that Fenton's highly oxidative species is the hydroxyl radical, generated by the catalytic decomposition of hydrogen peroxide in the presence of ferrous ions, in acidic medium, according to Eq. (1).

$$Fe^{2+} + H_2O_2 \rightarrow Fe^{3+} + HO^{\bullet} + OH^- \tag{1}$$

Fenton process can be applied to rapidly destroy organic compounds (RH), which are oxidized by the nonselective $HO^{\bullet}$ species in solution, according to Eq. (2) [81, 82]. Oxidation occurs by either hydrogen abstracting, addition to the double bonds, or electron transfer [83], originating new oxidized intermediates with lower molecular weight, which can be further decomposed until complete mineralization into H_2O and CO_2 (Eq. 3).

$$HO^{\bullet} + RH \rightarrow H_2O + \text{intermediates} \tag{2}$$

$$HO^{\bullet} + \text{intermediates} \rightarrow H_2O + CO_2 \tag{3}$$

Although Fenton's oxidation involves a complex free radical and chain mechanism, the traditionally accepted process can be simplified and described by Eqs. (4)–(10) in addition to Eq. (1), for which reaction rates are well reported in literature [84]. Of particular importance is the reaction associated with the ferrous ion regeneration, responsible for the cyclic continuity of the process (Eqs. 4 and 5). This process, commonly referred to as the Fenton-like stage, is much slower than the one depicted in Eq. (1) and involves the formation of hydroperoxyl radicals ($HO_2^{\bullet}$). Hydroperoxyl species also attack organic molecules, but they have significantly lower oxidative potential when compared to $HO^{\bullet}$ (cf. Table 1). Nonetheless, $HO_2^{\bullet}$ contributes to the additional regeneration of Fe^{2+}, according to Eq. (6).

$$Fe^{3+} + H_2O_2 \leftrightarrow Fe-OOH^{2+} + H^+ \tag{4}$$

$$Fe-OOH^{2+} \rightarrow Fe^{2+} + HO_2^{\bullet} \tag{5}$$

$$Fe^{3+} + HO_2^{\bullet} \rightarrow Fe^{2+} + O_2 + H^+ \tag{6}$$

Depending on the operating conditions, a wide number of competitive reactions may also take place; excess H_2O_2 or ferrous ions in solution are responsible for the scavenging of $HO^{\bullet}$ radicals, according to Eqs. (7) and (8), respectively, hindering/decreasing the Fenton process' overall efficiency [81].

$$HO^{\bullet} + H_2O_2 \rightarrow H_2O + HO_2^{\bullet} \tag{7}$$

$$HO^{\bullet} + Fe^{2+} \rightarrow Fe^{3+} + OH^- \tag{8}$$

Furthermore, radical-radical reactions that negatively affect the process are also reported to occur (Eqs. 9 and 10).

$$HO^{\bullet} + HO_2^{\bullet} \rightarrow O_2 + H_2O \qquad (9)$$

$$HO^{\bullet} + HO^{\bullet} \rightarrow H_2O_2 \qquad (10)$$

Operating at near-ambient temperature and atmospheric pressure, the Fenton process typically runs at much smaller energy consumption requirements when compared to alternative oxidation techniques [31], as the activation of the hydrogen peroxide requires no energy input. Other advantages of the Fenton process include the typically short reaction times and the use of easy-to-handle chemicals.

Several studies report the efficient application of this process for the treatment of coking [85], olive mill [86], textile [39, 74, 87], pharmaceutical [43], pulp and paper [88], cosmetic [89], and cork-processing wastewaters [90, 91], as well as numerous synthetic wastewaters containing target compounds, such as phenols, chlorobenzene, and toluene [92, 93], just to name a few.

However, despite the successful application in lab-scale tests, the process is still to be largely implemented at industrial scale, mainly because of

1. the ineffectiveness in reducing the level of certain organic refractory pollutants (acetic acid, carbon tetrachloride, acetone, oxalic acid, trichloroethane, etc.);
2. the high cost of the chemical reagents used (e.g., hydrogen peroxide and chemicals for pH adjustment, as the process has limited optimum pH value of ca. 3.0–3.5) [74];
3. the need of a subsequent process for the removal of accumulated iron sludge and (partial) catalyst recovery, which increases the global cost and complexity of the process; and
4. the need of high concentrations of homogeneous catalyst – concentration of iron ions is typically in the 50–80 ppm range for batch processes [94] – clearly above the European legal threshold of 2 ppm for safe wastewater discharge in natural waterbodies [95].

To overcome limitations of the Fenton process toward specific pollutants and/or to reduce costs associated with reagents, proposed methods include the coupling of Fenton oxidation with traditional processes like coagulation/flocculation, membrane filtration, and/or biological oxidation. Coupling or integration of Fenton's oxidation as a pretreatment stage with such processes is the most common practice for real industrial wastewater treatment facilities, since complete mineralization via AOPs alone is not cost-effective and often not necessary.

The enhancement of the Fenton process with the so-called hybrid methods, such as photo-Fenton-like, electro-Fenton-like, cavitation-Fenton-like, or microwave-Fenton-like processes, has been the focus of extensive research [96]. Also, alternatives to the classical Fe^{2+} catalyst, including low-valence transition metals such as Fe^{3+}, Cu^{2+}/Cu^{+}, and nano-zerovalent iron, are well-known sources of $HO^{\bullet}$ species that can be used for both homogeneous and heterogeneous Fenton-like processes.

More recently, the development of heterogeneous Fenton-like systems is gaining its importance, as the use of solid catalysts may reduce drawbacks related to the iron loss and Fe sludge management that the homogeneous process usually entails. Heterogeneous catalysts can be obtained from various sources, nature (natural iron-bearing materials such as pyrite, magnetite, hematite, etc.), artificial synthesis (e.g., Fe/AC, Fe/ZSM-5, Fe_2O_3/SBA-15, Cu-MCM41, etc.), or industrial by-products (e.g., fly ash, sludge, etc.), and their catalytic activity may be improved by a wide range of preparation/modification techniques [97–99].

The efforts on the development of highly active and stable catalysts as well as their application on heterogeneous Fenton-like systems for wastewater treatment are extensively reported in the literature. However, the vast majority of studies report operation in discontinuous reactors, which display clear disadvantages for industrial implementation when compared to systems operating continuously, as addressed in Sect. 2, which are the focus of this chapter. Table 3 provides an outlook of the studies found in the open literature in which heterogeneous materials have been used as support of metal catalysts (typically Fe-based) for the Fenton/Fenton-like oxidation of model compounds or wastewaters in continuous reactors. Operational conditions that affect the process efficiency, mainly in terms of degradation and/or mineralization of the model pollutant/wastewater, are also succinctly mentioned; the phenomenological comprehension of their effects is crucial for subsequent process optimization, although for such goal the use of statistically based tools might be an interesting alternative.

3.2 Operational Parameters of Influence

The key operational parameters affecting Fenton/Fenton-like processes are related with the chemicals, i.e., the catalyst and oxidant doses, in addition to the reaction characteristics: pH, temperature, and pollutant load (concentration and characteristics of organic and inorganic compounds present in solution). Additionally, for heterogeneous Fenton-like processes on continuous reactors, the amount of solid catalyst used (i.e., combination of support/matrix and low-valence transition metal) and feed flow rate are of great importance, since they influence the effluent's residence/contact time (depending on the reactor's configuration); catalyst nature/properties are in this case also of paramount importance. The effect of such operational conditions is discussed in the following sections, and some examples are provided to better illustrate their impact on the overall process efficiency.

3.2.1 Reaction pH

The performance of Fenton/Fenton-like oxidation processes is strongly dependent on the reaction medium's pH, particularly for homogeneous systems, due to the catalyst and oxidant speciation factors [23].

Table 3 Operational conditions and optimum catalytic performances of heterogeneous Fenton/Fenton-like systems, operating in continuous reactors, for the treatment of wastewater/model pollutants

Target pollutant	Type of reactor	Catalyst	Experimental conditions	Results obtained (steady-state conditions)	Remarks	Ref.
Pharmaceutical wastewater	Fixed-bed reactor	Fe_2O_3/SBA-15 nanocomposite (pellet, $d_p = 1.0$–1.6 mm) Fe (wt. %) = 14	$pH_0 = 3.0$, $T = 80°C$, $[WW]_{feed} = 860$ mg TOC L^{-1}, $[H_2O_2]_{feed} = 10.8$ g L^{-1}, $W_{catalyst} = 2.9$ g, $Q_{feed} = 0.25$ mL min^{-1}, $\tau = 3.8$ min	59% TOC removal, 81% COD, and 71% BOD_5 reduction	Activity of the catalyst is kept constant for at least 55 h of reaction; increase in the COD/BOD_5 ratio from 0.2 to 0.3	[43]
Textile wastewater	Packed-bed reactor	Fe/Norit RX3 Extra AC (pellet, $d_p = 0.8$–1.6 mm) Fe (wt. %) = 7	$pH_0 = 2.5$, $T = 50°C$, $[WW]_{feed} = 174.7$ mg TOC L^{-1}, $[H_2O_2]_{feed} = 30$ mM, $W_{catalyst}/Q = 3.3$ g min mL^{-1}	97% decolorization, 74% TOC removal, 66% COD, and 73% BOD_5 abatement, $t = 3$ h	Higher treatment efficiencies were obtained for the real wastewater when compared to the dye solution used for the parametric study; after 60 h of operation, only 1.25% cumulative Fe loss from the catalyst	[44]
Dye solution (CSB)	Packed-bed reactor	Fe/Norit RX3 Extra AC (pellet, $d_p = 0.8$–1.6 mm) Fe (wt. %) = 7	$pH_0 = 3.0$, $T = 50°C$, $[CSB]_{feed} = 0.012$ mM, $[H_2O_2]_{feed} = 2.25$ mM, $W_{catalyst}/Q = 4.1$ g min mL^{-1}	88% decolorization, 47% TOC removal, $t = 150$ min	Similar treatment efficiencies for at least three cycles; values of iron in solution after the treatment were below 0.4 ppm	[46]
Phenol	Fixed-bed reactor	Fe_2O_3/SBA-15 mesostructured ($d_p = 1.6$–2 mm) Fe (wt. %) = 14	$pH_0 = 5.5$, $T = 80°C$, $[phenol]_{feed} = 1.0$ g L^{-1}, $[H_2O_2]_{feed} = 5.1$ g L^{-1}, $Q_{feed} = 1$ cm^3 min^{-1}, $W_{catalyst} = 2.9$ g	Complete phenol degradation, 66% TOC removal	Increasing catalyst particle size in the range of 0.35–2 mm has not promoted a negative effect on the process efficiency	[47]

Pollutant	Reactor	Catalyst	Conditions	Results	Observations	Ref.
Phenol	Fixed-bed reactor	Fe_2O_3/Al_2O_3 (commercial alumina spheres – 2.5 mm) Fe (wt. %) = 4	$pH_0 = 6$, $T = 70°C$, [pollutant]$_{feed}$ = 5.0 g L^{-1}, [H_2O_2]$_{feed}$ = 0.67 mol L^{-1}, Q_{feed} = 5.4 mL min^{-1}, τ = 3.4 min	99% phenol degradation, 50–60% TOC removal, $t = 6$ h	Leaching levels were reduced from 25 to 11% by combining thermal and acidic treatments to the catalyst; oxidant consumption efficiency of 80% was attained for the optimum conditions tested	[48]
Dye solution (BB3)	Recirculating FBR	Natural magnetite (NM)	$pH_0 = 5.0$, T = room, [BB3]$_{feed}$ = 5 mg L^{-1}, [H_2O_2]$_{feed}$ = 4 mM, [BB3] = 2.27 g L^{-1}	84% decolorization, $t = 190$ min	Comparison between the efficiency of the FBR vs batch operation, under the same conditions, indicated that the FBR could improve BB3 degradation	[60]
Dye solution (AY36)	Recirculating FBR	Natural pyrite (NP)	$pH_0 = 3.7$, T = room, [AY36]$_{feed}$ = 5 mg L^{-1}, [H_2O_2]$_{feed}$ = 1.7 mmol L^{-1}, [NP] = 0.4 g L^{-1}, Q_{feed} = 680 L h^{-1}	Degradation efficiency (dye) = 99.2%	Central composite design has been used for optimization of the process. Data from the model was in in good agreement with the experimental data (R^2 = 0.964)	[64]
Dye solution (OII)	CSTR	Fe/ZSM-5 (d_p = 0.6–0.8 mm) Fe (wt. %) = 5.5	$pH_0 = 3.0$, $T = 70°C$, [OII]$_{feed}$ = 0.1 mM, [H_2O_2]$_{feed}$ = 6 mM, $W_{catalyst}/Q$ = 200 mg min mL^{-1}, $\tau = 90$ min	91% decolorization, 36% TOC removal	Only 0.13% iron leaching after the optimum run	[71]

(continued)

Table 3 (continued)

Target pollutant	Type of reactor	Catalyst	Experimental conditions	Results obtained (steady-state conditions)	Remarks	Ref.
Phenol	Plug flow	Al-Fe-pillared clay (extrudate 1 × 5 mm) Fe (wt. %) = 3	$pH_0 = 3.7$, $T = 25°C$, $[pollutant]_{feed} = 5$ g L^{-1}, $[H_2O_2]_{feed} = 0.1$ mM, $[catalyst] = 5$ g L^{-1}	80% phenol conversion, 60% TOC removal, $t = 4$ h	At recycling ratios higher than 10, the plug flow reactor has the behavior of a CSTR; iron leaching <2% of the initial Fe content, after 350 h	[100]
Phenol	Fixed-bed reactor	Fe_2O_3/γ-Al_2O_3 (spheres with $d_p = 0.45$–2.5 mm) Fe (wt. %) = ca. 10	$pH_0 = 4.5$, $T = 80°C$, $[PhenolII]_{feed} = 1$ g L^{-1}, $[H_2O_2]_{feed} = 5.1$ g L^{-1}, $Q = 1.0$ mL min^{-1}, $W_{catalyst} = 2.5$ g	100% phenol removal, 92% COD conversion	Short time to reach steady-state conditions (ca. 10 min); long-term studies showed that carboxylic acids formed as intermediates are responsible for the reduction of active sites on the catalyst	[101]
Phenol	Fixed-bed reactor	Fe_2O_3/MCM-41 mesoporous zeolite ($d_p = 2.5$ mm) Fe (wt. %) = 10	$pH_0 = 6.0$, $T = 80°C$, $[PhenolII]_{feed} = 1$ g L^{-1}, $[H_2O_2]_{feed} = 5.1$ g L^{-1}, $Q = 2$ mL min^{-1}, bed height = 4 cm	72.5% TOC removal, 99% phenol conversion	Almost no aromatic intermediates were found in the treated effluent when bed height was over 2 cm	[102]
Dye solution (RB19)	(Undisclosed)	Fe_3O_4-loaded chitosan hollow fibers (Fe_3O_4-CH HFs)	$pH_0 = 3.5$, $T = $ room, $[RB19]_{feed} = 50$ mg L^{-1}, $[H_2O_2]_{feed} = 10$ mM, $Q_{feed} = 8$ mL min^{-1}	89% dye removal, $t = 150$ min	In the 2nd run, only 74% dye removal was achieved for the same operational conditions	[103]
Phenol	Fixed-bed reactor	Fe-ZSM-5 ($d_p = 2$ mm) Fe (wt. %) = 9.4	$pH_0 = 6.0$, $T = 80°C$, $[PhenolII]_{feed} = 1$ g L^{-1}, $[H_2O_2]_{feed} = 5.1$ g L^{-1}, $Q = 2$ mL min^{-1}, bed height = 3.8 cm	77.7% TOC conversion, 99.2% phenol degradation	High oxidant conversion (93.1%)	[104]

Phenol	CSTR	Fe/AC (powder) Fe (wt. %) = 4	$pH_0 = 3$, $T = 50°C$, $[Phenol]_{feed} = 100\ mg\ L^{-1}$, $[H_2O_2]_{feed} = 500\ mg\ L^{-1}$, $[Fe/AC] = 5\ g\ L^{-1}$, $\tau = 200$ min, $Q_{feed} = 5\ mL\ min^{-1}$	For the first 20 h of operation: >90% phenol conversion and >80% TOC reduction; after that period, increasing loss of activity until stabilization at ca. 65% for phenol and 25% for TOC	Over 170 h of operation; H_2O_2 conversion values always >90%	[105]
Dyeing-finishing wastewater	Recirculating FBR	γ-FeOOH ($d_p = 0.6$ mm) Fe (wt. %) = 3.5	$pH_0 = 3.5–4.0$, T = room, $[COD]_{feed} = 185–210\ mg\ L^{-1}$, $[H_2O_2]_{feed} = 5.7–6.0$ mM, [γ-FeOOH] = 133 g L^{-1}, $\tau = 52$ min	COD removal between 29 and 40%	In the 3-month long-term operation treatment of the real wastewater, the catalytic performance was nearly unaffected	[106]
Tyrosol	(Undisclosed)	CuN_{He} (copper-supported Al-pillared clays) Cu (wt. %) = 1.93	$pH_0 = 5.6$, $T = 60°C$, $[Tyrosol]_{feed} = 500$ ppm $[CuN_{He}] = 0.5\ g\ L^{-1}$, $[H_2O_2]_{feed} = 6.8 \times 10^{-2}$ M, $W_{catalyst} = 1$ g, $Q_{feed} = 30\ mL\ h^{-1}$	Tyrosol and TOC conversions ca. 85 and 60%, respectively	The results obtained in the continuous reactor were similar to the batch ones	[107]
m-cresol	Fixed-bed reactor	HNO_3-SW (sewage sludge-derived carbonaceous material) ($d_p = 4–6$ mm) Fe (wt. %) = 1.74	$pH_0 = 7$, $T = 80°C$, $[m\text{-}cresol]_{feed} = 100\ mg\ L^{-1}$, $[H_2O_2]_{feed} = 15.7\ mmol\ L^{-1}$, LHSV = 1 h^{-1}, $[HNO_3\text{-}SW] = 0.5\ g\ L^{-1}$	*m*-cresol conversion reached 95%	*m*-cresol conversion reached 25% at 96 h, increased to 80% at 360 h, and was kept at 95% until 1,128 h of operation	[108]

For homogeneous systems, optimum pH values of the Fenton process are well known to be in the strict 3.0–3.5 range. In such conditions, maximum concentration of Fe^{2+} active species is achieved at the lowest rate of H_2O_2 parasitic decomposition [109]. Lower pH values are reported to slow down the formation of iron(III)-peroxo complexes (as intermediates, cf. Eq. 4), interfering with the Fe^{3+}/Fe^{2+} regeneration cycle [35]. Increasing the pH of the reaction leads to the precipitation of relatively inactive and insoluble ferric hydroxides ($Fe(OH)_3$), resulting in the drop of ferric ions' availability; thus the reaction depicted in Eq. (4) is again the rate-limiting step [33]. Also, hydrogen peroxide self-decomposition into molecular oxygen becomes significant under such conditions (Eq. 11), hindering the formation of oxidizing species [110, 111].

$$2H_2O_2 \rightarrow O_2 + 2H_2O \tag{11}$$

Drawbacks associated with the narrow optimum pH range of homogeneous systems may be avoided using solid catalysts, since heterogeneous Fenton-like catalysts are reported to show lower sensitivity to pH than iron ions in solution (for rather similar reaction conditions). Because iron species are restrained within the structure and inside the pore/interlayer of the solid catalyst, precipitation of iron hydroxide is reduced, resulting in the continuous ability for $HO^{\bullet}$ formation [45, 100, 112]. For instance, Chi et al. [113] showed that the modified polyacrylonitrile (PAN) catalyst, used on the oxidation of trace endocrine disrupting compounds (EDCs) and pharmaceutical and personal care products (PPCPs) in a pilot-scale continuous stirred-tank reactor, is catalytically active even at the natural pH of such effluents (pH range 6.6–8.0). The authors suggest that the catalytic activity in this wider pH range could be attributed to the strong ligands holding the metal ions, preventing the iron from precipitating (which usually happens in homogeneous systems when pH >4 – see Eq. 4). Lu et al. [101] also stated that the inexpensive Fe_2O_3/γ-Al_2O_3 catalyst, used in the continuous Fenton-like oxidation of a phenolic solution in a fixed-bed reactor, remained catalytically active in the pH range of 5–9. Phenol and COD removal for those runs were >99% and >90%, respectively (steady-state values), only dropping when pH = 10 (Fig. 9).

Nonetheless, other investigations on the use of continuous heterogeneous catalytic systems showed optimum pH values in a narrower acidic range. According to the work of Duarte et al. [44] on the degradation of a dye-containing solution (Alcian Blue-tetrakis – AB), using a fixed-bed reactor filled with an Fe on activated carbon catalyst (Fe/AC), optimum pH was suggested to be 2.5. The authors stated that a strongly acidified medium significantly improved discoloration and TOC abatement of the solution, as illustrated in Fig. 10. In fact, for the 2–4 range of pH values tested, pH = 2.0 allowed reaching maximum TOC removal (48.1%), while maintaining high discoloration percentage (>90%) at steady state. However, in respect to the leaching of iron from the activated carbon support, after 3 h of reaction Fe loss was ca. 5 times higher when compared to pH = 2.5. Despite that, total iron loss for that series of experiments was only 0.64% of the total Fe initially loaded into the column. It is anticipated that leaching >1% of active ferrous iron is

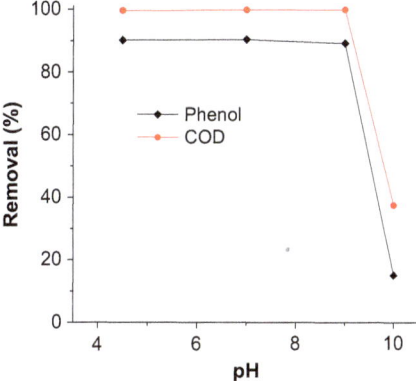

Fig. 9 Influence of the initial pH on the degradation of phenolic solutions at steady state. Experimental conditions: [Phenol]$_{feed}$ = 1 g L^{-1}, [H$_2$O$_2$]$_{feed}$ = 5.1 g L^{-1}, T = 80°C, $W_{catalyst}$ = 2.5 g, and Q_{feed} = 1.0 mL min^{-1}. Adapted from [101]

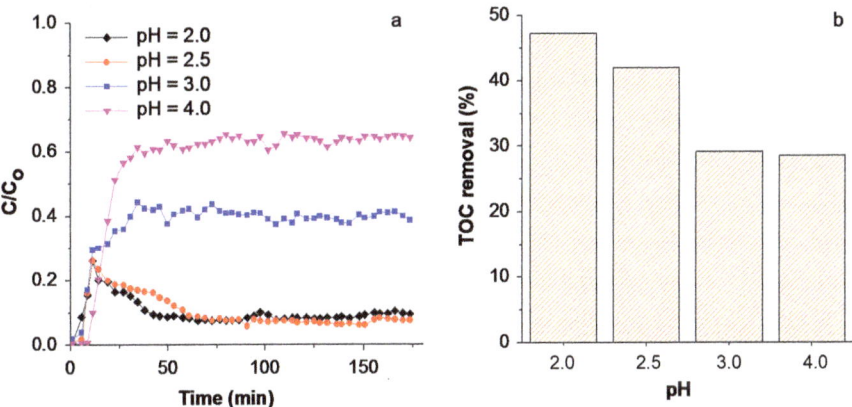

Fig. 10 Effect of medium pH in discoloration along time (**a**) and TOC removal in steady state (**b**) of the AB dye solution. Experimental conditions: [AB]$_{feed}$ = 0.01 mM, [H$_2$O$_2$]$_{feed}$ = 30 mM, T = 30°C, and $W_{catalyst}/Q$ = 3.3 g min mL^{-1}. Adapted from [44]

sufficient to cause catalytic activity in the homogeneous phase [114]. In such cases, the overall catalytic activity and efficiency of the process may be affected by the homogeneous contribution, but this depends of course on the catalyst used (namely, content of iron), target pollutant(s), and other operating conditions.

A rather similar behavior as reported by Duarte et al. [44] was found by Queirós et al. [71] on the oxidation studies of an azo dye solution (Orange II – OII) using an Fe/ZSM-5 zeolite as catalyst. The optimization of the Fenton-like process under continuous stirred-tank reactor operation (CSTR) showed that the 2–3 pH range allowed discoloration values between 65 and 75% (Fig. 11a). Operating at higher pH values leads to the significant reduction of the discoloration efficiency (for

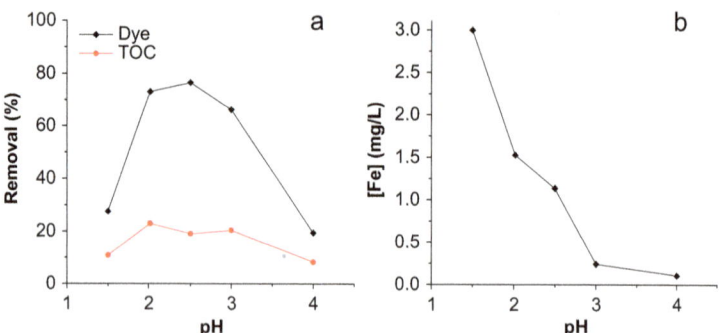

Fig. 11 Effect of medium pH in discoloration and TOC removal at steady state (**a**) and iron leaching from the catalyst support (**b**). Experimental conditions: $[OII]_{feed} = 0.1$ mM, $[H_2O_2]_{feed} = 6$ mM, $T = 50°C$, $W_{catalyst}/Q = 60$ mg min mL^{-1}, and $\tau = 30$ min. Adapted from [71]

pH = 4, color removal was ca. 20%; a similar decline in process performance was observed under very low pH values. In respect to the iron loss from the zeolite support, leaching levels were substantially higher as more acidic conditions were applied (Fig. 11b). To assess whether the contribution of the homogeneous Fenton process due to the iron loss from the support to the solution was influencing the efficiency of the process, an additional experiment was performed by feeding the reactor with the exact same Fe concentration that leached from the Fe/ZSM-5 zeolite in the run where pH = 3.0. For the same operational conditions, the homogeneous Fenton contribution was significantly less prominent than the heterogeneous one (9 vs 69% dye removal and 3 vs 21% TOC degradation), although the results confirm that iron leaching is a concern in such oxidative systems, since catalyst deactivation and homogenous oxidation coexist. This contribution is expected to be more significant for more acidic pH conditions, due to the higher Fe concentration in solution.

Operating at lower pH conditions entails several disadvantages, generally associated with the cost of acidification of the initial solution, need for post-treatment neutralization of the effluent, and the use of acid-resistant material for the reactor coating [115]. Besides, leaching of the iron catalyst from the support also causes secondary metal ion pollution in the treated effluent. The development of cost-effective catalysts with high catalytic activity and long-term stability in a broader pH range remains the greatest challenge for the effective application of Fenton-like heterogeneous systems for wastewater treatment.

3.2.2 Temperature

The reaction temperature also plays a critical role since oxidation proceeds at a faster rate when the reaction medium temperature is increased. This is due to the exponential dependence of kinetic constants (k), either for radical production, for iron regeneration, or for organics attack, from the temperature according with the Arrhenius law (Eq. 12),

$$k = k_0 \exp\left(-\frac{E_a}{RT}\right) \quad (12)$$

where k_0 is the global rate constant, R the ideal gas constant, E_a the activation energy, and T the temperature.

This dependence, for both discontinuous and continuous processes, is particularly evident between 10 and 40/50°C, as reported by several researchers [72, 116, 117]. Working with an Fe/AC catalyst for the catalytic oxidation of a dye solution in a fixed-bed reactor, Duarte et al. [44] stated that decolorization and TOC removal in the 10–50°C range were in accordance with the Arrhenius law dependence. Optimum temperature of 50°C leads to 74% decolorization and 37% TOC degradation (steady-state values). For $T = 70°C$, oxidation of the dye solution was no longer improved by the increase of reaction's temperature, and a slight mineralization efficiency decrease was even observed. This is probably related to the thermal decomposition of H_2O_2 into oxygen and water, promoted at higher temperatures, which affects the $HO^{\bullet}$ production, and consequently the degradation of organic matter. Also, the formation of gas bubbles observed in the reactor's column while operating at $T = 70°C$ may also have played an undesirable role on the mass transfer between the liquid and the catalyst phase, hindering the process efficiency [81].

A similar study was performed by Mesquita et al. [46], who also noticed the formation of gas bubbles inside the fixed-bed reactor when operating in the 50–70°C range. Even so, this phenomenon was not preponderant enough to hinder the degradation efficiency of the Chicago Sky Blue (CSB) dye solution (Fig. 12).

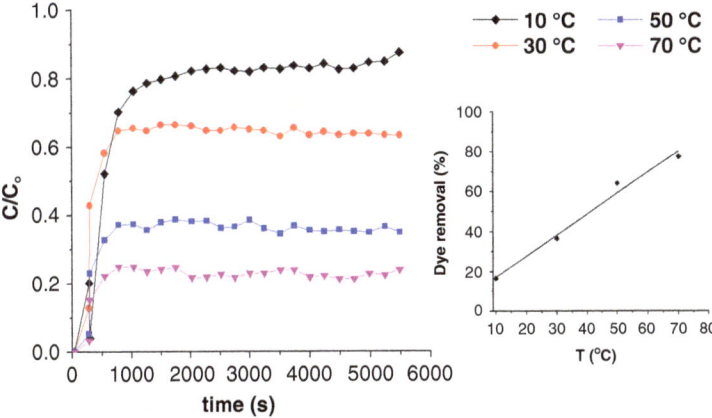

Fig. 12 Dye concentration history at the fixed-bed reactor outlet for different reaction temperatures; *inset* graphic shows steady-state dye conversion values. Experimental conditions: $[CSB]_{feed} = 0.012$ mM, $[H_2O_2]_{feed} = 2.25$ mM, pH = 3.0, and $W_{catalyst}/Q = 1.88$ g min mL^{-1}. Adapted from [46]

The study carried out by Liu et al. [102] pointed out similar results, but for the degradation of phenol over a Fe_2O_3/MCM-41 mesoporous zeolite catalyst. These results indicate that every catalytic system behaves differently depending on the reaction conditions, thus optimization of every single operational variable is crucial for a cost-efficient application of Fenton-like processes.

For industrial application, the reaction's temperature could be increased (up to a certain point) to improve degradation rates and consequently the process efficiency. In some specific circumstances, such is the case of effluents coming from textile industries, the heating of the system could eventually be avoided since textile wastewaters usually present high temperatures at the time of rejection [118, 119]. Nevertheless, for most applications, the associated energy consumption costs and the iron loss due to leaching phenomena, which is sometimes reported to be more prominent at higher temperatures [71, 96], make the use of lower temperatures (T ~30°C) more suitable for the majority of industrial wastewater treatment facilities.

3.2.3 H_2O_2 Dosage

For the vast majority of target compounds/wastewaters and operational conditions, an increase in the oxidant dose fed to the system leads to higher degradation rates. However, excess oxidant may cause unwanted parallel reactions, particularly via the so-called scavenging effect, where the $HO^{\bullet}$ is being ineffectively used to produce hydroperoxyl species ($HO_2^{\bullet}$) with lower oxidation potential (Eq. 7). Additionally, the enhanced production of $HO_2^{\bullet}$ origins undesired radical-radical reactions, also causing the parasitic consumption of $HO^{\bullet}$ (Eq. 9).

Mesquita et al. [46] showed that increasing the H_2O_2 dose fed to a fixed-bed reactor, which was loaded with Norit RX3 activated carbon impregnated with Fe (7 wt.% of iron), yields an increase in the CSB dye conversion. However, this was only true within a certain range of oxidant dosages (0.15–2.25 mM). Beyond that range, dye removal efficiency started to decrease, as depicted in Fig. 13. The existence of an optimum feed oxidant concentration was also reported in several other studies [45, 71, 72, 103].

The selection of the optimum oxidant concentration in the feed usually requires the knowledge of the H_2O_2 stoichiometric amount that is able to theoretically mineralize all the organic carbon present in solution. For most cases, where the wastewater to be treated has complex or unknown composition, this estimation can be achieved by the reaction presented in Eq. (13).

$$C + 2H_2O_2 \rightarrow CO_2 + 2H_2O \qquad (13)$$

Following the stoichiometry of the reaction, a H_2O_2/C molar ratio of 2, equivalent to a 5.7 mass ratio, is necessary to complete carbon depletion (as TOC content) [43]. Also, it can be referred to as the stoichiometric weight ratio between the

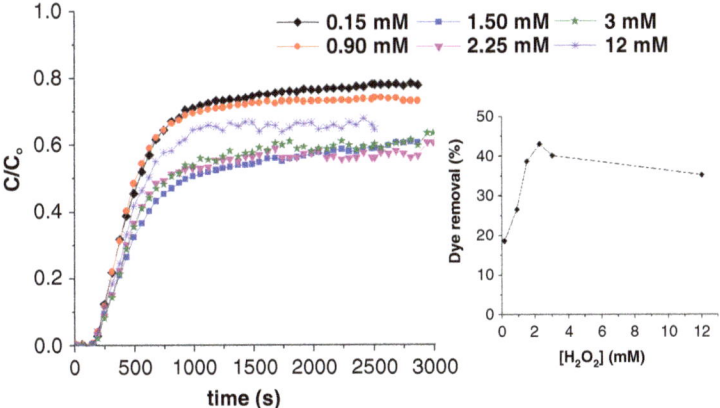

Fig. 13 Dye concentration at the fixed-bed reactor outlet for different H_2O_2 feed concentrations; the *inset* graphic represents dye conversion values at steady state. Experimental conditions: $[CBS]_{feed} = 0.012$ mM, $W_{catalyst}/Q = 1.88$ g min mL^{-1}, pH = 3, $T = 30°C$. Adapted from [46]

oxidant and effluent's initial COD – $R = H_2O_2/COD = 2.125$ (based on H_2O_2 complete conversion into O_2) [86]. In practice, because hydrogen peroxide participates in unwanted parallel reactions, as previously mentioned, the H_2O_2 dose applied is often higher than the theoretical stoichiometric one. For example, Queirós et al. [71] tested a range of hydrogen peroxide feed doses corresponding to 0.42–4.76 the stoichiometric concentration for the oxidation of an azo dye-containing solution, using an Fe/ZSM-5 zeolite as the catalyst. The best catalytic performance was observed at 1.42 times the stoichiometric dose. Esteves et al. [72] also reported the use of an oxidant feed concentration of 3.52 g L^{-1}, corresponding to two times the calculated theoretical stoichiometric concentration (R) for the heterogeneous Fenton oxidation of a simulated textile effluent, using an Fe/AC catalyst on a CSTR.

Melero et al. [43] tested a Fe_2O_3/SBA-15 nanocomposite material as a Fenton-like heterogeneous catalyst for the treatment of a pharmaceutical wastewater in a continuous upflow fixed-bed reactor. A stoichiometric amount of oxidant fed to the reactor (5,400 mg L^{-1}) yielded ca. 40% TOC degradation (steady-state value). Twice that amount improved the catalytic activity of the system, allowing a steady-state TOC conversion value of ca. 60%. However, further increase in the oxidant dosage did not influenced the efficiency of the process, as Fig. 14 shows, highlighting once again the existence of an optimum oxidant dose for Fenton/Fenton-like systems.

The efficient use of hydrogen peroxide is not only important from an economical point of view, since H_2O_2 is a relatively costly reactant, but also because the unused oxidant during the process may contribute to chemical oxygen demand (COD) of the treated effluent/solution [120]. Also, if Fenton oxidation is used as a pretreatment stage followed by a biological step, excess hydrogen peroxide has been linked to cause harmful effects on microorganisms, hindering biological depuration processes [121].

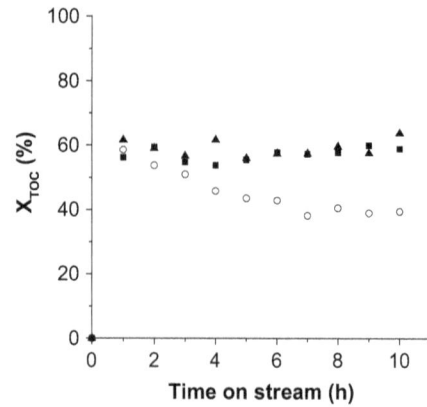

Fig. 14 Effect of hydrogen peroxide feed concentration in the effluent's (pharmaceutical wastewater) TOC degradation efficiency. Experimental conditions: $[H_2O_2]_{feed} = square$ 16,200 mg L^{-1}, *triangle* 10,800 mg L^{-1}, *circle* 5,400 mg L^{-1}; $T = 80°C$, $Q_{feed} = 0.25$ mL min^{-1}, $W_{catalyst} = 2.9$ g, and pH = 3.0. Adapted from [43]

3.2.4 Pollutant Concentration Fed

The nature and concentration of the target pollutant/wastewater also plays a key role for the efficient application of the majority of AOPs. High or close to complete mineralization via Fenton/Fenton-like processes is almost always favored at lower concentrations of the target compound(s) [23, 48].

The vast majority of studies reported in literature, and cited throughout this chapter, deal with the treatment of phenols and dye solutions. In fact, phenolic compounds are among the most abundant pollutants present in industrial wastewaters, owing to their widespread utilization in the synthesis of pesticides, solvents, lubricating oils, dyes, and resins, just to name a few. Long-term exposure to such compounds has been linked to cause several human health damages, as well as toxicity toward microorganisms and plants, even at very low concentrations [122]. Dyes and pigments are also frequently reported since they are commonly used as coloring agents in the textile, food, and pharmaceutical industries. The discharge of dye-containing effluents is known to cause significant environmental hazards related to the presence of organic compounds that are not easily degraded by conventional methods, particularly the azo dye category [123].

In the heterogeneous Fenton-like oxidation of Basic Blue 3 dye (BB3) with natural magnetite, Aghdasinia et al. [60] found that increasing the initial BB3 concentration fed in the 5–20 mg L^{-1} range led to a decline of the degradation efficiency, as illustrated in Fig. 15. Increasing the initial dye concentration may contribute to the occupation of active sites in the natural magnetite surface, which would deactivate accessible sites for H_2O_2, thus decreasing the generation of hydroxyl radicals, as the authors enlightened. Similar results were observed by Guo and Al-Dahhan [109] on the semicontinuous catalytic wet oxidation of phenolic solutions in the feed concentration range of 100–2,000 ppm ($T = 70°C$, in the presence of 0.3 M of H_2O_2 and 6.6 g L^{-1} of a Al-Fe-pillared clay catalyst). The elimination of phenol after 240 min of reaction was practically complete for feed concentrations in the 100–1,000 ppm range, although slower oxidation rates were

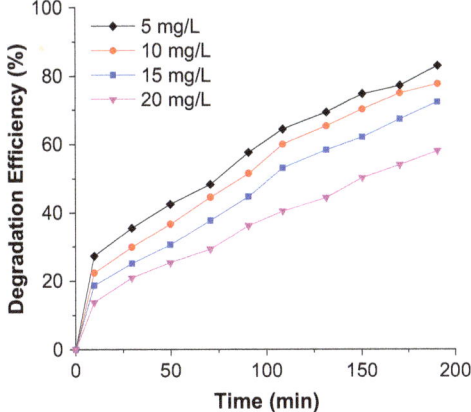

Fig. 15 Effect of BB3 initial concentration on the BB3 degradation efficiency in semicontinuous operation. Experimental conditions: $[H_2O_2]_{feed} = 4$ mM, $[NM]_{catalyst} = 2.27$ g L^{-1}, pH = 5, T = room. Adapted from [60]

observed when increasing the initial phenol concentration in that same range. For the maximum phenol feed dose tested (2,000 ppm), degradation efficiency was approximately half the one obtained previously, for the same reaction time.

In summary, Fenton-like processes have proved to be effective on the degradation of most refractory compounds, but technical and economic drawbacks arise when dealing with high organic loads, since H_2O_2 consumption requirements become substantially higher for such cases and blockage of the catalyst's surface may occur.

3.2.5 Catalyst Load

In heterogeneous Fenton/Fenton-like processes, ferric, ferrous, or other low-valence transition metal ions, situated in the surface sites of the solid material, take part in the reaction scheme yielding hydroxyl radical species. Similar to homogeneous systems, small amount of catalyst in solution will almost always result in lower efficiency of the process, since the availability to react with H_2O_2 is reduced. Also, when operating fixed-bed reactors, the catalyst load indirectly influences the residence time (t_R) of the liquid phase, according to Eq. (14),

$$t_R = \frac{V_L}{Q} = \frac{\epsilon_L V_B}{Q} \qquad (14)$$

where the liquid holdup, ϵ_L, is the ratio between the volume of the liquid phase, V_L, and the catalyst-bed volume, V_B (also referred to as catalyst bed height), and Q stands for the total feed flow rate. The liquid holdup is calculated according to parameters related to the packed bed, namely, porosity and external liquid saturation [47].

In fact, the catalytic volume/mass can be adjusted to modify the fluid residence time, as reported in the work of Yan et al. [104]. The influence of residence time on the catalytic wet hydrogen peroxide oxidation of phenol was addressed by modifying the catalyst bed height in the 1.1–3.8 cm range (corresponding to 2.5–9.7 g of Fe/ZSM-5 catalyst). As anticipated, higher phenol, TOC, and H_2O_2 conversions were obtained for the longer catalytic bed heights tested, either by the effect of the higher availability of active iron species or by the increased residence time of the solution.

In the attempt of improving the dye conversion of an azo dye-containing solution, Mesquita et al. [46] also increased the column load with a Fe/AC catalyst (7 wt.% of Fe) from 4.7 to 10.25 g. This increase led to ca. 70% dye conversion at steady state (at 30°C, for a feed containing 0.012 mM of the dye and 2.25 mM H_2O_2). The performance improvement, when compared to ca. 37% obtained when applying 4.7 g of Fe/AC, was attributed to the fact that longer catalyst beds lead to higher contact times, expressed as W_{cat}/Q (i.e., the amount of catalyst, W_{cat}, available inside the reactor is increased for the same feed flow rate, Q).

The work of Liu et al. [102] also showed that longer catalyst bed had a positive effect on phenol, TOC, and H_2O_2 conversion. By adjusting the catalytic bed height to 2, 3, and 4 cm, under the same operating conditions, phenol degradation reached >98% for all heights, TOC removal increased from 59 to 72.5%, and H_2O_2 conversion reached its higher value for 4 cm (92.1%), as shown in Fig. 16.

On the other hand, excessive loading may lead to the consumption of generated hydroxyl radicals by the excess catalyst (see Eq. 8). For example, Aghdasinia et al. [60] reported the existence of an optimum catalyst dose – 2.27 g L^{-1} of natural magnetite – for the Fenton-like oxidation of a dye-containing solution in a FBR; further increment in the catalyst applied to the system yields a dye conversion efficiency decrease. Besides the scavenger effect, the formation of iron complexes (iron + organics) is also reported to occur when excess amounts of catalyst are present in solution [116]. Martínez et al. [47] also reported the existence of an optimum dosage for the Fe_2O_3/SBA-15 catalyst used in the oxidation of phenolic solutions. TOC conversion enhancement is clear when catalyst mass (and consequently, the residence time) is increased from 1.4 to 2.9 g (48.1 vs 66.6% TOC removal); applying higher catalyst mass (3.9 g) leads to little to no variation in the mineralization degree, despite the high oxidant feed dose applied in that set of runs ([Phenol]$_{feed}$ = 1 g L^{-1}, [H_2O_2]$_{feed}$ = 5.1 g L^{-1}, T = 80°C). The presence of refractory compounds, formed during the oxidation of phenol, that are resistant to mineralization (namely, formic, acetic, and oxalic acids) were detected in the treated effluent.

3.2.6 Feed Flow Rate

Similarly to the catalyst dosage, variations in the feed flow rate also affect the residence/contact time (note: for perfectly mixed reactors operating with constant

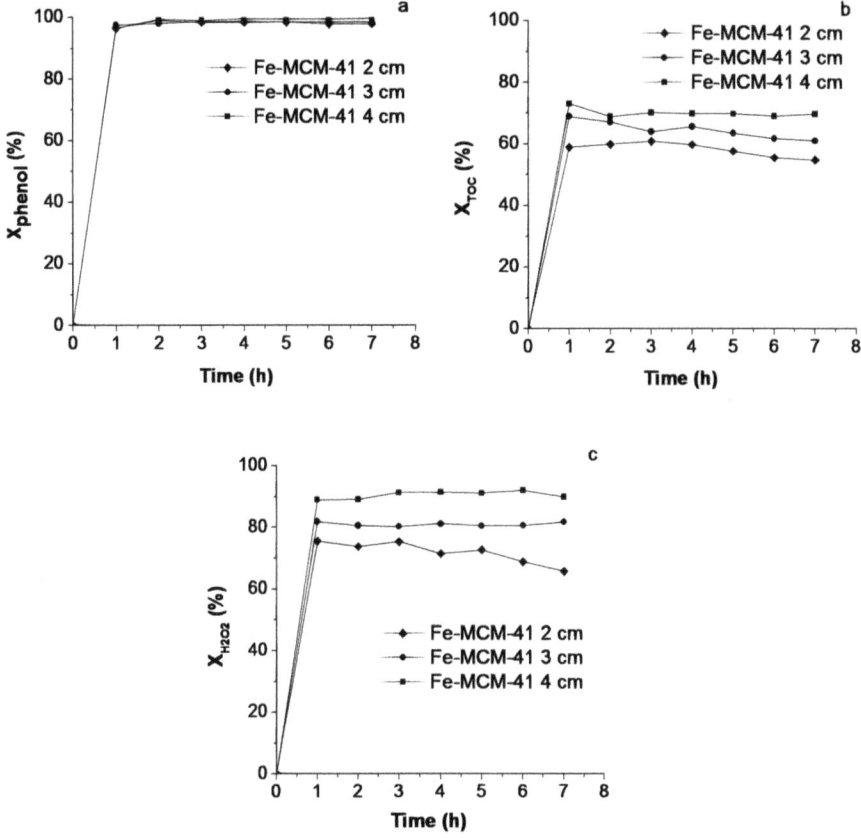

Fig. 16 Influence of catalyst bed height in the conversion efficiency of (**a**) phenol, (**b**) TOC, and (**c**) H_2O_2 of a phenolic solution. Experimental conditions: $[Phenol]_{feed} = 1.0$ g L^{-1}, $[H_2O_2]_{feed} = 5.1$ g L^{-1}, $Q = 2.0$ mL min^{-1}, $T = 80°C$. Adapted from [102]

density fluids, residence time is equivalent to the space-time, τ, which is calculated by the ratio between the total volume and the volumetric flow rate – $\tau = V/Q$).

Melero et al. [43] performed Fenton catalytic experiments for the degradation of a pharmaceutical wastewater using Fe_2O_3/SBA-15 as the heterogeneous catalyst, applying feed flow rates of 0.25, 0.5, and 1.0 mL min^{-1} (which provides residence times of 3.79, 1.90, and 0.95 min, respectively). TOC and H_2O_2 conversion histories at the outlet of the fixed-bed reactor are depicted in Fig. 17. Increasing the feed flow rate from 0.5 to 1.0 mL min^{-1} has the most impact in conversion profiles, as TOC degradation is almost inhibited by the greater feed flow rate tested. Likewise, H_2O_2 conversion is extremely low for $Q_{feed} = 1.0$ mL min^{-1}, owning to the lower time for oxidant/catalyst interaction. In terms of the catalytic system performance, maximum TOC removal of ca. 60% and H_2O_2 conversion of ca. 90% was attained at steady state for 0.25 and 0.5 mL min^{-1}. The similar behavior for the two feed flow rates could indicate the existence of refractory compounds, formed during the oxidation of the pharmaceutical wastewater.

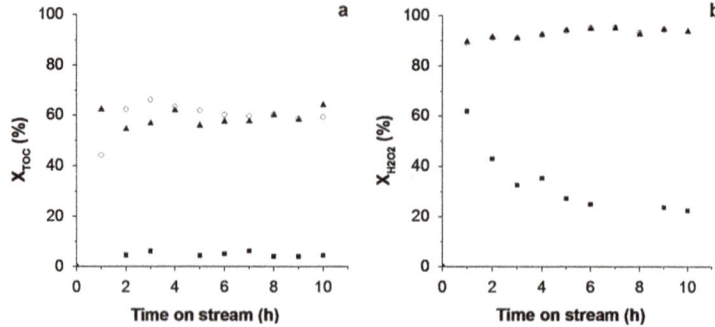

Fig. 17 Influence of the feed flow rate on TOC (**a**) and H_2O_2 (**b**) conversions with time on stream of a pharmaceutical wastewater. Experimental conditions: Q_{feed} = *square* 1.0, *circle* 0.5, and *triangle* 0.25 mL min^{-1}; $[TOC]_{feed}$ = 860 mg L^{-1}, $[H_2O_2]_{feed}$ = 10,800 mg L^{-1}, $W_{catalyst}$ = 2.9 g, T = 80°C, pH$_0$ = 3. Adapted from [43]

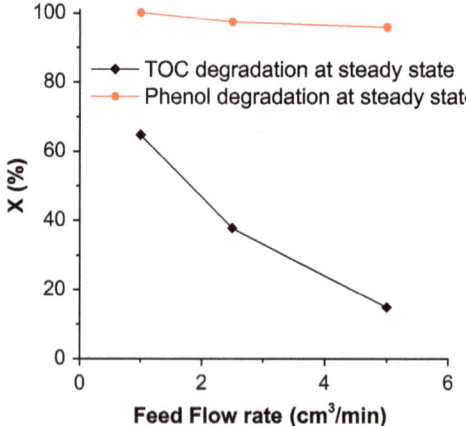

Fig. 18 Feed flow rate influence in the degradation efficiency of phenolic aqueous solutions at steady state. Experimental conditions: $[Phenol]_{feed}$ = 1.0 g L^{-1}, $[H_2O_2]_{feed}$ = 5.1 g L^{-1}, $W_{catalyst}$ = 2.9 g, T = 80°C, particle size = 1–1.6 mm. Adapted from [47]

In a previous work by the same research group [47], the same Fe_2O_3/SBA-15 catalyst was used for the degradation of a phenolic solution by CWHPO in the same reactor configuration. Varying the inlet feed flow rate of the aqueous phenolic effluent (1.0 g L^{-1} phenol and 5.1 g L^{-1} H_2O_2 solution) to 1.0, 2.5, and 5.0 cm^3 min^{-1} (corresponding to residence times of 3.6, 1.2, and 0.6 min, respectively), a clear relationship between mineralization degree and feed flow rate has been assessed, since higher residence times promote higher TOC degradation (Fig. 18). Rather similar results were found by Liu et al. [102] on the oxidation of phenol over a Fe_2O_3/MCM-41 catalyst (operational conditions were $[Phenol]_{feed}$ = 1.0 g L^{-1}, $[H_2O_2]_{feed}$ = 5.1 g L^{-1}, T = 80°C, and catalyst bed height = 4 cm). TOC conversion decreased from ca. 69 to 46% with the feed flow rate increase from 2 to 8 mL min^{-1}, and the same trend was observed in respect to the phenol degradation efficiencies (99.1–91.2%).

3.3 Influence of Catalyst's Physicochemical Properties

As already mentioned, the ideal catalyst for heterogeneous Fenton/Fenton-like processes should be a low-cost material with good catalytic activity and stability. Numerous parameters regarding heterogeneous catalysts are reported to directly affect the efficiency of Fenton's oxidation and the stability of the catalyst itself, namely, the type/source of support, catalyst's incorporation/synthesis method (e.g., nature of the precursor used and metal ion selected as the active phase), and the particle size applied to the catalytic system, among others.

Rodríguez et al. [124] compared the catalytic activity of Fe-based catalysts (Fe 4.2–4.9 wt.%) prepared over nine different supports (Table 4) for the degradation of the model acid Orange II dye (OII) by the heterogeneous Fenton-like process, although in a batch reactor. Since adsorption of organic molecules on the solid surface of the catalysts can enhance the efficiency of the process, additional experiments were also performed to evaluate whether adsorption or chemical reaction was occurring under identical experimental conditions. The study showed that overall TOC reduction of OII solutions by adsorption phenomena was more preponderant on carbon (AC, CNT, CNF) and hydrotalcite supports than in siliceous and clay mineral supports (e.g., sepiolite) – see Fig. 19. The Fe catalyst supported on sepiolite was found to be the most viable Fenton catalyst among all samples tested, with the overall highest color and TOC and COD conversion percentages, being the catalyst with lowest percentage of iron leaching among all nine materials tested.

The influence of the precursor used in the synthesis of heterogeneous Fe/AC catalysts (Fe 7 wt.%) was also evaluated by Duarte et al. [117]; this screening was

Table 4 Nomenclature of different iron catalysts and their main textural properties

Catalyst	Nomenclature	S_{BET} (m^2 g^{-1})	APS (Å)[a]
Hydrotalcite	Fe-HT	69.7	153.5
Zeolite USY	Fe-USY	172.9	19.0
Activated carbon	Fe-AC	673.9	24.3
Silica	Fe-S	392.7	17.7
Mesoporous silica	Fe-MCM	857.4	23.2
Silica xerogel	Fe-X	500.6	18.5
Carbon nanotubes	Fe-CNT	106.5	169.9
Carbon nanofibers	Fe-CNF	141.1	158.2
Sepiolite	Fe-SEP	201.9	85.2

Adapted from Rodríguez et al. [124]
[a]Average pore size

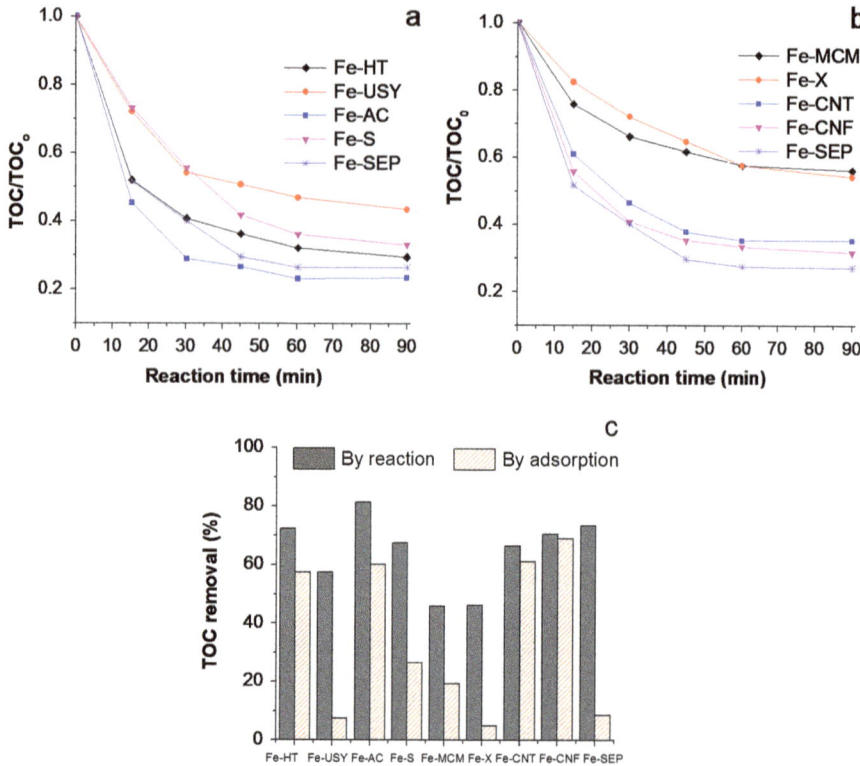

Fig. 19 (**a**, **b**) Evolution of TOC along time and (**c**) TOC removal after 120 min (by adsorption and heterogeneous Fenton reaction) for different Fe-based materials. Experimental conditions: $T = 30°C$, $W_{catalyst} = 0.5$ g (Fe 5 wt.%), $H_2O_2/AOII = 5$ w/w, $H_2O_2/Fe^{2+} = 3.5$ w/w, pH = 3. Adapted from [124]

carried out in a slurry batch reactor. Three Fe salts were tested, iron acetate (N-Ac$_2$Fe), iron sulfate (N-FeSO$_4$), and iron nitrate (N-Fe(NO$_3$)$_3$), and their effect on the degradation of the model Orange II azo dye by the heterogeneous Fenton-like process was assessed (N represents the support). It was observed that different precursors led to different iron particle sizes and location within the AC: for the N-Ac$_2$-Fe catalyst, a good iron dispersion with Fe-particles located inside the micropores was observed; with the N-FeSO$_4$ one, larger Fe-particles were found in the catalyst's surface (thus on larger micropores/mesopores); finally, an intermediate behavior was detected using N-Fe(NO$_3$)$_3$. Good iron dispersion appears to influence the mineralization efficiency, since the catalyst prepared with iron acetate showed the best TOC degradation percentages, while maintaining low levels of iron loss from the support.

Leaching of the active species is also closely related to the method and conditions of the catalyst preparation. Botas et al. [49] tested several catalysts that consisted of iron oxide, mostly crystalline hematite particles, over different silica supports. The authors found that samples prepared by post-synthesis incipient wetness impregnation displayed higher leaching levels in the outlet stream of the fixed-bed reactor (Fe concentrations above 100 mg L^{-1}) than the ones prepared by direct incorporation during the synthesis (i.e., co-condensation of iron and silica sources).

In another work of Duarte et al. [125], the influence of the particle size on adsorption and heterogeneous Fenton catalysis was also addressed. A commercial AC impregnated with iron (Fe content of 7 wt.%) was sieved and milled from the original pellet form (cylinders of ca. 3 × 5 mm) to obtain particle sizes in the intervals of 0.8–1.6, 0.25–0.8, and <0.15 mm (powder). The performance of the catalysts was evaluated regarding TOC removal of a dye-containing solution and Fe leaching from the AC support in batch runs. Results confirmed that the Fe-powder catalyst reached the higher levels of mineralization; however, leaching levels were also the highest for the smaller particle sizes tested. Taken into account the TOC mineralization degree and the iron leaching from the supports, the authors concluded that intermediate particle sizes in the 0.8–1.6 mm range were the most adequate for the operational conditions tested (see Fig. 20).

Lu et al. [101] also tested different sets of particle sizes of a Fe$_2$O$_3$/γ-Al$_2$O$_3$ catalyst in the intervals of 0.45–0.9, 0.9–1.5, and 1.5–2 mm. The catalytic performance of the continuous Fenton-like oxidation in a fixed-bed reactor was evaluated by COD reduction levels and phenol degradation ([Phenol]$_{feed}$ = 1 g L^{-1}). Phenol oxidation levels were ca. 100% for all particle sizes tested, and only a minor COD conversion decrease was observed for 1.5–2 mm. Hence, for the case in study, the effect of internal resistances to mass transfer (intraparticle diffusion) on the catalytic reaction can be disregarded for the smaller particle sizes tested. Although catalysts with smaller particle size are preferable to avoid internal diffusion problems, small particles experience a bigger pressure drop across the reactor and may eventually be cleared away by the liquid stream, so an intermediate range was selected as the optimum size.

The influence of the particle size was also addressed by Martínez et al. [47]. An extruded Fe$_2$O$_3$/SBA-15 catalyst was sieved and milled to obtain particle sizes in

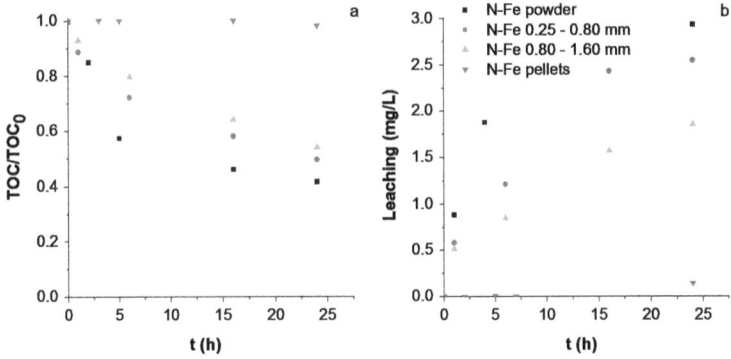

Fig. 20 (a) TOC reduction and (b) iron-leaching values along 24 h of reaction using Fe/AC catalysts with different particle sizes. Experimental conditions: T = 30°C, [Fe/AC]$_{catalyst}$ = 0.1 g L^{-1} (Fe 7 wt.%), [OII] = 0.1 mM, [H$_2$O$_2$] = 0.1 g L^{-1}, pH = 3. Adapted from [125]

the intervals of 0.35–0.7, 0.7–1, 1–1.6, and 1.6–2 mm. By changing the catalyst particle size within the packed bed ($W_{catalyst} = 2.9$ g), variations occur in the porosity and liquid holdup, which eventually modifies the residence time of the effluent in the reactor, as already addressed (higher residence times were obtained for larger particle sizes). Thus, and contrary to the expected, TOC conversions in steady state were slightly higher for the largest particle sizes tested (phenol degradation was always >99%). The authors also suggested that the contribution of the homogeneous Fenton oxidation reactions was significant since higher residence times for the liquid phase were expected for that catalyst particle size range. In that sense, the use of H_2O_2 as oxidant enables the degradation of phenol in the bulk liquid phase, mediated by its own oxidation or by the $OH^{\bullet}$ species (from the catalytic decomposition of the oxidant). This suggests that the oxidation of organic matter in the liquid phase plays an important role in the overall process efficiency.

3.4 Operational Issues Regarding Heterogeneous Catalysts: Stability

As discussed in Sect. 3.3, different materials show different activity to promote Fenton's reaction. Achieving a strong anchorage of the active species on any heterogeneous support, thus promoting extended catalytic activity and stability, is a key operational element for upscale viability of continuous heterogeneous Fenton-like systems.

Among all causes for catalyst deactivation, one of the most common is leaching of metal species from the solid support to the aqueous phase. As previously mentioned, a small concentration of the metal species in solution may act as homogenous catalyst (therefore, the catalytic activity may not be due exclusively to the action of the heterogeneous catalyst, but also to the effect of leached species in solution) but, more importantly, cause secondary metal ion pollution on the treated effluent. The most efficient way to assess the maximum productivity and stability of a catalyst is achieved by extending the reaction time with a fixed amount of catalyst until its complete or partial deactivation and, if necessary, elaborate reactivation procedures according to the data obtained during the process [126].

Long-term stability experiments with an Fe/AC catalyst (Fe 4 wt.%) were performed in a CSTR by Zazo et al. [105]. Operating with a total feed flow rate of 5 mL min^{-1} (100 mg L^{-1} phenol and 500 mg L^{-1} H_2O_2 solutions), phenol conversion during the initial 20 h of reaction was higher than 90%. Also, ca. 80% TOC degradation was observed during the same period (steady-state values). From that point on until the 170 h reaction mark, a progressive loss of activity was observed and removal percentages dropped to ca. 65% for phenol and 25% for TOC. Total iron loss from the AC support after the 170 h of reaction was almost 50% of the initial content.

The results of Melero et al. research team also stress out that the loss of activity is closely related to the features of the inlet solution/wastewater. For the same

Fe$_2$O$_3$/SBA-15 catalyst, and under the same operational conditions, the loss of activity reported by the parallel TOC reduction and H$_2$O$_2$ conversion levels was less prominent for the treatment of an industrial pharmaceutical wastewater [43], when compared to the degradation of a phenolic solution [47] (after 20 h of reaction).

Other known cause for catalyst deactivation is the formation of carbonaceous deposits that may reduce the specific surface area of the catalyst, as observed by Zazo et al. [105]. Characterization of fresh and used catalyst after a 170 h run in batch operation revealed a specific surface area reduction of ca. 40% (S$_{BET}$ of 781 vs 460 m^2 g^{-1}, respectively) and a total pore volume decrease from 0.67 to 0.36 cm^3 g^{-1} (not only microporous but also all porous structure). Reactivation procedures implemented by the authors included the washing of the used catalyst with 1 N NaOH, which allowed the recovery of most of the initial pore volume and S$_{BET}$, with no further loss of Fe; nevertheless, the catalytic activity of the recovered catalyst revealed a certain degree of irreversible deactivation, which is evidenced by the TOC reduction curves depicted in Fig. 21.

The poisoning of active catalytic sites within the heterogeneous surface is another known mechanism of catalyst deactivation, since the accessible active sites for hydrogen peroxide reduction, involved in the HO$^{\bullet}$ formation, become unavailable [47]. For three consecutive runs in a batch reactor using the same Fe-exchanged pillared beidellite catalyst, Catrinescu et al. [112] detected that the initial rate of phenol degradation was slowly diminishing. Then, when the catalyst was used after an intermediate calcination step, the initial oxidative capacity of the material was almost fully restored (without significant Fe loss), indicating that the organic species adsorbed on the active sites of the catalysts were causing the oxidation delay for short reaction times. Catalyst's stability investigations performed by Di Luca et al. [48] research group indicate that the use of a high calcination temperature (900°C), along with a step of immersion into an organic

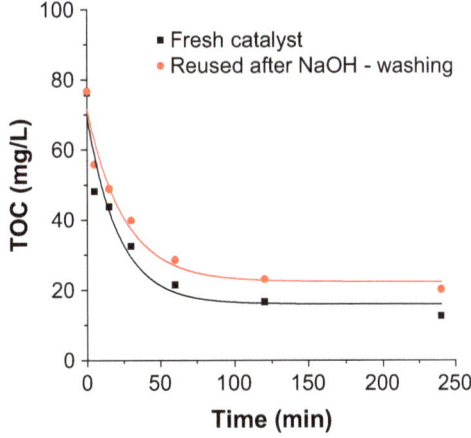

Fig. 21 Comparison of fresh vs used catalyst regenerated with 1 N NaOH-solution in batch runs. Experimental conditions: [Phenol]$_0$ = 100 mg L^{-1}, [H$_2$O$_2$]$_0$ = 500 mg L^{-1}, W$_{catalyst}$ = 500 mg L^{-1}, T = 50°C, pH$_0$ = 3.0. Adapted from [105]

acid solution, enhances the stability of a Fe_2O_3/Al_2O_3 catalyst by increasing the iron-support interactions.

Several efforts have been performed to overcome drawbacks associated with weak anchorage of metal ions on the catalyst, such as the modification of surface oxygenated groups by different treatment techniques [127] or the complexation of iron with organic ligands, which are linked to extend the pH range from acidic to basic regarding Fenton oxidation [128]; nevertheless, their deterioration occurs by the oxidative action of the process, which gradually destroys the ligands, with the consequent fragmentation of the structure over time.

Another major disadvantage of Fenton's heterogeneous catalysts is that pollutant's molecules in solution must diffuse to the surface of the material to reach active sites before they are oxidized. The detailed mechanism regarding the role of surface interactions and adsorption/desorption of metal ions and reactants in heterogeneous catalysts is still unclear; however, they are often considered rate-limiting steps of the process [35, 109].

3.5 Performance of Continuous vs Batch Reactors

An overwhelming number of studies regarding homogeneous and heterogeneous Fenton-like processes in discontinuous slurry reactors can be found in literature. However, not only is the operation of continuous reactors preferable for industrial applications, but it has also been demonstrated, mostly in lab-scale studies, that degradation efficiencies of several pollutants/wastewaters are very promising in continuous flow systems.

For example, Aghdasinia et al. [60] compared the catalytic performance of a heterogeneous Fenton-like system operating in a FBR with a batch one, under rather identical conditions, for the BB3-dye degradation. The higher dye removal level attained with the fluidized-bed reactor is probably related to the enhancement of contact between the natural magnetite catalyst particles and BB3, the high mass transfer coefficients in FBRs, and an effective and uniform mixing of the solution and catalyst inside the column.

The CWPO performance of sewage sludge carbonaceous materials on m-cresol (a toxic organic compound used in the manufacture of pharmaceuticals, pesticides, surfactants, etc.) degradation was evaluated by Yu et al. [108]. Results in batch and continuous fixed-bed reactor experiments showed that among all materials tested, the SW treated with HNO_3 presented higher conversion of the target compound. Conversion of m-cresol reached 98% and TOC depletion 67.1% in discontinuous mode (experimental conditions: $pH_0 = 7.0$, $T = 80°C$, [m-cresol] = 100 mg L^{-1}, [H_2O_2] = 15.7 mmol L^{-1}, [HNO_3-SW] = 0.5 g L^{-1}, 180 min of adsorption and 210 min of oxidation); conversion of m-cresol in the long-term continuous run (same conditions as the batch experiment and LHSV = 1 h^{-1}) was >95% from 500 to 1,200 h reaction time (Fig. 22). Nonetheless, different behavior was observed between the two configurations. For the batch operation, after the

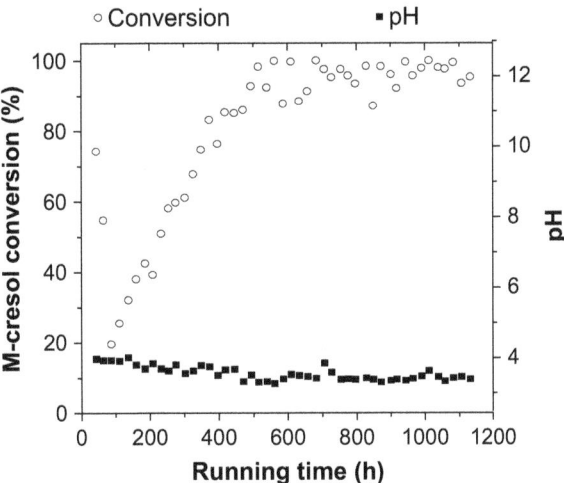

Fig. 22 Conversion of *m*-cresol and pH values along the reaction in continuous operation with HNO_3-SW catalyst. Experimental conditions: [*m*-cresol]$_{feed}$ = 100 mg L^{-1}, [H_2O_2]$_{feed}$ = 15.7 mmol L^{-1}, LHSV = 1 h^{-1}, T = 80°C, pH$_0$ = 7.0 Adapted from [108]

adsorption period, oxidation of *m*-cresol occurs almost instantly, whereas for the continuous run *m*-cresol conversion dropped along time in the first 100 h of operation, which was caused by both adsorption and oxidation phenomena. Once adsorption equilibrium was reached, the CWPO was the key process affecting the degradation and the efficiency of the process improved significantly. The behavior observed could have been caused by the catalyst's surface functional groups, since NO_2 groups (added during the HNO_3 treatment) may affect negatively the adsorption process; however, the loss of such groups was verified along the reaction, which may have led to the contact enhancement between *m*-cresol and the catalyst's active phase. Also, the lag observed could be due to the occurrence of different mass transfer rates: for the continuous system, the oxidant and pollutant passed through the granular catalysts; hence the mass transfer rate was lower than the one on the batch reactor (where the powder catalyst was mixed in solution with H_2O_2 and *m*-cresol).

4 Conclusions

Undoubtedly, homogeneous Fenton and Fenton-like processes are among the most interesting AOPs for the degradation of refractory organic compounds, increasingly common in industrially generated wastewaters. However, the added costs associated with the recovery of the catalyst (and the treatment of the sludge formed, which is a major concern especially for the treatment of large wastewater flow rates), the restricted pH range of operation, and the amount of dissolved catalyst needed in solution led to the exploration of cost-effective alternatives. The development of heterogeneous materials, capable of supporting the Fenton's reagent active phase (usually iron species), with consequent high catalytic activity and stability, became an interesting field of research over the last few decades. Numerous supports, preparation methods, transition metals, and other physicochemical characteristics regarding heterogeneous catalysts have been proposed and studied for the Fenton-like degradation of different pollutants and wastewaters. These studies, however, are mostly performed in discontinuous batch reactors, due to the easiness of operation and the faster screening of the key operational parameters influencing the oxidation process.

In recent years, more attention has been given toward the application of heterogeneous catalysts in continuous systems, namely, in fixed-bed reactors, fluidized-bed reactors, and CSTRs, which are understandably preferred for upscale industrial applications; each configuration presents its own advantages and disadvantages for the application of heterogeneous Fenton-like process, which are explained throughout this chapter.

Examples of the most up-to-date findings in this field of research are also provided, along with a comprehensive review of the process operational variables involved (e.g., pH, H_2O_2 feed dose, temperature, catalyst load, and properties) and their expected effect on the overall efficiency of this technology. We believe that such considerations provide helpful information for a faster and more effective optimization of heterogeneous Fenton-based processes.

Moreover, an overview of the main technological issue regarding this process – the catalysts deactivation – is made. Despite the recent progresses in this area, developing a cost-effective and strong anchorage of the metal ions in a solid matrix seems to be the limiting step toward the wider application of this catalytic process. Also, it is worth mentioning that despite the more-or-less successful application of this technology under lab-scale studies, especially for the treatment of model compounds, the successful upscaling of the process requires the development of more elaborated mechanisms and kinetic modeling and the study of more complex/real wastewaters.

The final section of this chapter illustrates the potential of heterogeneous Fenton-like continuous systems, when compared to the results obtained in batch processes under rather identical operating conditions.

Acknowledgments This work was the result of the Project POCI-01-0145-FEDER-006939 (Laboratory for Process Engineering, Environment, Biotechnology and Energy – UID/EQU/00511/2013) funded by the European Regional Development Fund (ERDF), through COMPETE2020 Programa Operacional Competitividade e Internacionalização (POCI), and by national funds, through Fundação para a Ciência e a Tecnologia (FCT), and Project NORTE-01-0145-FEDER-000005 – LEPABE-2-ECO-INNOVATION, supported by the North Portugal Regional Operational Programme (NORTE 2020), under the Portugal 2020 Partnership Agreement, through the European Regional Development Fund (ERDF).

Carmen Rodrigues is grateful to FCT for the financial support through the postdoctoral grant (SFRH/BPD/115879/2016).

References

1. Vandevivere PC, Bianchi R, Verstraete W (1998) Treatment and reuse of wastewater from the textile wet-processing industry: review of emerging technologies. J Chem Technol Biotechnol 72:289–302
2. Mohajerani M, Mehrvar M, Ein-Mozaffari F (2009) An overview of the integration of advanced oxidation technologies and other processes for water and wastewater treatment. Int J Eng 3:120–146
3. Oturan MA, Aaron J-J (2014) Advanced oxidation processes in water/wastewater treatment: principles and applications. A review. Crit Rev Environ Sci Technol 44:2577–2641
4. Eljarrat E, Barceló D (2003) Priority lists for persistent organic pollutants and emerging contaminants based on their relative toxic potency in environmental samples. Trends Anal Chem 22:655–665
5. Farré M, Pérez S, Kantiani L (2008) Fate and toxicity of emerging pollutants, their metabolites and transformation products in the aquatic environment. Trends Anal Chem 27:991–1007
6. Durmusoglu E, Taspinar F, Karademir A (2010) Health risk assessment of BTEX emissions in the landfill environment. J Hazard Mater 176:870–877
7. Toms L-ML, Hearn L, Mueller JF, Harden FA (2016) Assessing infant exposure to persistent organic pollutants via dietary intake in Australia. Food Chem Toxicol 87:166–171
8. Umbuzeiro GA, Freeman HS, Warren SH, Oliveira DP, Terao Y, Watanabe T, Claxton LD (2005) The contribution of azo dyes to the mutagenic activity of the Cristais River. Chemosphere 60:55–64
9. White PA, Claxton LD (2004) Mutagens in contaminated soil: a review. Mutat Res 567:227–345
10. Banat IM, Nigam P, Singh D, Marchant R (1996) Microbial decolorization of textile-dye containing effluents: a review. Bioresour Technol 58:217–227
11. Haritash AK, Kaushik CP (2009) Biodegradation aspects of polycyclic aromatic hydrocarbons (PAHs): a review. J Hazard Mater 169:1–15
12. Fu F, Chen R, Xiong Y (2006) Application of a novel strategy – coordination polymerization precipitation to the treatment of Cu^{2+}-containing wastewaters. Sep Purif Technol 52:388–393
13. Fu F, Wang Q (2011) Removal of heavy metal ions from wastewaters: a review. J Environ Manag 92:407–418
14. Cath TY, Childress AE, Elimelech M (2006) Forward osmosis: principles, applications, and recent developments. J Membr Sci 281:70–87
15. Busca G, Berardinelli S, Resini C, Arrighi L (2008) Technologies for the removal of phenol from fluid streams: a short review of recent developments. J Hazard Mater 160:265–288
16. Gabelman A, Hwang S-T (1999) Hollow fiber membrane contactors. J Membr Sci 159:61–106
17. Verma AK, Dash RR, Bhunia P (2012) A review on chemical coagulation/flocculation technologies for removal of colour from textile wastewaters. J Environ Manag 93:154–168
18. Duan J, Gregory J (2003) Coagulation by hydrolysing metal salts. Adv Colloid Interf Sci 100–102:475–502
19. Kurniawan TA, Chan GYS, Lo WH, Babel S (2006) Physico-chemical treatment techniques for wastewater laden with heavy metals. Chem Eng J 118:83–98
20. Kannan N, Sundaram MM (2001) Kinetics and mechanism of removal of methylene blue by adsorption on various carbons – a comparative study. Dyes Pigments 51:25–40
21. Babel S, Kurniawan TA (2003) Low-cost adsorbents for heavy metals uptake from contaminated water: a review. J Hazard Mater 97:219–243
22. Crini G (2006) Non-conventional low-cost adsorbents for dye removal: a review. Bioresour Technol 97:1061–1085
23. Babuponnusami A, Muthukumar K (2014) A review on Fenton and improvements to the Fenton process for wastewater treatment. J Environ Chem Eng 2:557–572

24. Levec J, Pintar A (2007) Catalytic wet-air oxidation processes: a review. Catal Today 124:172–184
25. Malato S, Blanco J, Vidal A, Richter C (2002) Photocatalysis with solar energy at a pilot-plant scale: an overview. Appl Catal B Environ 37:1–15
26. Scott JP, Ollis DF (1995) Integration of chemical and biological oxidation processes for water treatment: review and recommendations. Environ Prog 14:88–103
27. Saharan VK, Pinjari DV, Gogate PR, Pandit B (2014) Advanced oxidation technologies for wastewater treatment: an overview. Butterworth-Heinemann, Burlington, pp 141–191
28. Bowers AR, Gaddipati P, Eckenfelder WW, Monsen RM (1989) Treatment of toxic or refractory wastewaters with hydrogen peroxide. Water Sci Technol 21:477–486
29. García MT, Ribosa I, Guindulain T, Sánchez-Leal J, Vives-Rego J (2001) Fate and effect of monoalkyl quaternary ammonium surfactants in the aquatic environment. Environ Pollut 111:169–175
30. Lapertot M, Pulgarín C, Fernández-Ibáñez P, Maldonado MI, Pérez-Estrada L, Oller I, Gernjak W, Malato S (2006) Enhancing biodegradability of priority substances (pesticides) by solar photo-Fenton. Water Res 40:1086–1094
31. Bigda RJ (1995) Consider Fenton's chemistry for wastewater treatment. Chem Eng Prog 12:62–66
32. Huang CP, Dong C, Tang Z (1993) Advanced chemical oxidation: its present role and potential future in hazardous waste treatment. Waste Manag 13:361–377
33. Comninellis C, Kapalka A, Malato S, Parsons SA, Poulios I, Mantzavinos D (2008) Advanced oxidation processes for water treatment: advances and trends for R&D. J Chem Technol Biotechnol 83:769–776
34. Pera-Titus M, García-Molina V, Baños MA, Giménez J, Esplugas S (2004) Degradation of chlorophenols by means of advanced oxidation processes: a general review. Appl Catal B Environ 47:219–256
35. Pignatello JJ, Oliveros E, Mackay A (2006) Advanced oxidation processes for organic contaminant destruction based on the Fenton reaction and related chemistry advanced oxidation processes for organic contaminant destruction based on the Fenton. Crit Rev Environ Sci Technol 36:1–84
36. Poyatos JM, Muñio MM, Almecija MC, Torres JC, Hontoria E, Osorio F (2010) Advanced oxidation processes for wastewater treatment: state of the art. Water Air Soil Pollut 205:187–204
37. Froment GF, Bischoff KB (1990) Chemical reactor analysis and design. 2nd edn. Wiley, New York
38. Gogate PR, Pandit AB (2004) A review of imperative technologies for wastewater treatment I: oxidation technologies at ambient conditions. Adv Environ Res 8:501–551
39. Karthikeyan S, Titus A, Gnanamani A, Mandal AB, Sekaran G (2011) Treatment of textile wastewater by homogeneous and heterogeneous Fenton oxidation processes. Desalination 281:438–445
40. Punzi M, Mattiasson B, Jonstrup M (2012) Treatment of synthetic textile wastewater by homogeneous and heterogeneous photo-Fenton oxidation. J Photochem Photobiol A Chem 248:30–35
41. Liotta LF, Gruttadauria M, Di Carlo G, Perrini G, Librando V (2009) Heterogeneous catalytic degradation of phenolic substrates: catalysts activity. J Hazard Mater 162:588–606
42. Rey A, Zazo JA, Casas JA, Bahamonde A, Rodriguez JJ (2011) Influence of the structural and surface characteristics of activated carbon on the catalytic decomposition of hydrogen peroxide. Appl Catal A Gen 402:146–155
43. Melero JA, Martínez F, Botas JA, Molina R, Pariente MI (2009) Heterogeneous catalytic wet peroxide oxidation systems for the treatment of an industrial pharmaceutical wastewater. Water Res 43:4010–4018

44. Duarte F, Morais V, Maldonado-Hódar FJ, Madeira LM (2013) Treatment of textile effluents by the heterogeneous Fenton process in a continuous packed-bed reactor using Fe/activated carbon as catalyst. Chem Eng J 232:34–41
45. Fida H, Zhang G, Guo S, Naeem A (2016) Heterogeneous Fenton degradation of organic dyes in batch and fixed bed using La-Fe montmorillonite as catalyst. J Colloid Interface Sci 490:859–868
46. Mesquita I, Matos LC, Duarte F, Maldonado-Hódar FJ, Mendes A, Madeira LM (2012) Treatment of azo dye-containing wastewater by a Fenton-like process in a continuous packed-bed reactor filled with activated carbon. J Hazard Mater 237–238:30–37
47. Martínez F, Melero JA, Botas JA, Pariente MI, Molina R (2007) Treatment of phenolic effluents by catalytic wet hydrogen peroxide oxidation over Fe_2O_3/SBA-15 extruded catalyst in a fixed-bed reactor. Ind Eng Chem Res 46:4396–4405
48. Di Luca C, Massa P, Fenoglio R, Cabello FM (2014) Improved Fe_2O_3/Al_2O_3 as heterogeneous Fenton catalysts for the oxidation of phenol solutions in a continuous reactor. J Chem Technol Biotechnol 89:1121–1128
49. Botas JA, Melero JA, Martínez F, Pariente MI (2010) Assessment of Fe_2O_3/SiO_2 catalysts for the continuous treatment of phenol aqueous solutions in a fixed bed reactor. Catal Today 149:334–340
50. Pariente MI, Siles JA, Molina R, Botas JA, Melero JA, Martínez F (2013) Treatment of an agrochemical wastewater by integration of heterogeneous catalytic wet hydrogen peroxide oxidation and rotating biological contactors. Chem Eng J 226:409–415
51. Bello MM, Abdul Raman AA, Purushothaman M (2017) Applications of fluidized bed reactors in wastewater treatment – a review of the major design and operational parameters. J Clean Prod 141:1492–1514
52. Garcia-Segura S, Bellotindos LM, Huang Y-H, Brillas E, Lu M-C (2016) Fenton process as alternative wastewater treatment technology – a review. J Taiwan Inst Chem Eng 67:211–225
53. Yang W-C (2003) Handbook of fluidization and fluid-particle systems. Marcel Dekker, New York
54. Tisa F, Abdul Raman AA, Wan Daud WMA (2014) Applicability of fluidized bed reactor in recalcitrant compound degradation through advanced oxidation processes: a review. J Environ Manag 146:260–275
55. Shih YJ, Huang RL, Huang YH (2015) Adsorptive removal of arsenic using a novel akhtenskite coated waste goethite. J Clean Prod 87:897–905
56. Şen S, Demirer GN (2003) Anaerobic treatment of real textile wastewater with a fluidized bed reactor. Water Res 37:1868–1878
57. Bocos E, Pérez-Álvarez D, Pazos M, Rodríguez-Arguelles MC, Sanromán MA (2016) Coated nickel foam electrode for the implementation of continuous electro-Fenton treatment. J Chem Technol Biotechnol 91:685–692
58. Li H, Priambodo R, Wang Y, Zhang H, Huang Y-H (2015) Mineralization of bisphenol A by photo-Fenton-like process using a waste iron oxide catalyst in a three-phase fluidized bed reactor. J Taiwan Inst Chem Eng 53:68–73
59. Briones RM, De Luna MDG, Su C-C, Lu M-C (2014) Factors affecting Fenton oxidation of acetaminophen in a fluidized-bed reactor. J Environ Eng 140:77–83
60. Aghdasinia H, Khataee A, Sheikhi M, Takhtfiroozeha P (2017) Pilot plant fluidized-bed reactor for degradation of basic blue 3 in heterogeneous Fenton process in the presence of natural magnetite. Environ Prog Sustain Energy 2017:1–10
61. Chou S, Huang C, Huang YH (1999) Effect of Fe^{2+} on catalytic oxidation in a fluidized bed reactor. Chemosphere 39:1997–2006
62. Lyu C, Zhou D, Wang J (2016) Removal of multi-dye wastewater by the novel integrated adsorption and Fenton oxidation process in a fluidized bed reactor. Environ Sci Pollut Res 23:20893–20903
63. Hou L, Wang L, Royer S, Zhang H (2016) Ultrasound-assisted heterogeneous Fenton-like degradation of tetracycline over a magnetite catalyst. J Hazard Mater 302:458–467

64. Aghdasinia H, Bagheri R, Vahid B, Khataee A (2016) Central composite design optimization of pilot plant fluidized-bed heterogeneous Fenton process for degradation of an azo dye. Environ Technol 37:2703–2712
65. Luyben WL (2007) Chemical reactor design and control. Wiley, New York
66. Esposito G, Frunzo L, Panico A, Pirozzi F (2011) Modelling the effect of the OLR and OFMSW particle size on the performances of an anaerobic co-digestion reactor. Process Biochem 46:557–565
67. Nielsen HB, Ahring BK (2006) Responses of the biogas process to pulses of oleate in reactors treating mixtures of cattle and pig manure. Biotechnol Bioeng 95:95–105
68. Muñoz R, Kollner C, Guieysse B, Mattiasson B (2004) Photosynthetically oxygenated salicylate biodegradation in a continuous stirred tank photobioreactor. Biotechnol Bioeng 87:797–803
69. Sponza DT, Gök O (2010) Effect of rhamnolipid on the aerobic removal of polyaromatic hydrocarbons (PAHs) and COD components from petrochemical wastewater. Bioresour Technol 101:914–924
70. Martínez F, Molina R, Pariente MI, Siles JA, Melero JA (2016) Low-cost Fe/SiO_2 catalysts for continuous Fenton processes. Catal Today 280:176–183
71. Queirós S, Morais V, Rodrigues CSD, Maldonado-Hódar FJ, Madeira LM (2015) Heterogeneous Fenton's oxidation using Fe/ZSM-5 as catalyst in a continuous stirred tank reactor. Sep Purif Technol 141:235–245
72. Esteves BM, Rodrigues CSD, Boaventura RAR, Maldonado-Hódar FJ, Madeira LM (2016) Coupling of acrylic dyeing wastewater treatment by heterogeneous Fenton oxidation in a continuous stirred tank reactor with biological degradation in a sequential batch reactor. J Environ Manag 166:193–203
73. Zhang H, Choi H, Huang C (2006) Treatment of landfill leachate by Fenton's reagent in a continuous stirred tank reactor. J Hazard Mater 136:618–623
74. Ramirez JH, Duarte FM, Martins FG, Costa CA, Madeira LM (2009) Modelling of the synthetic dye Orange II degradation using Fenton's reagent: from batch to continuous reactor operation. Chem Eng J 148:394–404
75. Barreto-Rodrigues M, Silva FT, Paiva TCB (2009) Optimization of Brazilian TNT industry wastewater treatment using combined zero-valent iron and Fenton processes. J Hazard Mater 168:1065–1069
76. Hodaifa G, Ochando-Pulido JM, Rodriguez-Vives S, Martinez-Ferez A (2013) Optimization of continuous reactor at pilot scale for olive-oil mill wastewater treatment by Fenton-like process. Chem Eng J 220:117–124
77. Zhang H, Ran X, Wu X (2012) Electro-Fenton treatment of mature landfill leachate in a continuous flow reactor. J Hazard Mater 241–242:259–266
78. Zhang H, Fei C, Zhang D, Tang F (2007) Degradation of 4-nitrophenol in aqueous medium by electro-Fenton method. J Hazard Mater 145:227–232
79. Fenton HJH (1894) Oxidation of tartaric acid in presence of iron. J Chem Soc 65:899–910
80. Haber F, Weiss J (1934) The catalytic decomposition of hydrogen peroxide by iron salts. Proc R Soc Lond A 147:332–351
81. Walling C (1975) Fenton's reagent revisited. Acc Chem Res 8:125–131
82. Kuo WG (1992) Decolorizing dye wastewater with Fenton's reagent. Water Res 26:881–886
83. Bautista P, Mohedano AF, Casas JA, Zazo JA, Rodriguez JJ (2008) An overview of the application of Fenton oxidation to industrial wastewaters treatment. J Chem Technol Biotechnol 83:1323–1338
84. Sychev AY, Isak VG (1995) Iron compounds and the mechanisms of the homogeneous catalysis of the activation of O_2 and H_2O_2 and of the oxidation of organic substrates. Russ Chem Rev 64:1105–1129
85. Chu L, Wang J, Dong J, Liu H, Sun X (2012) Treatment of coking wastewater by an advanced Fenton oxidation process using iron powder and hydrogen peroxide. Chemosphere 86:409–414

86. Lucas MS, Peres JA (2009) Removal of COD from olive mill wastewater by Fenton's reagent: kinetic study. J Hazard Mater 168:1253–1259
87. Rodrigues CSD, Madeira LM, Boaventura RAR (2009) Treatment of textile effluent by chemical (Fenton's reagent) and biological (sequencing batch reactor) oxidation. J Hazard Mater 172:1551–1559
88. Tambosi JL, Di Domenico M, Schirmer WN, José HJ, Moreira RFPM (2006) Treatment of paper and pulp wastewater and removal of odorous compounds by a Fenton-like process at the pilot scale. J Chem Technol Biotechnol 81:1426–1432
89. Bautista P, Mohedano AF, Gilarranz MA, Casas JA, Rodriguez JJ (2007) Application of Fenton oxidation to cosmetic wastewaters treatment. J Hazard Mater 143:128–134
90. Peres JA, Heredia JB, Domínguez JR (2004) Integrated Fenton's reagent – coagulation/flocculation process for the treatment of cork processing wastewaters. J Hazard Mater 107:115–121
91. Guedes A, Madeira LM, Boaventura RAR, Costa CA (2003) Fenton oxidation of cork cooking wastewater – overall kinetic analysis. Water Res 37:3061–3069
92. Wang W, Liu Y, Li T, Zhou M (2014) Heterogeneous Fenton catalytic degradation of phenol based on controlled release of magnetic nanoparticles. Chem Eng J 242:1–9
93. Watts RJ, Jones AP, Chen P-H, Kenny A (1997) Mineral-catalyzed Fenton-like oxidation of sorbed chlorobenzenes. Water Environ Res 69:269–275
94. Idel-Aouad R, Valiente M, Yaacoubi A, Tanouti B, Lópz-Mesas M (2011) Rapid decolourization and mineralization of the azo dye C.I. Acid Red 14 by heterogeneous Fenton reaction. J Hazard Mater 186:745–750
95. EC Directive 2000/60/EC of the European parliament and of the council of October 23 (2000) Establishing a framework for community action in the field of water policy
96. Rokhina EV, Virkutyte J (2010) Environmental application of catalytic processes: heterogeneous liquid phase oxidation of phenol with hydrogen peroxide. Crit Rev Environ Sci Technol 41:125–167
97. Barroso-Bogeat A, Alexandre-Franco M, Fernández-González C, Gómez-Serrano V (2016) Activated carbon surface chemistry: changes upon impregnation with Al(III), Fe(III) and Zn (II)-metal oxide catalyst precursors from NO_3^- aqueous solutions. Arab J Chem, in press. doi: 10.1016/j.arabjc.2016.02.018
98. Oliveira LCA, Silva CN, Yoshida MI, Lago RM (2004) The effect of H_2 treatment on the activity of activated carbon for the oxidation of organic contaminants in water and the H_2O_2 decomposition. Carbon N Y 42:2279–2284
99. Zazo JA, Bedia J, Fierro CM, Pliego G, Casas JA, Rodriguez JJ (2012) Highly stable Fe on activated carbon catalysts for CWPO upon $FeCl_3$ activation of lignin from black liquors. Catal Today 187:115–121
100. Tatibouët JM, Guélou E, Fournier J (2005) Catalytic oxidation of phenol by hydrogen peroxide over a pillared clay containing iron. Active species and pH effect. Top Catal 33:225–232
101. Lu M, Yao Y, Gao L, Mo D, Lin F, Lu S (2015) Continuous treatment of phenol over an $Fe_2O_3/\gamma\text{-}Al_2O_3$ catalyst in a fixed-bed reactor. Water Air Soil Pollut 226:87
102. Liu P, He S, Wei H, Wang J, Sun C (2015) Catalytic wet peroxide oxidation of phenol over Fe_2O_3/MCM-41 in a fixed bed reactor. Ind Eng Chem Res 54:130–136
103. Seyed Dorraji MS, Mirmohseni A, Carraro M, Gross S, Simone S, Tasselli F, Figoli A (2015) Fenton-like catalytic activity of wet-spun chitosan hollow fibers loaded with Fe_3O_4 nanoparticles: batch and continuous flow investigations. J Mol Catal A Chem 398:353–357
104. Yan Y, Jiang S, Zhang H (2014) Efficient catalytic wet peroxide oxidation of phenol over Fe-ZSM-5 catalyst in a fixed bed reactor. Sep Purif Technol 133:365–374
105. Zazo JA, Casas JA, Mohedano AF, Rodríguez JJ (2006) Catalytic wet peroxide oxidation of phenol with a Fe/active carbon catalyst. Appl Catal B Environ 65:261–268

106. Chou S, Huang C, Huang Y (2001) Heterogeneous and homogeneous catalytic oxidation by supported γ-FeOOH in a fluidized-bed reactor: kinetic approach. Environ Sci Technol 35:1247–1251
107. Achma RB, Ghorbel A, Dafinov A, Medina F (2008) Copper-supported pillared clay catalysts for the wet hydrogen peroxide catalytic oxidation of model pollutant tyrosol. Appl Catal A Gen 349:20–28
108. Yu Y, Wei H, Yu L, Wang W, Zhao Y, Gu B, Sun C (2016) Sewage-sludge-derived carbonaceous materials for catalytic wet hydrogen peroxide oxidation of m-cresol in batch and continuous reactors. Environ Technol 37:153–162
109. Guo J, Al-Dahhan M (2003) Catalytic wet oxidation of phenol by hydrogen peroxide over pillared clay catalyst. Ind Eng Chem Res 42:2450–2460
110. Jung YS, Lim WT, Park J, Kim Y (2009) Effect of pH on Fenton and Fenton-like oxidation. Environ Technol 30:183–190
111. Ramirez JH, Lampinen M, Vicente MA, Madeira LM (2008) Experimental design to optimize the oxidation of Orange II dye solution using a clay-based Fenton-like catalyst. Ind Eng Chem Res 47:284–294
112. Catrinescu C, Teodosiu C, Macoveanu M (2003) Catalytic wet peroxide oxidation of phenol over Fe-exchanged pillared beidellite catalytic wet peroxide oxidation of phenol over Fe-exchanged pillared beidellite. Water Res 37:1154–1160
113. Chi GT, Churchley J, Huddersman KD (2013) Pilot-scale removal of trace steroid hormones and pharmaceuticals and personal care products from municipal wastewater using a heterogeneous Fenton's catalytic process. Int J Chem Eng 2013:1–10
114. Kuznetsova EV, Savinov EN, Vostrikova LA, Parmon VN (2004) Heterogeneous catalysis in the Fenton-type system FeZSM-5/H_2O_2. Appl Catal B Environ 51:165–170
115. Muruganandham M, Suri RPS, Jafari S, Sillanpää M, Lee G-J, Wu JJ, Swaminathan M (2014) Recent developments in homogeneous advanced oxidation processes for water and wastewater treatment. Int J Photoenergy 1:1–21
116. Ramirez JH, Maldonado-Hódar FJ, Pérez-Cadenas AF, Moreno-Castilla C, Costa CA, Madeira LM (2007) Azo-dye Orange II degradation by heterogeneous Fenton-like reaction using carbon-Fe catalysts. Appl Catal B 75:312–323
117. Duarte F, Maldonado-Hódar FJ, Madeira LM (2013) Influence of the iron precursor in the preparation of heterogeneous Fe/activated carbon Fenton-like catalysts. Appl Catal A Gen 458:39–47
118. Rache ML, García AR, Zea HR, Silva AMT, Madeira LM, Ramirez JH (2014) Azo-dye orange II degradation by the heterogeneous Fenton-like process using a zeolite Y-Fe catalyst – kinetics with a model based on the Fermi's equation. Appl Catal B Environ 146:192–200
119. Rodrigues CSD, Boaventura RAR, Madeira LM (2014) Technical and economic feasibility of polyester dyeing wastewater treatment by coagulation/flocculation and Fenton's oxidation. Environ Technol 35:1307–1319
120. Lin SH, Lo CC (1997) Fenton process for treatment of desizing wastewater. Water Res 31:2050–2056
121. Ito K, Jian W, Nishijima W, Baes AU, Shoto E, Okada M (1998) Comparison of ozonation and AOPs combined with biodegradation for removal of thm precursors in treated sewage effluents. Water Sci Technol 38:179–186
122. Nieto LM, Hodaifa G, Rodriguez S, Giménez JA, Ochando J (2011) Degradation of organic matter in olive-oil mill wastewater through homogeneous Fenton-like reaction. Chem Eng J 173:503–510
123. Ramirez JH, Costa CA, Madeira LM (2005) Experimental design to optimize the degradation of the synthetic dye Orange II using Fenton's reagent. Catal Today 107–108:68–76
124. Rodríguez A, Ovejero G, Sotelo JL, Mestanza M, García J (2010) Heterogeneous Fenton catalyst supports screening for mono azo dye degradation in contaminated wastewaters. Ind Eng Chem Res 49:498–505

125. Duarte F, Maldonado-Hódar FJ, Madeira LM (2012) Influence of the particle size of activated carbons on their performance as Fe supports for developing Fenton-like catalysts. Ind Eng Chem Res 51:9218–9226
126. Navalon S, Alvaro M, Garcia H (2010) Heterogeneous Fenton catalysts based on clays, silicas and zeolites. Appl Catal B Environ 99:1–26
127. Figueiredo JL, Pereira MFR, Freitas MMA, Órfão JJM (1999) Modification of the surface chemistry of activated carbons. Carbon 37:1379–1389
128. Yao Y, Wang L, Sun L, Zhu S, Huang Z, Mao Y, Lu W, Chen W (2013) Efficient removal of dyes using heterogeneous Fenton catalysts based on activated carbon fibers with enhanced activity. Chem Eng Sci 101:424–431

Disinfection by Chemical Oxidation Methods

Luis-Alejandro Galeano, Milena Guerrero-Flórez,
Claudia-Andrea Sánchez, Antonio Gil, and Miguel-Ángel Vicente

Abstract Poor quality in drinking water is primary cause of pathogen transmission and responsible of varied infectious diseases. Methods of water treatment for human consumption must pay special attention on microbiological safe disinfection. Indeed, from the past few years laws all around the world have included new, more stringent water quality parameters. Chlorination and other mainly used conventional disinfection processes usually do not achieve full inactivation of all microorganisms present in real water supplies, whereas the presence of even low concentrations of organic matter can lead to form harmful disinfection by-products. Protozoan parasites *Giardia* sp. and *Cryptosporidium* sp. are some of the microorganisms that cannot be completely inactivated via chlorination under the same contact times typical of bacteria or virus elimination. It has increased toxicological and microbiological risks as well as operational costs. Disinfection by the advanced oxidation process more intensively studied in the past few years has been reviewed including

---

L.-A. Galeano (✉) and Claudia-Andrea Sánchez
Grupo de Investigación en Materiales Funcionales y Catálisis (GIMFC), Departamento de Química, Facultad de Ciencias Exactas y Naturales, Universidad de Nariño, Pasto-Nariño, Colombia
e-mail: alejandrogaleano@udenar.edu.co; clasanchezr@unal.edu.co

M. Guerrero-Flórez
Grupo de Investigación en Materiales Funcionales y Catálisis (GIMFC), Departamento de Química, Facultad de Ciencias Exactas y Naturales, Universidad de Nariño, Pasto-Nariño, Colombia

Present Address: Universidad Nacional de Colombia, Bogotá, DC, Colombia
e-mail: milenague@udenar.edu.co

A. Gil
Departamento de Química Aplicada, Edificio de los Acebos, Universidad Pública de Navarra, Pamplona, Spain
e-mail: andoni@unavarra.es

Miguel-Ángel Vicente
GIR-QUESCAT, Departamento de Química Inorgánica, Universidad de Salamanca, Salamanca, Spain
e-mail: mavicente@usal.es

Fenton and photo-Fenton processes and photocatalytic and electro-catalytic variants; this vibrant topic still remains partially uncovered in the available scientific background, which has motivated many recent researches and publications. This chapter is then devoted to briefly review the most recent reports studying the disinfecting potential displayed by mentioned AOPs with respect to widely and currently used conventional techniques. Revision of the inactivation of water-borne pathogens including *E. coli*, total coliforms, parasites as *Giardia* and *Cryptosporidium,* and virus such as coliphages has focused on advantages and disadvantages in application of every particular AOP, their disinfecting mechanisms, and the main parameters affecting the disinfection response.

Keywords Advanced oxidation processes, Chemical oxidation, Disinfection, Drinking water, Oxidative stress, Water microbiology

Contents

1 Introduction 259
2 Some General Issues About Pathogens and Resistance to Water Disinfection 260
3 Disinfection Methods More Widely Used in Production of Drinking Water 264
   3.1 Chlorine 264
   3.2 Ozone (O_3) 265
   3.3 Hydrogen Peroxide (H_2O_2) 266
4 Advanced Oxidation Processes: AOPs 267
   4.1 Disinfecting Potential of AOPs 267
   4.2 Fenton Process 269
   4.3 Photo-Fenton Process 271
   4.4 Heterogeneous TiO_2 Photocatalysis 275
   4.5 Electrochemical Oxidation 281
5 Conclusions 284
References 285

Abbreviations

AOPs	Advanced oxidation processes
BDD electrode	Boron-doped diamond electrode
CoA	Coenzyme A
COWPs	*Cryptosporidium* oocyst wall proteins
CWPO	Catalytic wet peroxide oxidation
DBP-Cl	Chlorinated DBP
DBP-N	Nitrogenated DBP
DBPs	Disinfection by-products
DNA	Deoxyribonucleic acid
DOC	Dissolved organic carbon
EPS	Exopolysaccharide
GalNAc	β-1,3-linked *N*-acetylgalactosamine

HAAs	Haloacetic acids
IEP	Isoelectric point
IR	Infrared radiation
LPS	Lipopolysaccharide
NOM	Natural organic matter
PG	Peptidoglycan
PRE	Photoreactivation enzyme
RCS	Reactive chlorine species
RNA	Ribonucleic acid
ROS	Reactive oxygen species
RPA	Replication protein A
SEM	Scanning electron microscopy
SOD	Superoxide dismutase
SODIS	Simulated solar disinfection
STP	Sewage treatment plants
TEM	Transmission electron microscopy
THMs	Trihalomethanes
UPEC	*E. coli* with uropathogenic characteristics
UV	Ultraviolet radiation
VGs	Virulence genes

1 Introduction

Water is one of the most important natural resources for human beings and ecosystems. However, such a valuable resource is undergoing constant spoilage caused by the sustained growth of population and wastes in both industry and agriculture, as main factors responsible of the contamination exerted on the aquatic environments [1, 2]. Entities in charge for the surveillance and protection of the environment, social development, and health care are imposing regulations more and more stringent in order to achieve extended and deeper water purification [3, 4].

Currently, one of the most significant challenges is water production with optimal parameters of quality for human consumption. Its availability would be at the same time a strong contribution to reduce spreading diseases related to environmental factors. Water disinfection is a critical process for protection of public health; either pathogen's inactivation or elimination plays a key role on water's quality [5–10]. Some pathogens often transmitted by water are displayed in Table 1.

Safe disinfection of water supplies is so much important that the employment of high-performance, top efficient methods should always be completely guaranteed. Therefore, we share a brief revision about microbial resistance and disinfecting potential of some chemical oxidation methods, mainly focused in AOPs, as technological alternatives currently available to achieve safer water disinfection than traditional methods.

Table 1 Pathogens commonly transmitted by water [11–13]

Bacteria	*Campylobacter* spp., *E. coli*, *Leptospira* spp., *Salmonella* spp., *Shigella* spp., *Vibrio cholerae*, *Yersinia enterocolitica*
Protozoa	*Balantidium coli*, *Cryptosporidium parvum*, *Cyclospora cayetanensis*, *Entamoeba histolytica*, *Giardia intestinalis*, *Toxoplasma gondii*
Virus	Adenovirus, enterovirus, hepatitis virus A, hepatitis virus E, norovirus, sapovirus, rotavirus
Helminths	*Ascaris lumbricoides*, *Enterobius vermicularis*, *Fasciola hepatica*, *Hymenolepis nana*, *Taenia saginata*, *Taenia solium*, *Trichuris trichiura*

2 Some General Issues About Pathogens and Resistance to Water Disinfection

Effectiveness of any disinfectant depends on several properties of targeted water like pH, temperature, presence of suspended particles, and chemical composition (loading of natural organic material, agricultural and industrial chemicals). The direct biocidal action depends on the *CT* value, parameter calculated as the product of disinfectant concentration and contact time, specific for each microorganism and disinfectant [14].

Disinfectants as chlorine can form reactive chlorine species (RCS) and reactive oxygen species (ROS) which can cause oxidation of proteins, lipids, and nucleic acids. In the last one either, changes in A, T, C, G, or U nitrogenous bases, breakup of the phosphodiester bond between nucleotide chains of DNA and RNA, and cross-linking with other molecules lead to formation of pyrimidine dimers, changes in tridimensional structure, alteration of DNA replication process, and consequent death as main effects on cells [11, 15, 16].

Oxidative damages accumulated in molecules as RNA can affect protein synthesis and may generate lack of cultivability in vitro [16]. RNA repair is a mechanism not well understood because it is considered that damaged RNA is degraded rather than repaired [17]. RNA is more susceptible to oxidative damage by exposure to UV than DNA; besides, both nucleic acids can suffer damage at the same rate [16].

In cytoplasm of eukaryotic cells, the natural formation of ROS and the important role of enzymes superoxide dismutase (SOD), peroxidase, and catalase to keep the balance inducing detoxification are well established. When the normal concentration of ROS in cells increases, it leads to toxic effect in response to accumulation of reactive species; cell may lack functions, viability, and integrity, eventually causing death [16, 18].

Microorganisms have biochemical and structural differences. The cell wall is a complex multilayered structure based on variety of proteins, lipids, and polysaccharides protecting the microorganism from environmental conditions [19]. Gram-positive bacteria are surrounded by cell walls composed by PG (peptidoglycan) and teichoic and lipoteichoic acids, whereas Gram-negative bacteria contain a thin PG cell wall, which itself is surrounded by an outer membrane formed by lipopolysaccharide (LPS). Fungus cell walls instead of PG are composed by chitin, glucans, and proteins. Plasma membrane differs from most subcellular membranes because high fractions of sterols, sphingolipids, and glycerophospholipids form lipid rafts clusters into liquid-ordered domains instead of giving rise to homogeneous mixtures of glycerophospholipids mainly associated with proteins (e.g., acylated and glycosylphosphatidylinositol [GPI]-anchored proteins) [20]. Bacteria can generate new structural macromolecules to resist hostile disinfection conditions, but in fungus this process may also involve more virulence [21].

In bacteria, several mechanisms of resistance to biocides can be devised: (1) decreasing biocide penetration (e.g., changing membrane's composition and porin regulation), (2) depleting intracellular biocide concentration (e.g., either expression of efflux pumps or activation of detoxifying enzymes SOD or catalase) [22], (3) changing metabolic pathways [23], and (4) DNA repair mechanisms [24] helped by lyases. Other worth mentioning include EPS (exopolysaccharide) production and cell aggregation, which allow to decrease the concentration of disposable biocides and cell protection, respectively, important factors in bacterial biofilm resistance [25].

The disinfecting potential of AOPs to produce drinking water of course should consider the resistance mechanisms, survival potential, environmental resilience, as well as profiles of genetic variability and distribution of microorganisms through the water resources. Currently, these are responsible of treatment troubles and derived risks for public health; parasites as *Cryptosporidium parvum* and *Giardia lamblia* have developed new virulence profiles and new genotypes [26]. In the other hand, there are also viruses capable to infect new hosts and to change their own genomic configuration to generate new types, variants, and viral strains with high virulence power.

The genomic variations and mechanisms of DNA repair can lead to reactivation or regrowth of the microorganisms and subsequent low efficiency of any disinfecting treatment. Currently there are new approaches attempting to explain the cell and molecular response allowing some water-borne pathogens to resist mainly the UV radiation and Fenton process [27–30].

Oocysts of *Cryptosporidium* have shown strong resistance to conventional chlorination treatments, since they require disinfectant doses and contact times near to four orders of magnitude higher than the ones typically used for bacteria and virus inactivation [14, 31]. The resistance obeys in part to the outer wall, a common structure in *Apicomplexan* parasites that produce environmentally stable oocysts. The accurate erection of the oocyst's wall and the mechanisms regulating the process remain unclear. *Cryptosporidium* oocyst wall proteins (COWPs) are an integral part of this structure, encoded by at least nine genes [32]. This composition is related with a major resistance and survival, but other mechanisms of genomic remodeling and response to DNA damage can be also implicated. Meanwhile, the DNA replication machinery of *C. parvum* includes two unique replication protein A (RPA) heterotrimeric complexes, expressed differentially during the parasite life cycle in response to DNA damage [32]. This response may occur via resistance to water disinfectants, since it has been related with response to stressing environmental conditions.

Determining the changes in the gene expression profile of oocysts in response to ultraviolet (UV) irradiation may help to understand how the parasite responds to stress, of course with more practical relevance to UV exposure encountered in external environment as well as UV irradiation commonly used to treat drinking and surface waters [33]. Oocysts of *Cryptosporidium* exposed to UV radiation (UV_{254}) have shown complex and dynamic regulations in gene expression, activation of DNA repair, and intracellular trafficking mechanisms. In response to long-time UV exposure (5 h), high activities of genes coding for protein degradation, amino acid recycling, proteasome, ubiquitin-associated genes, and encoding hypothetical or membrane proteins have been found to be displayed. Among the stress-related genes, TCP-1 family members and some thioredoxin-associated genes appear to play important roles in the recovery of UV-induced damages. There are more than 1,924 genes positively expressed in oocysts after exposition to environmental stress conditions, representing near 51% of the genes in *C. parvum* genome [33]: 1,048 hypothetical genes with undefined functions (vs. 876 functional annotated ones), actively expressed at different levels in the oocysts (e.g., cgd3_1570, unknown, possible sporozoite antigen) and proteins containing apicomplexan-specific domains (e.g., cgd2_200), or predicted membrane proteins (e.g., cgd6_780). Transcriptomic analyses have shown that CpArray15K could be used both to study profiling of gene expression parasite oocysts as to explain responses to UV irradiation [34]. Future research on these genes may help to elucidate their functional roles in survival and stress resistance by parasite oocysts in the environment [33]. The effect and oxidation mechanisms on *Cryptosporidium* oocyst by AOPs are still unclear.

In the case of *Giardia*, the host is protected outside against hypotonic lysis in the environment by a hard cyst wall (60% carbohydrate plus 40% protein) [35]. Whereas the walls of plants and fungi contain several sugar homopolymers (cellulose, chitin, and β-1,3-glucans) along with dozens of proteins, the cyst wall of *Giardia* contains an homopolymer of β-1,3-linked *N*-acetylgalactosamine (GalNAc) and at least three cyst wall proteins (CWPs), each one composed by Leu-rich repeats and a C-terminal Cys-rich region [36]. Then, the efficient disinfection of this parasite depends at least in part to excystation state; during this process the fibrils of GalNAc homopolymer and how deproteinated fibrils of GalNAc homopolymer are degraded for *Giardia* cysts but not trophozoites are important. The behavior showed during excystation, host, and *Giardia* proteases appear to degrade bound CWPs, exposing GalNAc homopolymer fibrils, digested by a stage-specific glycohydrolase. In contrast, cyst walls of *Giardia* treated with hot alkali to deproteinate fibrils of the GalNAc homopolymer were thick (~1.2 μm), resistant to sonication, and permeable [37]. These findings about a new curled fiber and lectin model of the intact *Giardia* cyst wall protease and glycohydrolase model of excystation have contributed to understand how this parasite can resist to disinfection processes and should be carefully taken into account for future research in this field.

The regrowth of bacteria after treatment by AOPs has been reported, which has suggested one or more possible mechanisms there involved [38–41]. Studying various stages of sewage treatment plants (STPs), numerous virulence genes

(VGs), resistant genes, and phylogenetic groups responsible of *E. coli* strain survival have been established. When used separately either UV radiation or chlorination as disinfecting treatment, *E. coli* strains carrying VGs showed better survivability against UV treatment than chlorination [42–44]. When both disinfection methods were combined, greater elimination efficiency of *E. coli* with uropathogenic characteristics (UPEC) was achieved, preventing then pathogenic *E. coli* strains to be released into the environment from STPs.

UV irradiation has shown to be less effective against fecal coliforms like *E. coli* with capacity to repair their UV damages for both photoreactivation or light-free mechanisms (dark repair) [27, 45–48]. In photoreactivation, a complex is formed between the photoreactivation enzyme (PRE) and the nucleotide dimer to be repaired followed by prereleasing and the DNA repairing; DNA repair enzymes photo-lyases participate in the process. This process depends on the light source (visible, IR, UV) between 380 and 480 nm in the visible spectrum and the radiant intensity. On the other hand, the light-independent, bacterial dark repair is governed by expression of recA protein which allows to identify damaged DNA to form repairing complex, cleavage and elimination of defective DNA strand, filling by polymerization of new nucleotides (by the DNA polymerase), and final binding by corresponding ligase [38]. Other mechanism involved in disinfection resistance of bacteria is the capacity to use and share the genetic material with other microorganisms previously inactivated, as a new tool to take and recombine DNA and even also as new source of energy to survive [16].

In studies of virus disinfection performed between the 1960s and the 1980s, scientists used scintillation spectroscopy and electron microscopy techniques to detect modifications in viral genomes and proteins, concluding that the virus inactivation is the result of either protein damages or genome damages [49–52]. Nowadays, the study of the physical/chemical inactivation of virus has become simpler than 10 years ago thanks to cornerstone advances in genomic, sequencing, protein mass spectrometry and structural virology techniques [53]. The structure and composition of a virus is capable to elucidate the damage or injury by disinfection process, but it is not enough to understand and predict the mechanism of inactivation [11, 54].

The effect that genetic modification has on viral functions (e.g., host cell binding, genome entry, etc.) has improved the fundamental understanding of biological functionality of viral domains, virus-host cell interactions, virus assembly, capsid-RNA interaction sites, and RNA release. This knowledge could also provide valuable insight into critical structures and biological functions that should be targeted in virus neutralization or inactivation strategies [53].

In drinking water production, the biological safety offered by any disinfection method and topics related with control in a context of global climate change are fully relevant. It is also necessary to improve the surveillance on potential mutagenicity related with either microbial regrowth or reactivation as well as potential formation of DBPs because of the biological risk that may impose high costs on human and animal health [55].

3 Disinfection Methods More Widely Used in Production of Drinking Water

Conventional methods for water disinfection can be divided into two categories: physical and chemical [56, 57]. In the first case, heat and pasteurization are widely used in household disinfection, as well as UV lamps emitting radiation at 254 nm, but microbes differ in sensitivity to inactivation with heat and UV irradiation; in general, heat has been established to be more effective against vegetative bacteria [13].

Chemical disinfection has been made by using oxidizing agents relatively stable such as chlorine and its derivatives: molecular chlorine (Cl_2), hypochlorite (ClO^-), monochloramine (NH_2Cl), and chlorine dioxide (ClO_2) with remarkable toxicity disadvantages and formation of DBP-N and DBP-Cl [58]. Hydrogen peroxide (H_2O_2) with recognized biocide effect [18], ozone (O_3), as the second more frequently used method of disinfection only after chlorination, and other agents occasionally used are potassium permanganate ($KMnO_4$), bromine, and iodine and their derivatives [59, 60].

3.1 Chlorine

Chlorination features many practical advantages like simple application, high efficiency, and low cost. However, its efficiency also depends on the type of microorganism to be inactivated [61, 62] and a specific CT value for virus, bacteria, and protozoa [14]. A free chlorine concentration ranging 0.1 mg/L has been estimated to be required for inactivation of bacteria, 1.0 mg/L for virus (poliovirus 1), and 5.0 mg/L for parasites (*Giardia* sp. and *Entamoeba histolytica*) during contact times of 0.4, 1.7, and 18 min, respectively [14]. The microbiocide effect occurs when chlorine contacts with the pathogen's surface; it reacts causing oxidation of lipids in membrane changing its permeability, destabilizing the cytoplasm, inhibiting the enzymatic activity, and causing irreversible damage at the level of the nucleic acids.

After water treatment and before the drinking water distribution systems, a residual concentration of free chlorine is required to be dosed in order to or minimize the risk of regrowth and formation of biofilms in the distribution pipes. Evidences about subsequent profiles of bacterial regrowth and bacterial communities' structure after chlorine disinfection along with using new generation sequencing technologies have allowed to realize that most chlorine-tolerant bacterial groups belong to orders Sphingomonadales and Rizhobiales, previously reported as inducers of biofilm formation [63]. Biofilms represent a serious problem for water quality affecting efficiency and costs in disinfection by chlorine units, both in pipes and distribution networks [58].

A disadvantage derived of disinfection only based in chlorine is the ability to react with the organic matter present in most natural surface waters, conducting to formation of halogenated, highly toxic, and hazardous disinfection by-products

(DBPs). Increased dosages or time of contacts with chlorine used in disinfection processes also conduces to higher potentials of formation of toxic trihalomethanes (THMs), halo-acetic acids (HAAs), and some other emergent halogenated nitrogenous by-products DBPs-N, which represent till 100 times more carcinogenic risk compared to typical chlorinated by-products of public health importance [64]. Sequencing analysis of antibiotic resistance genes in assay of bacteria exposure to different N-DBPs has shown mutagenicity of both *Pseudomonas aeruginosa* PAO1 and *E. coli*, representing important genotoxicologic and epidemiologic risks through treated water [55].

DBPs have cytotoxic, genotoxic, apoptotic, and even carcinogenic effects even under very low concentrations [64, 65]. They could affect either the central nervous system or the kidney and liver functions in humans and animals, seriously threatening public health [2, 66–68]. Despite its limited efficiency to inactivate parasites as *Cryptosporidium* spp., and the potential formation of toxic by-products, chlorination for production of drinking water is currently accepted and widely used all around the world [11, 69]. Since the potential toxic risk on both humans and animal's health derived from the use of chlorine in water facilities it has been recently and extensively discussed in both, academic and industrial scenarios, as it has motivated the exploration on novel technological approaches too. They must be at least as cost-effective but safer than chlorination to decrease its use or even substitute it completely in the mid-/long-term.

3.2 Ozone (O_3)

Ozone displays high oxidation potential to decompose organic and inorganic contaminants, as well as also to inactivate microorganisms [67, 70, 71]. It displays a stronger bactericidal effect than chlorine-derived agents, performs efficiently at low concentrations, and eliminates a greater number of pathogens, including viruses, cysts, and protozoa. One of the main species responsible for such an effect is the hydroxyl radical ($HO^{\bullet}$), strongly oxidizing intermediate generated through the ozonation process, that may destroy enzymes and coenzymes needed for the life and survival of the microorganisms present in water [2, 59, 67]. Since ozone is highly unstable, it cannot be either transported or stored in big amounts, and for practical purposes it is produced in situ [2, 57, 59]. Production of large amounts (let's say in the scale of kg/h) requires the use of specialized equipment that involves expensive operation making it difficult to be implemented at big scales [5, 11]. Other considerations that must be kept in mind are both the high rate of reaction and no residual effect; then, combined use with other oxidizing residual agents should be considered in order to keep controlled regrowth of the pathogens along prolonged times [2, 57, 71]. A general disadvantage of ozonation is the similar behavior to chlorine once in contact with bromides, aldehydes, and ketones present in water; brominated compounds, oxides, peroxides, and cyanides are some by-products that may be formed, also displaying toxic and other undesired effects for humans and animals [11, 55, 72, 73].

3.3 Hydrogen Peroxide (H_2O_2)

Hydrogen peroxide is extensively used as a biocide, mainly in applications where its decomposition into nontoxic by-products is important [18]. Along with its high disinfecting efficiency, there is still little understanding about its biocidal mechanism. Hydrogen peroxide displays higher oxidizing power in comparison with chlorine and slightly below than ozone. It can inactivate bacteria (*Escherichia coli and Salmonella* spp.), bacterial spores (*Bacillus anthracis, Bacillus subtilis, and Geobacillus stearothermophilus*) [74, 75], fungus (*Histoplasma capsulatum, Blastomyces dermatitidis, and Coccidioides immitis*) [76, 77], viruses like bacteriophage MS2 [78], and even prions [79].

In disinfection with H_2O_2, the cytotoxic effect has not been ascribed to the oxidizing properties of the molecular form but to oxidizing stress caused by the hydroxyl radicals ($HO^•$), singlet oxygen (O_2^*), and superoxide species ($O_2^{-•}$) coming from its natural decomposition. These species may attack a wide variety of organic compounds as lipids, membrane proteins, and nucleic acids of cells [74, 80]. The temperature can promote the diffusion of the hydrogen peroxide through the cell's membrane; once inside, it can generate reactive species of oxygen like the hydroperoxyl and the hydroxyl radicals, which react more easily at high temperature with a wide spectrum of organic molecules with its subsequent significant oxidative stress [61].

The effect of the pH on the bactericidal action has been also documented; H_2O_2 activity does not vary significantly against pH within a range of 2.0–10.0 [61, 81], whereas other authors have mentioned that its action is higher under acidic conditions [82]. Raffellini et al. [83] found the most efficient conditions to inactivate *E. coli* under increased temperature and concentration together with low pH: peroxide concentrations close to 3.00% w/v and pH 3.0 at 25 °C depleted in 5-log the concentration of *E. coli* in 1–9 min of reaction. Meanwhile, Labas et al. [77] found low levels of *E. coli* inactivation at pH 7.0, $T = 20$ °C, H_2O_2 concentrations of 15–300 mg/L, and prolonged times of exposure in comparison with other disinfection technologies. As advantage, hydrogen peroxide has associated low risk by formation of toxic DBPs; standard side products of decomposition of any remaining concentration once the oxidizing treatment is finished are just water and oxygen; thus, it has been widely employed for disinfection of foods, fruits, vegetables, and food processing equipment [74, 84]. However, its use in disinfection of surface streams for drinking water production is more rather scarce and not extended.

4 Advanced Oxidation Processes: AOPs

These modern technologies offer a cost-effective and high-performing response to the challenges imposed by the current high levels of contamination in water, the problems of public health related with the consumption of unclean water, and the current laws on drinking water quality [9, 85–90]. AOPs could improve the systems of treatment producing drinking water [91–94].

AOPs are a set of methods taking advantage on the high oxidizing power of the HO$^{\bullet}$ radical that basically differ one to each other in the way used to generate it [95, 96]. Several combinations of methods to produce hydroxyl radicals have been investigated: for example, O_3/H_2O_2, O_3/UV, catalytic peroxidation assisted by Fenton catalysts, photocatalytic activation by semiconductors (mostly TiO_2), water sonolysis, sono-photocatalytic activation, catalytic wet peroxide oxidation (CWPO or CWHPO), and combinations of them [57, 97–99].

The hydroxyl radicals produced by the AOPs are capable to mineralize a wide variety of organic compounds to CO_2, water, and inorganic ions, without significant limitations exerted by concentration or phase change [86, 87, 100], whereas a high fraction of microorganisms could be also inactivated as outstanding advantage. The HO$^{\bullet}$ radical is a ROS, and it enables processes more thermodynamically favored as well as rates of oxidation pretty higher in comparison with most of the other known oxidizing methods or species. They may react 10^2 to 10^3 times faster than alternative oxidizers like ozone [5, 101]. The AOPs may release significantly lower charge of toxic DBPs in comparison with other disinfection methodologies, if some at all; in addition, they frequently enhance organoleptic features of the so-treated waters and show quite low selectivity in the degradation of wide variety of organic compounds [62, 89, 100], as extensively reviewed elsewhere [91].

4.1 Disinfecting Potential of AOPs

Disinfection by AOPs has suggested high inactivation potential against a wide range of microorganisms like spore-forming bacteria, yeasts, fungus, protozoa, and virus, which relies primarily on the formation of ROS including singlet oxygen, triplet oxygen, anion-radical superoxide ($O_2^{-\bullet}$), hydroxyl radical (HO$^{\bullet}$), hydroperoxyl radical (HO$_2^{\bullet}$), and hydrogen peroxide itself. Altogether they may exert a toxic effect and loss of structural integrity on such microorganisms affecting their viability and survival [9, 11, 15, 30, 69, 74]. The mechanism of action of each ROS on every type of microorganism is not still completely established, and it remains being an enthusiastic seed of emerging investigations [2, 102].

The damage on microorganisms attributed to the oxidative stress caused by ROS can take place when their concentrations are high enough to exceed the defense capacity of the cell. ROS species like hydroxyl radical feature short half-live times. In bacteria once got this barrier over, the components of the cell's membrane are attacked by peroxidation of the lipid chains, oxidation of polyunsaturated fat acids and proteins, in turn increasing the membrane's permeability and altering the transmembrane gradients of concentration, essential to preserve the equilibrium and appropriate function of the cell. The long-lived endogenous ROS like the H_2O_2 may straightforwardly damage the cell or penetrate it giving rise to other ROS; it may primarily lead to attack on the nucleic acids by either modifying nucleic bases and breaking of the DNA's double helix or simply activating cell injury processes [9, 15, 61, 80, 86, 90, 95, 102].

The time of exposure with microorganisms may vary according with the conditions of reaction of every AOP and its characteristic rate of ROS formation. During the oxidation reaction, the microorganisms can activate their mechanisms of defense before they truly undergo oxidative stress [103]; cells may then die by accumulation of several damages [90, 95].

Regarding toxicity of the HO• radicals, Storz et al. [104] found that such radicals might produce chemical modifications on the nitrogenous bases and breaks on the DNA chains of *Salmonella typhimurium*; a little bit latter, Anzai et al. [105] determined that hydroxyl radicals attack lipids of the cell's membrane leading to oxidative degradation. More recently, according to Spuhler et al. [90], Flores et al. [84], and Pablos et al. [2], the action of the HO• radicals on the cell's components is complete and irreversible, with the pathogens unable to recover its viability. However, the microbial resistance against an AOP treatment may sharply vary depending on the structure and particular chemical composition of every microorganism [106–108].

The kinetics of the microbial inactivation has been mostly studied by using spiked distilled water in order to avoid the interference of ions and organic compounds during the oxidation process. But it is still required elucidation of the mechanism of action exerted by the AOP under real conditions, since the dissolved ionic species often available in real aqueous influents may significantly affect the concentrations of Ca^{2+} and Mg^{2+} between the inside and outside of the cell. It would lead to stressful conditions of inner osmotic pressure and, then, weakening of the cell's wall followed by higher susceptibility to the action of the AOP [109].

Rubio et al. [96] compared the efficiency of photo-Fenton and H_2O_2/UV_{254} AOPs in the elimination of *E. coli* by using three types of water. They found a rate of bacterial inactivation in the next order as a function of the water matrix: Milli-Q water > Leman lake water > artificial sea water. The evolution is probably due to presence of organic material and inorganic ions such as chlorides, fluorides, sulfates, phosphates, bromides, and bicarbonates, which may compete against hydrogen peroxide for the binding sites on the cell's membrane, react with the HO• radicals, or generate other compounds making difficult the pass of light inside the water or preventing microorganisms against the attack of ROS generated during treatment with AOPs. Assessment of AOP's disinfection power on real drinking water facilities should be primarily made on the effluent of conventional physicochemical treatments (coagulation-flocculation), as the much suitable stream; it might enable easier recording of the inactivation process and the microbial elimination but also because this combined and sequential set of treatments – physicochemical followed by AOP – should be more favorable than any single one.

Compared with the potential of the AOPs in the abatement of organic matter and other highly refractory pollutants, the microbiological disinfection remains a field more rather scarcely explored. The photocatalytic approaches have been the most studied AOPs in microbial inactivation, but some other studies focused in simultaneously elucidating microorganisms inactivation and organic matter elimination are still strongly necessary. Thus, from now on a brief description of AOPs studied in the inactivation of microorganisms is displayed.

4.2 Fenton Process

This is a cheap and effective homogeneous catalytic system, where ROS are produced by interaction of H_2O_2 and dissolved salts of Cu^{2+} or more often Fe^{2+}. In this case, the decomposition of peroxide mostly to hydroxyl and hydroperoxyl radicals happens very fast, thanks to the catalyzed homolytic breakdown of the peroxide bond by the dissolved metal ions in homogeneous phase [80, 84, 91]. Hence, the overall disinfecting efficiency displayed by the Fenton process will depend on the rate of ROS generation but in turn also on the rate of reaction between the catalyzing metal ion and the hydrogen peroxide [11]. Besides to action of ROS, disinfection may occur by direct chemical and physical interactions between the reagents and microorganisms, where the straightforward attack of H_2O_2 is probably the most significant [110]. Microorganisms like *E. coli* may retain iron into the periplasmic space but also in the intracellular spaces that allow either Fe complexes or Fe-storing proteins to form. Once inside, iron may react with peroxide molecules to carry out what could be so-called intracellular Fenton process [88, 90, 111]. Since the Fe^{3+} formed in the very first step of reaction is rapidly back-converted to Fe^{2+} in the second stage, a cyclic, sustained transformation of iron between Fe^{2+} and Fe^{3+} is verified throughout the Fenton process that in turn enable continuous production of ROS [reactions (1) and (2)] [112]:

$$Fe^{2+} + H_2O_2 \rightarrow Fe^{3+} + HO^{\bullet} + OH^- \quad (1)$$
$$Fe^{3+} + H_2O_2 \rightarrow Fe^{2+} + HO_2^{\bullet} + H^+ \quad (2)$$

Some reports related with the effects of the Fenton reaction on microorganisms are displayed in Table 2.

Linley et al. [18] also described the mechanism of cytotoxic activity generally reported to be based on the production of highly reactive hydroxyl radicals from the interaction of the superoxide ($O_2^{\bullet-}$) radical and H_2O_2, a reaction first proposed by Haber and Weiss (Eq. 3):

Table 2 Microorganisms inactivated or eliminated by Fenton treatment

Microorganism		Reference(s)
Bacteria	*Enterococcus* sp.	[113, 114]
	E. coli	[60, 110, 115–119]
	Pseudomonas aeruginosa	[117]
	Salmonella sp.	[90]
Virus	MS2 coliphage	[110, 120]
Biofilms[a]		[111]

[a]*Assemblage of microbial cells that is irreversibly associated (not removed by gentle rinsing) with a surface and enclosed in a matrix of primarily polysaccharide material* [121]

$$O_2^{\cdot -} + H_2O_2 \rightarrow O_2 + OH^- + HO^{\cdot} \quad (3)$$

The production of extremely short-lived hydroxyl radicals within the cell by the Haber-Weiss cycle is catalyzed in vivo by the presence of transition metal ions (iron-II) via Fenton chemistry (Eq. 1). Investigations about DNA oxidation have suggested that the oxidizing radical is the ferryl radical formed from DNA-associated iron but not the hydroxyl one [18]. Before, Kim et al. [110] measured the Fenton inactivation of MS2 coliphages hosted in *E. coli* resistant to antibiotics. The authors reported that MS2 can be inactivated via the Fenton process by means of H_2O_2 initiated by either Fe^{2+} or Fe^{3+}. However, since the reaction with the former ion occurred faster, the coliphages became inactivated even from the first minute, whereas for the second form of ion, the inactivation was significantly slower. Such a higher rate of peroxide activation by iron in its lower oxidation state has been widely reported based on the elimination of different biorefractory organic molecules, where indeed it has been realized that an induction period takes place when the reaction is started by Fe^{3+} [91]. Under several conditions of pH (6.0–8.0) and presence of Fe-chelating agents like o-phenanthroline and bipyridine, inactivation of coliphages occurred, thanks to hydroxyl radicals formed during the Fenton process. It is somewhat confusing to understand the mechanism at elevated values of pH without undergoing significant precipitation of the catalyzing metal ions. Apparently, the potential inactivating effect exerted by the carbonate buffer on both the walls and membranes of the host bacteria *E. coli* before every assay of inactivation has not been taken into account; these factors might at least partially affect the recorded results of the process [110].

Rather scarce studies can be found about disinfection kinetics on microorganisms after Fenton treatment. Among the parameters affecting the rate of reaction, iron concentration, peroxide concentration, and pH of the reacting mixture can be cited. Concerning the pH, alkaline conditions lead the dissolved Fe^{2+} to get transformed in colloidal ferric forms, which break the peroxide primarily in water and oxygen instead than releasing the active radicals. In other words, it more rather promotes a parasite reaction that consumes hydrogen peroxide without production of hydroxyl radicals, as already said the main responsible of the microbial inactivation [110, 111]. The optimum reaction pH for the homogeneous Fenton process has been reported to be below 3.0 [122–124], which is a disadvantage for use in production of drinking water since many real polluted effluents feature pH values close to neutral or sometimes slightly higher. Therefore, recommended pH values in AOPs should be near neutral [91, 125].

The pH range of operation of the Fenton process can be significantly increased when the reaction is catalyzed in heterogeneous phase. Here, activation of hydrogen peroxide to generate HO$^{\bullet}$ radicals has been performed with various transition metals like for instance Fe, Cu, and Mn, immobilized on several solid supports as natural aluminosilicates, zeolites, alumina, silica, and activated carbon. This process has been referred to by some authors as catalytic wet peroxide oxidation – CWPO [91, 97, 100, 126, 127]. This AOP has relevant advantages in the context of water treatment such as the high efficiency even under conditions of low, ambient pressures and temperatures, low-energy consumption, prolonged recycling of the catalyst, and pretty low cost for scaling up [91].

The removal of 98% of DOC after 4 h of reaction and full removal of true color from real surface river water in less than 45 min of reaction by Al/Fe-mixed pillared clay (Al/Fe-PILC) activated CWPO under pH 3.7 and very slight conditions of room temperature and atmospheric pressure (291 K, 72 kPa) has been reported [85]. Nieto-Juarez et al. [120] studied the inactivation and removal of MS2 coliphage by heterogeneous Fenton-like process catalyzed by iron (hydr)oxide particles in batch reactor at circumneutral pH, containing 200 mg/L of various commercial iron (hydr)oxide particles of similar particle size: hematite (α-Fe_2O_3), goethite (α-FeOOH), magnetite (Fe_3O_4), and amorphous iron(III) hydroxide ($Fe(OH)_3$). The authors concluded that the CWPO process can inactivate and physically remove the MS2 coliphage from water through two pathways: the particle-mediated (photo-)Fenton-like process and the adsorption process which could be mediated by hydrophobic interactions or Van der Waals forces. All particles showed similar adsorption capacity, in the order FeOOH > Fe_2O_3 > Fe_3O_4 ≈ $Fe(OH)_3$. Adsorption on three types of the solid particles studied, α-FeOOH, $Fe(OH)_3$ and Fe_3O_4, caused virus inactivation of 7.0%, 14.0%, and 22.0%, respectively. Exposure of particle-adsorbed viruses to sunlight and H_2O_2 resulted in highly efficient additional inactivation. Only Fe_3O_4 caused the inactivation of MS2 coliphage via dark Fenton-like process.

4.3 Photo-Fenton Process

The photo-Fenton process (Fe/H_2O_2/UV) is a photo-assisted homogeneous catalytic treatment, also based in Fenton chemistry [57, 128]. Its enhanced efficiency with respect to the process in absence of UV radiation (UV, UV-A, UV-B, or UV-C) can be ascribed to the extra generation of active radicals by photo-reduction of Fe^{3+} (see reaction 4) of course besides those taking place in reactions (1), (2), and (3). It is noteworthy that this extra-pathway does not consume hydrogen peroxide either [129].

$$[Fe^{III}(OH)]^{2+} + H_2O_2 + h\nu\ (UV) \rightarrow Fe^{2+} + HO^\bullet \tag{4}$$

The outstanding performance by the photo-Fenton inactivation of several microorganisms has been studied on bacteria, fungi, and some protein pathogen particles (see Table 3).

The mechanism of action of disinfection by $Fe/H_2O_2/UV$ process of course is induced by the ROS oxidative stress but also by the action of the UV radiation itself producing peroxide and superoxide species in situ as well as by causing direct irreversible damages on the DNA of the pathogens. The ROS may be produced from H_2O_2 previously diffused through the cell's membrane that may react with free dissolved iron available into within the cell [113, 135, 139]. In absence of iron, H_2O_2 and O^{2-} species formed starting from self-oxidation of flavin adenine dinucleotide $FADH_2$ may in turn cause oxidation of iron sulfide *clusters* [4Fe-4S] releasing Fe^{3+} ions [147]. These ions may promote the overall Fenton reaction, and the UV light can also attack enterobactin or ferritin proteins, delivering Fe^{2+} ions [28, 90, 95].

Fe^{2+} ions can also induce harmful effects on the targeted microorganisms because of its capability to diffuse across the cell's membrane and then react with the metabolic H_2O_2 [3, 28, 90, 95, 142, 148]. A diagram with several possible pathways for generation of ROS in the photo-inactivation of *E. coli* and related damages is shown in Fig. 1.

The mechanism varies according to the structure and chemical composition of cell membranes and other organelles [114]. It was found that *Fusarium solani* showed higher resistance in comparison with *E. coli* [95], which was attributed to the chemical composition and structure of the fungus wall that in general is more rigid due to presence of polysaccharides like glucans, mannans, chitin, proteins, and glycoproteins [139]. Chitarra et al. [149] found ROS production favored during

Table 3 Some microorganisms inactivated by photo-Fenton oxidation

Microorganism		Reference
Bacteria	Endospores of *Bacillus subtilis*	[130, 131]
	Enterococcus faecalis	[113, 132]
	Enterococcus sp.	[133]
	E. coli	[3, 60, 90, 95, 113, 114, 118, 125, 128, 134–138]
	Pseudomonas aeruginosa	[134]
	Salmonella sp.	[80, 136]
	Total coliforms	[3]
Fungi	*Fusarium solani*	[95, 139–143]
	Phytophthora capsici	[142]
Virus	MS2 coliphage	[120, 133]
Helminth	*Ascaris suum* eggs	[130, 144, 145]
Recombinant prion[a]		[146]

[a]Pathogenic protein particles

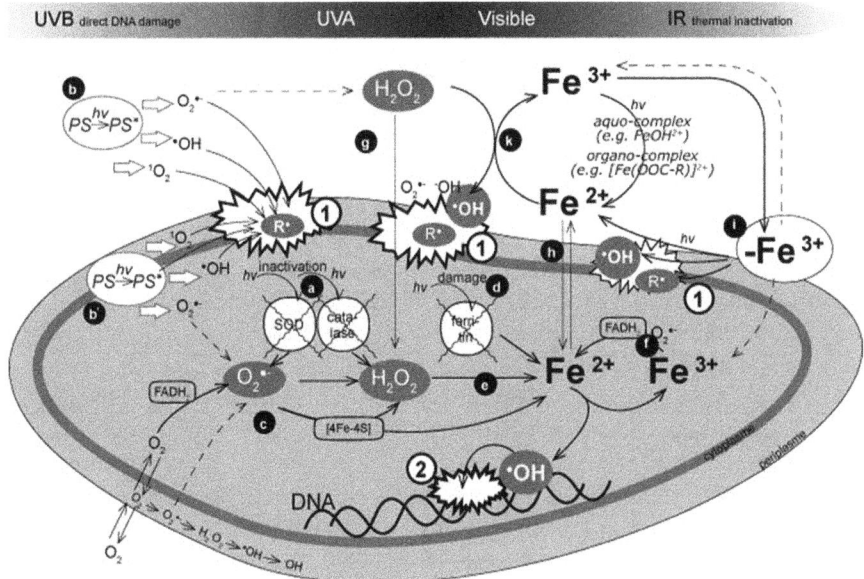

Fig. 1 Possible pathways involved in the photo-inactivation of *E. coli* in the presence of Fe^{2+}, Fe^{3+}, and H_2O_2. Reproduced with permission of Elsevier from [90]

germination of the spores and start of the metabolic activity displayed by *Fusarium solani*, since throughout that stage spores absorb water to hydrate its nucleus, making then easier the access of iron toward the inner structure.

Polo-López et al. [142] studied the capability of the photo-Fenton system to eliminate zoospores of *Phytophthora capsici* in distilled water, reaching the highest inactivation rate with 5.0 mg/L of either Fe^{3+} or Fe^{2+} along with 10 mg/L H_2O_2, under 2.5 kJ/L of solar UV dosage (60 min exposure). Under low concentrations of iron (1.0 mg/L, 2.5 mg/L), the diffusion of H_2O_2 and iron inside of the cell was limited, and it did not generate enough abundance of $HO^{\bullet}$ radicals for inactivation of the targeted zoospores. A little bit earlier, it was determined that the effect of the photo-Fenton process strongly depends on the structure of either the cell or the spores [107]; thereafter, optimal UV, H_2O_2, and Fe dosages as well as the times of exposure are target-specific parameters for every microorganism as a function of its intrinsic resistance against each AOP. The best photo-Fenton inactivation of *E. coli* was achieved by Rodríguez-Chueca et al. [60] with 20 mg/L of H_2O_2, 5.0 mg/L of dissolved Fe, pH 5.0, under irradiation with wavelengths in the range 320–800 nm (light intensity of 500 W/m^2); the irradiation and the interaction $[Fe^{3+}]:[H_2O_2]$ were chosen as experimental, more influencing factors in the factorial, statistical experimental design of the Fenton and photo-Fenton disinfecting treatments.

The simultaneous photo-Fenton microbial disinfection and elimination of organic matter were also recently studied by Spuhler et al. [90]. The higher performance in both *E. coli* inactivation and resorcinol degradation as model refractory molecule was found (0.6 mg/L in Fe^{2+} or Fe^{3+}, 10 mg/L of H_2O_2, pH in the range 5.0–5.5, radiation below 300 nm UV-C/UV-B, and radiation intensity 550 W/m^2). Resorcinol competed with bacteria for available Fe^{2+} or Fe^{3+} then limiting the contact of the microorganism and then disfavoring intracellular diffusion. The efficiency of the system was more affected in the presence of Fe^{2+} than with Fe^{3+}. However, Fe^{2+} still showed a higher bactericidal effect. Concerning similar system, Moncayo-Lasso et al. [40] realized that upon addition of 0.6 mg/L of Fe^{3+} and 10 mg/L of H_2O_2 under pH 6.5, the mineralization of 55% of the starting 5.3 mg/L of total dissolved organic carbon present in a river's water along with simultaneous *E. coli* inactivation without any regrowth evidenced over 24 h later under darkness was attained.

A little bit later, Ortega-Gómez et al. [57] used photooxidation of resorcinol as a model of natural organic matter (NOM) and inactivation of *Enterococcus faecalis*; both decreased as the dosage of hydrogen peroxide and iron raised (H_2O_2/Fe^{2+}: 50/20 mg/L), thanks to the subsequent production of active free radical in the system. The photo-Fenton inactivation of the Gram-positive microorganism was much more efficient when the temperature was increased to 37–40 °C, which apparently arose to the very close conditions of temperature with respect to the optimal one in the growth of *E. faecalis*; the disinfection limits were accomplished at 80, 65, and 40 min under 20, 30, and 40 °C, respectively. Apparently under such conditions of temperature, bacteria display complete metabolic activity, becoming much more vulnerable to the attack by the ROS generated throughout the Fe/H_2O_2/UV process.

The photo-Fenton in elimination of *E. coli* using two light sources (lamp for light emission at 254 nm and SunTest simulator) has been also recently studied [96]. Under UV_{254} irradiation, the presence of organic matter produced a negative effect on the rate of disinfection ascribed to the formation of triplet-state dissolved organic matter due to absorption of UV/visible light (UV/visible), whose deactivation should be mediated by reaction with oxygen, leading to subsequent formation of singlet oxygen that can later react enhancing overall photocatalytic treatment. As indicated, UV_{254} could be filtrated by the NOM then decreasing the rate of *E. coli* inactivation. Development of Fe-NOM complexes improved recycling of the transition metal in both oxidation states (Fe^{3+} and Fe^{2+}) as well as the rate of formation of $HO^{\bullet}$ radicals.

Heaselgrave and Kilvington [150] studied the antimicrobial effect of simulated solar disinfection-SODIS in presence and absence of riboflavin on various protozoa and helminths: *Ascaris, Giardia, Acanthamoeba, Naegleria, Entamoeba,* and *Cryptosporidium*. The authors found that SODIS can successfully disinfect water samples containing cyst of *Naegleria, Entamoeba,* and *Giardia*; they demonstrated that inactivation of highly resistant cysts of *Acanthamoeba* could be significantly enhanced by the addition of riboflavin (SODIS-R). However, *Ascaris larvae* continued developing inside the ova even after exposure to SODIS, whereas *Cryptosporidium* remained impermeable to propidium iodide staining, probably indicating that it might remain infectious.

This suggests that protozoan disinfection remains being the most challenging issue for both conventional and also modern methods of treatment. Moreover, it also drives forward research looking for improved disinfecting conditions displayed by AOPs. In addition, there is still a lack of information concerning the combined analysis of the kinetic parameters featuring both the photo-Fenton reaction itself and the microbial growing, to more clearly establish the set of conditions able to maximize the performance of this AOP as well as the mechanism(s) implicated in the disinfection of a range of types and viable forms of several hazardous microorganisms as suspended in real water systems.

4.4 Heterogeneous TiO_2 Photocatalysis

The heterogeneous photocatalysis by using titanium dioxide as catalytic active surface has been reported as one of the most attractive AOPs for treatment of surface waters in production of drinking water. It is mainly due to its ability, under certain conditions, for taking direct advantage of the sunlight as primary source of energy and to carry out a simultaneous degradation of the organic matter and the microbial inactivation without demanding addition of other oxidizing agents [8, 62, 73, 91, 93, 151]. Probably, the disinfecting potential of the photocatalytic oxidation was first described by Matsunaga et al. [20]; since then many researchers have made important steps forward [152–155]. There are now accumulated evidences about Gram-positive and Gram-negative bacteria, protozoa, microalgae, and fungi that could be inactivated by combination of heterogeneous photocatalysis with TiO_2 [119, 156]. A wide variety of microorganisms, whose photocatalytic inactivation has been investigated along the past three decades, is summarized in Table 4.

The effect of the TiO_2 photocatalytic treatment on cellular lines as HeLa (cervical carcinoma) [174] and Ls-174-t (human colon carcinoma) [175] has been also studied. Photocatalytic killing effect of the TiO_2 on human colon carcinoma cells with concentrations of TiO_2 above 200 µg/mL under UV-A irradiation at room temperature was evidenced. In the case of the HeLa cancer cells, Abdulla et al. [174] reported removals of 20% and 25% when employing Ag metal core-TiO_2 shell composite nanocluster as photocatalytic system with 10 min of irradiation and concentrations of 5.0 and 7.0 mM of TiO_2, respectively. It was enhanced to 100% for molar ratios Ag:TiO_2 of 1:5 and 1:7. Most efficient photocatalysts were prepared starting from $AgNO_3$ (1 mM) and titanium (IV) (triethanolaminato)-isopropoxide solutions in concentrations of 5.0 mM or 7.0 mM, respectively. The killing of HeLa cells was more efficient in the presence of the Ag/TiO_2 shell composite photocatalyst than with either TiO_2 or Ag alone, thanks to a synergistic response.

As displayed in Table 4, the role played by the TiO_2-assisted heterogeneous photocatalytic inactivation has been assessed for a wide range of targeted microorganisms. However, their corresponding mechanisms of action have not been fully elucidated so far [10, 15, 62, 157]. It is particularly interesting to get deeper insight in the study of the TiO_2-assisted, heterogeneous photocatalytic inactivation of hazardous protozoa, mainly (oo)cyst of parasites *Giardia* spp. and *Cryptosporidium* spp.

It is proven that ROS can play a central role in fast disinfection processes. However, linking it to the formation of superoxide anion might significantly further

Table 4 Some microorganisms inactivated by TiO_2 photocatalytic treatment

Microorganism		Reference(s)
Bacteria	*Clostridium perfringens*	[5, 157]
	Endospores of *Bacillus subtilis*	[130, 158, 159]
	Enterococcus faecalis	[160]
	Enterococcus sp.	[86]
	E. coli	[5, 15, 20, 96, 108, 118, 125, 153–155, 159–167]
	Listeria monocytogenes	[15]
	Lactobacillus acidophilus	[20]
	Pseudomonas aeruginosa	[159]
	Salmonella typhimurium	[15]
Fungus	*Candida albicans*	[159]
	Fusarium anthophilum	[107, 108]
	Fusarium equiseti	[107]
	Fusarium oxysporum	[107]
	Fusarium solani	[107, 108, 141, 159, 163, 168]
	Fusarium verticillioides	[107]
	Endospores of *Aspergillus niger*	[169]
Yeast	*Saccharomyces cerevisiae*	[20, 59, 156]
Protozoa	*Acanthamoeba castellani*	[94, 159]
	Acanthamoeba polyphaga	[159]
	Giardia intestinalis	[94]
	Cryptosporidium parvum	[158]
Algae	*Chlorella vulgaris*	[59]
Virus	Hepatitis B virus	[170]
	Herpes simplex virus	[171]
	MS2 coliphage	[106, 172]
	Poliovirus 1	[173]

improve cellular oxidative state [156]. Throughout photocatalytic processes, ROS can oxidize lipids, proteins, and amino acids; subsequent lipid peroxidation and formation of secondary products as for instance hydroperoxides, which carry a major oxidative damage, secondary attack of amino acids, and lack of normal function of membrane may also take place.

Kim et al. [176] stated almost a decade ago that DNA, RNA, proteins, and lipids are the main biological targets of the ROS generated in the photocatalytic treatment. An approximation to the photocatalytic generation of ROS and its consequent effect on several microorganisms is shown in Fig. 2 [177]. As it can be seen, resistance to oxidative damages increases from viruses to bacteria and finally yeasts and molds. According to Kim et al. [15], ROS may damage the DNA after only 60 to 90 s of treatment by the TiO_2 – photocatalytic process.

The progressive absorption of photons by cellular envelopes may lead to the loss of its semi-permeability, then making easier the photooxidation of the inner components of the cell [11, 16, 86, 153]. Likewise, the direct photocatalytic oxidation of

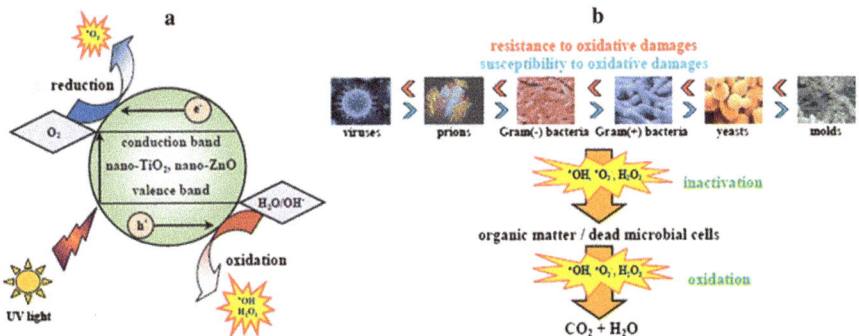

Fig. 2 (**a**) Mechanism of generation of reactive oxygen species (ROS) on the surface of photocatalyst nanoparticles of titanium or zinc oxides and (**b**) effects of ROS on several infectious agents. Reproduced with permission of Springer from [177]

the intracellular coenzyme A (CoA) to play a role in the disinfection process has been considered, since it can involve a decrease in respiratory activities, decreasing too in turn the cell viability [11, 20, 59, 109, 153, 178]. Dalrymple et al. [11] studied mechanistic models based on peroxidation of the membrane's lipid layer as well as the interaction between the microorganisms and the photocatalyst's particles, very often ignored; according to this and other authors, this process can be mediated by the orientation of the TiO_2 particles, the shape and size of the microorganisms, and the pH-controlled electric charge of the surface [16, 167].

Electron microscopy has revealed the interaction between catalyst and microorganisms. For instance, in the organization of TiO_2 particles on yeast *Saccharomyces cerevisiae* as observed by scanning electron microscopy (SEM) and transmission electron microscopy (TEM), yeast exposed separately to catalyst and photocatalytic treatment was surrounded by nanoparticles causing serious structural damages on yeast's cells, but no direct contact between catalyst and cell wall neither entry of nanoparticles into the cellular compartments was observed. In other study [16], TEM showed the structural damages caused by TiO_2 and photocatalytic treatment on *E. coli*; after 30 min of contact, *E. coli* envelope demonstrated lower resistance in comparison to yeast *S. cerevisiae*; it was attributed to polysaccharides and glycoproteins forming a reticular tridimensional structure surrounding yeasts capable of generating protection against stress conditions [179] (see more details in Fig. 3).

TiO_2 is for sure the most widely used active solid in photocatalytic studies because of its high activity, stability, and safety [86, 180]. The most used is the commercial form Degussa P-25 (mixture of approx. 70% of the anatase phase and 30% of the rutile phase), which has exhibited the highest efficiency in degradation of organic pollutants [72, 151, 181]. The phase content of the TiO_2 photocatalyst has been a variable considered in evaluation of disinfecting capability. Gumy et al. [182] studied the catalytic activity of 13 commercial powders of TiO_2 with specific surface areas varying from 9 to 335 m^2/g and isoelectric points (IEP) from 3.0 to 7.5 in *E. coli* inactivation. The most efficient material in the photocatalytic disinfection was TiO_2

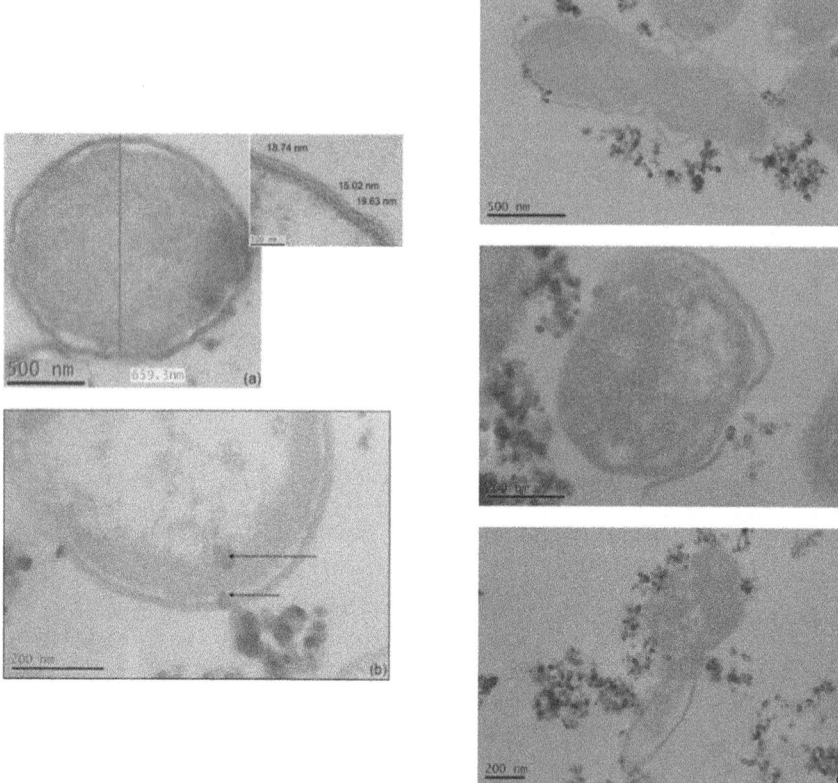

Fig. 3 TEM pictures of *E. coli* cells in aqueous suspension: (*Left*) (**a**) In the dark without TiO_2 (the insert is a zoom picture in the bacterial envelope); (**b**) after 30 min of contact with TiO_2 P-25 (0.25 g/L) in the dark. (*Right*) Exposed to UV-A photocatalysis with TiO_2 P-25 (0.25 g/L) and $I_{UV-A} = 3.45$ mW/cm^2: (**a**) after 80 min of exposure, (**b**) and (**c**) after 180 min of exposure. Reproduced with permission of Elsevier from [16]

Degussa P-25. Likewise, minor bacterial inactivation for lower IEP values was observed, whereas the disinfection showed to be somehow independent of the catalyst's specific surface area, behavior observed with most catalyst tested. Guillard et al. [153] also compared two catalysts in disinfection of two strains of *E. coli* (K12 PHL849 and K12 PHL1273) differing in the number of bacterial fimbriae: one commercial TiO_2 Degussa P-25 (anatase/rutile = 80/20, S_{BET} = 50 m^2/g, size particle = 30 nm) and one industrial named Millennium PC500 from Millennium Inorganic Chemicals (anatase, S_{BET} higher than 300 m^2/g, size particle ≈ 5 nm, IEP lower than TiO_2 P-25). In the presence of Degussa P-25, both strains were inactivated in a similar way, but without significant effects, mainly due to alterations in bacteria surface with respect to fimbriae number. Likewise, with the second catalyst, the bacterial strain PHL1273, containing a higher fimbriae number, was inactivated faster than strain PHL849, probably due to stronger adherence of first strain to catalyst's surface.

Thus, the selection of active catalyst in photocatalytic inactivation of microorganisms should consider particle size, crystalline phase, isoelectric point, and surface area, but mainly surface charge and particular electrostatic interaction of targeted infectious agent with TiO_2 particles. More investigations are still required in order to improve the understanding about the effects exerted by the main operational parameters (type of catalyst, type of catalysis reaction) in the photocatalytic inactivation of wide diversity of water-borne pathogens.

Regarding inactivation kinetics, Agulló-Barceló et al. [5] recently studied the heterogeneous photocatalytic reaction in presence of 100 mg/L of TiO_2 on four microbial indicators: *E. coli*, spores of sulfite-reducing clostridia, somatic coliphages, and F-specific RNA bacteriophages. These authors realized several disinfection kinetics for every target. The viral indicators were those more susceptible to the treatment, followed by *E. coli* that showed similar behavior, whereas the higher relative resistance was exhibited by the spores of sulfite-reducing clostridia, in similar trend latter reported elsewhere [177]. Nevertheless, viruses had shown to be more resistant to the action of either H_2O_2, UV (>295 nm), or UV/H_2O_2 in comparison with bacteria [183]; apparently TiO_2 leads to increase susceptibility of the virus as advantage of this AOP. It must be stressed that anyway the effect strongly depends on the size and type of virion, as well as on the type of nucleic acid [172]. The relatively high resistance exhibited by the microbial spores could be caused by structural components of multiple layers covering them, mainly composed by proteins, glycoproteins, PG, calcium, and dipicolinic acid that provide them enhanced impermeability against toxic molecules and greater resistance under hostile conditions [169, 184]. In this regard, Cho et al. [106] previously showed the biocidal modes of action of ROS to be very different depending on the specific microorganism involved; whereas the MS-2 phage was inactivated mainly by the free hydroxyl radical in the solution bulk, *E. coli* was inactivated by both the free and the surface-bound hydroxyl radicals, but probably also by other ROS species generated along the TiO_2 photocatalytic reaction.

Similar to bacterial endospores, (oo)cystic forms of protozoa parasites have displayed significantly higher resistance to disinfection [185, 186]. Sokmen et al. [94] assessed inactivation of cysts of two protozoans by the TiO_2/UV system (titanium dioxide as anatase 99.9%, low-pressure Hg lamp, 300 W, 254 nm, and light intensity 5.8 W/m^2). In the case of *Giardia intestinalis*, 52.5% of the cysts were inactivated after 25 min of reaction, while the complete disinfection of the effluent was achieved over 30 min; irreversible damages on the cell's wall were evidenced and hence no chances of regrowing. However, the same system was not effective in the elimination of *Acanthamoeba castellani*. Meanwhile, the increase in concentration of H_2O_2 in the range of 25–100 mg/L did not produce any significant effect on the inactivation of *E. coli* [181]. These evidences have allowed to elucidate the composition of walls playing a central role involved in generating resistance to disinfection by (oo)cystic forms of protozoa parasites [187].

Along with the particular type of targeted microorganism, experimental parameters of the TiO_2-assisted heterogeneous photocatalytic disinfection such as pH, type, and intensity of UV radiation, phase of the TiO_2 photocatalyst, initial concentration of microorganism, temperature, turbidity, and presence of either electrolytes and/or organic compounds have shown to participate in overall inactivation of pathogens [5, 154, 161–163, 182].

External sources of UV radiation or dope-sensitizing by various ways have increased the absorption of low-energy UV or visible-photons by the catalysts toward only solar-driven photocatalytic treatment of real streams [167, 188, 189]. The photo-response of TiO_2 in the visible region has been varied by the doping of the photocatalyst with metal and nonmetals: e.g., N-doped TiO_2 [190], N, S co-doped TiO_2 [191], and carbon-sensitized $TiO_{2-x}N_x$ [103]. Using external sources of UV irradiation, the efficiency of the catalytic disinfection process can be modified. UV-A radiation induces less deleterious effects than UV-B and UV-C on cell's survival [119, 155], whereas UV-C radiation is capable to generate the lack of integrity in cell's envelope after only 10 min of exposure, but it was also observed that the lack of total bacterial cultivability is faster in the presence of UV-C with respect to UV-A radiation. Pigeot-Rémy et al. [155] and Kim et al. [176] determined by independent studies that UV-C range of radiation is more efficient than UV-A and UV-B fractions in elimination of pathogenic microorganisms. Likewise, the ability of UV-C to affect essential functions of the cells such as the DNA replication, due to induced formation of ROS, is well-known.

Regarding the way the UV light is applied, it can be continuous, intermittent, average irradiation, direct, or diffuse. This parameter must be carefully considered for comparisons between results obtained by employing different photoactivated systems [12, 15, 73, 154, 192]. High values of pH (7.31 ± 0.30) disfavor the interaction of the microorganisms with the particles of the photocatalyst, whereas under pH values below 6.5 the proteins of the cell's membrane get destabilized along with increasing both ROS concentration and then disinfecting performance. Throughout the process, the reaction mixture tends to acidify as a product of carboxylic functionalities as intermediates on the oxidizing treatment. This induces positive charges on the catalyst's surface facilitating electrostatic interaction of catalyst pathogens to accelerate the microbial inactivation [5, 182].

In real water sources, ions introduced by anthropogenic activities may exert an important effect on the disinfection kinetics; phosphate anions, carbonates, or bicarbonates may inhibit the adsorption of amino acids on the TiO_2 particles. The chloride anion (Cl^-) increases the time required for photocatalytic bacterial inactivation, because it slows down oxidation mediated by radical species like $HO^{\bullet}$ and $HO_2^{\bullet}$ and blocks photoactivation of sites on the catalyst's surface [154]. Other oxidizable species establish a competition with microorganisms, reducing overall disinfecting efficiency of the photocatalytic system [5, 60, 109]. Rincon et al. [154] found that even low concentrations (~ 0.2 mmol/L) of SO_4^{2-} or HCO_3^- can affect bacteria inactivation. Finally, the time of reaction required for full disinfection is a proper indirect parameter in order to compare the efficiency of several AOPs; prolonged time of treatment may suppose very often complete elimination of the microorganisms, but if partial, some undesirable events of microbial resistance or genetic modifications could be also involved, leading to reactivation or persistence of the microorganisms in water. There is scarce knowledge about the mutagenic and cytotoxic potential of AOPs.

4.5 Electrochemical Oxidation

This AOP has claimed recent interest as an alternative treatment of various water matrices including drinking water, ballast water, and wastewaters [87, 193–199]. Some related advantages are environmental friendly, easily automated, simple operation, lower consumption of additional reagents, fast treatment, and more severe injuries compared with chlorination [198, 200]. Among the disadvantages, high costs of operation compared with other AOPs in terms of energy consumption, short life cycle of the required electrodes, and possible formation of hazardous DBPs like the THMs must be cited [196, 201]. Thus, it is important to determine how cost-effective could be this family of processes in the real-scale application on every type of targeted influent streams. Studies about the electrochemical disinfection on microorganisms have been reported (see Table 5).

The bactericidal properties of the electrooxidation techniques arise to the synergistic effect of the oxidizing species directly produced on the anode's surface together with the indirect oxidation by substances in the fluid aqueous phase [208]. Hence, as a function of the conditions of reaction and the chemical composition of the electrodes, several reactive species of oxygen can be produced like molecular hydrogen peroxide (H_2O_2), ozone (O_3), hydroxyl radicals (HO$^{\bullet}$), free chlorine Cl_2 (in the presence of chlorides), peroxodisulfate $S_2O_8^{2-}$ (in the presence of sulfates), and peroxydicarbonate $C_2O_6^{2-}$ (in the presence of carbonates) [87, 115, 193, 196, 197, 199]. The role of every oxidizing species during the operation of this AOP is in some cases still scarcely understood, and then there are clear efforts motivating more investigations in order to explain it [203, 204].

Diao et al. [115] compared the effects of chlorination, ozonation, Fenton reaction, and electrooxidation in the inactivation of *E. coli* (Fig. 4). Similar damages can be observed on the cells treated by Fenton and electrooxidation, along with substantial leakage of intracellular materials, less evident upon ozonation and almost imperceptible after chlorination. Both results involve the action of chemical species with comparable oxidizing power to the hydroxyl radicals of the Fenton reaction; it means higher oxidizing potential compared to chlorine. Thus, probably the mechanism involved in the electrochemical disinfection was primarily related with the action of ROS; it could be linked with the hypothesis of dissolved chloride ions being not only precursors to form free chlorine Cl_2 but also catalyzing a set of reactions believed to increase the lifetime of the hydroxyl radicals, consequently enhancing their disinfecting action.

Concerning the short-lived ROS, in order to enhance the efficiency of their production and hence their effectiveness destroying pathogen cells, it would be better to use a direct electrocatalyst instead of the electrochemical generation of the oxidizing species in solution [203, 204].

Table 5 Microorganisms inactivated or eliminated by electrochemical oxidation

Microorganisms		Reference(s)
Bacteria	Bacillus subtilis	[202]
	Endospores of Bacillus subtilis	[203]
	Enterococcus faecalis	[193, 196, 204]
	E. coli	[87, 115, 193–196, 198, 200, 202, 204–206]
	Legionella	[207]
	Pseudomonas aeruginosa	[199]
	Staphylococcus aureus	[202]
Virus	MS2 coliphage	[195, 196]
	Recombinant adenovirus serotype 5	[196]

The mechanism proposed to explain the high disinfecting potential of the electrooxidation claims that it is strongly related with the increased membrane's permeability caused by the exposure to the electric field that facilitates the diffusion of the oxidizing species into the cell, in turn leading to alterations of the enzymes, oxidation of the N-terminal groups in the cell-wall proteins, lysis, and final cellular death. It has been also found that throughout the electrochemical treatment, the oxidation of the coenzyme A may take place, even without prior rupture of the membrane, seriously modifying the metabolism of the cell [115, 193, 200, 204, 208].

Several factors that must be accounted for to increase the efficiency of the electroch4emical disinfection include the material of the electrode, electrolyte composition, applied current, pH, temperature, and type of electrolysis [199, 197, 204–206, 209]. High temperature is a condition under which the microbial metabolism responds faster to the oxidizing action of chlorine [209], and then it should display the same trend in the presence of ROS generated by the AOPs.

Electric current presents an inverse relationship with the contact time required for the disinfection process. Electrochemical treatment of *E. coli* under 25 mA/cm^2 yields complete inactivation of this microorganism in only 0.5 min of contact, whereas disinfection reached 99.98% for 2 min of reaction under 16 mA/cm^2 (Fig. 4, [115]). These results showed a clearly superior performance of the electrochemical oxidation in comparison with the conventional disinfection treatments by chlorination, which in 30 min of reaction attained 88.94% of efficiency measured under identical experimental conditions. In the other hand, the material of the electrodes used in the electrochemical oxidation has been considered one of the more relevant factors on microbial inactivation. The materials more frequently used thanks to their high oxygen overpotential have been lead dioxide, dimension-stable anodes based on SnO_2, and boron-doped diamond BDD electrodes [197, 198]. The BDD electrodes display great electrochemical stability along with high overpotential for water oxidation, allowing the efficient electro-generation of hydroxyl radicals and other ROS directly into the aqueous solution, but are significantly less effective producing free chlorine from the dissolved chlorides available in the target stream. Accordingly, such a type of electrodes is more suitable for treatment of water with

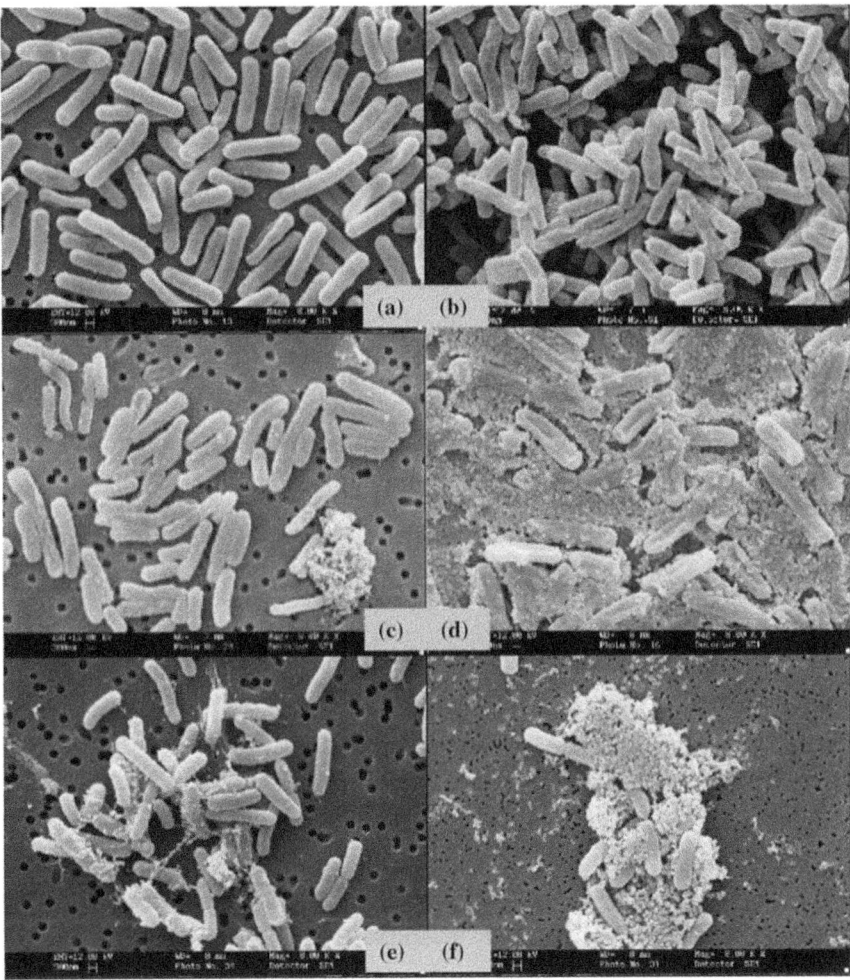

Fig. 4 SEM micrographs of *E. coli* in (**a**) fresh culture and after either (**b**) chlorination at 5 mg/L for 30 min, (**c**) ozonation at 10 mg/L for 5 min, (**d**) Fenton reaction (8.5 mg/L H_2O_2 and 0.85 mg/L Fe^{2+} at pH 4.0) for 10 min, (**e**) electrochemical disinfection (16 mA/cm^2) for 2 min, or (**f**) electrochemical disinfection at 25 mA/cm^2 for 2 min. Reproduced with permission of Elsevier from [115]

low contents of chloride [87, 204, 207]. In contrast, Pt and RuO_2 electrodes exhibit higher effectiveness to treat water containing this anion [204, 209]. From the results reported by Jeong et al. [197], the production of active chlorine as a function of the electrode's material decreases in the following order: Ti/IrO_2 > Ti/RuO_2 > Ti/Pt-IrO_2 > BDD > Pt; it is a trend quite different to that observed for electrochemical production of ROS. In this sense, BDD electrode produced tenfold the amount of hydroxyl radicals formed with the Ti/RuO_2 and the Pt electrodes. BDD electrode was also much more efficient in the production of O_3, effect attributed to the distinct electrostatic activities of the materials. Finally, it must be stressed that the study and application of this AOP at big scale is still very incipient. Then, further studies to figure out other factors favoring its entire efficiency and to clarify the disinfectant potential are both interesting issues in the field to be addressed in the very short-term.

The disinfecting potential of other AOPs and their combinations with either other AOPs or different disinfection systems is focus of a vibrant, currently developing field of research in water science. The literature reports are still incipient, and for this reason this review is circumscribed to the most studied AOPs so far.

5 Conclusions

The disinfection process plays an essential role in the production of drinking water. According with the most up-to-date available references, AOPs are technological alternatives to conventional disinfection methods widely used in drinking water facilities that have not been still fully developed and exploited. Those AOPs activated by catalytic routes probably are going to play the central role in the short term. Moreover, in order to speed up their applicability at real scale, at least the following aspects should be considered:

1. The mechanisms of inactivation and/or elimination of the main and more hazardous water-borne microorganisms must be elucidated taking into account not only the attack by the powerful radical oxidizing species but also their particular interaction with either the surface of the solid (photo)-catalyst/ electrocatalyst (in the case of the heterogeneously activated) and the implicated radiation (photoactivated processes).
2. The increasing variability of species, types, or strains of parasites, virus, and bacteria, as well as phages or prions present in water, must be carefully accounted in the analysis; besides, their evolutionary and adaptive mechanisms to environmental conditions and capability to form new structures with varied composition under determined stressing conditions must also be considered.
3. The presence of spores, cyst forms, endospores, plasmids, and other genomic rearrangements, aggregations, or biofilms are featured by high resistance to disinfection.
4. Potential formation of intermediate metabolites alongside the oxidizing process.
5. Presence and type of NOM that may affect the overall performance or delay the oxidation and contribute to the pathogen's resistance.
6. Dosage, particle size, type, and surface charge of solid catalyst.
7. Realistic potential for the scaling up of the process in terms of both technical and economic issues. Finally, other critical aspects concern the biological safety displayed by every AOP; in this regard, potential cytotoxic and/or genotoxic effects produced by AOPs on every type of microorganisms including cyst,

oocyst, spores, endospores, and plasmids that can be present in varied surface waters must be carefully examined.

Acknowledgment Authors acknowledge funding received from Nariño University (VIPRI), EMPOPASTO S.A. E.S.P. – Specific Cooperation Agreement 09/2012 and Research project Agua Potable Nariño – SGR, CT&I Fund of the SGR, Departamento de Nariño, Colombia (BPIN 2014000100020). C.A. Sánchez also thanks Young Researchers scholarship granted by COLCIENCIAS (2012). MAV and AG thank the support from the Spanish Ministry of Economy and Competitiveness (MINECO) and the European Regional Development Fund (FEDER) (projects MAT2013-47811-C2-R and MAT2016-78863-C2-R).

References

1. Bichai F, Polo-Lopez MI, Fernandez Ibanez P (2012) Solar disinfection of wastewater to reduce contamination of lettuce crops by Escherichia coli in reclaimed water irrigation. Water Res 46(18):6040–6050
2. Pablos C, Marugan J, van Grieken R, Serrano E (2013) Emerging micropollutant oxidation during disinfection processes using UV-C, UV-C/H_2O_2, UV-A/TiO_2 and UV-A/TiO_2/H_2O_2. Water Res 47(3):1237–1245
3. Ortega-Gomez E, Esteban Garcia B, Ballesteros Martin MM, Fernandez Ibanez P, Sanchez Perez JA (2014) Inactivation of natural enteric bacteria in real municipal wastewater by solar photo-Fenton at neutral pH. Water Res 63:316–324
4. Primo Martínez O (2008) Mejoras en el tratamiento de lixiviados de vertedero de RSU mediantes procesos de oxidación avanzada. PhD thesis. Universidad de Cantabria, Spain
5. Agulló-Barceló M, Polo-López MI, Lucena F, Jofre J, Fernández-Ibáñez P (2013) Solar advanced oxidation processes as disinfection tertiary treatments for real wastewater: implications for water reclamation. Appl Catal B-Environ 136:341–350
6. Gomez-Couso H, Fontan-Sainz M, Navntoft C, Fernandez-Ibanez P, Ares-Mazas E (2012) Comparison of different solar reactors for household disinfection of drinking water in developing countries: evaluation of their efficacy in relation to the waterborne enteropathogen Cryptosporidium parvum. Trans R Soc Trop Med Hyg 106(11):645–652
7. Guzzella L, Di Caterino F, Monarca S, Zani C, Feretti D, Zerbini I, Nardi G, Buschini A, Poli P, Rossi C (2006) Detection of mutagens in water-distribution systems after disinfection. Mutat Res 608(1):72–81
8. Malato S, Fernández-Ibáñez P, Maldonado MI, Blanco J, Gernjak W (2009) Decontamination and disinfection of water by solar photocatalysis: recent overview and trends. Catal Today 147(1):1–59
9. McGuigan KG, Conroy RM, Mosler H-J, Md P, Ubomba-Jaswa E, Fernandez-Ibañez P (2012) Solar water disinfection (SODIS): a review from bench-top to roof-top. J Hazard Mater 235:29–46
10. McLoughlin OA, Kehoe SC, McGuigan KG, Duffy EF, Touati FA, Gernjak W, Alberola IO, Rodríguez SM, Gill LW (2004) Solar disinfection of contaminated water: a comparison of three small-scale reactors. Sol Energy 77(5):657–664
11. Dalrymple OK, Stefanakos E, Trotz MA, Goswami DY (2010) A review of the mechanisms and modeling of photocatalytic disinfection. Appl Catal B-Environ 98(1):27–38
12. Méndez-Hermida F, Ares-Mazás E, McGuigan KG, Boyle M, Sichel C, Fernández-Ibáñez P (2007) Disinfection of drinking water contaminated with Cryptosporidium parvum oocysts under natural sunlight and using the photocatalyst TiO_2. J Photochem Photobiol B 88(2):105–111
13. Guidelines for drinking-water quality (2011) 4th edn. WHO Press, Geneva, Switzerland

14. Hoff JC, Akin EW (1986) Microbial resistance to disinfectants: mechanisms and significance. Environ Health Perspect 69:7–13
15. Kim S, Ghafoor K, Lee J, Feng M, Hong J, Lee D-U, Park J (2013) Bacterial inactivation in water, DNA strand breaking, and membrane damage induced by ultraviolet-assisted titanium dioxide photocatalysis. Water Res 47(13):4403–4411
16. Pigeot-Rémy S, Simonet F, Erraruziz-Cerda E, Lazzaroni JC, Atlan D, Guillard C (2011) Photocatalysis and disinfection of water: identification of potential bacterial targets. Appl Catal B-Environ 104(3):390–398
17. Emadoldin F, Ottar S, Marianne Pedersen W, Per Arne A, Cathrine BV, Marit O, Geir S, Hans EK (2007) RNA base damage and repair. Curr Pharm Biotechnol 8(6):326–331
18. Linley E, Denyer SP, McDonnell G, Simons C, Maillard J-Y (2012) Use of hydrogen peroxide as a biocide: new consideration of its mechanisms of biocidal action. J Antimicrob Chemother 67(7):1589–1596
19. Silhavy TJ, Kahne D, Walker S (2010) The bacterial cell envelope. Cold Spring Harb Perspect Biol 2(5):a000414
20. Matsunaga T, Tomoda R, Nakajima T, Wake H (1985) Photoelectrochemical sterilization of microbial cells by semiconductor powders. FEMS Microbiol Lett 29(1–2):211–214
21. Alvarez FJ, Douglas LM, Konopka JB (2007) Sterol-rich plasma membrane domains in fungi. Eukaryot Cell 6(5):755–763
22. Maillard JY (2007) Bacterial resistance to biocides in the healthcare environment: should it be of genuine concern? J Hosp Infect 65:60–72
23. Webber MA, Randall LP, Cooles S, Woodward MJ, Piddock LJV (2008) Triclosan resistance in Salmonella enterica serovar Typhimurium. J Antimicrob Chemother 62(1):83–91
24. Finnegan M, Linley E, Denyer SP, McDonnell G, Simons C, Maillard JY (2010) Mode of action of hydrogen peroxide and other oxidizing agents: differences between liquid and gas forms. J Antimicrob Chemother 65(10):2108–2115
25. Martin NL, Bass P, Liss SN (2015) Antibacterial properties and mechanism of activity of a novel silver-stabilized hydrogen peroxide. PLoS One 10(7):e0131345
26. Abdou AG, Harba NM, Afifi AF, Elnaidany NF (2013) Assessment of Cryptosporidium parvum infection in immunocompetent and immunocompromised mice and its role in triggering intestinal dysplasia. Int J Infect Dis 17(8):e593–e600
27. Hallmich C, Gehr R (2010) Effect of pre- and post-UV disinfection conditions on photoreactivation of fecal coliforms in wastewater effluents. Water Res 44(9):2885–2893
28. Imlay JA (2008) Cellular defenses against superoxide and hydrogen peroxide. Annu Rev Biochem 77:755–776
29. Coulon C, Collignon A, McDonnell G, Thomas V (2010) Resistance of Acanthamoeba spp. cysts to disinfection treatments used in health care settings. J Clin Microbiol 48(8):2689–2697
30. Zhang Y, Zhang Y, Zhou L, Tan C (2014) Factors affecting UV/H_2O_2 inactivation of Bacillus atrophaeus spores in drinking water. J Photochem Photobiol B 134:9–15
31. Okoh A, Odjadjare EE, Igbinosa EO, Osode AN (2007) Wastewater treatment plants as a source of microbial pathogens in receiving watersheds. Afr J Biotechnol 6(25):2932–2944
32. Rider SD Jr, Zhu G (2010) Cryptosporidium: genomic and biochemical features. Exp Parasitol 124(1):2–9
33. Zhang H, Guo F, Zhou H, Zhu G (2012) Transcriptome analysis reveals unique metabolic features in the Cryptosporidium parvum Oocysts associated with environmental survival and stresses. BMC Genomics 13:647
34. Rider SD Jr, Zhu G (2008) Differential expression of the two distinct replication protein a subunits from Cryptosporidium parvum. J Cell Biochem 104(6):2207–2216
35. Ankarklev J, Jerlström-Hultqvist J, Ringqvist E, Troell K, Svärd SG (2010) Behind the smile: cell biology and disease mechanisms of Giardia species. Nat Rev Microbiol 8(6):413–422
36. Sun C-H, McCaffery JM, Reiner DS, Gillin FD (2003) Mining the Giardia lamblia genome for new Cyst Wall proteins. J Biol Chem 278(24):21701–21708

37. Chatterjee A, Carpentieri A, Ratner DM, Bullitt E, Costello CE, Robbins PW, Samuelson J (2010) Giardia cyst wall protein 1 is a lectin that binds to curled fibrils of the GalNAc homopolymer. PLoS Pathog 6(8):e1001059
38. Giannakis S, Darakas E, Escalas-Cañellas A, Pulgarin C (2014) Elucidating bacterial regrowth: effect of disinfection conditions in dark storage of solar treated secondary effluent. J Photoch Photobio A 290:43–53
39. Gilboa Y, Friedler E (2008) UV disinfection of RBC-treated light greywater effluent: kinetics, survival and regrowth of selected microorganisms. Water Res 42(4–5):1043–1050
40. Moncayo-Lasso A, Sanabria J, Pulgarin C, Benitez N (2009) Simultaneous E. coli inactivation and NOM degradation in river water via photo-Fenton process at natural pH in solar CPC reactor. A new way for enhancing solar disinfection of natural water. Chemosphere 77(2):296–300
41. Sanabria J, Wist J, Pulgarin C (2011) Photocatalytic disinfection treatments: viability, cultivability and metabolic changes of E. coli using different measurements methods. Dyna 78(166):150–157
42. Anastasi EM, Wohlsen TD, Stratton HM, Katouli M (2013) Survival of Escherichia coli in two sewage treatment plants using UV irradiation and chlorination for disinfection. Water Res 47(17):6670–6679
43. Anastasi EM, Matthews B, Gündoğdu A, Vollmerhausen T, Ramos NL, Stratton H, Ahmed W, Katouli M (2010) Prevalence and persistence of Escherichia coli strains with uropathogenic virulence characteristics in sewage treatment plants. Appl Environ Microbiol 76(17):5882–5886
44. Anastasi EM, Matthews B, Stratton HM, Katouli M (2012) Pathogenic Escherichia coli found in sewage treatment plants and environmental waters. Appl Environ Microbiol 78(16):5536–5541
45. Guo M, Huang J, Hu H, Liu W, Yang J (2012) UV inactivation and characteristics after photoreactivation of Escherichia coli with plasmid: health safety concern about UV disinfection. Water Res 46(13):4031–4036
46. Guo M, Huang J, Hu H, Liu W (2011) Growth and repair potential of three species of bacteria in reclaimed wastewater after UV disinfection. Biomed Environ Sci 24(4):400–407
47. Hijnen WA, Beerendonk EF, Medema GJ (2006) Inactivation credit of UV radiation for viruses, bacteria and protozoan (oo)cysts in water: a review. Water Res 40(1):3–22
48. Locas A, Demers J, Payment P (2008) Evaluation of photoreactivation of Escherichia coli and enterococci after UV disinfection of municipal wastewater. Can J Microbiol 54(11):971–975
49. Dennis WH, Olivieri VP, Krusé CW (1979) The reaction of nucleotides with aqueous hypochlorous acid. Water Res 13(4):357–362
50. Kim CK, Gentile DM, Sproul OJ (1980) Mechanism of ozone inactivation of bacteriophage f2. Appl Environ Microbiol 39(1):210–218
51. O'Brien RT, Newman J (1979) Structural and compositional changes associated with chlorine inactivation of polyviruses. Appl Environ Microbiol 38(6):1034–1039
52. Roy Y, Wong PKY, Engelbrecht RS, Chian ESK (1981) Mechanism of enteroviral inactivation by ozone. Appl Environ Microbiol 41(3):718–723
53. Wigginton KR, Kohn T (2012) Virus disinfection mechanisms: the role of virus composition, structure, and function. Curr Opin Virol 2(1):84–89
54. Page MA, Shisler JL, Marinas BJ (2010) Mechanistic aspects of adenovirus serotype 2 inactivation with free chlorine. Appl Environ Microbiol 76(9):2946–2954
55. Lv X, Lu Y, Yang X, Dong X, Ma K, Xiao S, Wang Y, Tang F (2015) Mutagenicity of drinking water sampled from the Yangtze River and Hanshui River (Wuhan section) and correlations with water quality parameters. Sci Rep 5:9572
56. Hindiyeh M, Ali A (2010) Investigating the efficiency of solar energy system for drinking water disinfection. Desalination 259(1–3):208–215
57. Ortega-Gómez E, Fernández-Ibáñez P, Ballesteros Martín MM, Polo-López MI, Esteban García B, Sánchez Pérez JA (2012) Water disinfection using photo-Fenton: effect of temperature on enterococcus faecalis survival. Water Res 46(18):6154–6162
58. Lin W, Yu Z, Zhang H, Thompson IP (2014) Diversity and dynamics of microbial communities at each step of treatment plant for potable water generation. Water Res 52:218–230

59. Venieri D, Fraggedaki A, Kostadima M, Chatzisymeon E, Binas V, Zachopoulos A, Kiriakidis G, Mantzavinos D (2014) Solar light and metal-doped TiO_2 to eliminate water-transmitted bacterial pathogens: photocatalyst characterization and disinfection performance. Appl Catal B-Environ 154:93–101
60. Rodríguez-Chueca J, Mosteo R, Ormad MP, Ovelleiro JL (2012) Factorial experimental design applied to Escherichia coli disinfection by Fenton and photo-Fenton processes. Sol Energy 86(11):3260–3267
61. Raffellini S, Schenk M, Guerrero S, Alzamora SM (2011) Kinetics of Escherichia coli inactivation employing hydrogen peroxide at varying temperatures, pH and concentrations. Food Control 22(6):920–932
62. Pariente MI, Martínez F, Melero JA, Botas JÁ, Velegraki T, Xekoukoulotakis NP, Mantzavinos D (2008) Heterogeneous photo-Fenton oxidation of benzoic acid in water: effect of operating conditions, reaction by-products and coupling with biological treatment. Appl Catal B-Environ 85(1):24–32
63. Acharya SM, Kurisu F, Kasuga I, Furumai H (2016) Chlorine dose determines bacterial community structure of subsequent regrowth in reclaimed water. J Water Environ Technol 14 (1):15–24
64. Chu W, Li X, Gao N, Deng Y, Yin D, Li D, Chu T (2015) Peptide bonds affect the formation of haloacetamides, an emerging class of N-DBPs in drinking water: free amino acids versus oligopeptides. Sci Rep 5:14412
65. Zhang SH, Miao DY, Liu AL, Zhang L, Wei W, Xie H, WQ L (2010) Assessment of the cytotoxicity and genotoxicity of haloacetic acids using microplate-based cytotoxicity test and CHO/HGPRT gene mutation assay. Mutat Res Genet Toxicol Environ Mutagen 703 (2):174–179
66. Ye B, Wang W, Yang L, Wei J, Xueli E (2009) Factors influencing disinfection by-products formation in drinking water of six cities in China. J Hazard Mater 171(1–3):147–152
67. Molnar JJ, Agbaba JR, Dalmacija BD, Klasnja MT, Dalmacija MB, Kragulj MM (2012) A comparative study of the effects of ozonation and TiO_2-catalyzed ozonation on the selected chlorine disinfection by-product precursor content and structure. Sci Total Environ 425:169–175
68. Tian C, Liu R, Liu H, Qu J (2013) Disinfection by-products formation and precursors transformation during chlorination and chloramination of highly-polluted source water: significance of ammonia. Water Res 47(15):5901–5910
69. Sciacca F, Rengifo-Herrera JA, Wéthé J, Pulgarin C (2011) Solar disinfection of wild Salmonella sp. in natural water with a 18L CPC photoreactor: detrimental effect of non-sterile storage of treated water. Sol Energy 85(7):1399–1408
70. Chin A, Berube PR (2005) Removal of disinfection by-product precursors with ozone-UV advanced oxidation process. Water Res 39(10):2136–2144
71. Yang X, Peng J, Chen B, Guo W, Liang Y, Liu W, Liu L (2012) Effects of ozone and ozone/peroxide pretreatments on disinfection byproduct formation during subsequent chlorination and chloramination. J Hazard Mater 239–240:348–354
72. Coleman HM, Marquis CP, Scott JA, Chin SS, Amal R (2005) Bactericidal effects of titanium dioxide-based photocatalysts. Chem Eng J 113(1):55–63
73. Guillard C, Pigeot-Rémy S, Simonet F, Atlan D, Lazzaroni JC (2012) Antimicrobial activity and mode of action of photochemistry and photocatalysis. Book of abstracts, 7th international conference on environmental catalysis. Lyon, France
74. Malik DJ, Shaw CM, Rielly CD, Shama G (2013) The inactivation of Bacillus subtilis spores at low concentrations of hydrogen peroxide vapour. J Food Eng 114(3):391–396
75. Rogers JV, Sabourin CL, Choi YW, Richter WR, Rudnicki DC, Riggs KB, Taylor ML, Chang J (2005) Decontamination assessment of Bacillus anthracis, Bacillus subtilis, and Geobacillus stearothermophilus spores on indoor surfaces using a hydrogen peroxide gas generator. J Appl Microbiol 99(4):739–748

76. Hall L, Otter JA, Chewins J, Wengenack NL (2008) Deactivation of the dimorphic fungi Histoplasma capsulatum, Blastomyces dermatitidis and Coccidioides immitis using hydrogen peroxide vapor. Med Mycol 46(2):189–191
77. Labas MD, Zalazar CS, Brandi RJ, Cassano AE (2008) Reaction kinetics of bacteria disinfection employing hydrogen peroxide. Biochem Eng J 38(1):78–87
78. Pottage T, Richardson C, Parks S, Walker JT, Bennett AM (2010) Evaluation of hydrogen peroxide gaseous disinfection systems to decontaminate viruses. J Hosp Infect 74(1):55–61
79. Fichet G, Antloga K, Comoy E, Deslys JP, McDonnell G (2007) Prion inactivation using a new gaseous hydrogen peroxide sterilisation process. J Hosp Infect 67(3):278–286
80. Sciacca F, Rengifo-Herrera JA, Wethe J, Pulgarin C (2010) Dramatic enhancement of solar disinfection (SODIS) of wild Salmonella sp. in PET bottles by H_2O_2 addition on natural water of Burkina Faso containing dissolved iron. Chemosphere 78(9):1186–1191
81. Block SS (2001) Disinfection, sterilization, and preservation, 5th edn. Lippincott Williams & Wilkins, Philadelphia
82. Cords BR, Burnett SL, Hilgren J, Finley M, Magnuson J (2005) Sanitizers: halogens, surface-active agents, and peroxides. In: Davidson PM, Sofos JN, Branen AL (eds) Antimicrobials in food. Taylor & Francis, Boca Raton, FL
83. Raffellini S, Guerrero S, Alzamora SM (2008) Effect of hydrogen peroxide concentration and pH on inactivation kinetics of Escherichia coli. J Food Saf 28(4):514–533
84. Flores MJ, Brandi RJ, Cassano AE, Labas MD (2012) Chemical disinfection with H_2O_2 – the proposal of a reaction kinetic model. Chem Eng J 198–199:388–396
85. Galeano LA, Bravo PF, Luna CD, Vicente MÁ, Gil A (2012) Removal of natural organic matter for drinking water production by Al/Fe-PILC-catalyzed wet peroxide oxidation: effect of the catalyst preparation from concentrated precursors. Appl Catal B-Environ 111–112:527–535
86. Lanao M, Ormad MP, Mosteo R, Ovelleiro JL (2012) Inactivation of Enterococcus sp. by photolysis and TiO_2 photocatalysis with H_2O_2 in natural water. Sol Energy 86(1):619–625
87. Lopez-Galvez F, Posada-Izquierdo GD, Selma MV, Perez-Rodriguez F, Gobet J, Gil MI, Allende A (2012) Electrochemical disinfection: an efficient treatment to inactivate Escherichia coli O157:H7 in process wash water containing organic matter. Food Microbiol 30(1):146–156
88. Ndounla J, Spuhler D, Kenfack S, Wéthé J, Pulgarin C (2013) Inactivation by solar photo-Fenton in pet bottles of wild enteric bacteria of natural well water: absence of re-growth after one week of subsequent storage. Appl Catal B-Environ 129:309–317
89. Primo O, Rivero MJ, Ortiz I (2008) Photo-Fenton process as an efficient alternative to the treatment of landfill leachates. J Hazard Mater 153(1–2):834–842
90. Spuhler D, Andrés Rengifo-Herrera J, Pulgarin C (2010) The effect of Fe^{2+}, Fe^{3+}, H_2O_2 and the photo-Fenton reagent at near neutral pH on the solar disinfection (SODIS) at low temperatures of water containing Escherichia coli K12. Appl Catal B-Environ 96(1):126–141
91. Galeano L-A, Vicente MÁ, Gil A (2014) Catalytic degradation of organic pollutants in aqueous streams by mixed Al/M-pillared clays (M = Fe, Cu, Mn). Catal Rev 56(3):239–287
92. Mohajerani M, Mehrvar M, Ein-Mozaffari F (2009) An overview of the integration of advanced oxidation technologies and other processes for water and wastewater treatment. Int J Eng 3(2):120–146
93. Rincón A-G, Pulgarin C (2007) Absence of E. coli regrowth after Fe^{3+} and TiO_2 solar photoassisted disinfection of water in CPC solar photoreactor. Catal Today 124(3):204–214
94. Sokmen M, Degerli S, Aslan A (2008) Photocatalytic disinfection of Giardia intestinalis and Acanthamoeba castellani cysts in water. Exp Parasitol 119(1):44–48
95. García-Fernández I, Polo-López MI, Oller I, Fernández-Ibáñez P (2012) Bacteria and fungi inactivation using Fe^{3+}/sunlight, H_2O_2/sunlight and near neutral photo-Fenton: a comparative study. Appl Catal B-Environ 121:20–29
96. Rubio D, Nebot E, Casanueva JF, Pulgarin C (2013) Comparative effect of simulated solar light, UV, UV/H_2O_2 and photo-Fenton treatment (UV-Vis/H_2O_2/$Fe^{2+,3+}$) in the Escherichia coli inactivation in artificial seawater. Water Res 47(16):6367–6379

97. Galeano LA, Gil A, Vicente MA (2010) Effect of the atomic active metal ratio in Al/Fe-, Al/Cu- and Al/(Fe–Cu)-intercalating solutions on the physicochemical properties and catalytic activity of pillared clays in the CWPO of methyl orange. Appl Catal B-Environ 100 (1–2):271–281
98. Joseph CG, Li Puma G, Bono A, Krishnaiah D (2009) Sonophotocatalysis in advanced oxidation process: a short review. Ultrason Sonochem 16(5):583–589
99. Malato S, Blanco J, Alarcón DC, Maldonado MI, Fernández-Ibáñez P, Gernjak W (2007) Photocatalytic decontamination and disinfection of water with solar collectors. Catal Today 122(1–2):137–149
100. Galeano LA, Gil A, Vicente MA (2011) Strategies for immobilization of manganese on expanded natural clays: catalytic activity in the CWPO of methyl orange. Appl Catal B-Environ 104 (3–4):252–260
101. Li B, Xu X, Zhu L, Ding W, Mahmood Q (2010) Catalytic ozonation of industrial wastewater containing chloro and nitro aromatics using modified diatomaceous porous filling. Desalination 254(1–3):90–98
102. Flores MJ, Brandi RJ, Cassano AE, Labas MD (2012) Chemical disinfection with H_2O_2 – the proposal of a reaction kinetic model. Chem Eng J 198:388–396
103. Liu Y, Li J, Qiu X, Burda C (2007) Bactericidal activity of nitrogen-doped metal oxide nanocatalysts and the influence of bacterial extracellular polymeric substances (EPS). J Photoch Photobio A 190(1):94–100
104. Storz G, Christman MF, Sies H, Ames BN (1987) Spontaneous mutagenesis and oxidative damage to DNA in Salmonella typhimurium. Proc Natl Acad Sci U S A 84:8917–8921
105. Anzai K, Hamasuna M, Kadono H, Lee S, Aoyagi H, Kirino Y (1991) Formation of ion channels in planar lipid bilayer membranes by synthetic basic peptides. Biochim Biophys Acta-Biomembr 1064(2):256–266
106. Cho M, Chung H, Choi W, Yoon J (2005) Different inactivation behaviors of MS-2 phage and Escherichia coli in TiO_2 photocatalytic disinfection. Appl Environ Microbiol 71(1):270–275
107. Sichel C, de Cara M, Tello J, Blanco J, Fernández-Ibáñez P (2007) Solar photocatalytic disinfection of agricultural pathogenic fungi: Fusarium species. Appl Catal B-Environ 74 (1):152–160
108. Sichel C, Tello J, de Cara M, Fernández-Ibáñez P (2007) Effect of UV solar intensity and dose on the photocatalytic disinfection of bacteria and fungi. Catal Today 129(1):152–160
109. Ubomba-Jaswa E, Navntoft C, Polo-López MI, Fernández-Ibáñez P, McGuigan KG (2009) Solar disinfection of drinking water (SODIS): an investigation of the effect of UV-A dose on inactivation efficiency. Photochem Photobiol Sci 8(5):587–595
110. Kim JY, Lee C, Sedlak DL, Yoon J, Nelson KL (2010) Inactivation of MS2 coliphage by Fenton's reagent. Water Res 44(8):2647–2653
111. Gosselin F, Madeira LM, Juhna T, Block JC (2013) Drinking water and biofilm disinfection by Fenton-like reaction. Water Res 47(15):5631–5638
112. Benatti CT, Tavares CR, Guedes TA (2006) Optimization of Fenton's oxidation of chemical laboratory wastewaters using the response surface methodology. J Environ Manage 80 (1):66–74
113. Rodríguez-Chueca J, Polo-López MI, Mosteo R, Ormad MP, Fernández-Ibáñez P (2014) Disinfection of real and simulated urban wastewater effluents using a mild solar photo-Fenton. Appl Catal B-Environ 150:619–629
114. Rodríguez-Chueca J, Ormad MP, Mosteo R, Ovelleiro JL (2015) Kinetic modeling of Escherichia coli and Enterococcus sp. inactivation in wastewater treatment by photo-Fenton and H_2O_2/UV–vis processes. Chem Eng Sci 138:730–740
115. Diao HF, Li XY, JD G, Shi HC, Xie ZM (2004) Electron microscopic investigation of the bactericidal action of electrochemical disinfection in comparison with chlorination, ozonation and Fenton reaction. Process Biochem 39(11):1421–1426

116. Mackulak T, Nagyova K, Faberova M, Grabic R, Koba O, Gal M, Birosova L (2015) Utilization of Fenton-like reaction for antibiotics and resistant bacteria elimination in different parts of WWTP. Environ Toxicol Pharmacol 40(2):492–497
117. Park SC, Kim NH, Yang W, Nah JW, Jang MK, Lee D (2016) Polymeric micellar nanoplatforms for Fenton reaction as a new class of antibacterial agents. J Control Release 221:37–47
118. Rincón A-G, Pulgarin C (2006) Comparative evaluation of Fe^{3+} and TiO_2 photoassisted processes in solar photocatalytic disinfection of water. Appl Catal B-Environ 63(3):222–231
119. Rodriguez C, Di Cara A, Renaud FNR, Freney J, Horvais N, Borel R, Puzenat E, Guillard C (2014) Antibacterial effects of photocatalytic textiles for footwear application. Catal Today 230:41–46
120. Nieto-Juarez JI, Kohn T (2013) Virus removal and inactivation by iron (hydr)oxide-mediated Fenton-like processes under sunlight and in the dark. Photochem Photobiol Sci 12(9):1596–1605
121. Donlan RM (2002) Biofilms: microbial life on surfaces. Emerg Infect Dis 8(9):881–890
122. Bozzi A, Yuranova T, Mielczarski E, Mielczarski J, Buffat PA, Lais P, Kiwi J (2003) Superior biodegradability mediated by immobilized Fe-fabrics of waste waters compared to Fenton homogeneous reactions. Appl Catal B-Environ 42(3):289–303
123. Rodriguez ML, Timokhin VI, Contreras S, Chamarro E, Esplugas S (2003) Rate equation for the degradation of nitrobenzene by "Fenton-like" reagent. Adv Environ Res 7(2):583–595
124. Garrido-Ramírez EG, Theng BKG, Mora ML (2010) Clays and oxide minerals as catalysts and nanocatalysts in Fenton-like reactions – a review. Appl Clay Sci 47(3–4):182–192
125. Ruales-Lonfat C, Benítez N, Sienkiewicz A, Pulgarín C (2014) Deleterious effect of homogeneous and heterogeneous near-neutral photo-Fenton system on Escherichia coli. Comparison with photocatalytic action of TiO_2 during cell envelope disruption. Appl Catal B-Environ 160:286–297
126. Rey A, Faraldos M, Casas JA, Zazo JA, Bahamonde A, Rodríguez JJ (2009) Catalytic wet peroxide oxidation of phenol over Fe/AC catalysts: influence of iron precursor and activated carbon surface. Appl Catal B-Environ 86(1–2):69–77
127. Bautista P, Mohedano AF, Menéndez N, Casas JA, Rodriguez JJ (2010) Catalytic wet peroxide oxidation of cosmetic wastewaters with Fe-bearing catalysts. Catal Today 151(1–2):148–152
128. Moncayo-Lasso A, Torres-Palma RA, Kiwi J, Benítez N, Pulgarin C (2008) Bacterial inactivation and organic oxidation via immobilized photo-Fenton reagent on structured silica surfaces. Appl Catal B-Environ 84(3–4):577–583
129. Tokumura M, Morito R, Hatayama R, Kawase Y (2011) Iron redox cycling in hydroxyl radical generation during the photo-Fenton oxidative degradation: dynamic change of hydroxyl radical concentration. Appl Catal B-Environ 106(3–4):565–576
130. Bandala ER, Corona-Vasquez B, Guisar R, Uscanga M (2009) Deactivation of highly resistant microorganisms in water using solar driven photocatalytic processes. Int J Chem React Eng 7: A7
131. Bandala ER, Perez R, Velez-Lee AE, Sanchez-Salas JL, Quiroz MA, Méndez-Rojas MA (2011) Bacillus Subtilis spore inactivation in water using photo-assisted Fenton reaction. Sustain Environ Res 21(5):285–290
132. Ortega-Gómez E, Martín MMB, García BE, Pérez JAS, Ibáñez PF (2016) Wastewater disinfection by neutral pH photo-Fenton: the role of solar radiation intensity. Appl Catal B-Environ 181:1–6
133. Ortega-Gómez E, Ballesteros Martín MM, Carratalà A, Fernández Ibañez P, Sánchez Pérez JA, Pulgarín C (2015) Principal parameters affecting virus inactivation by the solar photo-Fenton process at neutral pH and μM concentrations of H_2O_2 and $Fe^{2+/3+}$. Appl Catal B-Environ 174:395–402
134. Bandala ER, González L, de la Hoz F, Pelaez MA, Dionysiou DD, Dunlop PSM, Byrne JA, Sanchez JL (2011) Application of azo dyes as dosimetric indicators for enhanced photocatalytic solar disinfection (ENPHOSODIS). J Photoch Photobio A 218(2):185–191

135. Barreca S, Velez Colmenares JJ, Pace A, Orecchio S, Pulgarin C (2015) Escherichia coli inactivation by neutral solar heterogeneous photo-Fenton (HPF) over hybrid iron/montmorillonite/alginate beads. J Environ Chem Eng 3(1):317–324
136. Ndounla J, Pulgarin C (2014) Evaluation of the efficiency of the photo Fenton disinfection of natural drinking water source during the rainy season in the Sahelian region. Sci Total Environ 493:229–238
137. Ruales-Lonfat C, Barona JF, Sienkiewicz A, Bensimon M, Vélez-Colmenares J, Benítez N, Pulgarín C (2015) Iron oxides semiconductors are efficients for solar water disinfection: a comparison with photo-Fenton processes at neutral pH. Appl Catal B-Environ 166:497–508
138. Ruales-Lonfat C, Barona JF, Sienkiewicz A, Vélez J, Benítez LN, Pulgarín C (2016) Bacterial inactivation with iron citrate complex: a new source of dissolved iron in solar photo-Fenton process at near-neutral and alkaline pH. Appl Catal B-Environ 180:379–390
139. Aurioles-López V, Polo-López MI, Fernández-Ibáñez P, López-Malo A, Bandala ER (2016) Effect of iron salt counter ion in dose–response curves for inactivation of Fusarium solani in water through solar driven Fenton-like processes. Phys Chem Earth A/B/C 91:46–52
140. Dubey JP, Huong LT, Sundar N, Su C (2007) Genetic characterization of toxoplasma gondii isolates in dogs from Vietnam suggests their South American origin. Vet Parasitol 146(3–4):347–351
141. Polo-López MI, Castro-Alférez M, Oller I, Fernández-Ibáñez P (2014) Assessment of solar photo-Fenton, photocatalysis, and H_2O_2 for removal of phytopathogen fungi spores in synthetic and real effluents of urban wastewater. Chem Eng J 257:122–130
142. Polo-López MI, Oller I, Fernández-Ibáñez P (2013) Benefits of photo-Fenton at low concentrations for solar disinfection of distilled water. A case study: Phytophthora capsici. Catal Today 209:181–187
143. Polo-López MI, García-Fernández I, Velegraki T, Katsoni A, Oller I, Mantzavinos D, Fernández-Ibáñez P (2012) Mild solar photo-Fenton: an effective tool for the removal of Fusarium from simulated municipal effluents. Appl Catal B-Environ 111:545–554
144. Ramírez Zamora RM, Galván García M, Gallardo IR, Rigas F, Durán-Moreno A (2006) Viability reduction of parasites (Ascaris spp.) in water with photo-Fenton reaction via response surface methodology. Water Pract Technol 1(2):wpt2006043
145. Bandala ER, Gonzalez L, Sanchez-Salas JL, Castillo JH (2012) Inactivation of Ascaris eggs in water using sequential solar driven photo-Fenton and free chlorine. J Water Health 10 (1):20–30
146. Paspaltsis I, Berberidou C, Poulios I, Sklaviadis T (2009) Photocatalytic degradation of prions using the photo-Fenton reagent. J Hosp Infect 71(2):149–156
147. Imlay JA (2003) Pathways of oxidative damage. Annu Rev Microbiol 57:395–418
148. Cartron ML, Maddocks S, Gillingham P, Craven CJ, Andrews SC (2006) Feo – transport of ferrous iron into bacteria. Biometals 19(2):143–157
149. Chitarra GS, Breeuwer P, Rombouts FM, Abee T, Dijksterhuis J (2005) Differentiation inside multicelled macroconidia of Fusarium culmorum during early germination. Fungal Genet Biol 42(8):694–703
150. Heaselgrave W, Kilvington S (2011) The efficacy of simulated solar disinfection (SODIS) against Ascaris, Giardia, Acanthamoeba, Naegleria, Entamoeba and Cryptosporidium. Acta Trop 119(2–3):138–143
151. Rubio D, Casanueva JF, Nebot E (2013) Improving UV seawater disinfection with immobilized TiO_2: study of the viability of photocatalysis (UV_{254}/TiO_2) as seawater disinfection technology. J Photoch Photobio A 271:16–23
152. Dunlop PSM, Byrne JA, Manga N, Eggins BR (2002) The photocatalytic removal of bacterial pollutants from drinking water. J Photoch Photobio A 148(1):355–363
153. Guillard C, Bui T-H, Felix C, Moules V, Lina B, Lejeune P (2008) Microbiological disinfection of water and air by photocatalysis. C R Chim 11(1–2):107–113
154. Rincón A-G, Pulgarin C (2004) Bactericidal action of illuminated TiO_2 on pure Escherichia coli and natural bacterial consortia: post-irradiation events in the dark and assessment of the effective disinfection time. Appl Catal B-Environ 49(2):99–112

155. Pigeot-Remy S, Simonet F, Atlan D, Lazzaroni JC, Guillard C (2012) Bactericidal efficiency and mode of action: a comparative study of photochemistry and photocatalysis. Water Res 46 (10):3208–3218
156. Thabet S, Simonet F, Lemaire M, Guillard C, Cotton P (2014) Impact of photocatalysis on fungal cells: depiction of cellular and molecular effects on Saccharomyces cerevisiae. Appl Environ Microbiol 80(24):7527–7535
157. Dunlop PSM, McMurray TA, Hamilton JWJ, Byrne JA (2008) Photocatalytic inactivation of Clostridium perfringens spores on TiO_2 electrodes. J Photoch Photobio A 196(1):113–119
158. Cho M, Yoon J (2008) Measurement of OH radical CT for inactivating Cryptosporidium parvum using photo/ferrioxalate and photo/TiO_2 systems. J Appl Microbiol 104(3):759–766
159. Lonnen J, Kilvington S, Kehoe SC, Al-Touati F, McGuigan KG (2005) Solar and photocatalytic disinfection of protozoan, fungal and bacterial microbes in drinking water. Water Res 39(5):877–883
160. van Grieken R, Marugán J, Pablos C, Furones L, López A (2010) Comparison between the photocatalytic inactivation of Gram-positive E. faecalis and Gram-negative E. coli faecal contamination indicator microorganisms. Appl Catal B-Environ 100(1):212–220
161. Benabbou AK, Derriche Z, Felix C, Lejeune P, Guillard C (2007) Photocatalytic inactivation of Escherichia coli: effect of concentration of TiO_2 and microorganism, nature, and intensity of UV irradiation. Appl Catal B-Environ 76(3):257–263
162. Castillo-Ledezma JH, Sánchez Salas JL, López-Malo A, Bandala ER (2011) Effect of pH, solar irradiation, and semiconductor concentration on the photocatalytic disinfection of Escherichia coli in water using nitrogen-doped TiO_2. Eur Food Res Technol 233(5):825–834
163. García-Fernández I, Fernández-Calderero I, Polo-López MI, Fernández-Ibáñez P (2015) Disinfection of urban effluents using solar TiO_2 photocatalysis: a study of significance of dissolved oxygen, temperature, type of microorganism and water matrix. Catal Today 240:30–38
164. Khraisheh M, Wu L, Al-Muhtaseb AH, Al-Ghouti MA (2015) Photocatalytic disinfection of Escherichia coli using TiO_2 P25 and Cu-doped TiO_2. J Ind Eng Chem 28:369–376
165. Pratap Reddy M, Phil HH, Subrahmanyam M (2008) Photocatalytic disinfection of Escherichia coli over titanium (IV) oxide supported on Hβ zeolite. Catal Lett 123(1):56–64
166. Rincón AG, Pulgarin C, Adler N, Peringer P (2001) Interaction between E. coli inactivation and DBP-precursors – dihydroxybenzene isomers – in the photocatalytic process of drinking-water disinfection with TiO_2. J Photoch Photobio A 139(2):233–241
167. Rizzo L, Sannino D, Vaiano V, Sacco O, Scarpa A, Pietrogiacomi D (2014) Effect of solar simulated N-doped TiO_2 photocatalysis on the inactivation and antibiotic resistance of an E. coli strain in biologically treated urban wastewater. Appl Catal B-Environ 144:369–378
168. Fernández-Ibáñez P, Sichel C, Polo-López MI, de Cara-García M, Tello JC (2009) Photocatalytic disinfection of natural well water contaminated by Fusarium solani using TiO_2 slurry in solar CPC photo-reactors. Catal Today 144(1):62–68
169. Pigeot-Remy S, Real P, Simonet F, Hernandez C, Vallet C, Lazzaroni JC, Vacher S, Guillard C (2013) Inactivation of Aspergillus niger spores from indoor air by photocatalytic filters. Appl Catal B-Environ 134–135:167–173
170. Zan L, Fa W, Peng T, Gong ZK (2007) Photocatalysis effect of nanometer TiO_2 and TiO_2-coated ceramic plate on hepatitis B virus. J Photochem Photobiol B 86(2):165–169
171. Hajkova P, Spatenka P, Horsky J, Horska I, Kolouch A (2007) Photocatalytic effect of TiO_2 films on viruses and bacteria. Plasma Processes Polym 4(S1):S397–S401
172. Misstear DB, Gill LW (2012) The inactivation of phages MS2, PhiX174 and PR772 using UV and solar photocatalysis. J Photochem Photobiol B 107:1–8
173. Watts RJ, Kong S, Orr MP, Miller GC, Henry BE (1995) Photocatalytic inactivation of coliform bacteria and viruses in secondary wastewater effluent. Water Res 29(1):95–100
174. Abdulla-Al-Mamun M, Kusumoto Y, Zannat T, Islam MS (2011) Synergistic cell-killing by photocatalytic and plasmonic photothermal effects of Ag@TiO_2 core–shell composite nanoclusters against human epithelial carcinoma (HeLa) cells. Appl Catal A-Gen 398(1):134–142
175. Zhang A-P, Sun Y-P (2004) Photocatalytic killing effect of TiO_2 nanoparticles on Ls-174-t human colon carcinoma cells. World J Gastroenterol 10(21):3191–3193

176. Kim TY, Lee Y-H, Park K-H, Kim SJ, Cho SY (2005) A study of photocatalysis of TiO_2 coated onto chitosan beads and activated carbon. Res Chem Intermediat 31(4):343–358
177. Bogdan J, Zarzyńska J, Pławińska-Czarnak J (2015) Comparison of infectious agents susceptibility to photocatalytic effects of nanosized titanium and zinc oxides: a practical approach. Nanoscale Res Lett 10(1):1023
178. Cheng YW, Chan RC, Wong PK (2007) Disinfection of legionella pneumophila by photocatalytic oxidation. Water Res 41(4):842–852
179. Klis FM, Boorsma A, De Groot PW (2006) Cell wall construction in Saccharomyces cerevisiae. Yeast 23(3):185–202
180. Gogate PR, Pandit AB (2004) A review of imperative technologies for wastewater treatment I: oxidation technologies at ambient conditions. Adv Environ Res 8(3):501–551
181. Paleologou A, Marakas H, Xekoukoulotakis NP, Moya A, Vergara Y, Kalogerakis N, Gikas P, Mantzavinos D (2007) Disinfection of water and wastewater by TiO_2 photocatalysis, sonolysis and UV-C irradiation. Catal Today 129(1):136–142
182. Gumy D, Morais C, Bowen P, Pulgarin C, Giraldo S, Hajdu R, Kiwi J (2006) Catalytic activity of commercial of TiO_2 powders for the abatement of the bacteria (E. coli) under solar simulated light: influence of the isoelectric point. Appl Catal B-Environ 63(1):76–84
183. Mamane H, Shemer H, Linden KG (2007) Inactivation of E. coli, B. subtilis spores, and MS2, T4, and T7 phage using UV/H_2O_2 advanced oxidation. J Hazard Mater 146(3):479–486
184. Dunlop PSM, Sheeran CP, Byrne JA, McMahon MAS, Boyle MA, McGuigan KG (2010) Inactivation of clinically relevant pathogens by photocatalytic coatings. J Photochem Photobiol A 216(2):303–310
185. Ali MA, Al-Herrawy AZ, El-Hawaary SE (2004) Detection of enteric viruses, Giardia and Cryptosporidium in two different types of drinking water treatment facilities. Water Res 38(18):3931–3939
186. Castro-Hermida JA, Gonzalez-Warleta M, Mezo M (2015) Cryptosporidium spp. and Giardia duodenalis as pathogenic contaminants of water in Galicia, Spain: the need for safe drinking water. Int J Hyg Environ Health 218(1):132–138
187. Jenkins MB, Eaglesham BS, Anthony LC, Kachlany SC, Bowman DD, Ghiorse WC (2010) Significance of wall structure, macromolecular composition, and surface polymers to the survival and transport of Cryptosporidium parvum oocysts. Appl Environ Microbiol 76(6):1926–1934
188. Chen E, Su H, Tan T (2011) Antimicrobial properties of silver nanoparticles synthesized by bioaffinity adsorption coupled with TiO_2 photocatalysis. J Chem Technol Biotechnol 86(3):421–427
189. Zhang N, Hu K, Shan B (2014) Ballast water treatment using UV/TiO_2 advanced oxidation processes: an approach to invasive species prevention. Chem Eng J 243:7–13
190. Li Q, Xie R, Li YW, Mintz EA, Shang JK (2007) Enhanced visible-light-induced photocatalytic disinfection of E. coli by carbon-sensitized nitrogen-doped titanium oxide. Environ Sci Technol 41(14):5050–5056
191. Rengifo-Herrera JA, Pierzchała K, Sienkiewicz A, Forró L, Kiwi J, Pulgarin C (2009) Abatement of organics and Escherichia coli by N, S co-doped TiO_2 under UV and visible light. Implications of the formation of singlet oxygen ($^1O_2$) under visible light. Appl Catal B-Environ 88(3–4):398–406
192. Merwald H, Klosner G, Kokesch C, Der-Petrossian M, Honigsmann H, Trautinger F (2005) UVA-induced oxidative damage and cytotoxicity depend on the mode of exposure. J Photochem Photobiol B 79(3):197–207
193. Chen S, Hu W, Hong J, Sandoe S (2016) Electrochemical disinfection of simulated ballast water on PbO_2/graphite felt electrode. Mar Pollut Bull 105(1):319–323
194. Cui X, Quicksall AN, Blake AB, Talley JW (2013) Electrochemical disinfection of Escherichia coli in the presence and absence of primary sludge particulates. Water Res 47(13):4383–4390
195. Hong X, Wen J, Xiong X, Hu Y (2016) Silver nanowire-carbon fiber cloth nanocomposites synthesized by UV curing adhesive for electrochemical point-of-use water disinfection. Chemosphere 154:537–545

196. Huang X, Qu Y, Cid CA, Finke C, Hoffmann MR, Lim K, Jiang SC (2016) Electrochemical disinfection of toilet wastewater using wastewater electrolysis cell. Water Res 92:164–172
197. Jeong J, Kim C, Yoon J (2009) The effect of electrode material on the generation of oxidants and microbial inactivation in the electrochemical disinfection processes. Water Res 43 (4):895–901
198. Lacasa E, Tsolaki E, Sbokou Z, Rodrigo MA, Mantzavinos D, Diamadopoulos E (2013) Electrochemical disinfection of simulated ballast water on conductive diamond electrodes. Chem Eng J 223:516–523
199. Rajab M, Heim C, Letzel T, Drewes JE, Helmreich B (2015) Electrochemical disinfection using boron-doped diamond electrode – the synergetic effects of in situ ozone and free chlorine generation. Chemosphere 121:47–53
200. Long Y, Ni J, Wang Z (2015) Subcellular mechanism of Escherichia coli inactivation during electrochemical disinfection with boron-doped diamond anode: a comparative study of three electrolytes. Water Res 84:198–206
201. Schaefer CE, Andaya C, Urtiaga A (2015) Assessment of disinfection and by-product formation during electrochemical treatment of surface water using a Ti/IrO$_2$ anode. Chem Eng J 264:411–416
202. Li H, Zhu X, Ni J (2011) Comparison of electrochemical method with ozonation, chlorination and monochloramination in drinking water disinfection. Electrochim Acta 56(27):9789–9796
203. Jeong J, Kim JY, Cho M, Choi W, Yoon J (2007) Inactivation of Escherichia coli in the electrochemical disinfection process using a Pt anode. Chemosphere 67(4):652–659
204. Polcaro AM, Vacca A, Mascia M, Palmas S, Pompei R, Laconi S (2007) Characterization of a stirred tank electrochemical cell for water disinfection processes. Electrochim Acta 52(7):2595–2602
205. Cano A, Cañizares P, Barrera-Díaz C, Sáez C, Rodrigo MA (2012) Use of conductive-diamond electrochemical-oxidation for the disinfection of several actual treated wastewaters. Chem Eng J 211:463–469
206. Hussain SN, de Las Heras N, Asghar HM, Brown NW, Roberts EP (2014) Disinfection of water by adsorption combined with electrochemical treatment. Water Res 54:170–178
207. Furuta T, Tanaka H, Nishiki Y, Pupunat L, Haenni W, Rychen P (2004) Legionella inactivation with diamond electrodes. Diamond Relat Mater 13(11):2016–2019
208. Feng C, Suzuki K, Zhao S, Sugiura N, Shimada S, Maekawa T (2004) Water disinfection by electrochemical treatment. Bioresour Technol 94(1):21–25
209. Schmalz V, Dittmar T, Haaken D, Worch E (2009) Electrochemical disinfection of biologically treated wastewater from small treatment systems by using boron-doped diamond (BDD) electrodes – contribution for direct reuse of domestic wastewater. Water Res 43 (20):5260–5266

Inactivation of *Cryptosporidium* by Advanced Oxidation Processes

Abidelfatah M. Nasser

Abstract *Cryptosporidium*, a protozoan parasite, was found responsible for numerous water- and foodborne outbreaks. The high risk introduced by the presence of *Cryptosporidium* in water is attributed to its low infectious dose and its resistance to environmental stress and conventional disinfection processes. Since *Cryptosporidium* oocysts are highly resistant to chlorine, the most applied water disinfectant, alternative disinfectants were proposed and applied to reduce the health risks of *Cryptosporidium* in water, among them advanced oxidation processes (AOPs), based on highly reactive oxidants, mainly hydroxyl radicals. AOPs proved to be efficient in reducing the concentration of micropollutants. The data presented here proved also that AOPs are effective in the inactivation of *Cryptosporidium* and other waterborne pathogens. Therefore AOPs can be applied as a barrier for reducing the health risks of waterborne *Cryptosporidium*.

Keywords Advanced oxidation processes, *Cryptosporidium*, Disinfection, Inactivation, Water

Contents

1 Importance of *Cryptosporidium* for Public Health 298
2 Challenges in Determining the Viability and Infectivity of *Cryptosporidium* 299
3 Inactivation of *Cryptosporidium* by Chlorine, Monochloramine, Chlorine Dioxide, and Ozone 300
4 Sequential Inactivation of *Cryptosporidium* by Ozone as Primary and by Chlorine or Monochloramine as Secondary Disinfectants 300

A.M. Nasser (✉)
Institute of Natural Sciences, Beit Berl Academic College of Education, Kfar Saba, Israel
e-mail: abid@beitberl.ac.il

5	Photocatalytic and Advanced Oxidation Processes for the Inactivation of *Cryptosporidium*	302
6	Conclusions	305
References		306

1 Importance of *Cryptosporidium* for Public Health

Cryptosporidium is an enteric protozoan parasite which causes severe diarrhea in human and domestic animals. Although *Cryptosporidium* was first recognized as a waterborne pathogen during an outbreak in Braun station, Texas, in 1984, in which more than 2,000 individuals have developed cryptosporidiosis, only 10 years later, following the largest waterborne outbreak in Milwaukee, Wisconsin, its importance as waterborne pathogen was recognized worldwide [1, 2]. The occurrence of the outbreaks which involved thousands of individuals stimulated research for the development of sensitive and specific methods for monitoring drinking water and drinking water sources for the presence of low levels of *Cryptosporidium*. On the other hand, intensive efforts were allocated for the development of efficient treatment methods for the reduction of *Cryptosporidium* from drinking water sources.

Cryptosporidium is an intracellular protozoan parasite with an infectious dose of as low as 120 oocysts [3]. The thick outer wall of the oocysts enables the persistence of *Cryptosporidium* for long time in the environment which may facilitate its transmission by contaminated water and food [4]. Additional routes of transmission of *Cryptosporidium* may include person to person and exposure to infected animals and to recreational waters [5, 6]. Numerous waterborne outbreaks of *Cryptosporidium* have been reported in developed countries [4]. Most of the waterborne outbreaks resulted from the consumption of drinking water, contaminated either by anthropogenic or animal wastes, which received inadequate treatment. It is worth noting that in some cryptosporidiosis outbreaks, the suspected drinking water complied with the fecal coliform guideline, since *Cryptosporidium* oocysts are very resistant to chlorine [7].

The source of *Cryptosporidium* oocysts in the environment can be from the discharge of untreated human and animal wastes. Studies have shown that *Cryptosporidium* oocysts are highly prevalent in raw wastewater and often poorly removed by wastewater treatment processes [8, 9]. Therefore, surface-water sources have been found to be contaminated with various levels of *Cryptosporidium* oocysts [10, 11]. The thick wall surrounding the oocysts of *Cryptosporidium* makes them resistant to natural decay. Oocysts of *Cryptosporidium* may survive for months in soil at cold temperature in dark [11]. Fayer et al. have shown that oocysts of *Cryptosporidium* remained infectious for up to 24 weeks in water at 20°C; the low infectious dose of *Cryptosporidium* and its high persistence in natural water may result in the occurrence of cryptosporidiosis [12]. Therefore, efficient water treatment processes must be applied to prevent the transmission of *Cryptosporidium* oocysts by drinking water.

2 Challenges in Determining the Viability and Infectivity of *Cryptosporidium*

The difference between the inactivation of *Cryptosporidium* and bacteria or viruses by oxidizers stem from that the disinfectant must penetrate through oocysts wall to inactivate the sporozoites. To measure the inactivation efficiency of *Cryptosporidium*, the analysis should be able to determine whether the sporozoites are alive or dead. Viability assays for *Cryptosporidium* oocysts based on the inclusion or exclusion of fluorogenic vital dyes, 4′,6-diamidino-2-phenylindole (DAPI), and propidium iodide (PI) were reported by Campbell et al. [13] (Table 1). Inclusion of DAPI and/or PI indicated that oocysts of *Cryptosporidium* lost their viability. In addition to the dye inclusion-exclusion method, differential interphase contrast (DIC) is also applied to distinguish between empty and full oocysts [14]. An additional method for viability measurement of *Cryptosporidium* is by in vitro excystation (the ability of oocysts to release sporozoites) which was found to be highly correlated with the dye inclusion-exclusion method [13]. Both dye-based methods and in vitro excystation cannot measure infectivity.

Assessment of the infectivity of *Cryptosporidium* oocysts relayed initially on animal hosts, which is time consuming and very costly [15]. The development of an in vitro method for the detection of infectious *Cryptosporidium* in cell culture enables accurate evaluation of the persistence of the parasites in environmental settings and its inactivation by disinfection methods [16]. Rochelle et al. compared the sensitivity of in vitro cell culture with mouse assay for measuring the infectivity of *Cryptosporidium* [17]. These authors demonstrated good correlation between the infectivity for HCT-8 cells and the results for CD-1 mice for untreated oocysts and oocysts exposed to ozone and UV irradiation. Consequently, the authors concluded that in vitro cell culture is equivalent to the "gold standard" mouse infectivity and should be considered as practical and accurate method to assess oocysts' infectivity

Table 1 Detection of *Cryptosporidium* oocysts by viability and infectivity methods

Method	Detection viability/ infectivity	Parameter detected	Reference
DAPI[a]	Viability	Permeability of oocysts wall and membrane	[13]
PI[b]	Viability	Permeability of oocysts wall and membrane	[13]
DIC[c]	Viability	Empty or full oocyst	[14]
Excystation	Viability	Ability to release sporozoites	[13]
Animal	Infectivity	Ability to multiply in animal intestine	[15]
Cell culture	Infectivity	Ability to multiply in cell culture	[16, 17]
Cell culture PCR	Infectivity	Ability to multiply in cell culture	[18, 19]

[a]*DAPI* 4′,6-diamidino-2-phenylindole
[b]*PI* propidium iodide
[c]*DIC* differential interphase contrast

and inactivation. An improved sensitivity was accomplished by integrating cell culture real-time polymerase chain reaction (CC-RT-PCR), where single infectious oocyst was detected [18, 19].

3 Inactivation of *Cryptosporidium* by Chlorine, Monochloramine, Chlorine Dioxide, and Ozone

Chlorine is the most commonly used disinfectant in water and wastewater treatment. Studies using mouse infectivity to measure the inactivation rate of *Cryptosporidium* oocysts by free chlorine demonstrated low reduction efficiency [20–22]. Korich et al. compared the inactivation efficiency of *Cryptosporidium* by ozone, chlorine dioxide, chlorine, and monochloramine using mouse infectivity to determine oocysts' viability [20]. A $C \times T$ (concentration × time) of five of ozone was sufficient to cause 90% inactivation of *Cryptosporidium*, whereas a $C \times T$ of 7,200 was required to reach the same inactivation level for chlorine and monochloramine (Table 2). A *CT* of up to 7,200 mg min/L of free chlorine was required to obtain an inactivation of 2 log of *Cryptosporidium* (Table 2). Inactivation of *Cryptosporidium* oocysts by monochloramine was inefficient also, and a *CT* of 80,000 mg min/L of monochloramine was required for the inactivation of 2 log [22]. Chlorine dioxide was found more efficient than free chlorine and chloramine for the inactivation of *Cryptosporidium*, where a *CT* of 111 mg min/L resulted in 2-log inactivation as measured by mouse infectivity [23]. However, Chauret et al. reported a *CT* of 1,000 mg min/L to obtain 2-log inactivation of *Cryptosporidium* using cell culture to determine infectivity [24]. Ozone was found the most efficient disinfectant compared to monochloramine, free chlorine, and chlorine dioxide, and a *CT* of 5 mg min/L resulted in 2-log inactivation of *Cryptosporidium*. It is worth noting that an agreement was observed between the inactivation results measured by mouse infectivity and excystation (Table 2).

4 Sequential Inactivation of *Cryptosporidium* by Ozone as Primary and by Chlorine or Monochloramine as Secondary Disinfectants

Since the *CT* values required for the inactivation of *Cryptosporidium* are large, free chlorine may not be suitable for disinfecting drinking water under most treatment sittings. Therefore, sequential inactivation using two disinfectants was proposed, in an attempt to improve pathogens inactivation efficiency in drinking water treatment plants [26–28]. Sequential disinfection using ozone as a primary disinfectant and either free chlorine or monochloramine as secondary disinfectants have been studied (Table 3). Corona-Vasquez et al. reported that ozone pretreatment results

Table 2 Inactivation of *Cryptosporidium* by chlorine, monochloramine, chlorine dioxide, and ozone

Matrix	Disinfectant	CT (mg min/L)	Method of detection	Log/% inactivation	Reference
DF[a] 0.01 M PBS[b]	Free chlorine	7,200	Mouse infectivity	2.0	[20]
ODF[c] PBS	Free chlorine	1,200	Mouse infectivity	1.5	[21]
DF water, pH 6, 22°C	Free chlorine	3,200	Mouse infectivity	1.0	[22]
DF 0.01 M PBS	Monochloramine	7,200	Mouse infectivity	1.0	[20]
DF water	Monochloramine	80,000	Mouse infectivity	2.0	[23]
DF water	Monochloramine	11,400	Excystation	2.0	[25]
DF 0.01 M PBS	Chlorine dioxide	78	Mouse infectivity	90%	[20]
DF water	Chlorine dioxide	111	Mouse infectivity	2.0	[22]
Deionized water	Chlorine dioxide	1,000	Cell culture	99%	[24]
Ozone DF water	Ozone	5	Mouse infectivity	99%	[20]
DF 0.01 M PBS	Ozone	5	Mouse infectivity	1.5	[22]
DF 0.01 M PBS	Ozone	5	Excystation	2.0	[26]

[a]*DF* demand free
[b]*PBS* phosphate-buffered saline
[c]*ODF* oxidant demand free

in the disappearance of the lag phase often observed when free chlorine is used to inactivate *Cryptosporidium* [26]. Although temperature dependent, after pretreatment with 1.8 mg min/L ozone, only a CT of 40 mg min/L of free chlorine was required to inactivate 2 log of *Cryptosporidium* at 20°C. Similar results were observed for the sequential inactivation of *Cryptosporidium* by chloramine. After pretreatment of ozone at CT of 22.5 mg min/L, a CT of 900 of chloramine was required to inactivate 2 log of *Cryptosporidium* at 20°C. Sequential inactivation using ozone and either free chlorine or monochloramine was found to be temperature dependent, increasing temperature, in the range 1–25°C, resulting in decreased inactivation efficiency of *Cryptosporidium*. Li et al. determined the sequential inactivation of ozone as primary disinfectant and free chlorine as secondary disinfectant by animal infectivity using neonatal CD-1 mice [22] (Table 3). The same authors reported also that temperature is critical for both single and sequential inactivation. After a primary kill of 1.6 log by ozone, the free chlorine ($C_{avg}\ T$ products) for inactivation of 3.0 log units were 1,000, 2,000, and 3,300 mg min/L for 22°C, 10°C, and 1°C, respectively [22].

Table 3 Sequential inactivation of *Cryptosporidium* by ozone as primary and chlorine or monochloramine as secondary disinfectants

Matrix	Disinfectant (CT = mg min/L)		Detection method	Log/% inactivation	Reference
	Primary	Secondary			
0.01 M PBS, pH 7.0, 5°C	Ozone (28)	F. chlorine[a] (20)	Excystation	1.0	[25]
0.01 M PBS, pH 7.0, 20°C	Ozone (1.8)	F. chlorine (40)	Excystation	2.0	[25]
0.01 M PBS, pH 8.0, 20°C	Ozone (22.5)	MCA[b] (900)	Excystation	2.0	[30]
0.01 M PBS, pH 8.0, 5°C	Ozone (11.5)	MCA (1400)	Excystation	2.0	[30]
0.01 M PBS, pH 6.0, 20°C	Ozone (1.4)	F. chlorine (600)	Excystation	2.0	[28]
0.01 M PBS, pH 8.0, 20°C	Ozone (1.4)	MCA (2000)	Excystation	2.0	[28]
0.05 M PB, pH 6, 22°C	Ozone (5)	F. chlorine (1000)	Mouse infectivity	4.6	[22]

[a]*F. chlorine* free chlorine
[b]*MCA* monochloramine

No synergy was observed in the inactivation rate of *Cryptosporidium* when sequential inactivation of chlorine dioxide and free chlorine or monochloramine was applied [29]. The authors explained that the absence of synergy for chlorine dioxide/free chlorine and chlorine dioxide/monochloramine could be the result of chlorine dioxide reacting with oocyst chemical groups that are mostly different from those reacting with ozone, free chlorine, and monochloramine. Driedger et al. proposed that secondary inactivation step may be the result of permeation within the oocyst wall layers, whereas primary disinfection may be a permeation process within more reactive oocyst wall layers [27].

5 Photocatalytic and Advanced Oxidation Processes for the Inactivation of *Cryptosporidium*

Titanium dioxide (TiO_2) photocatalysis is an advanced oxidation process (AOP) which can be used for water treatment. When irradiated with supra band gap energy, electron-hole pairs are generated within the titanium dioxide particles. Redox reactions with water and dissolved oxygen at the particle solution interface can result in the formation of reactive oxygen species (ROS). ROS formed by photocatalysis include the hydroxyl radical, a powerful indiscriminate oxidant, and superoxide anions, formed by reduction of dissolved oxygen [30, 31]. Photocatalysis can degrade a wide range of organic compounds and inactivate pathogens in water [32]. The ROS generated at the TiO_2 surface have been proposed to be the responsible agent for the disruption of cell membranes leading to loss of respiration and microbial inactivation [33]. DNA damage has also been reported following exposure of cells to superoxide anions and/or hydroxyl radicals [34].

Holes in the valence band are strongly oxidizing agents able to oxidize water to highly aggressive hydroxyl radicals as well as superoxide, and other active oxygen species are responsible for disinfection in water by attacking cellular membranes as well as cytoplasmic proteins. When UV energy is combined with ozone, it creates a chemical reaction resulting in hydroxyl radicals, short lived, but very powerful oxidizer that can kill pathogens as well as destroy inorganic contaminants in the water. The more ozone molecules in the system, the more hydroxyl radicals are created [35].

King et al. used cell culture infectivity to measure solar inactivation of *Cryptosporidium* [36]. A reduction of 90% of *Cryptosporidium* infectivity was recorded within 1 h on a high UV index summer day, whereas up to 6.4 h were needed to obtain 90% inactivation on an overcast day of low UV index rating. *Cryptosporidium* inactivation was attributed to the UV-B and UV-A fractions of the solar radiation. These results are in agreement with those reported by Connelly et al., who also used cell culture infectivity to evaluate the inactivation of *Cryptosporidium* by solar radiation [37]. These authors reported that exposures of *Cryptosporidium* to 32 and 66 kJ/m^2 artificial UV-B significantly decreased oocyst infectivity by an average of 58 and 98%, respectively. Exposure of oocysts to approximately half and full intensity of solar spectrum (all wavelengths) for a period of less than 1 day (10 h) in midsummer reduced mean infectivity by an average of 67% and of >99.99%, respectively [37]. McGuigan et al. observed that after 6 h of solar disinfection system (SODIS) exposure 60% of infective *Cryptosporidium parvum* were permeable to DAPI and from 10 h, all oocysts were PI positive [38] (Table 4). Ten hours of SODIS exposure was also sufficient to fully inactivate *Cryptosporidium* oocysts as measured by Swiss CD-1 mice. Inactivation of *Cryptosporidium* by sunlight may be the result of optical and thermal process, and a strong synergistic effect occurs at temperatures exceeding 45°C [38, 39]. Sunlight may be absorbed by photosensitizers present in water which reacts with oxygen and then produce highly reactive superoxide and hydrogen peroxide.

To enhance the inactivation efficiency of waterborne pathogenic microorganisms by SODIS, semiconductor photocatalysis is used to produce highly reactive species [40]. As indicated above, ROS formed by photocatalysis include hydroxyl radical and superoxide anions [41]. The generation of OH radicals by photo/ferrioxalate and photo/TiO$_2$ systems was evaluated for the inactivation of *C. parvum* [42] (Table 4). In vitro excystation of *Cryptosporidium* was used to evaluate the inactivation efficiency. The study revealed that CT values of OH radicals to achieve 2-log *C. parvum* inactivation were 9.3×10^{-5} and 7.9×10^{-5} mg min/L from photo/ferrioxalate and photo/TiO$_2$ systems, respectively [42] (Table 4). OH radical is one of the most effective oxidants (oxidation potential: OH radical 2.7 V, ozone 2.07 V, chlorine dioxide 1.91 V, and free chlorine) 1.36 V [32]. Sunnotel et al. demonstrated a reduction of 78.4 and 73.7% after 180 min exposure to photocatalytic activity of using TiO$_2$ films [30]. These authors assessed the oocysts viability by dye exclusion,

Table 4 Photocatalytic and AOP inactivation of *Cryptosporidium*

Treatment	*Cryptosporidium* detection method	Inactivation efficiency (log/%)	Reference
1 h high UV index summer day	Cell culture	90%	[36]
6.4 h low UV index rating	Cell culture	90%	[36]
SODIS[a], 10 h	Mouse infectivity	100%	[38]
SODIS[a], 6 h	DAPI	60%	[38]
Photo/TiO_2, $CT = 7.9 \times 10^{-5}$	In vitro excystation	2 log	[42]
Photo/ferrioxalate, $CT = 9.3 \times 10^{-5}$	In vitro excystation	2 log	[42]
Photocatalysis with TiO_2 film	Dye exclusion and in vitro excystation	78.4%	[30]
10% H_2O_2, 2 h	Cell culture	99%	[44]
TiO_2 and UV 4 mJ/cm^2	CC-PCR	2 log	[44]
TiO_2 and UV 11 mJ/cm^2	CC-PCR	3 log	[44]
SODIS and 8 mg/L FAC	Cell culture	>2 log	[46]

[a]*SODIS* solar disinfection system

excystation, and a gene expression assay based on lactate dehydrogenase 1 (LDH1) expression level. Scanning electron microscopy analysis showed a large number of empty oocysts following treatment confirming rupture of the oocyst wall suture line [30]. The ROS generated at the TiO_2 surface have been proposed to be the responsible agents for the disruption of the cell membrane leading to loss of respiration and microbial inactivation [33].

Although the evaluation of *Cryptosporidium* viability was performed by the inclusion/exclusion of the fluorogenic vital dye propidium iodide (PI), it has been demonstrated that the effectiveness of solar disinfection against *C. parvum* was enhanced when TiO_2 was incorporated. Incorporation of TiO_2 resulted in decreasing the time needed for *Cryptosporidium* inactivation by solar radiation [43].

Reduction of dissolved oxygen generates additional reactive oxygen species including superoxide radical anion ($O_2^{\cdot -}$), hydroperoxyl radical ($HO_2^{\cdot -}$), and hydrogen peroxide (H_2O_2). These species can feed into the radical attack mechanisms responsible for the oxidation of organic matter and inactivation of pathogenic microorganisms present in the water. Delling et al. reported that exposure of *Cryptosporidium* for 2 h to 10% H_2O_2 resulted in 99% inactivation as measured by CC-PCR [44] (Table 4).

Ryu et al. studied the inactivation efficiency of *Cryptosporidium* by the synergistic effect of UV irradiation and TiO_2 [45]. Whereas low inactivation (<0.28 log) was recorded by long-wave UV radiation, the synergistic effect of germicidal UV and TiO_2 resulted in 2-log and 3-log oocysts inactivation with 4.0 and 11.0 mJ/cm^2, respectively.

Zhou et al. demonstrated that solar radiation of water containing free available chlorine (FAC) dramatically enhances the inactivation of *C. parvum* compared to solar radiation and FAC alone [46]. Exposure to simulated sunlight for 60 min in the presence of 8 mg/L free chlorine resulted in photo decomposition of free chlorine to 1 mg/L and inactivation of >2 log of *Cryptosporidium*. The enhanced *Cryptosporidium* inactivation was due to in situ ROS and ozone production via FAC photolysis.

Kniel et al. recorded 50% inhibition of protease activity by treatment of *C. parvum* with H_2O_2 [47]. These authors proposed several mechanisms for the inhibition of infectivity and excystation of *Cryptosporidium* oocysts by H_2O_2. These mechanisms include oxidation of cell wall proteins or lipids, chelation of cations necessary for infection or hydroxyl radical-induced DNA damage in sporozoites, or both.

Abeledo-Lameiro et al. demonstrated enhancement in the effectiveness of solar disinfection procedures against *C. parvum*, when TiO_2 or TiO_2/H_2O_2 were incorporated, decreasing the time needed for oocysts inactivation and resulting in 97.7% and 99.1% inactivation, respectively [48].

Cho and Yoon used *Bacillus subtilis* spores as surrogate for *Cryptosporidium* inactivation by O_3/H_2O_2 followed by Cl_2 [49]. The authors demonstrated a 2.7 log inactivation in system of O_3/H_2O_2 followed by Cl_2 at 20°C which was induced by OH radical enhancement and synergistic effect of sequential disinfection. The presence of OH radical achieved 1-log inactivation. In addition, the sequential treatment with O_3/H_2O_2 followed by Cl_2 achieved an additional 1.4 log inactivation with the extra synergistic effect.

Barbee et al. reported that exposure of *Cryptosporidium* to 3% H_2O_2 for 20 min at 20°C, resulted in 1-log inactivation, whereas exposure to 6% resulted in over 3-log inactivation [50].

Rodríguez-Chueca et al. demonstrated high inactivation efficiency of *Escherichia coli* by peroxone system (O_3/H_2O_2) and catalytic ozonation (O_3/TiO_2) [51]. Inactivation of 6.80 log after 10 min of contact time was observed for the peroxone system and 6.22 log for the catalytic ozonation.

6 Conclusions

Disinfection is an important barrier to prevent the infection of *Cryptosporidium* by water consumption. Since *Cryptosporidium* oocysts are highly resistant to chlorine, alternative disinfection methods are applied to efficiently inactivate *Cryptosporidium*. Advanced oxidation processes, especially reactive oxygen species and hydroxyl radicals, have been shown to be efficient for the inactivation of *Cryptosporidium* oocysts and other waterborne pathogens.

References

1. D'Antonio RG, Winn RE, Taylor JP, Gustafson TL, Current WL, Rhodes MM, Gary GW Jr, Zajac RA (1985) A waterborne outbreak of cryptosporidiosis in normal hosts. Ann Intern Med 103(6_Part_1):886–888
2. MacKenzie W, Hoxie N, Proctor M, Gradus M, Blari K, Peterson D, Kazmierczak J, Davis J (1994) A massive outbreak in Milwaukee of *Cryptosporidium* infection transmitted through the public water supply. N Engl J Med 331(3):161–167
3. DuPont HL, Chappell CL, Sterling CR, Okhuysen PC, Rose JB, Jakubowski W (1995) The infectivity of *Cryptosporidium parvum* in healthy volunteers. N Engl J Med 332:855–859
4. Baldursson S, Karanis P (2011) Waterborne transmission of protozoan parasites: review of worldwide outbreaks – an update 2004-2010. Water Res 45(20):6603–6614
5. Rose JB, Huffman DE, Gennaccaro A (2002) Risk and control of waterborne cryptosporidiosis. FEMS Microbiol Rev 26(2):113–123
6. Long SM, Adak GK, O'Brien SJ, Gillespie IA (2002) General outbreaks of infectious intestinal disease linked with salad vegetables and fruit, England and Wales, 1992-2000. Commun Dis Public Health 5(2):101–105
7. Craun GF, Hubbs SA, Frost F, Calderon RL, Via SH (1998) Waterborne outbreaks of cryptosporidiosis. J Am Water Works Assoc 90(9):81–91
8. Benshoshan M, Vaizel-Ohayon D, Aharoni A, Nitzan Y, Rebhun M, Nasser AM (2015) Prevalence and fate of *Cryptosporidium* genotypes in two wastewater treatment plants. J Environ Sci Health A 50(12):1265–1273
9. Nasser AM (2016) Removal of *Cryptosporidium* by wastewater treatment processes, a review. J Water Health 14(1):1–13
10. Robertson LJ, Campbell AT, Smith HV (1992) Survival of *Cryptosporidium parvum* oocysts under various environmental pressures. Appl Environ Microbiol 58(11):3494–3500
11. Nasser AM, Tweto E, Nitzan Y (2007) Die-off of *C. parvum* in soil and wastewater effluents. J Appl Microbiol 102:169–176
12. Fayer R, Trout JM, Jenkins MC (1998) Infectivity of *Cryptosporidium parvum* oocysts stored in water at environmental temperatures. J Parasitol 84:1165–1169
13. Campbell AT, Robertson LJ, Smith HV (1992) Viability of *Cryptosporidium parvum* oocysts: correlation of in vitro excystation with inclusion or exclusion of fluorogenic vital dyes. Appl Environ Microbiol 58:3488–3493
14. U.S. Environmental Protection Agency Method 1623.1: *Cryptosporidium* and *Giardia* in water by IFA/IMS/FA (2012) EPA 821-R-01-025. Office of Water, U.S. Environmental Protection Agency, Washington
15. Tzipori S, Angus KW, Campbell I, Gray EW (1982) Experimental infection of lambs with *Cryptosporidium* isolated from a human patient with diarrhea. Gut 23(1):71–74
16. Slifko TR, Friedman D, Rose JB, Jakubowski W (1997) An in vitro method for detecting infectious *Cryptosporidium* oocysts with cell culture. Appl Environ Microbiol 63(9):3669–3675
17. Rochelle PA, Marshall MM, Mead JR, Johnson AM, Korich DG, Rosen JS, De Leon R (2002) Comparison of in vitro cell culture and a mouse assay for measuring infectivity of *Cryptosporidium parvum*. Appl Environ Microbiol 68(8):3809–3817
18. Rochelle PA, Ferguson DM, Handojo TJ, De Leon R, Stewart MH, Wolfe RL (1997) An assay combining cell culture with reverse transcriptase PCR to detect and determine the infectivity of waterborne *Cryptosporidium parvum*. Appl Environ Microbiol 63:2029–2037
19. Di Giovanni GD, Hashemi FH, Shaw NJ, Abrams FA, LeChevallier MW, Abbaszadegan M (1999) Detection of infectious *Cryptosporidium parvum* oocysts in surface and filter backwash water samples by immunomagnetic separation and integrated cell culture-PCR. Appl Environ Microbiol 65(8):3427–3432
20. Korich DG, Mead JR, Madore MS, Sinclair NA, Sterling CR (1990) Effects of ozone, chlorine dioxide, chlorine, and monochloramine on *Cryptosporidium parvum* oocyst viability. Appl Environ Microbiol 56(5):1423–1428

21. Venczel LV, Arrowood M, Hurd M, Sobsey MD (1997) Inactivation of *Cryptosporidium parvum* oocysts and *Clostridium perfringens* spores by a mixed-oxidant disinfectant and by free chlorine. Appl Environ Microbiol 63(4):1598–1601
22. Li H, Finch GR, Smith DW, Belosevic M (2001) Sequential inactivation of *Cryptosporidium parvum* using ozone and chlorine. Water Res 35(18):4339–4348
23. Gyurek LL, Finch GR, Belosevic M (1997) Modeling chlorine inactivation of *Cryptosporidium parvum* oocysts. J Environ Eng 9:865–875
24. Chauret CP, Radziminski CZ, Lepuil M, Creason R, Andrews RC (2001) Chlorine dioxide inactivation of *Cryptosporidium parvum* oocysts and bacterial spore indicators. Appl Environ Microbiol 67(7):2993–3001
25. Driedger AM, Rennecker JL, Mariñas BJ (2001) Inactivation of *Cryptosporidium parvum* oocysts with ozone and monochloramine at low temperature. Water Res 35(1):41–48
26. Corona-Vasquez B, Samuelson A, Rennecker JL, Mariñas BJ (2002) Inactivation of *Cryptosporidium parvum* oocysts with ozone and free chlorine. Water Res 36(16):4053–4063
27. Driedger AM, Rennecker JL, Mariñas BJ (2000) Sequential inactivation of *Cryptosporidium parvum* oocysts with ozone and free chlorine. Water Res 34(14):3591–3597
28. Rennecker JL, Driedger AM, Rubin SA, Mariñas BJ (2000) Synergy in sequential inactivation of *Cryptosporidium parvum* with ozone/free chlorine and zone/monochloramine. Water Res 34(17):4121–4130
29. Corona-Vasquez B, Rennecker JL, Driedger AM, Mariñas BJ (2002) Sequential inactivation of *Cryptosporidium parvum* oocysts with chlorine followed by free chlorine or monochloramine. Water Res 36:178–188
30. Sunnotel O, Verdoold R, Dunlop PS, Snelling WJ, Lowery CJ, Dooley JS, Moore JE, Byrne JA (2010) Photocatalytic inactivation of *Cryptosporidium parvum* on nanostructured titanium dioxide films. J Water Health 8(1):83–91
31. Cho M, Lee Y, Chung H, Yoon J (2004) Inactivation of *Escherichia coli* by photochemical reaction of ferrioxalate at slightly acidic and near-neutral pHs. Appl Environ Microbiol 70 (2):1129–1134
32. Egerton TA, Kosa SA, Christensen PA (2006) Photoelectrocatalytic disinfection of *E. coli* suspensions by iron doped TiO_2. Phys Chem Chem Phys 21(3):398–406
33. Maness P-C, Smolinski S, Blake DM, Huang Z, Wolfrum EJ, Jacoby WA (1999) Bactericidal activity of photocatalytic TiO_2 reaction: toward an understanding of its killing mechanism. Appl Environ Microbiol 65:4094–4098
34. Su M, Yang Y, Yang G (2006) Quantitative measurement of hydroxyl radical induced DNA double-strand breaks and the effect of N-acetyl-L-cysteine. FEBS Lett 580(17):4136–4142
35. Rincón A-G, Pulgarin C (2004) Field solar *E. coli* inactivation in the absence and presence of TiO_2: is UV solar dose an appropriate parameter for standardization of water solar disinfection? Solar Energy 77(5):635–648
36. King BJ, Hoefel D, Daminato DP, Fanok S, Monis PT (2008) Solar UV reduces *Cryptosporidium parvum* oocyst infectivity in environmental waters. J Appl Microbiol 104(5):1311–1323
37. Connelly SJ, Wolyniak EA, Williamson CE, Jellison KL (2007) Artificial UV-B and solar radiation reduce in vitro infectivity of the human pathogen *Cryptosporidium parvum*. Environ Sci Technol 41(20):7101–7106
38. McGuigan KG, Méndez-Hermida F, Castro-Hermida JA, Ares-Mazás E, Kehoe SC, Boyle M, Sichel C, Fernández-Ibáñez P, Meyer BP, Ramalingham S, Meyer EA (2006) Batch solar disinfection inactivates oocysts of *Cryptosporidium parvum* and cysts of *Giardia muris* in drinking water. J Appl Microbiol 101(2):453–463
39. Theitler DJ, Nasser A, Gerchman Y, Kribus A, Mamane H (2012) Synergistic effect of heat and solar UV on DNA damage and water disinfection of *E. coli* and bacteriophage MS2. J Water Health 10(4):605–618
40. Byrne JA, Fernandez-Ibañez PA, Dunlop PSM, Alrousan DMA, Hamilton JWJ (2011) Photocatalytic enhancement for solar disinfection of water: a review. Int J Photoenergy 2011:12. Article ID: 798051

41. Ireland JC, Klostermann P, Rice EW, Clark RM (1993) Inactivation of *Escherichia coli* by titanium dioxide photocatalytic oxidation. Appl Environ Microbiol 59(5):1668–1670
42. Cho M, Yoon J (2008) Measurement of OH radical CT for inactivating *Cryptosporidium parvum* using photo/ferrioxalate and photo/TiO_2 systems. J Appl Microbiol 104:759–766
43. Abeledo-Lameiro MJ, Ares-Mazás E, Gómez-Couso H (2016) Evaluation of solar photocatalysis using TiO_2 slurry in the inactivation of *Cryptosporidium parvum* oocysts in water. J Photochem Photobiol B 163:92–99
44. Delling C, Holzhausen I, Daugschies A, Lendner M (2016) Inactivation of *Cryptosporidium parvum* under laboratory conditions. Parasitol Res 115(2):863–866
45. Ryu H, Gerrity D, Crittenden JC, Abbaszadegan M (2008) Photocatalytic inactivation of *Cryptosporidium parvum* with TiO_2 and low-pressure ultraviolet irradiation. Water Res 42 (6–7):1523–1530
46. Zhou P, Di Giovanni GD, Meschke JS, Dodd MC (2014) Enhanced inactivation of *Cryptosporidium parvum* oocysts during solar photolysis of free available chlorine. Environ Sci Technol Lett 1(11):453–458
47. Kniel KE, Sumner SS, Pierson MD, Zajac AM, Hackney C, Fayer RR, Lindsay DS (2004) Effect of hydrogen peroxide and other protease inhibitors on *Cryptosporidium parvum* excystation and in vitro development. J Parasitol 90(4):885–888
48. Abeledo-Lameiro MJ, Reboredo-Fernández A, Polo-López MI, Fernández-Ibáñez P, Ares-Mazás E, Gómez-Couso H (2017) Photocatalytic inactivation of the waterborne protozoan parasite *Cryptosporidium parvum* using TiO_2/H_2O_2 under simulated and natural solar conditions. Catal Today 280:132–138
49. Cho M, Yoon Y (2006) Enhanced bactericidal effect of O_3/H_2O_2 followed by Cl_2. J Ozone Sci Eng 28(5):335–340
50. Barbee SL, Weber DJ, Sobsey MD, Rutala WA (1999) Inactivation of *Cryptosporidium parvum* oocyst infectivity by disinfection and sterilization processes. Gastrointest Endosc 49 (5):605–611
51. Rodríguez-Chueca J, Ormad Melero MP, Mosteo Abad R, Esteban Finol J, Ovelleiro Narvión JL (2015) Inactivation of *Escherichia coli* in fresh water with advanced oxidation processes based on the combination of O_3, H_2O_2 and TiO_2. Kinetic modeling. Environ Sci Pollut Res Int 22(13):10280–10290

Cost-Effective Catalytic Materials for AOP Treatment Units

Shahryar Jafarinejad

Abstract Catalysts (homogeneous or heterogeneous) can be utilized to improve the performance of conventional advanced oxidation processes (AOPs). In general, catalyst activity, selectivity, stability, simplicity of preparation, preparation time, cost, nontoxicity, availability, recycling capability, environmental suitability, etc. can be the important parameters in the catalyst selection. High costs, cumbersome preparations, and environmental unsuitability can usually hinder the industrial applicability of a catalyst. In this chapter, catalytic AOPs (Fenton-based processes, catalytic ozonation, heterogeneous photocatalysis, catalytic wet air oxidation, and catalytic supercritical water oxidation), related catalytic materials, and cost-effective catalytic materials used in these processes are discussed.

Keywords Advanced oxidation processes (AOPs), Catalyst, Cost, Treatment

Contents

1 Introduction 310
2 Overview of AOPs 310
3 Catalytic AOPs and Related Catalytic Materials 311
    3.1 Fenton-Based Processes 313
    3.2 Catalytic Ozonation 318
    3.3 Heterogeneous Photocatalysis 322
    3.4 Catalytic Wet Air Oxidation (CWAO) 325
    3.5 Catalytic Supercritical Water Oxidation (CSCWO) 328
4 Conclusions 330
References 330

S. Jafarinejad (✉)
Chemical Engineering Division, College of Environment, UoE, Karaj, Iran
e-mail: jafarinejad83@gmail.com

1 Introduction

Generally, various techniques have been applied for treatment of industrial and municipal wastewaters, purification of drinking water, ultrapurification of water for particular applications, etc. In some cases, conventional methods are not suitable to attain the degree of purity needed by international or local regulations or by the subsequent utilization of the effluent. In these situations, advanced oxidation processes (AOPs) can be useful as novel techniques for water and wastewater treatment that allow the total or partial degradation of compounds resistant to conventional treatment methods [1]. In real, AOPs can cause a total mineralization, converting recalcitrant compounds into inorganic substances (CO_2 and H_2O), or partial mineralization, converting them into more biodegradable substances [2]. In other words, they not only destruct pollutants but also prevent consequent generation of toxic residues, whereas conventional methods (i.e., nondestructive physical separation methods) only eliminate the contaminants, transferring them to other phases, thereby producing concentrated residues [3, 4].

AOPs have appeared as powerful processes and are extensively applied in drinking water purification for the destruction of all kinds of constituents, including endocrine disrupting chemicals [5, 6]. Also, they are largely utilized for the elimination of recalcitrant organic compounds from industrial and domestic wastewaters [7].

The existence of other water and wastewater contaminants such as natural organic matter, dissolved and suspended solids, alkalinity, pH, and temperature can affect the performance of AOPs [6, 8]. The operational cost of many AOPs can be relatively high [9]. Therefore, the costs of materials and equipment, energy requirements, and efficiency should be considered when evaluating the overall performance of AOPs [6].

In this chapter, overview of AOPs, catalytic AOPs (Fenton-based processes, catalytic ozonation, heterogeneous photocatalysis, catalytic wet air oxidation, and catalytic supercritical water oxidation), related catalytic materials, and cost-effective catalytic materials used in these processes are discussed.

2 Overview of AOPs

AOPs were first proposed for drinking water treatment by Glaze [10] in 1987 [11], which are referred to clean technologies based on physicochemical processes to generate the extremely reactive and nonselective hydroxyl radicals ($HO^{\bullet}$), with very high oxidative power ($E_0 = 2.8$ V) for toxic organic compounds degradation in a medium [1, 3, 4, 12–14]. AOPs may be applied for various purposes [1, 3, 8–22].

According to Litter and Quici [1], Mota et al. [3], and Jafarinejad [4], the electrophilic addition of a hydroxyl radical ($HO^{\bullet}$) to π systems (unsaturated or aromatic) leading to the formation of organic radicals (Eq. 1), the hydrogen abstraction by reacting the hydroxyl radical (HO˙) with a saturated aliphatic compound (Eq. 2), and electron transfer reactions (Eq. 3) may be the possible reaction pathway in the AOPs [1, 3, 4]. It is necessary to note that some chemical species in water such as carbonate and bicarbonate ions may react with the hydroxyl radicals, hence competing with the organic compounds through the hydroxyl radicals [3, 4].

$$HO^{\bullet} + \text{Unsaturated or Aromatic} \rightarrow \text{Unsaturated} - OH \text{ or Aromatic} - OH \quad (1)$$

$$HO^{\bullet} + R - H \rightarrow R^{\bullet} + H_2O \quad (2)$$

$$HO^{\bullet} + R - X \rightarrow [R - X]^{+\bullet} + HO^{-} \quad (3)$$

Depending on the number of phases involved, the AOPs can be divided into homogeneous and heterogeneous processes [3, 4, 23, 24]. They can be also categorized based on light usage in them as non-photochemical and photochemical processes [3, 4, 24]. According to these classifications, the types of AOPs are listed in Table 1.

3 Catalytic AOPs and Related Catalytic Materials

A catalyst is a material which can be added in the balance of a chemical reaction to speed up the attainment of the chemical equilibrium between reactants and products without affecting the thermodynamic equilibrium of the process. In most cases, the rate of chemical reactions can be increased by lowering their activation energy with a catalyst. Generally, consumption of catalysts does not occur during the reaction, and they can be found unchanged after completion of the reaction (deactivation of the catalyst may occur). In homogeneous catalysis, the catalyst, reactants, and products are in the same phase, usually the liquid phase. If the catalyst and reactants are in different phases, usually the catalyst is a solid and the reactants and the products are in the liquid or vapor phase; this catalyst will be called heterogeneous [34].

Nowadays, due to the environmental issues, catalysis seems to be even more important than before and to constitute one of the main sources of developments in our society [35–37]. Improvements in the catalytic activity and selectivity are essential to make catalytic processes more efficient [37].

Due to the very short lifetime of hydroxyl radicals, they are only in situ generated during AOPs, including a combination of oxidizing agents, irradiation, and/or catalysts [14, 23]. Large consumption of chemicals, acidic pH, high cost of oxidizing agents, generation of sludge (e.g., iron sludge), and necessity of posttreatment can be some limitations of the homogeneous AOPs. Thus, heterogeneous catalysts should be used to improve the performance of conventional AOPs [38]. In real, catalysts can be applied in combination with some of AOPs to enhance the oxidation process [2, 39]. Activity, selectivity, stability [40], recycling capability [11], cost, etc. of catalysts are important. In addition, the use of suitable catalysts

Table 1 Various advanced oxidation processes (AOPs) [1, 3, 4, 22, 25–33]

Non-photochemical processes	Photochemical processes
Homogeneous processes	
Ozonation in alkaline media (O_3/HO^-)	Photolysis of water in vacuum ultraviolet (VUV)
Ozonation with hydrogen peroxide (O_3/H_2O_2)	US/UV
Fenton (Fe^{2+} or Fe^{3+}/H_2O_2)	UV/H_2O_2
Electro-Fenton	UV/O_3
Electrochemical oxidation	UV/O_3/H_2O_2
Electrohydraulic discharge – ultrasound (US)	Photo-Fenton (Fe^{2+} or Fe^{3+}/H_2O_2/UV)
US/Fenton	US/photo-Fenton
O_3/US	
H_2O_2/US	
Microwave (MW)/H_2O_2	
Gamma-radiolysis and electron beam treatment	
Nonthermal plasma (surface corona discharge)	
Ferrate(VI) oxidation	
Wet air oxidation (WAO)	
Supercritical water oxidation (SCWO)	
Heterogeneous processes	
Catalytic wet air oxidation (CWAO)	Heterogeneous photocatalysis: ZnO/UV, SnO_2/UV, TiO_2/UV, TiO_2/H_2O_2/UV
Zerovalent iron (ZVI) and related processes	ZVI/UV
Catalytic ozonation	Photocatalytic ozonation (O_3/TiO_2/UV)
Heterogeneous Fenton	Heterogeneous photo-Fenton
Catalytic supercritical water oxidation (CSCWO)	
US/H_2O_2/catalyst	
O_3/TiO_2/electron beam irradiation	

for AOPs not only decreases the severity of reaction conditions but also more easily destructs even the most refractory contaminants, thereby diminishing capital and operational cost [21, 41]. In this section, catalytic AOPs (Fenton-based processes, catalytic ozonation, heterogeneous photocatalysis, catalytic wet air oxidation, and catalytic supercritical water oxidation), related catalytic materials, and cost-effective catalytic materials used in these processes are discussed.

3.1 Fenton-Based Processes

3.1.1 Principles

The Fenton process is one of the cost-effective and environmentally friendly oxidation methods, which was discovered by Henry J. Fenton in 1894 when he strongly improved tartaric acid oxidation by the use of ferrous ion (Fe^{2+}) and H_2O_2 [4, 12, 42, 43]. This process was not utilized for the degradation of organic pollutants until late 1960s [43, 44]; nowadays, it is widely used in wastewater treatment. In real, Fenton's reagent is a solution of H_2O_2 and ferrous ions and this process has a complex mechanism, which primarily can involve the following reactions [3, 4, 12–14, 45, 46]:

$$H_2O_2 + Fe^{2+} \rightarrow Fe^{3+} + OH^- + HO^{\bullet} \quad (4)$$

$$\text{Organic matter} + HO^{\bullet} \rightarrow \text{Oxidation intermediates} \quad (5)$$

$$\text{Oxidation intermediates} + HO^{\bullet} \rightarrow CO_2 + H_2O \quad (6)$$

$$H_2O_2 + Fe^{3+} \rightarrow Fe^{2+} + H^+ + HO_2^{\bullet} \quad (7)$$

$$H_2O_2 + HO^{\bullet} \rightarrow H_2O + HO_2^{\bullet} \quad (8)$$

$$Fe^{2+} + HO^{\bullet} \rightarrow Fe^{3+} + HO^- \quad (9)$$

$$Fe^{3+} + HO_2^{\bullet} \rightarrow Fe^{2+} + O_2H^+ \quad (10)$$

$$Fe^{2+} + HO_2^{\bullet} + H^+ \rightarrow Fe^{3+} + H_2O_2 \quad (11)$$

$$HO^{\bullet} + HO^{\bullet} \rightarrow H_2O_2 \quad (12)$$

$$2HO_2^{\bullet} \rightarrow H_2O_2 + O_2 \quad (13)$$

Briefly, the reaction between ferrous ions and H_2O_2 generates hydroxyl radicals with high oxidative power (Eq. 4), which attack the organic compounds present in the water (Eq. 5). Unfortunately, some parallel reactions take place, and so the hydroxyl radicals are not only consumed to degrade the organic matter but also to produce other radicals, with less oxidative power, or other species (scavenging effect of $HO^{\bullet}$). Besides, this leads to the undesired consumption of H_2O_2 (Eq. 8) [4, 12, 13]. Thus, a suitable molar ratio of iron ion to H_2O_2 requires to be experimentally determined for minimization of the unwanted scavenging [14]. According to Mota et al., although a [Fe^{2+}]/[H_2O_2] ratio of 1/2 has a higher destruction rate of organic matters, it is generally recommended to use the 1/5 ratio, which yields similar results and requires fewer reagents [3]. Besides, the ideal pH in Fenton's reaction is at an acidic pH condition (3) [3, 4]. Fe^{3+} can generate iron sludge at typical water and wastewater treatment situations [14]. In other words, the generation of a substantial amount of $Fe(OH)_3$, which precipitates and causes additional water pollution, by the homogeneous catalyst added as an iron salt that cannot be retained in the process, can restrict the application of this system [4, 14, 47].

On the other hand, Eqs. (14) and (15) demonstrate a production of Fe^{2+} by the reaction between H_2O_2 and Fe^{3+} (Fenton-like process); this way ferrous ion is restored, acting as catalyst in the overall process [4, 12, 13, 48].

$$Fe^{3+} + H_2O_2 \leftrightarrow \left[Fe^{III}(HO_2)\right]^{2+} + H^+ \quad (14)$$

$$\left[Fe^{III}(HO_2)\right]^{2+} \rightarrow Fe^{2+} + HO_2^{\bullet} \quad (15)$$

In real, according to Andreozzi et al., and Ribeiro et al., the reduction of Fe^{3+} to Fe^{2+} is often called "Fenton-like" process [48, 49]. This concept is also extensively used to refer to variants of the Fenton process. For example, in the $Fe°/H_2O_2/H_2SO_4$ Fenton-like process, zerovalent iron ($Fe°$) is applied directly, and the efficiency of the process can rely on the gradual production of Fe^{2+} from $Fe°$ (Eq. 16), followed by Fenton reaction (Eq. 4) [48, 50].

$$Fe° + H_2SO_4 \rightarrow Fe^{2+} + SO_4^{2-} + H_2 \quad (16)$$

The Fenton-based processes can be divided into homogeneous and heterogeneous [1, 38, 43, 48, 51–55]. Generally, there are many variations of the Fenton process [55] and Fenton-based processes may be classified as:

- Classical Fenton process [48]
- Photo-Fenton-like processes
- Electro-Fenton-like processes [48, 53]
- Cavitation-Fenton-like processes
- Microwave (another physical field)-Fenton-like processes [53]
- Fenton-like reactions using nano-zerovalent iron (nZVI) [52]

In the photo-Fenton-like systems, UV and/or visible light may be used to decrease the catalyst loading and/or to enhance the catalytic capacity of the catalyst in both heterogeneous and homogeneous Fenton-like processes with a major goal of increasing the UV-induced reduction of dissolved Fe^{3+} to Fe^{2+} [14, 48, 53]. In the electro-Fenton-like processes, either or both of the Fenton reagents may be produced by electrochemical methods [14]. An effective alternative for organic degradation may be provided by acoustic cavitation (produced by the passage of high-frequency sound waves, i.e., ultrasonic) and hydrodynamic cavitation (generated by the pressure variations in the fluid because of a sudden alteration in the linear flow rate of the fluid). The utilization of microwave irradiation in the Fenton-like processes has offered evident superiority in comparison with the Fenton-like process alone [53].

Zerovalent state metals (such as $Fe°$, $Zn°$, $Sn°$, and $Al°$) can be effective agents for the treatment of polluted water [53, 56, 57]. In heterogeneous Fenton reaction, an alternative mean of inducing Fenton oxidation can be supplied by oxidation of nZVI [53, 58, 59] as shown in Eq. (17) to overcome the disadvantages of homogeneous Fenton processes [53]:

$$\text{Fe}^\circ + \text{O}_2 + 2\text{H}^+ \to \text{Fe}^{2+} + \text{H}_2\text{O}_2 \tag{17}$$

Ferrous ion concentration, H_2O_2 concentration, operating pH [52, 53], catalyst type and dosage [53], initial concentration of pollutant, and operating temperature [52] are parameters which may affect the efficiency of the Fenton-based processes.

3.1.2 Catalytic Materials Used in Fenton-Like Processes

Despite its high performance and simplicity [53, 60, 61] and its low toxicity [53, 62] (H_2O_2 can decompose into environmentally friendly compounds such as water and oxygen), Fenton process has some drawbacks such as high operating cost, narrow optimum pH range, large volume of metal-containing sludge generation, difficulties in recycling of the homogeneous catalyst (Fe^{2+}) [53, 63], and the required iron ion concentration range of 50–80 ppm for batch processes which is above the standards (e.g., 2 ppm limit forced by the European Union (EU) directives for direct discharge of wastewater into the environment) [53, 64]. To resolve some of the above disadvantages, heterogeneous Fenton-like processes have been developed [65].

Replacing Fe^{2+} in the Fenton reagent with a solid catalyst can establish heterogeneous Fenton-like systems, while homogeneous Fenton-like systems are because of a combination of other metal ions/metal ion-organic ligand complexes and H_2O_2 [53, 66]. In the homogeneous process, the catalysis can take place in the whole liquid phase; meanwhile in the heterogeneous process, the catalysis always takes place on the surface of the catalyst [53, 67, 68]. The surface characteristics and the pore structure of the solid catalyst [33], stability, reusability, diffusion, and adsorption processes of H_2O_2 and other reactants to the surface of catalyst [53, 67, 68] are important in the solid catalyzed Fenton reactions.

The removal rate of the organics in the wastewater can sometimes be improved by enhancing the catalyst dosage. The excessive loading of catalyst in both hetero- and homogeneous Fenton-like systems may have a negative effect on the wastewater treatment due to the consumption of the generated hydroxyl radicals by the excess catalyst. In addition, it may increase the treatment cost. Therefore, optimization of the catalyst loading in both hetero- and homogeneous Fenton-like systems is essential [53].

In recent years, the application of solid catalysts in the so-called catalytic wet peroxide oxidation (CWPO) or heterogeneous Fenton oxidation has extensively been investigated [43]. Heterogeneous Fenton-like catalysts can be solids including cations of transition metals [65, 69, 70]. The catalytic activity can be directly influenced by metals and the selected support because they are important parameters in the stability of the supported metal and its dispersion [65, 71, 72]. Various solid catalysts such as metal oxide (CuO and Cu/Al_2O_3) [73], Fe-clay [74], metal oxide impregnated activated carbon (Fe_2MO_4, M:Fe and Mn) [75], iron molybdate $Fe_2(MoO_4)_3$ [76], CuFeZSM-5 zeolite [77], Fe exchanged/supported on zeolite [78–80], goethite [81], Fe°/Fe_3O_4 composites [82, 83], nanomagnetite [84], iron-containing SBA-15 [85–87], Al-Fe-pillared clays [88–90], Fe_3O_4 supported on γ-Al_2O_3 and activated carbon as well as bare carbon materials (graphite) [91, 92], and nZVI [93] manganese functionalized mesoporous molecular sieves Ti-HMS [20] have been tested in Fenton reactions.

Different supports such as nafion [94], zeolite [95–99], silica [100], clay [101], resins, and activated carbon [102, 103] have been applied to prepare heterogeneous Fenton catalysts [33]. In real, carbon materials such as activated carbon, MWCNTs, and reduced graphene oxide ((r)GO) and silicate materials such as mesoporous silica, quartz, zeolite, and saponite may be applied as support for efficient heterogeneous Fenton catalysts because of the good chemical stability, high mechanical strength, and low cost [54].

Munoz et al. reviewed magnetite-based catalysts and their application in heterogeneous Fenton oxidation [43]. In addition, Ezzatahmadi et al. [104] recently reviewed clay-supported nZVI composite materials for the remediation of contaminated aqueous solutions. Besides, Wang et al. [53] reviewed catalyst types for organic wastewater treatment in various Fenton-like processes.

Based on the reactive content and physicochemical properties, the iron-based heterogeneous Fenton catalysts may be divided into five primary categories: (1) zerovalent iron, (2) iron minerals and iron (hydr)oxides [magnetite (Fe_3O_4), goethite (α-FeOOH), etc.], (3) multimetallic iron-based materials, (4) supported iron-based materials (carbon-based materials as support, silicate-based materials as support, and organic compound as support), and (5) horseradish peroxidase (HRP)-based materials. Interfacial mechanisms of heterogeneous Fenton reactions catalyzed by iron-based materials have been reviewed by He et al. [54].

Some researches have been carried out to develop non-Fe catalysts (e.g. Mn-, Co-containing materials) as the substitute for Fe-based catalysts in Fenton reactions [105–108]. For example, according to Song et al. [20], Mn contained (Ti-)HMS catalyst may introduce a new type of Fenton-like catalyst, which can be environmentally friendly and cost-effective and has superior efficiency for the reactive hydroxyls generated in AOPs [20].

3.1.3 Cost-Effective Catalytic Materials Used in Fenton-Like Processes

Heterogeneous Fenton-like processes are cost-effective in overcoming the drawbacks of homogeneous Fenton process [38]. The development of efficient and low-cost catalytic materials to promote sufficient treatment in heterogeneous Fenton-like systems is essential [7, 109]. The wastewater treatment cost can be influenced by the source (nature, industrial waste, and artificial synthesis), preparation and modification method, activity, and stability of the heterogeneous catalysts. On the other hand, prevention of precipitation/flocculation of the homogeneous catalyst is critical due to the deactivation of the catalyst that can adversely affect the decreasing of the wastewater treatment cost [53].

High Fenton activity, low cost, negligible toxicity, and easy recovery are advantages of iron-based materials [54, 110, 111]. The use of natural minerals as catalysts in CWPO is interesting because of their wide availability and low cost [43]. Among iron-oxide minerals, magnetite is extensively used as catalyst in heterogeneous Fenton oxidation because of its relatively high abundance, low cost, and easy magnetic separation from the reaction medium [43, 110]. In addition, ZVI can be applied as low-cost catalytic material in Fenton reaction [112]. High production cost of ZVI is an important drawback which can be reduced by producing ZVI from waste iron oxide generated in steel works [113]. For example, Martins et al. [112] tested low-cost materials such as red volcanic rock, sepiolite, and iron shavings (coming from Fe processing industries) without modification as catalysts during the Fenton-like process on the treatment of simulated and actual olive mill wastewaters. They claimed that iron shavings can be promising low-cost catalytic materials applied in Fenton's reaction for the treatment of bio-refractory effluents in batch or continuous operations [112]. In other works, Pereira and Freire [114] used the recycled waste ZVI powder (iron particles discarded from a manufacturing process) as an environmentally friendly and low-cost material for azo dye degradation. Reduction in color and total organic carbon were reported to be 95% and 70%, respectively, under the optimum operational conditions (pH = 3, [Fe] = 5 g/L, and iron particle size $\leq$250 µm) [114].

Manu et al. applied a novel iron catalyst source laterite soil as a cost-effective material to degrade paracetamol in aqueous solutions by Fenton oxidation process. At [H_2O_2]:[Laterite iron] = 40:1, pH of 3, 10 mg/L initial concentration of paracetamol, and a reaction time of 60 min, the paracetamol and chemical oxygen demand (COD) removals were reported to be 75 and 63, respectively. In addition, paracetamol degradation of 100% was reported with HPLC analysis in 240 min [115].

Due to the wide availability of clay minerals in nature, they can be considered as low-cost alternatives to zeolites in many applications [116]. In recent years, the pillared clays have been widely applied as catalysts for Fenton-like advanced oxidation, specifically wet hydrogen peroxide catalytic oxidation (WHPCO) of organic contaminants due to their high activity, reusability, easy recovery, environmental compatibility, low cost, high stability of the active phase in the reaction medium, and conducting oxidation at room temperature and atmospheric pressure. According to Sanabria et al. [65], the combination of the biological and catalytic oxidation with AlFe-pillared interlayered clays (PILCs) could attain a 96.7% reduction of COD in coffee wet processing wastewater.

Song et al. [20] claimed that manganese functionalized mesoporous molecular sieves Ti-HMS (Mn contained (Ti-)HMS) catalyst is a novel Fenton-like catalyst that could be environmentally friendly and cost-effective and has superior efficiency for the reactive hydroxyls generated in AOPs [20].

Because of high surface area, well-developed porous structure, variable surface composition, good chemical resistance, and acceptable cost, activated carbon has mostly been used as support [43, 117, 118]. Also, alumina (Al_2O_3) is extensively used as catalyst support because of its mechanical, electrical, and chemical stability and relatively low cost. Alumina-supported catalysts have demonstrated a considerably higher stability in CWPO in comparison with carbon-based ones [43, 119].

3.2 Catalytic Ozonation

3.2.1 Principles

Because of the strong oxidative properties of ozone (O_3), it can oxidize organic compounds in aqueous solution by direct or indirect pathways [4, 120, 121]. Despite the high demand of energy to generate ozone (high production cost) [3, 4, 13, 17, 121, 122] and efficiency dependence of ozone on gas-liquid mass transfer [3, 4, 13, 17, 123], its use in wastewater treatment has numerous advantages such as no sludge remaining, minimal danger, easy performance, little space requirement, one-step degradation, and easy conversion of residual ozone to oxygen and water [121]. Having a high reduction potential (2.07 V), it can react with an organic substrate (R in Eq. 18) [3, 4, 13, 17, 121, 124]. However, the application of ozone is only defined as an AOP when it can decompose for hydroxyl radical generation (Eq. 19) and can be catalyzed by hydroxyl ions in alkaline medium or by transition metal cations (indirect type of ozonation) [3, 4, 13, 17, 120, 122, 124].

$$O_3 + R \rightarrow RO + O_2 \qquad (18)$$
$$2O_3 + 2H_2O \rightarrow 2HO^\bullet + 2HO_2^\bullet + O_2 \qquad (19)$$

The ozone decomposition mechanism in water is very complicated which may be explained by the chain mechanism of Weiss as follows [120, 125, 126]:

$$O_3 + H_2O \rightarrow 2HO^\bullet + O_2 \qquad (20)$$
$$O_3 + OH^- \rightarrow O_2^{\bullet -} + HO_2^\bullet \qquad (21)$$
$$O_3 + HO^\bullet \rightarrow HO_2^\bullet + O_2 \qquad (22)$$
$$O_3 + HO_2^\bullet \rightarrow HO^\bullet + 2O_2 \qquad (23)$$
$$2HO_2^\bullet \rightarrow H_2O_2 + O_2 \qquad (24)$$

The efficiency of ozone in destructing organic pollutants can be enhanced in combination with catalyst, H_2O_2, radiolysis, UV radiation, or ultrasound [3, 4, 13, 127, 128].

Ozonation reactions can be improved by applying heterogeneous or homogeneous catalysts [4, 127]. Homogeneous catalytic ozonation is based on ozone activation by metal ions present in the reaction medium (aqueous solution), while in heterogeneous catalytic ozonation, the major catalysts are metal oxides or metals/metal oxide on supports [120, 128]. The possible mechanisms for homogeneous and heterogeneous catalytic ozonation have been given in some references [120, 128, 129]. The removal of organic compounds from aqueous solutions can be improved by homogeneous catalytic ozonation, but the costs of water treatment can be enhanced due to the introduction of ions which result in the secondary pollution. Heterogeneous catalytic ozonation can extensively be applied in water treatment due to the higher stability, reusability, and lower loss of heterogeneous catalysts that improve the efficiency of ozone decomposition [129]. In the heterogeneous system, ozonation can occur not only in the bulk solution but also in all the interfaces involved between the different phases present in the reaction media [116, 120, 130]. Simplified scheme of the catalytic ozonation pilot is shown in Fig. 1, which includes an ozone generator, a continuous stirred tank reactor containing the suspended catalyst powder with an ozone distribution system, a recirculation loop with a membrane module for the catalyst separation from the treated effluent, and residual ozone destruction unit (an oven at 475 °C) [131, 132]. The catalyst, its surface properties, the pH of the solution (influences the characteristics of the surface active sites and ozone decomposition reactions in wastewater), etc. can affect the efficiency of the catalytic ozonation process [129]. Selection of catalysts is the most important parameter for heterogeneous systems [120].

Addition of hydrogen peroxide to ozone (this mixture is called peroxone process) can start the decomposition cycle of ozone [4, 127], in which hydroxyl radicals can be formed as follows [3, 4]:

$$O_3 + H_2O_2 \rightarrow HO^\bullet + HO_2^\bullet + O_2 \quad (25)$$

UV radiation can readily be absorbed at 254 nm wavelength by ozone, generating as an intermediate hydrogen peroxide (Eq. 26), which can then be decomposed to hydroxyl radicals under exposure of UV radiation as shown by Eq. (27) [3, 4, 127]:

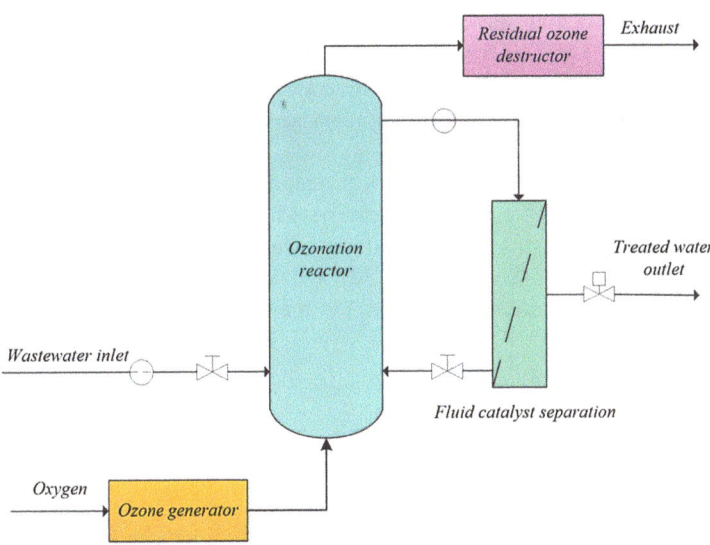

Fig. 1 Simplified schematic of the catalytic ozonation pilot (modified from [131, 132])

$$O_3 + H_2O \xrightarrow{h\nu} H_2O_2 + O_2 \quad (26)$$

$$H_2O_2 \xrightarrow{h\nu} 2HO^{\bullet} \quad (27)$$

The production of hydroxyl radicals may also be accelerated by combined use of ultrasound and ozone due to the reduction of mass transfer limitations resulting from the turbulence generation by the acoustic current induced by ultrasound [3, 4, 123].

3.2.2 Catalytic Materials Used in Catalytic Ozonation

In homogeneous catalytic ozonation, several transition metals such as Fe^{2+}, Fe^{3+}, Mo^{6+}, Mn^{2+}, Ni^{2+}, Co^{2+}, Cd^{2+}, Cu^{2+}, Ag^+, Cr^{3+}, and Zn^{3+} have been used as homogeneous solutions in ozonation media [4, 116, 120, 127–129, 133], whose nature determines the reaction rate, selectivity of ozone oxidation system, ozone consumption, and the efficiency of ozone application [120, 134, 135]. The highest catalytic performances have been demonstrated by bivalent cations and more especially by Fe^{2+} ions [116, 136, 137].

In heterogeneous catalytic ozonation, metal oxides (MnO_2, TiO_2, Al_2O_3, FeOOH, ZnO, Fe_2O_3, WO_3, CuO, Ni_2O_3, CoO, V_2O_5, Cr_2O_3, MoO_3, CeO, and CeO_2), metals (Cu, Ru, Pt, Pb, Pd, Co, etc.) or metal oxides (Fe_2O_3, TiO_2, Co_3O_4, CuO, etc.) on supports (SiO_2, Al_2O_3, TiO_2, CeO_2, ZrO_2, activated carbon, clay, silica gel, ceramic honeycomb, and so on), zeolites and related catalysts, carbon-based materials as catalysts or catalyst supports (activated carbon, granulated activated carbon (GAC), carbon black powder and graphite, multiwalled carbon nanotubes (MWCNTs), etc.), modified activated carbon, mixed hydroxide catalysts, and natural minerals have been applied as catalysts [4, 116, 120, 127–129, 133, 135, 138–141].

Shahidi et al. reviewed oxidative water treatments with emphasis on catalytic ozonation [116]. In addition, Fe-based catalysts for heterogeneous catalytic ozonation of emerging pollutants in water and wastewater have been reviewed by Wang and Bai [142]. Besides, Xiao et al. [138] and Mehrjouei et al. [143] reviewed heterogeneous photocatalytic ozonation for the removal of contaminants from wastewater.

3.2.3 Cost-Effective Catalytic Materials Used in Catalytic Ozonation

Catalyst activity, simplicity of preparation, preparation time, cost [144], nontoxicity, availability, environmental suitability [145], etc. can be important parameters in the catalyst selection. High costs, cumbersome preparations, and environmental unsuitability can usually hinder the industrial applicability of a catalyst [145]. TiO_2 is a common and low-cost catalyst extensively applied in photo-oxidative processes, especially photocatalytic ozonation [116, 146]. Due to relatively low-cost, nontoxicity, and availability of WO_3, it may be an interesting alternative to TiO_2 for the photocatalytic ozonation under visible or solar light radiation [147]. Fe_2O_3 has been proposed as an attractive and efficient catalyst for treatment, especially for the removal of color from distillery wastewater, due to its cost-effectiveness and nontoxicity [148]. High catalytic capacity, low cost, and low toxicity of ZnO make it promising catalyst for catalytic ozonation [149, 150]. Copper oxide may be cost-competitive, chemically stable, and nontoxic in the catalytic ozonation process [151].

Clay minerals with extensive abundance in nature can be low-cost alternatives to zeolites in different applications, especially in catalytic ozonation [116, 152, 153]. Natural raw mixtures of clay minerals, volcanic ashes, silicas, carbonates, and miscellaneous are different types of clays [116, 152]. The clays from smectite mineral family, more specifically montmorillonite, the major component of bentonites, are attractive and low-cost aluminosilicates [116].

Recycling/reusing of particular wastes to manufacture a catalyst for the ozonation process can be cost-effective and "wastes-treat-wastes" or "green" strategy [145, 154]. Cobalt doped red mud (Co/RM) (RM from the alumina industry waste as the catalyst supporter) has been applied in the catalytic ozonation for bezafibrate destruction in water; better performance and detoxification than using un-doped RM has been reported. The higher catalytic activity in Co/RM can be due to the presence of Fe, Al, Si, and Ti oxides and oxyhydroxides in RM as potential reaction sites and the surface modification by cobalt oxide [155, 156]. Sewage sludge was reused as a catalyst in the ozonation process by Wen et al. [157]. Sludge-based catalyst (SBC) was prepared with $ZnCl_2$ as activation agent. The removal efficiency of oxalic acid was reported to be improved by 45.4% with simultaneous application of ozone and SBC in comparison with that of single SBC adsorption plus ozone alone, which demonstrates a strong synergistic effect [157]. The spent fluid catalytic cracking catalysts (sFCCc), a typical petrochemical waste, were recycled and applied for the advanced treatment of petrochemical wastewater by Chen et al. [145]. Al_2O_3 and various types of zeolites are the main components of sFCCc that can have excellent adsorptive and catalytic performance. In addition, the metal oxides in the waste which poisoned FCCc could be the active components of catalytic ozonation. Hydroxyl radical generation and hydroxyl radical-mediated oxidation could be promoted by multivalent vanadium (V^{4+} and V^{5+}), iron (Fe^{2+} and Fe^{3+}), and nickel (Ni^{2+}) oxides which were highly distributed on the surface of sFCCc. The sFCCc assisted catalytic ozonation (sFCCc-O) showed high potential for the treatment of bio-refractory wastewaters and enhanced to threefold the COD removal compared to single ozonation. Besides, nonbiodegradable polar contaminants were remarkably destructed [145]. Wu et al. used recycled waste iron shavings as a catalyst for the catalytic ozonation of organic pollutants from actual bio-treated dyeing and finishing wastewater (BDFW) [158]. The effluent COD level reported to meet the discharge limitation of 80 mg/L. The iron shavings with their high stability and excellent reusability at a low cost and without the requirement for separation from the treated effluent can be appropriate for both successive batch process and continuous (flow) process [158].

3.3 Heterogeneous Photocatalysis

3.3.1 Principles

For the first time, the term photocatalysis was introduced in the 1930s as a new branch of catalysis [4, 37, 159]. Promotion or acceleration of a chemical reaction under the action of light in the presence of photocatalysts which absorb light quanta and are involved in the chemical transformations of the reaction components is called photocatalysis [37, 160, 161]. In the heterogeneous photocatalysis, the catalyst occurs in a different phase than the reaction medium [162]. Since 1972, when Fujishima and Honda [163] reported the photochemical splitting of water into hydrogen and oxygen in the presence of TiO_2, heterogeneous photocatalysis has been interestingly investigated [4, 159, 164–166]. Since the 1980s, this process has been evaluated and examined with great success by various scientists for environmental (air, water, and soil) applications [3, 4, 13, 159, 167].

According to Mota et al. [3], heterogeneous photocatalysis principle (Fig. 2) is based on the activation of a semiconductor material such as TiO_2 by the action of radiation with a suitable wavelength (e.g., with the light of $\lambda < 390$ nm). With the absorption of photons by the semiconductor particulate possessing enough energy to promote the conduction of an electron (e^-) from its valence band (VB) to the conduction band (CB) (a transition called band gap energy), activation is attained generating holes in the valence band (h^+) that will act as oxidizing sites [3, 4, 13]. In other words, illumination onto a photocatalyst excites to produce electron and hole pair (e^-/h^+) with high-energy state, which move to the particle surface, where they take part in redox reactions with adsorbed species and, thus, form superoxide radical anion ($O_2^{\bullet-}$) and hydroxyl radical as follows [4, 121, 159, 168–172]:

$$TiO_2 \xrightarrow{hv} TiO_2(e^- + h^+) \tag{28}$$

$$e^- + O_2 \rightarrow O_2^{\bullet-} \tag{29}$$

$$h^+ + H_2O \rightarrow OH^{\bullet} + H^+ \tag{30}$$

$$h^+ + OH^- \rightarrow OH^{\bullet} \tag{31}$$

The generated radicals are applied as powerful oxidizing agents to transform organic contaminants into H_2O, CO_2, and less toxic by-products [4, 13, 18, 168, 169, 171]. The generated hydroxyl radical attacks organic compounds as follows [173]:

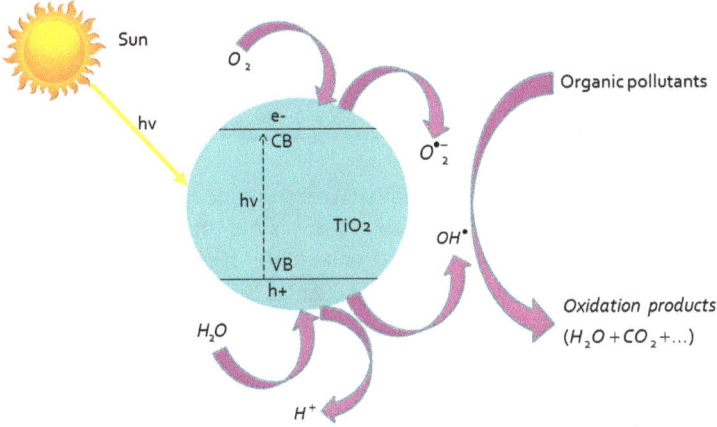

Fig. 2 Typical mechanism of TiO$_2$ in solar photocatalysis process (modified from [4, 166, 171])

$$\text{OH}^{\cdot} + \text{Organic (R)} \rightarrow \text{Intermediates} \rightarrow \text{CO}_2 + \text{H}_2\text{O} \qquad (32)$$

According to Bockelmann et al. [174] and Mota et al. [3], the heterogeneous photocatalytic process may also be favored by the addition of H$_2$O$_2$, given that, like O$_2$, it may act as the acceptor of electrons present in the system, producing hydroxyl radicals as follows [3, 4, 174]:

$$e^{-} + \text{H}_2\text{O}_2 \rightarrow \text{HO}^{\cdot} + \text{HO}^{-} \qquad (33)$$

In the photocatalytic systems, artificial sources (e.g., UV polychromatic lamps) or the sun may be utilized as radiation sources. Solar light-based photocatalytic degradation systems are of interest because of economical aspects [4, 121, 170, 175].

The major factors that can affect the performance of heterogeneous photocatalytic process in water and wastewater treatment are the initial organic load or initial concentration of substrate, catalyst and state of catalyst, amount of catalyst, reactor's design, irradiation source, UV irradiation time, temperature, solution's pH, light intensity, and the presence of ionic species [4, 7, 121, 170].

Wastewater treatment photocatalytic reactors can be classified based on their design characteristics such as the state of the photocatalyst, type of illumination, and position of the irradiation source [4, 170]. The reactors may be categorized according to the state of the photocatalyst into two main groups: (1) a suspension/slurry type and (2) thin film type [4, 18, 165, 170]. The slurry reactor needs an additional downstream separation unit for the photocatalyst particles' recovery while the second one permits a continuous operation [176]. Efficient illumination of the catalyst can be the major challenge in the design of a photocatalytic reactor [4, 166]. The annular slurry photoreactor, cascade photoreactor, downflow contactor reactor, etc. have been applied in the photocatalytic water and wastewater treatment [176].

3.3.2 Catalytic Materials Used in Heterogeneous Photocatalysis

An ideal photocatalyst should be photoactive, biologically and chemically inert, stable toward photo-corrosion, suitable for visible or near UV light energy harnessing, low cost, nontoxic, and environmentally friendly [4, 166].

Various materials such as oxides (TiO_2, ZnO, SiO_2, SnO_2, WO_3, ZrO_2, CeO, Nb_2O_3, Fe_2O_3, MgO, etc.) [4, 121, 165, 177, 178], mixed oxides ($SrTiO_3$, Co-TiO_2 nanocatalysts [179], $FeTiO_3$ [180]), sulfides (CdS, ZnS, etc.), GaP [177], semiconductor-conjugated CP composites [e.g., AgI/UiO-66 (zirconium-based metal-organic framework) photocatalyst] [177, 181], Ag/$BiVO_4$ composite [182], 10:100 (w/w) of oxidized multiwalled carbon nanotube ($MWCNT_{ox}$)-anatase (10-$MWCNT_{ox}$-TiO_2) composite [183], graphitic carbon nitride (g-C_3N_4)/semiconductor (CNS) nanocomposites [184], etc. have been applied in photocatalysis.

TiO_2 is an excellent photocatalyst due to its high resistance to photo-corrosion and desirable bandgap energy, availability in the market, chemical inertness and durability, and nontoxicity [4, 165]. Anatase and rutile are the most common crystalline forms of TiO_2, while brookite is an uncommon form which is unstable. Having more open structure in comparison with rutile, anatase is more efficient than rutile in photocatalytic utilizations. The commercially available form of TiO_2 is Degussa P25 which includes two forms of ca. 25% rutile and 75% anatase [4, 166].

Supports can play a major role in immobilizing active catalyst, increasing the surface area of catalytic material, sintering reduction, and improving hydrophobicity, thermal, hydrolytic, and chemical stability of the catalytic material [4, 166]. Activated carbon, fiber-optic cables, fiberglass, glass, glass beads, glass wool, membranes, quartz sand, zeolites, silica gel, stainless steel, Teflon, clays, polymeric materials, pumice stone, monoliths, cellulose, and others have been studied as TiO_2 supports [4, 170, 176, 185].

3.3.3 Cost-Effective Catalytic Materials Used in Heterogeneous Photocatalysis

If the photocatalyst is nonrecyclable, cost efficiency will be a concern. Photocatalysts which apply rare metals can be expensive due to their low abundance. Treatment cost can be low when photocatalysts are recyclable [186, 187].

Among the photocatalysts studied for photodegradation, TiO_2 and ZnO have intensively been used due to their low cost, high efficiency and quantum yield, resistance to photo-corrosion, and safe handling [188, 189]. TiO_2 is the most extensively applied catalyst in heterogeneous photocatalysis to destruct a large range of contaminants due to its wide bandgap (3.03 eV for rutile, 3.18 eV for anatase) [183, 190], spectral overlap with sunlight emission (about 5%) [183], photostability, nontoxicity, low cost [183, 191], availability, high photoactivity, environmentally friendly [192], and water insolubility under most environmental conditions [191].

Natural clays (bentonite, sepiolite, montmorillonite, and kaolinite) and zeolites have been widely reported in the literature as supports for TiO_2 due to their high adsorption capacity and cost-effectiveness [176].

Singh et al. developed an effective and low-cost TiO_2/polystyrene floating photocatalyst using solvent-cast method for environmental application and the color removal efficiency by the optimized photocatalyst on reuse reported to be in the range of 99–100% [193].

García-Muñoz et al. assessed ilmenite mineral ($FeTiO_3$) as low-cost catalyst in different AOPs (photocatalysis, CWPO, and CWPO-photoassisted processes) using phenol as target compound. Results revealed a scarce activity of ilmenite in photocatalytic process. But, ilmenite mineral has demonstrated to be active and highly stable catalyst for CWPO-photoassisted process; a complete phenol and H_2O_2 conversion and a TOC conversion higher than 95% were reported to be attained at pH 3 and 25 °C with 500 mg/L of ilmenite and the stoichiometric amount of H_2O_2 [180].

3.4 Catalytic Wet Air Oxidation (CWAO)

3.4.1 Principles

Zimmermann [194] originally developed the WAO process and the first industrial utilizations of WAO appeared in the late 1950s [195–198]. Generally, the oxidation of soluble or suspended organic or inorganic substances in an aqueous solution by means of oxygen or air as the oxidizing agent at elevated temperatures and pressures is called WAO process [4, 197, 199–201]. The temperature usually varies from approximately 175 to 320 °C [4, 199] or 150 to 320 °C [4, 200]. Hydroxyl radicals are formed in this process [3, 4]. The products of WAO system can be determined by the degree of oxidation. For high degrees of oxidation, organic matter can mainly be transformed to CO_2 and H_2O [4, 199]. The following reaction can explain the typical material balance for the WAO process where the heat value is close to 435 kJ/(mole O_2 reacted) [4, 202]:

$$C_m H_n O_k Cl_w N_x S_y P_z + [m + 0.25(n - 3x) - 0.5k + 2(y + z)]O_2 \rightarrow \\ mCO_2 + 0.5(n - 3x)H_2O + xNH_3 + wCl^- + ySO_4^{2-} + zPO_4^{3-} + \text{heat} \quad (34)$$

Temperature, pressure, inert gas flow rate, pH evolution during oxidation, etc. are factors which can affect the efficiency of a WAO system and should be considered during reactor design [4, 202].

More than 200 WAO treatment plants have been built around the world, the majority to treat sewage sludge. Other major fields of utilization include the regeneration of activated carbon and the treatment of industrial wastewater (e.g., petrochemical, chemical, and pharmaceutical wastewaters) [197, 198, 203].

Application of WAO units can be very useful for the treatment of a variety of refractory organic contaminants in wastewater, but the need of high pressure and high temperature for its operation restricts their practical utilization; thus, CWAO has been suggested to relax the oxidation conditions [204, 205]. CWAO is an improvement over the WAO process by the introduction of a catalyst (homogeneous or heterogeneous) in the reaction [3, 4, 201, 206].

The development of commercial CWAO units began as early as the mid-1950s in the USA [198, 207, 208]. In Japan, several companies developed CWAO technologies based on heterogeneous catalysts including precious metals deposited on titania or titania-zirconia oxides. In Europe, the interest was more on homogeneous CWAO. In recent years, various homogeneous CWAO processes based on this concept (e.g., Ciba-Geigy, LOPROX, WPO, ORCAN, and ATHOS processes) have been developed [198, 208, 209]. A scheme of the Bayer LOPROX process (with Fe^{2+} ions as catalyst) is shown in Fig. 3. NS-LC (with $Pt-Pd/TiO_2-ZrO_2$ honeycomb catalyst), Osaka gas (with ZrO_2 or TiO_2 with noble or base metal catalyst), Kurita (with supported Pt catalyst), and CALIPHOX (with a metal oxide catalyst, e.g., $CuO-ZnO-Al_2O_3$) processes are examples of industrial heterogeneous CWAO processes [198, 210].

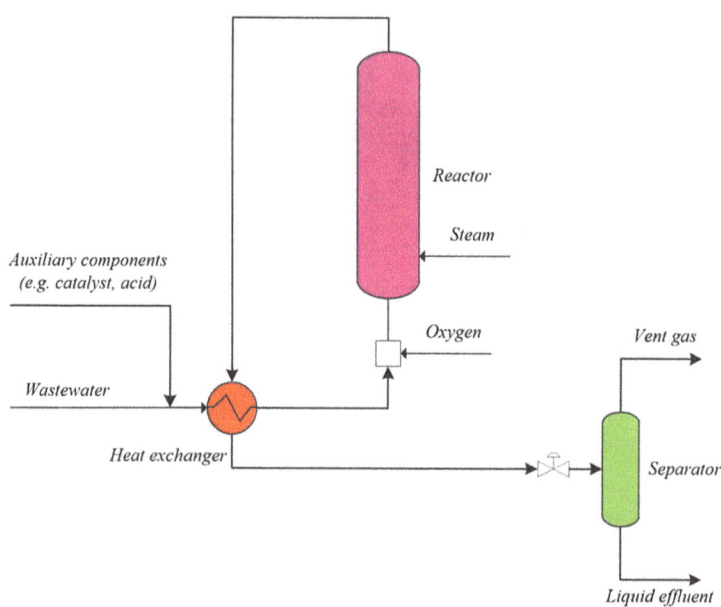

Fig. 3 Schematic of the Bayer LOPROX process (modified from [211])

3.4.2 Catalytic Materials Used in CWAO

Homogeneous and heterogeneous catalysts can be applied in CWAO process. Selection of the suitable catalyst for CWAO system not only can reduce the severity of reaction conditions, but it can also increase the degradation of refractory organic contaminants, resulting in diminishing investment and operational cost [209, 212, 213]. Homogeneous catalysts such as dissolved salts of Cu, Fe, Ni, Co, Mn, etc. have successfully been applied in CWAO process [214, 215]. Generally, the homogeneous catalysts, such as dissolved copper salts, are regarded as effective, but an additional separation step (e.g., precipitation or membrane separation) is required for the removal or recovery of metal ions from the treated effluent because of their toxicity, which substantially enhances operational costs [213, 216]. Note that homogeneous catalyst utilization should be assessed in the early process design stage based on the existing discharge standards or regulations either in the liquid or solid phases [198, 208].

Various heterogeneous catalysts including supported noble metals and metal oxides (as well as mixed metal oxides) have been widely studied to enhance the efficiency of CWAO [4, 21, 198, 211, 217–220]. Heterogeneous catalysts show very high activity and stability. They can be easily separated by filtration from the treated solution [216]. However, deactivation phenomena and leaching of active metals to the liquid phase can exist. For this reason, metal-free carbon materials have been evaluated as catalysts in various CWAO utilizations, including activated carbons [220–222], carbon xerogels [223, 224], multiwalled carbon nanotubes [225–227], carbon foams and fibers enriched with nitrogen [220, 228], and graphene oxide (GO) and chemically reduced graphene oxides [229]. Stuber et al. reviewed the application of carbon materials as catalytic supports or direct catalyst in CWAO of organic pollutants [221]. The employed heterogeneous catalysts in CWAO are listed in Table 2.

Tailoring of catalyst for each particular application is essential and inexpensive materials should be used for its preparation. Oxides of Zr, Ce, and Ti may be applied as stable supports. Metal oxide catalysts can be very active but unstable due to leaching. The catalytically active compounds must be incorporated into a lattice of catalyst support to decrease leaching. If this is not practicable, precious metals should be applied in the catalytic active phase [208].

3.4.3 Cost-Effective Catalytic Materials Used in CWAO

Soluble transition metal catalysts based on iron or copper salts have been used in various commercial WAO plants, which were successfully employed in the treatment of industrial effluents and sludge [198, 208].

High catalytic activity, long life cycle, and strong adaptability are demonstrated by precious metals such as Pt and Pd; however, their utilizations are restricted due to high cost [231, 232]. Thus, a catalyst for CWAO comparable to noble metal

Table 2 Some of the employed heterogeneous catalysts in CWAO process [4, 21, 198, 211, 217–230]

Type of catalyst	Description
Noble metals	Pt supported on γ-Al_2O_3; Ru, Ir, Pd, Ag, base metals supported on CeO_2, TiO_2, ZrO_2; Pt supported on C; Pt supported on resin; Pt supported on SDB resin; Rh supported on TiO_2, CeO_2, C; Pt-Ru supported on C; Pd, Pd-Pt supported on ALON^{TM}; Pd-Pt-Ce supported on γ-Al_2O_3
Metal oxides (as well as mixed metal oxides)	Cu/Cr oxides; Cu/Cr/Ba/Al oxides; Co, Fe, Mn, Zn oxides with Cu oxides supported on γ-Al_2O_3; CuO supported on γ-Al_2O_3; Cu/Ni/Al oxides; CuO/ZnO/CoO supported on cement; CuO/ZnO supported on γ-Al_2O_3; CuO/CeO_2; K-MnO_2/CeO_2; MnO_2/CeO_2; CuO supported on C; MnO_2, Co_2O_3; Ni-oxide; Cu/Zn/Cr/Ba/Al oxides; Fe_2O_3; Cu/Mn/La supported on ZnO-Al_2O_3; Ce/Zr/Cu oxides or Ce/Zr/Mn; Cu-Fe-La supported on γ-Al_2O_3
Carbon materials	Activated carbons; carbon xerogels; multiwalled carbon nanotubes; carbon foams and fibers enriched with nitrogen; graphene oxide and chemically reduced graphene oxides

catalysts is highly wanted. Transition metal catalysts are low cost, and introduction and removal of the lattice oxygen in transition metal oxides can be easily done. They are extensively employed due to their high catalytic activity [230, 233–235]. In other words, according to Kim and Ihm [21], with regard to cost and resistance to poisoning by halogen-containing compound, base metal oxide catalysts can be more desirable than noble metals, though their catalytic activities are still lower than noble metals. The most futuristic species of them to challenge with noble metals can be Cu, Mn, and Ce [21].

3.5 Catalytic Supercritical Water Oxidation (CSCWO)

3.5.1 Principles

A promising process which takes place in an aqueous medium at temperatures and pressures above the critical point of water ($T_c = 374$ °C and $P_c = 22.1$ MPa) is called SCWO [4, 15, 16, 19, 236–245]. A smaller amount of hydrogen bonds, a lower dielectric constant, a lower viscosity, and a higher diffusion coefficient compared to normal water are characteristics of supercritical water [4, 240]. Having unique properties under these conditions, supercritical water can be utilized as a reaction medium to produce nanoparticles, destruct organic compounds, etc. In general, operational temperature and pressure ranges in SCWO process are 400–600 °C and of 24–28 MPa, respectively [4, 16, 19].

Degradation of organic wastes is the primary use of SCWO technology [4, 16, 246]. Formation of a single phase environment by dissolving organic substances and oxidants (such as air, or pure oxygen, or H_2O_2) into supercritical water can overcome the interphase mass transfer resistance and enhance the whole reaction rate. In this process, organic matter can be destructed into small molecular compounds such as CO_2, N_2, H_2O, etc., and heteroatoms may also be converted into their mineral acids [4, 240]. Removal rates higher than 99% can be attained with residence times shorter than 1 min. This technology will not generate a secondary pollution [4, 237, 246].

The technology of waste treatment by SCWO was suggested by Medoll in the 1980s [247]. Since then, a wide variety of toxic and hazardous industrial wastes has been successfully treated by SCWO which has been reviewed by some researchers [4, 237, 240, 246]. In real, commercial-scale SCWO plants have been employed in the USA, the UK, Japan, China, Sweden, etc. [4, 237, 240, 246]. The main disadvantages of SCWO which still exist and even make some commercial-scale SCWO plants inactive are corrosion, plugging, and high running costs [4, 237, 240, 248].

The oxidation rate enhancement, the residence time and operational temperature reduction, and possibly a better selectivity for competing reaction pathways can be attained by catalysts [244, 249]. CSCWO is an amelioration over the SCWO system by the introduction of a catalyst (homogeneous or heterogeneous) in the reaction. Effective use of a catalyst in CSCWO process has the ability to reduce reactor temperature and the energy consumption of the device and oxidant concentration and improve the CO_2 formation efficiency, which makes this process to be much more feasible for practical use in the aspects of device safety, durability, and economy [241, 250, 251].

3.5.2 Catalytic Materials Used in CSCWO

The applied catalysts in CSCWO may be generally divided into noble metals, metal oxides, and transition metal salts [241]. In real, $H_4Si_{12}O_{40}$ [252], activated carbon [31, 253], CuO/ZnO/CoO supported on porous cement [254], MnO_2/CeO_2, V_2O_5/Al_2O_3 [255], Carulite 150, a commercial catalyst [256], TiO_2 [257], MnO_2 [244, 258], CuO, CuO/Al_2O_3 [30], MnO_2/CuO [259], Mn_2O_3/Ti-Al oxide composite catalyst [245], Ni/Al_2O_3 [260], $Pt/\gamma-Al_2O_3$ [261], NaOH [262], $CuSO_4$ [241, 263], $CuNO_3$, $FeCl_3$, $(Fe)_2(SO_4)_3$, V_2O_5, $MnSO_4$, $FeSO_4$ [263], $ZnSO_4$ [264], etc. have been studied as catalysts in the CSCWO of organic components.

3.5.3 Cost-Effective Catalytic Materials Used in CSCWO

Some frequently used homogeneous catalysts in CSCWO such as Cu^{2+} have shown to be an efficient catalyst [241, 265, 266]. Transition metal catalysts can be generally low cost [230, 233–235].

4 Conclusions

In this chapter, overview of AOPs, catalytic AOPs (Fenton-based processes, catalytic ozonation, heterogeneous photocatalysis, catalytic wet air oxidation, and catalytic supercritical water oxidation), related catalytic materials, and cost-effective catalytic materials used in these processes have been discussed.

The use of natural minerals as catalysts in CWPO is interesting because of their wide availability and low cost. In addition, ZVI can be applied as low-cost catalytic material in Fenton reaction. Activated carbon has mostly been used as support. Also, alumina-supported catalysts have demonstrated a considerably higher stability in CWPO in comparison with carbon-based ones.

TiO_2 is a common and low-cost catalyst extensively applied in photo-oxidative processes, especially photocatalytic ozonation. Due to relatively low cost, nontoxicity, and availability of WO_3, it may be an interesting alternative to TiO_2 for the photocatalytic ozonation under visible or solar light radiation. Clay minerals with extensive abundance in nature can be low-cost alternatives to zeolites in different applications, especially in catalytic ozonation. Recycling/reusing of particular wastes (such as SBC, sFCCc, etc.) to manufacture a catalyst for the ozonation process can be cost-effective and "wastes-treat-wastes" or "green" strategy.

Among the photocatalysts studied for photodegradation, TiO_2 and ZnO have intensively been used due to their low cost, high efficiency and quantum yield, resistance to photo-corrosion, and safe handling. Natural clays (bentonite, sepiolite, montmorillonite, and kaolinite) and zeolites have been widely reported in the literature as supports for TiO_2 due to their high adsorption capacity and cost-effectiveness.

Soluble transition metal catalysts based on iron or copper salts have been used in various commercial WAO plants, which were successful in the treatment of industrial effluents and sludge.

Some frequently used homogeneous catalysts in CSCWO such as Cu^{2+} have shown to be an efficient catalyst. Transition metal catalysts can be generally low cost.

References

1. Litter MI, Quici N (2010) Photochemical advanced oxidation processes for water and wastewater treatment. Rec Pat Eng 4:217–241
2. Covinich LG, Bengoechea DI, Fenoglio RJ, Area MC (2014) Advanced oxidation processes for wastewater treatment in the pulp and paper industry: a review. Am J Environ Eng 4(3):56–70
3. Mota ALN, Albuquerque LF, Beltrame LTC, Chiavone-Filho O, Machulek A Jr, Nascimento CAO (2008) Advanced oxidation processes and their application in the petroleum industry: a review. Brazil J Petrol Gas 2(3):122–142
4. Jafarinejad S (2017) Petroleum waste treatment and pollution control. 1st edn. Elsevier Inc., Butterworth-Heinemann, USA

5. Badriyha BN, Song W, Ravindran V, Pirbazari M (2007) Advanced oxidation processes for destruction of endocrine disrupting chemicals in water treatment: comparison of free-radical reaction mechanisms, pathways and kinetics. 2007 AIChE Annual Meeting
6. Hofman-Caris CHM, Harmsen DJH, Beerendonk EF (2010) Advanced oxidation processes, degradation of priority compounds by UV and UV-oxidation. TECHNEAU, deliverable number D 2.4.1.2b, Dec 2010
7. Stasinakis AS (2008) Use of selected advanced oxidation processes (AOPs) for wastewater treatment – a mini review. Global NEST J 10(3):376–385
8. Ikehata K, Jodeiri Naghashkar N, Gamal El-Din M (2006) Degradation of aqueous pharmaceuticals by ozonation and advanced oxidation processes: a review. Ozone Sci Eng 28(6):353–414
9. Remya N, Lin JG (2011) Current status of microwave application in wastewater treatment – a review. Chem Eng J 166:797–813
10. Glaze WH (1987) Drinking-water treatment with ozone. Environ Sci Technol 21(3):224–230
11. Fu J, Kyzas GZ (2014) Wet air oxidation for the decolorization of dye wastewater: an overview of the last two decades. Chin J Catal 35:1–7
12. Santos MSF, Alves A, Madeira LM (2011) Paraquat removal from water by oxidation with Fenton's reagent. Chem Eng J 175:279–290
13. Jafarinejad S (2015) Recent advances in determination of herbicide paraquat in environmental waters and its removal from aqueous solutions: a review. Inter Res J Appl Basic Sci 9(10):1758–1774
14. Deng Y, Zhao R (2015) Advanced oxidation processes (AOPs) in wastewater treatment. Curr Pollution Rep 1:167–176
15. Jafarinejad S, Abolghasemi H, Golzary A, Moosavian MA, Maragheh MG (2010) Fractional factorial design for the optimization of hydrothermal synthesis of lanthanum oxide under supercritical water condition. J Super Fluid 52:292–297
16. Jafarinejad S (2014) Supercritical water oxidation (SCWO) in oily wastewater treatment. National e-conference on advances in basic sciences and engineering (AEBSCONF), Iran
17. Jafarinejad S (2015) Ozonation advanced oxidation process and place of its use in oily sludge and wastewater treatment. 1st international conference on environmental engineering (eiconf), Tehran, Iran
18. Jafarinejad S (2015) Heterogeneous photocatalysis oxidation process and use of it for oily wastewater treatment. 1st international conference on environmental engineering (eiconf), Tehran, Iran
19. Sabet JK, Jafarinejad S, Golzary A (2014) Supercritical water oxidation for the recovery of dysprosium ion from aqueous solutions. Inter Res J Appl Basic Sci 8(8):1079–1083
20. Song H, You JA, Chen C, Zhang H, Ji XZ, Li C, Yang Y, Xu N, Huang J (2016) Manganese functionalized mesoporous molecular sieves Ti-HMS as a Fenton-like catalyst for dyes wastewater purification by advanced oxidation processes. J Environ Chem Eng 4:4653–4660
21. Kim KH, Ihm SK (2011) Heterogeneous catalytic wet air oxidation of refractory organic pollutants in industrial wastewaters: a review. J Hazard Mater 186:16–34
22. Mahamuni NN, Adewuyi YG (2010) Advanced oxidation processes (AOPs) involving ultrasound for waste water treatment: a review with emphasis on cost estimation. Ultrason Sonochem 17:990–1003
23. Huang CP, Dong C, Tang Z (1993) Advanced chemical oxidation: its present role and potential future in hazardous waste treatment. Waste Manag 13:361–377
24. Loures CCA, Alcântara MAK, Filho HJI, Teixeira ACSC, Silva FT, Paiva TCB, Samanamud GRL (2013) Advanced oxidative degradation processes: fundamentals and applications. Inter Rev Chem Eng 5(2):102–120
25. Jelonek P, Neczaj E (2012) The use of advanced oxidation processes (AOP) for the treatment of landfill leachate. Inżynieria i Ochrona Środowiska 15(2):203–217
26. Trapido M (2008) Ozone-based advanced oxidation processes, ozone science and technology. Encyclopedia of Life Support Systems (EOLSS), Developed under the Auspices of UNESCO, 1–17. http://www.eolss.net/sample-chapters/c07/e6-192-07a-00.pdf

27. Kalra SS, Mohan S, Sinha A, Gurdeep Singh G (2011) Advanced oxidation processes for treatment of textile and dye wastewater: a review. 2nd international conference on environmental science and development IPCBEE, vol 4. IACSIT Press, Singapore, pp 271–275
28. Burgos AJ, Rodriguez PU, Lopez JS (2015) Advanced oxidation processes (AOPs), series: advanced treatments, technology fact sheets for effluent treatment plants of textile industry, INDITEX, FS-AVA-001, 1-27
29. Sharma S, Ruparelia JP, Patel ML (2011) A general review on advanced oxidation processes for waste water treatment. Institute of Technology, Nirma University, Ahmedabad – 382 481, 08-10 December, 1–7
30. Yu J, Savage PE (2000) Phenol oxidation over CuO/Al_2O_3 in supercritical water. Appl Catal Environ 28(3–4):275–288
31. Matsumura Y, Urase T, Yamamoto K, Nunoura T (2002) Carbon catalyzed supercritical water oxidation of phenol. J Super Fluids 22(2):149–156
32. Adewuyi YG, Peters RW (2013) Fundamental developments and economic feasibility of AOPs involving ultrasound for environmental remediation. Core Programming Area at the 2013 AIChE Annual Meeting: Global Challenges for Engineering a Sustainable Future, San Francisco, CA, USA, 3–8 Nov 2013. https://www3.aiche.org/Proceedings/content/Annual-2013/extended-abstracts/P342115.pdf
33. Blanco M, Martinez A, Marcaide A, Aranzabe E, Aranzabe A (2014) Heterogeneous Fenton catalyst for the efficient removal of Azo dyes in water. Am J Anal Chem 5:490–499
34. Sahu O, Paul D, Chaudhari PK (2014) A comparatively study on thermal and advance oxidation wastewater treatment process: review. J Chem Eng Chem Res 1(6):353–364
35. Corma A (1997) From microporous to mesoporous molecular sieve materials and their use in catalysis. Chem Rev 97:2373–2420
36. Somorjai GA, Rioux RM (2005) High technology catalysts towards 100% selectivity. Fabrication, characterization and reaction studies. Catal Today 100:201–215
37. Fechete I, Wang Y, Vedrine JC (2012) The past, present and future of heterogeneous catalysis. Catal Today 189:2–27
38. Buthiyappan A, Aziz ARA, Daud WMAW (2016) Recent advances and prospects of catalytic advanced oxidation process in treating textile effluents. Rev Chem Eng 32(1):1–47
39. Abramov VO, Abramov OV, Gekhman AE, Kuznetsov VM, Price GJ (2006) Ultrasonic intensification of ozone and electrochemical destruction of 1,3-dinitrobenzene and 2,4-dinitrotoluene. Ultrason Sonochem 13:303–307
40. Arena F, Chio RD, Gumina B, Spadaro L, Trunfio G (2015) Recent advances on wet air oxidation catalysts for treatment of industrial wastewaters. Inorg Chim Acta 431:101–109
41. Luck F (1996) A review of industrial catalytic wet air oxidation processes. Catal Today 27:195–202
42. Fenton HJH (1894) Oxidation of tartaric acid in presence of iron. J Chem Soc Trans 65:899–910
43. Munoz M, de Pedro ZM, Casas JA, Rodriguez JJ (2015) Preparation of magnetite-based catalysts and their application in heterogeneous Fenton oxidation – a review. Appl Catal Environ 176–177:249–265
44. Eisenhauer HR (1964) Oxidation of phenolic wastes. J Water Pollut Control Fed 36:1116–1128
45. Sun JH, Sun SP, Fan MH, Guo HQ, Qiao LP, Sun RX (2007) A kinetic study on the degradation of p-nitroaniline by Fenton oxidation process. J Hazard Mater 148:172–177
46. Jiang C, Pang S, Ouyang F, Ma J, Jiang J (2010) A new insight into Fenton and Fenton like processes for water treatment. J Hazard Mater 174:813–817
47. Awaleh MO, Soubaneh YD (2014) Waste water treatment in chemical industries: the concept and current technologies. Hydrol Current Res 5(1):1–12
48. Ribeiro AR, Nunes OC, Pereira MFR, Silva AMT (2015) An overview on the advanced oxidation processes applied for the treatment of water pollutants defined in the recently launched directive 2013/39/EU. Environ Int 75:33–51

49. Andreozzi R, Caprio V, Insola A, Marotta R (1999) Advanced oxidation processes (AOP) for water purification and recovery. Catal Today 53:51–59
50. Mackul'ak T, Prousek J, Švorc LU (2011) Degradation of atrazine by Fenton and modified Fenton reactions. Monatsh Chem 142:561–567
51. Muruganandham M, Suri RPS, Jafari S, Sillanpää M, Lee GJ, Wu JJ, Swaminathan M (2014) Recent developments in homogeneous advanced oxidation processes for water and wastewater treatment. Int J Photoenergy 2014:21 p, Article ID 821674. Hindawi Publishing Corporation. http://dx.doi.org/10.1155/2014/821674
52. Babuponnusami A, Muthukumar K (2014) A review on Fenton and improvements to the Fenton process for wastewater treatment. J Environ Chem Eng 2:557–572
53. Wang N, Zheng T, Zhang G, Wang P (2016) A review on Fenton-like processes for organic wastewater treatment. J Environ Chem Eng 4:762–787
54. He J, Yang X, Men B, Wang D (2016) Interfacial mechanisms of heterogeneous Fenton reactions catalyzed by iron-based materials: a review. J Environ Sci 39:97–109
55. Luiz DB, Jose HJ, Moreira RFPM (2012) A discussion paper on challenges and proposals for advanced treatments for potabilization of wastewater in the food industry. In: Valdez B (ed) Scientific, health and social aspects of the food industry. InTech, Rijeka, Croatia
56. Powell RM, Puls RW, Hightower SK, Sabatini DA (1995) Coupled iron corrosion and chromate reduction: mechanisms for subsurface remediation. Environ Sci Technol 29:1913–1922
57. Warren KD, Arnold RG, Bishop TL, Lindholm LG, Betterton EA (1995) Kinetics and mechanism of reductive dehalogenation of carbon tetrachloride using zerovalence metals. J Hazard Mater 41:217–227
58. Joo SH, Feitz AJ, Waite TD (1995) X oxidative degradation of the carbothioate herbicide, molinate using nanoscale zero-valent iron. Environ Sci Technol 38:2242–2247
59. Babuponnusami A, Muthukumar K (2012) Removal of phenol by heterogenous photo electro Fenton-like process using nano-zero valent iron. Separ Purif Tech 98:130–135
60. Bigda RJ (1995) Consider Fenton's chemistry for wastewater treatment. Chem Eng Prog 91:62–66
61. Mesquita I, Matos LC, Duarte F, Maldonado-Hódar FJ, Mendes A, Madeira LM (2012) Treatment of azo dye-containing wastewater by a Fenton-like process in a continuous packed-bed reactor filled with activated carbon. J Hazard Mater 237–238:30–37
62. Duarte F, Maldonado-Hódar FJ, Madeira LM (2011) Influence of the characteristics of carbon materials on their behaviour as heterogeneous Fenton catalysts for the elimination of the azo dye Orange II from aqueous solutions. Appl Catal B 103:109–115
63. Yuan SH, Gou N, Alshawabkeh AN, Gu AZ (2013) Efficient degradation of contaminants of emerging concerns by a new electro-Fenton process with Ti/MMO cathode. Chemosphere 93:2796–2804
64. Sabhi S, Kiwi J (2001) Degradation of 2,4-dichlorophenol by immobilized iron catalysts. Water Res 35:1994–2002
65. Sanabria NR, Molina R, Moreno S (2012) Development of pillared clays for wet hydrogen peroxide oxidation of phenol and its application in the posttreatment of coffee wastewater. Int J Photoenergy 2012:17 p. Article ID 864104. Hindawi Publishing Corporation. doi: 10.1155/2012/864104
66. Zhou T, Lim TT, XH W (2011) Sonophotolytic degradation of azo dye reactive black 5 in an ultrasound/UV/ferric system and the roles of different organic ligands. Water Res 45:2915–2924
67. Yang XJ, Tian PF, Zhang XM, Yu X, Wu T, Xu J, Han YF (2014) The generation of hydroxyl radicals by hydrogen peroxide decomposition on FeOCl/SBA-15 catalysts for phenol degradation. AIChE J 61:166–176
68. Zhang C, Zhou MH, Ren GB, Yu XM, Ma L, Yang J, Yu FK (2015) Heterogeneous electro-Fenton using modified iron-carbon as catalyst for 2,4-dichlorophenol degradation: influence factors, mechanism and degradation pathway. Water Res 70:414–424

69. Soon AN, Hameed BH (2011) Heterogeneous catalytic treatment of synthetic dyes in aqueous media using Fenton and photo-assisted Fenton process. Desalination 269(1–3):1–16
70. Moreno S, Sanabria N, Molina R (2008) Chapter 4. Recent tendencies in the synthesis of pillared clays for phenol oxidation. In: Heikkine E (ed) Focus on water resource research. Nova Science Publisher, New York, NY, USA, pp 185–209
71. Rao TSRP, Dhar GM (1998) Recent advanced in basic and applied aspects of industrial catalysis. Elsevier Science B.V., Amsterdam, The Netherlands
72. Sanabria NR, Molina R, Moreno S (2012) Raschig rings based on pillared clays: efficient reusable catalysts for oxidation of phenol. J Advan Oxid Technol 15(1):117–124
73. Kim JK, Martinez F, Metcalfe IS (2007) The beneficial role of use of ultrasound in heterogeneous Fenton-like system over supported copper catalysts for degradation of p-chlorophenol. Catal Today 124(3–4):224–231
74. Hassan H, Hameed BH (2011) Fe-clay as effective heterogeneous Fenton catalyst for the decolorization of reactive blue 4. Chem Eng J 171(3):912–918
75. Nguyen TD, Phan NH, Do MH, Ngo KT (2011) Magnetic Fe_2MO_4 (M:Fe, Mn) activated carbons: fabrication, characterization and heterogeneous Fenton oxidation of methyl orange. J Hazard Mater 185(2–3):653–661
76. Tian SH, YT T, Chen DS, Chen X, Xiong Y (2011) Degradation of acid Orange II at neutral pH using $Fe_2(MoO4)_3$ as a heterogeneous Fenton-like catalyst. Chem Eng J 169(1–3):31–37
77. Dukkanci M, Gunduz G, Yilmaz S, Prihod'ko RV (2010) Heterogeneous Fenton-like degradation of Rhodamine 6G in water using CuFeZSM-5 zeolite catalyst prepared by hydrothermal synthesis. J Hazard Mater 181(1–3):343–350
78. Idel-aouad R, Valiente M, Yaacoubi A, Tanouti B, Lopez-Mesas M (2011) Rapid decolourization and mineralization of the azo dye C.I. Acid red 14 by heterogeneous Fenton reaction. J Hazard Mater 186(1):745–750
79. Ku˘si'c H, Koprivanac N, Selanec I (2006) Fe-exchanged zeolite as the effective heterogeneous Fenton-type catalyst for the organic pollutant minimization: UV irradiation assistance. Chemosphere 65(1):65–73
80. Kuznetsova EV, Savinov EN, Vostrikova LA, Parmon VN (2004) Heterogeneous catalysis in the Fenton-type system FeZSM-5/H_2O_2. Appl Catal B 51(3):165–170
81. de la Plata GBO, Alfano OM, Cassano AE (2010) Decomposition of 2-chlorophenol employing goethite as Fenton catalyst II: reaction kinetics of the heterogeneous Fenton and photo-Fenton mechanisms. Appl Catal B 95(1–2):14–25
82. Costa RCC, Moura FCC, Ardisson JD, Fabris JD, Lago RM (2008) Highly active heterogeneous Fenton-like systems based on Fe^0/Fe_3O_4 composites prepared by controlled reduction of iron oxides. Appl Catal B 83(1–2):131–139
83. Moura FCC, Araujo MH, Costa RCC et al (2005) Efficient use of Fe metal as an electron transfer agent in a heterogeneous Fenton system based on Fe0/Fe3O4 composites. Chemosphere 60(8):1118–1123
84. Sun SP, Lemley AT (2011) P-Nitrophenol degradation by a heterogeneous Fenton-like reaction on nano-magnetite: process optimization, kinetics, and degradation pathways. J Molec Catal A 349(1–2):71–79
85. Mart'ınez F, Calleja G, Melero JA, Molina R (2005) Heterogeneous photo-Fenton degradation of phenolic aqueous solutions over iron-containing SBA-15 catalyst. Appl Catal B 60(3–4):181–190
86. Molina R, Mart'ınez F, Melero JA, Bremner DH, Chakinala AG (2006) Mineralization of phenol by a heterogeneous ultrasound/Fe-SBA-15/H_2O_2 process: multivariate study by factorial design of experiments. Appl Catal B 66(3–4):198–207
87. Shukla P, Wang S, Sun H, Ang HM, Tad'e M (2010) Adsorption and heterogeneous advanced oxidation of phenolic contaminants using Fe loaded mesoporous SBA-15 and H_2O_2. Chem Eng J 164(1):255–260

88. Galeano LA, Vicente MA, Gil A (2011) Treatment of municipal leachate of landfill by fenton-like heterogeneous catalytic wet peroxide oxidation using an Al/Fe-pillared montmorillonite as active catalyst. Chem Eng J 178:146–153
89. Luo M, Bowden D, Brimblecombe P (2009) Catalytic property of Fe-Al pillared clay for Fenton oxidation of phenol by H_2O_2. Appl Catal B 85(3–4):201–206
90. Molina CB, Casas JA, Zazo JA, JJ R'ı (2006) A comparison of Al-Fe and Zr-Fe pillared clays for catalytic wet peroxide oxidation. Chem Eng J 118(1–2):29–35
91. Munoz M, Dominguez CM, de Pedro ZM, Quintanilla A, Casas JA, Rodriguez JJ (2016) Degradation of imidazolium-based ionic liquids by catalytic wet peroxide oxidation with carbon and magnetic iron catalysts. J Chem Technol Biotechnol 91(11):2882–2887
92. Munoz M, de Pedro ZM, Menendez N, Casas JA, Rodriguez JJ (2013) Appl Catal Environ 136–137:218–224
93. Xu L, Wang J (2011) A heterogeneous Fenton-like system with nanoparticulate zero-valent iron for removal of 4-chloro-3-methyl phenol. J Hazard Mater 186:256–264
94. Parra S, Henao L, Mielczarski E, Mielczarski J, Albers P, Suvorova E (2004) Synthesis, testing, and characterization of a novel nafion membrane with superior performance in photo assisted immobilized Fenton catalysis. Langmuir 20:5621–5629. http://dx.doi.org/10.1021/la049768d
95. Aleksić M, Kušić H, Koprivanac N, Leszczynska D, Božić AL (2010) Heterogeneous Fenton type processes for the degradation of organic dye pollutant in water-the application of zeolite assisted AOPs. Desalination 257:22–29. http://dx.doi.org/10.1016/j.desal.2010.03.016
96. Gonzalez-Olmos R, Holzer F, Kopinke FD, Georgi A (2011) Indications of the reactive species in a heterogeneous Fenton-like reaction using Fe-containing zeolites. Appl Catal A Gen 398:44–53. http://dx.doi.org/10.1016/j.apcata.2011.03.005
97. Gonzalez-Olmos R, Martin MJ, Georgi A, Kopinke FD, Oller I, Malato S (2012) Fe-zeolites as heterogeneous catalysts in solar Fenton-like reactions at neutral pH. Appl Catal Environ 125:51–58. http://dx.doi.org/10.1016/j.apcatb.2012.05.022
98. Tekbas M, CengizYatmaz H, Bektas N (2008) Heterogeneous photo-Fenton oxidation of reactive Azo dye solutions using iron exchanged zeolite as a catalyst. Micropor Mesopor Mat 115:594–602. http://dx.doi.org/10.1016/j.micromeso.2008.03.001
99. Pirkanniemi K, Sillanpää M (2002) Heterogeneous water phase catalysis as an environmental application: a review. Chemosphere 48:1047–1060. http://dx.doi.org/10.1016/S0045-6535(02)00168-6
100. Soon AN, Hameed BH (2013) Degradation of acid blue 29 in visible light radiation using iron modified mesoporous silica as heterogeneous photo-Fenton catalyst. Appl Catal A Gen 450:96–105. http://dx.doi.org/10.1016/j.apcata.2012.10.025
101. Ramirez JH, Vicente MA, Madeira LM (2010) Heterogeneous photo-Fenton oxidation with pillared clay-based catalysts for wastewater treatment: review. Appl Catal Environ 98:10–26. http://dx.doi.org/10.1016/j.apcatb.2010.05.004
102. Martínez F, Pariente MI, Ángel J, Botas JA, Melero JA, Rubalcaba A (2012) Influence of preoxidizing treatments on the preparation of iron-containing activated carbons for catalytic wet peroxide oxidation of phenol. J Chem Technol Biotechnol 87:880–886. http://dx.doi.org/10.1002/jctb.2744
103. Santos A, Yustos P, Rodríguez S, Garcia-Ochoa F, de Gracia M (2007) Decolorization of textile dyes by wet oxidation using activated carbon as catalyst. Ind Eng Chem Res 46:2423–2427. http://dx.doi.org/10.1021/ie0614576
104. Ezzatahmadi N, Ayoko GA, Millar GJ, Speight R, Yan C, Li J, Li S, Zhu J, Xi Y (2017) Clay-supported nanoscale zero-valent iron composite materials for the remediation of contaminated aqueous solutions: a review. Chem Eng J 312:336–350
105. Tušar NN, Maucec D, Rangus M, Arcon I, Mazaj M, Cotman M, Pintar A, Kaucic V (2012) Manganese functionalized silicate nanoparticles as a Fenton-type catalyst for water purification by advanced oxidation processes (AOP). Adv Funct Mater 22:820–826

106. Yao Y, Cai Y, Wu G, Wei F, Li X, Chen H, Wang S (2015) Sulfate radicals induced from peroxymonosulfate by cobalt manganese oxides ($CoxMn_3$-xO_4) for Fenton-like reaction in water. J Hazard Mater 296:128–137
107. Rhadfi T, Piquemal JY, Sicard L, Herbst F, Briot E, Benedetti M, Atlamsani A (2010) Polyol-made Mn_3O_4 nanocrystals as efficient Fenton-like catalysts. Appl Catal A Gen 386:132–139
108. Karthikeyan S, Boopathy R, Sekaran G (2015) In situ generation of hydroxyl radical by cobalt oxide supported porous carbon enhance removal of refractory organics in tannery dyeing wastewater. J Colloid Interface Sci 448:163–174
109. Comninellis C, Kapalka A, Malato S, Parsons SA, Poulios I, Mantzavinos D (2008) Advanced oxidation processes for water treatment: advances and trends for R&D. J Chem Technol Biotechnol 83:769–776
110. Pereira MC, Oliveira LCA, Murad E (2012) Iron oxide catalysts: Fenton and Fenton-like reactions – a review. Clay Miner 47:285–302
111. Rahim Pouran S, Abdul Raman AA, Wan Daud WMA (2014) Review on the application of modified iron oxides as heterogeneous catalysts in Fenton reactions. J Clean Prod 64:24–35
112. Martins RC, Henriques LR, Quinta-Ferreira RM (2013) Catalytic activity of low cost materials for pollutants abatement by Fenton's process. Chem Eng Sci 100:225–233
113. Lee H, Kim BH, Park YK, Kim SJ, Jung SC (2015) Application of recycled zero-valent iron nanoparticle to the treatment of wastewater containing nitrobenzene. J Nanomater 2015:8 p. Article ID 392537. Hindawi Publishing Corporation. http://dx.doi.org/10.1155/2015/392537
114. Pereira WS, Freire RS (2006) Azo dye degradation by recycled waste zero-valent iron powder. J Braz Chem Soc 17(5):832–838
115. Manu B, Mahamood S, Vittal H, Shrihari S (2011) A novel catalytic route to degrade paracetamol by Fenton process. Int J Res Chem Environ 1(1):157–164
116. Shahidi D, Roy R, Azzouz A (2015) Advances in catalytic oxidation of organic pollutants-prospects for thorough mineralization by natural clay catalysts. Appl Catal Environ 174:277–292
117. Zazo JA, Casas JA, Mohedano AF, Rodriguez JJ (2006) Catalytic wet peroxide oxidation of phenol with a Fe/active carbon catalyst. Appl Catal Environ 65:261–268
118. Rey A, Faraldos M, Casas JA, Zazo JA, Bahamonde A, Rodriguez JJ (2009) Catalytic wet peroxide oxidation of phenol over Fe/AC catalysts: influence of iron precursor and activated carbon surface. Appl Catal Environ 86:69–77
119. Bautista P, Mohedano AF, Menendez N, Casas JA, Rodriguez JJ (2010) Catalytic wet peroxide oxidation of cosmetic wastewaters with Fe-bearing catalysts. Catal Today 151:148–152
120. Guo Y, Yang L, Cheng X, Xiangtao Wang X (2012) The application and reaction mechanism of catalytic ozonation in water treatment. J Environ Anal Toxicol 2:150. doi:10.4172/2161-0525.1000150
121. Krzemińska D, Neczaj E, Borowski G (2015) Advanced oxidation processes for food industrial wastewater decontamination. Rev Article J Ecol Eng 16(2):61–71
122. Pera-Titus M, Garcia-Molina V, Banos MA, Gimenez J, Esplugas S (2004) Degradation of chlorophenols by means of advanced oxidation processes: a general review. Appl Catal Environ 47:219–256
123. Gogate PR, Pandit AB (2004) A review of imperative technologies for wastewater treatment I: oxidation technologies at ambient conditions. Adv Environ Res 8(3–4):501–551
124. Augugliaro V, Litter M, Palmisano L, Soria J (2006) The combination of heterogeneous photocatalysis with chemical and physical operations: a tool for improving the photoprocess performance. J Photochem Photobiol C Photochem Rev 7(4):127–144
125. Weiss J (1935) Investigations on the radical HO_2 in solution. Trans Faraday Soc 31:668–681
126. Masschelein WJ (1992) Unit processes in drinking water treatment. Marcel Dekker, New York

127. Munter R (2001) Advanced oxidation processes – current status and prospects. Proc Est Acad Sci Chem 50(2):59–80
128. Wang JL, LJ X (2012) Advanced oxidation processes for wastewater treatment: formation of hydroxyl radical and application. Crit Rev Environ Sci Tech 42:251–325
129. Kasprzyk-Hordern B, Ziółek M, Nawrocki J (2003) Catalytic ozonation and methods of enhancing molecular ozone reactions in water treatment. Appl Catal Environ 46:639–669
130. Legube B, Leitner NKV (1999) Catalytic ozonation: a promising advanced oxidation technology for water treatment. Catal Today 53:61–72
131. Abouzlam M, Ouvrard R, Mehdi D, Pontlevoy F, Gombert B, Leitner NKV, Boukari S (2013) An optimal control of a wastewater treatment reactor by catalytic ozonation. Control Eng Practice 21:105–112
132. Abouzlam M, Ouvrard R, Mehdi D, Pontlevoy F, Gombert B, Leitner NKV, Boukari S (2015) A H_∞ control for optimizing the advanced oxidation processes-case of a catalytic ozonation reactor. Control Eng Practice 44:1–9
133. Pirgalıoglu S, Ozbelge TA (2009) Comparison of non catalytic and catalytic ozonation processes of three different aqueous single dye solutions with respect to powder copper sulfide catalyst. Appl Catal A: General 363:157–163
134. Munoz MSG (2010) Catalytic ozonation of pharmaceuticals in aqueous solution, PhD thesis, Alcala University
135. Nawrocki J, Kasprzyk-Hordern B (2010) The efficiency and mechanisms of catalytic ozonation. Appl Catal Environ 99:27–42
136. Azzouz A, Kotbi A, Niquette P, Sajin T, Ursu AV, Rami A, Monette F, Hausler R (2010) Ozonation of oxalic acid catalyzed by ion-exchanged montmorillonite in moderately acidic media. React Kinet Mech Catal 99:289–302
137. Liotta LF, Gruttadauria M, DiCarlo G, Perrini G, Librando V (2009) Heterogeneous catalytic degradation of phenolic substrates: catalysts activity. J Hazard Mater 162:588–606
138. Xiao J, Xie Y, Cao H (2015) Organic pollutants removal in wastewater by heterogeneous photocatalytic ozonation. Chemosphere 121:1–17
139. Pocostales P, Álvarez P, Beltrán FJ (2011) Catalytic ozonation promoted by alumina-based catalysts for the removal of some pharmaceutical compounds from water. Chem Eng J 168:1289–1295
140. Liu Y, Wang S, Gong W, Chen Z, Liu H, Bu Y, Zhang Y (2017) Heterogeneous catalytic ozonation of p-chloronitrobenzene (pCNB) in water with iron silicate doped hydroxylation iron as catalyst. Catal Commun 89:81–85
141. Li B, Xu X, Zhu L, Ding W, Mahmood Q (2010) Catalytic ozonation of industrial wastewater containing chloro and nitro aromatics using modified diatomaceous porous filling. Desalination 254:90–98
142. Wang J, Bai Z (2017) Fe-based catalysts for heterogeneous catalytic ozonation of emerging contaminants in water and wastewater. Chem Eng J 312:79–98
143. Mehrjouei M, Müller S, Möller D (2015) A review on photocatalytic ozonation used for the treatment of water and wastewater. Chem Eng J 263:209–219
144. Orge CA, Órfão JJM, Pereira MFR (2012) Carbon xerogels and ceria–carbon xerogel materials as catalysts in the ozonation of organic pollutants. Appl Catal Environ 126:22–28
145. Chen C, Yu J, Yoza BA, Li QX, Wang G (2015) A novel "wastes-treat-wastes" technology: role and potential of spent fluid catalytic cracking catalyst assisted ozonation of petrochemical wastewater. J Environ Manage 152:58–65
146. Gilbert E (2002) Influence of ozone on the photocatalytic oxidation of organic compounds. Ozone Sci Eng 24:75–82
147. Rey A, Mena E, Chávez AM, Beltrán FJ, Medina F (2015) Influence of structural properties on the activity of WO_3 catalysts for visible light photocatalytic ozonation. Chem Eng Sci 126:80–90
148. Sreethawong T, Chavadej S (2008) Color removal of distillery wastewater by ozonation in the absence and presence of immobilized iron oxide catalyst. J Hazard Mater 155:486–493

149. Jung H, Choi H (2006) Catalytic decomposition of ozone and parachlorobenzoic acid (pCBA) in the presence of nanosized ZnO. Appl Catal Environ 66:288–294
150. Gharbani P, Mehrizad A (2014) Heterogeneous catalytic ozonation process for removal of 4-chloro-2-nitrophenol from aqueous solutions. J Saudi Chem Soc 18:601–605
151. Hu E, Wu X, Shang S, Tao X, Jiang S, Gan L (2016) Catalytic ozonation of simulated textile dyeing wastewater using mesoporous carbon aerogel supported copper oxide catalyst. J Clean Prod 112:4710–4718
152. Bergaya F, Theng BKG, Lagaly G (2006) Handbook of clay science. Elsevier, Oxford, pp 541–546
153. Shahidi D, Roy R, Azzouz A (2014) Total removal of oxalic acid via synergistic parameter interaction in montmorillonite catalyzed ozonation. J Environ Chem Eng 2:20–30
154. Aly AA, Hasan YNY, Al-Farraj AS (2014) Olive mill wastewater treatment using a simple zeolite-based low-cost method. J Environ Manage 145:341–348
155. Li H, Xu B, Qi F, Sun D, Chen Z (2014) Degradation of bezafibrate in wastewater by catalytic ozonation with cobalt doped red mud: efficiency, intermediates and toxicity. Appl Catal Environ 152:342–351
156. Xu B, Qi F, Zhang J, Li H, Sun D, Robert D, Chen Z (2016) Cobalt modified red mud catalytic ozonation for the degradation of bezafibrate in water: catalyst surface properties characterization and reaction mechanism. Chem Eng J 284:942–952
157. Wen G, Pan ZH, Ma J, Liu ZQ, Zhao L, Li JJ (2012) Reuse of sewage sludge as a catalyst in ozonation – efficiency for the removal of oxalic acid and the control of bromate formation. J Hazard Mater 239–240:381–388
158. Wu J, Ma L, Chen Y, Cheng Y, Liu Y, Zha X (2016) Catalytic ozonation of organic pollutants from bio-treated dyeing and finishing wastewater using recycled waste iron shavings as a catalyst: removal and pathways. Water Res 92:140–148
159. Kamboj ML (2009) Studies on the degradation of industrial wastewater using heterogeneous photocatalysis master thesis. Thapar University, Patiala
160. Braslavsky SE, Houk KN (1988) Glossary of terms used in photochemistry. Pure Appl Chem 60:1055–1106
161. Verhoven JW (1996) Glossary of terms used in photochemistry. Pure Appl Chem 68:2223–2286
162. Cesaro A, Belgiorno V (2015) Removal of endocrine disruptors from urban wastewater by advanced oxidation processes (AOPs): a review. Open Biotechnol J 9:1–28
163. Fujishima A, Honda K (1972) Electrochemical photolysis of water at a semiconductor electrode. Nature 238:37–38
164. Renge VC, Khedkar SV, Thanvi NJ (2012) Photocatalytic oxidation and reactors – a review. Int J Adv Eng Technol 3(4):31–35
165. Kaan CC, Aziz AA, Ibrahim S, Matheswaran M, Saravanan P (2012) Heterogeneous photocatalytic oxidation an effective tool for wastewater treatment – a review, studies on water management issues, Dr. Muthukrishnavellaisamy Kumarasamy (ed) ISBN: 978-953-307-961-5, InTech. [Online] http://www.intechopen.com/books/studies-on-water-managementissues/heterogeneous-photocatalytic-oxidation-an-effective-tool-for-wastewater-treatment-a-review
166. Ibhadon AO, Fitzpatrick P (2013) Heterogeneous photocatalysis: recent advances and applications. Catalysts 3:189–218
167. Fujishima A, Zhang X, Tryk DA (2007) Heterogeneous photocatalysis: from water photolysis to applications in environmental cleanup. Int J Hydro Energy 32:2664–2672
168. Hoffmann HR, Martin ST, Choi W, Bahnemann DW (1995) Environmental applications of semiconductor photocatalysis. Chem Rev 69:95–101
169. Lee JC, Kim MS, Kim CK, Chung CH, Cho SM, Han GY, Yoon KJ, Kim BW (2003) Removal of paraquat in aqueous suspension of TiO_2 in an immersed UV photoreactor. Korean J Chem Eng 20(5):862–868

170. Mok NB (2009) Photocatalytic degradation of oily wastewater: effect of catalyst concentration load, irradiation time and temperature. Bachelor thesis, Faculty of Chemical & Natural Resources Engineering, University Malaysia Pahang
171. Zhang T, Wang X, Zhang X (2014) Recent progress in TiO_2-mediated solar photocatalysis for industrial wastewater treatment. Int J Photoenergy 12 p. Article ID 607954. Hindawi Publishing Corporation. [Online] http://dx.doi.org/10.1155/2014/607954
172. Khataee AR, Zarei M, Ordikhani-Seyedlar R (2011) Heterogeneous photocatalysis of a dye solution using supported TiO_2 nanoparticles combined with homogeneous photoelectrochemical process: molecular degradation products. J Molec Catal A Chem 338:84–91
173. Ahmed S, Rasul MG, Brown R, Hashi MA (2011) Influence of parameters on the heterogeneous photocatalytic degradation of pesticides and phenolic contaminants in wastewater: a short review. J Environ Manage 92:311–330
174. Bockelmann D, Weichgrebe D, Goslich R, Bahnemann D (1995) Concentrating versus non-concentrating reactors for solar water detoxication. Sol Energ Mater Sol Cell 38:441–251
175. Banu JR, Anandan S, Kaliappan S, Yeom IY (2008) Treatment of dairy wastewater using anaerobic and solar photocatalytic methods. Sol Energy 82:812–819
176. Chong MN, Jin B, Chow CWK, Saint C (2010) Recent developments in photocatalytic water treatment technology: a review. Water Res 44:2997–3027
177. Zhang H, Liu G, Shi L, Liu H, Wang T, Ye J (2016) Engineering coordination polymers for photocatalysis. Nano Energy 22:149–168
178. Kumar P, Kumar S, Bhardwaj NK, Kumar S (2011) Titanium dioxide photocatalysis for the pulp and paper industry wastewater treatment. Ind J Sci Technol 4(3):327–332
179. Chen Q, Ji F, Guo Q, Fan J, Xu X (2014) Combination of heterogeneous Fenton-like reaction and photocatalysis using Co-TiO_2 nanocatalyst for activation of $KHSO_5$ with visible light irradiation at ambient conditions. J Environ Sci 26:2440–2450
180. García-Muñoz P, Pliego G, Zazo JA, Bahamonde A, Casas JA (2016) Ilmenite ($FeTiO_3$) as low cost catalyst for advanced oxidation processes. J Environ Chem Eng 4:542–548
181. Sha Z, Sun J, Chan HSO, Jaenicke S, Wu J (2015) Enhanced photocatalytic activity of the AgI/UiO-66(Zr) composite for Rhodamine B degradation under visible-light irradiation. Chem Plus Chem 80(8):1321–1328. http://dx.doi.org/10.1002/cplu.201402430
182. Booshehri AY, Polo-Lopez MI, Castro-Alférez M, He P, Xu R, Rong W, Malato S, Fernández-Ibáñez P (2017) Assessment of solar photocatalysis using Ag/$BiVO_4$ at pilot solar compound parabolic collector for inactivation of pathogens in well water and secondary effluents. Catal Today 281:124–134
183. Martínez C, Canle LM, Fernández MI, Santaballa JA, Faria J (2011) Aqueous degradation of diclofenac by heterogeneous photocatalysis using nanostructured materials. Appl Catal Environ 107:110–118
184. Lam SM, Sin JC, Mohamed AR (2016) A review on photocatalytic application of g-C3N4/semiconductor (CNS) nanocomposites towards the erasure of dyeing wastewater. Mater Sci Semicond Process 47:62–84
185. Shan AY, Ghaz TIM, Rashid SA (2010) Immobilisation of titanium dioxide onto supporting materials in heterogeneous photocatalysis: a review. Appl Catal A General 389:1–8
186. Rajeshwar K, Chenthamarakshan CR, Goeringer S, Djukic M (2001) Titania based heterogeneous photocatalysis: materials mechanistic issues and implications for environmental remediation. Pure Appl Chem 73(12):1849–1860
187. Murphy S (2012) Photocatalytic degradation of pharmaceuticals in aqueous solutions and development of new dye sensitised photocatalytic materials, PhD thesis, Dublin City University
188. Osarumwense JO, Amenaghawn NA, Aisien FA (2015) Heterogeneous photocatalytic degradation of phenol in aqueous suspension of periwinkle shell ash catalyst in the presence of UV from sunlight. J Eng Sci Technol 10(12):1525–1539

189. Ahmed S, Rasul MG, Martens WN, Brown R, Hashib MA (2011) Advances in heterogeneous photocatalytic degradation of phenols and dyes in wastewater: a review. Water Air Soil Pollut 215(1–4):3–29
190. Jin YX, Li GH, Zhang Y, Zhang YX, Zhang LD (2001) Photoluminescence of anatase TiO_2 thin films achieved by the addition of $ZnFe_2O_4$. J Phys Condens Matter 13:L913–L918
191. Belgiorno V, Rizzo L, Fatta D, Rocca CD, Lofrano G, Nikolaouc A, Naddeo V, Meric S (2007) Review on endocrine disrupting-emerging compounds in urban wastewater: occurrence and removal by photocatalysis and ultrasonic irradiation for wastewater reuse. Desalination 215:166–176
192. Hassan M, Zhao Y, Xie B (2016) Employing TiO_2 photocatalysis to deal with landfill leachate: current status and development. Chem Eng J 285:264–275
193. Singh S, Singh PK, Mahalingam H (2015) An effective and low-cost TiO_2/polystyrene floating photocatalyst for environmental remediation. Int J Environ Res 9(2):535–544
194. Zimmermann FJ (1954) Waste disposal. US Patent No. 2 665 249, US Patent Office 6 (10):630–631
195. Zimmermann FJ, Diddams DG (1960) The Zimmermann process and its application in the pulp and paper industry. TAPPI 43:710–715
196. Zou LY, Li Y, Hung YT (2007) Wet air oxidation for waste treatment. In: Wang LK, Hung YT, Shammas NK (eds) Advanced physicochemical treatment technologies. Handbook of Environmental Engineering, vol 5. Humana Press. https://doi.org/10.1007/978-1-59745-173-4_13
197. Zou L, Zhu B (2006) Literature review report for smart water project "improving recycled water aesthetic quality by removing colour and trace organics." Oxidation processes for degradation of organic pollutants in water. Institute of Sustainability and Innovation Victoria University
198. Roy S, Vashishtha M, Saroha AK (2010) Catalytic wet air oxidation of oxalic acid using platinum catalysts in bubble column reactor: a review. J Eng Sci Technol Rev 3(1):95–107
199. Copa WM, Dietrich MJ (1988) Wet air oxidation of oils, oil refinery sludges, and spent drilling muds. ZIMPRO/PASSAVANT INC., Apr 1988. [Online] http://infohouse.p2ric.org/ref/25/24892.pdf
200. Siemens Water Technologies Corp (2011) Can you treat the most difficult wastewater with only air?, Zimpro® wet air oxidation systems: the cleanest way to treat the dirtiest water, answers for industry, Siemens Water Technologies Corp, GIS-WAO-BR-0111. [Online] http://www.energy.siemens.com/hq/pool/hq/industries-utilities/oil-gas/water-solutions/Zimpro-Wet-Air-Oxidation-System-The-Cleanest-Way.pdf
201. Gaikwad RW, Malik I, Kulkarni V, Mhaske S, Badadhe S (2016) Review on catalytic wet air oxidation. Int J Environ Natural Sci 9:1–8
202. Debellefontaine H, Foussard JN (2000) Wet air oxidation for the treatment of industrial wastes. Chemical aspects, reactor design and industrial applications in Europe. Waste Manag 20:15–25
203. Kolaczkowski ST, Plucinski P, Beltran FJ, Rivas FJ, McLurgh DB (1999) Wet air oxidation: a review of process technologies and aspects in reactor design. Chem Eng J 73:143–160
204. Ovejero G, Sotelo JL, Garcia J, Rodrıguez A (2005) Catalytic removal of phenol from aqueous solutions in a trickle bed reactor. J Chem Technol Biotechnol 80:406–412
205. Rodr'ıguez A, Ovejero G, Romero MD, Diaz C, Barreiro M, Garcia J (2008) Catalytic wet air oxidation of textile industrial wastewater using metal supported on carbon nanofibers. J Super Fluids 46:163–172
206. Hong TY (2013) Catalytic wet air oxidation of wastewater containing acetic acid. BSc thesis, Faculty of Chemical and Natural Resources Engineering, Universiti Malaysia Pahang
207. Moses DV, Smith EA (1954) Wet air oxidation of aqueous wastes, US Patent 2,690,425
208. Levec J, Pintar A (2007) Catalytic wet-air oxidation processes: a review. Catal Today 124:172–184
209. Luck F (1999) Wet air oxidation: past, present and future. Catal Today 53:81–91

210. Hosseini AM (2013) Intensification of wet oxidation of industrial process wastewater. Department of Chemical and Environmental Process Engineering of BME and Centre of Energy Research of Hungarian Academy of Sciences
211. Trunfio G Catalyst development for the catalytic wet air oxidation (CWAO) of phenol. Thesis for the degree of Doctor of Philosophy in "Chemical Technologies and Innovative Processes," University of Messina, Area 03-Scienze Chimiche (CHIM 04), CYCLE XXI (2006–2008)
212. Oliviero L, Barbier J, Duprez D (2003) Wet air oxidation of nitrogen-containing organic compounds and ammonia in aqueous media. Appl Catal Environ 40:163–184
213. Erjavec B, Kaplana R, Djinovic P, Pintar A (2013) Catalytic wet air oxidation of bisphenol a model solution in a trickle-bed reactor over titanate nanotube-based catalysts. Appl Catal Environ 132-133:342–352
214. Modi RR, Vyas DS, Patel SM (2016) Catalytic wet air oxidation of dye industry wastewater using metallic catalyst. IJARIIE 2(3):461–469
215. Jani HR (2008) Catalytic wet air oxidation of pulp and paper mills effluent. PhD thesis, School of Applied Sciences, Science, Engineering, and Technology Portfolio, RMIT University
216. Gomes HT, Figueiredo JL, Faria JL (2007) Catalytic wet air oxidation of olive mill wastewater. Catal Today 124:254–259
217. Imamura S (1999) Catalytic and noncatalytic wet oxidation. Ind Eng Chem Res 38:1743–1753
218. Matatov-Meytal Y, Sheintuch M (1998) Catalytic abatement of water pollutants. Ind Eng Chem Res 37:309–326
219. Eftaxias A (2002) Catalytic wet air oxidation of phenol in a trickle bed reactor: kinetics and reactor modelling, PhD thesis, Escola T'ecnica Superior de Enginyeria Qu'ımica, Departament d'Enginyeria Qu'ımica, Universitat Rovira i Virgili, Tarragona
220. Katsoni A, Gomes HT, Pastrana-Martínez LM, Faria JL, Figueiredoc JL, Mantzavinos D, Silva AMT (2011) Degradation of trinitrophenol by sequential catalytic wet air oxidation and solar TiO_2 photocatalysis. Chem Eng J 172:634–640
221. Stüber F, Font J, Fortuny A, Bengoa C, Eftaxias A, Fabregat A (2005) Carbon materials and catalytic wet air oxidation of organic pollutants in wastewater. Topics Catal 33:3–50
222. Morales-Torres S, Silva AMT, Pérez-Cadenas AF, Faria JL, Maldonado-Hódar FJ, Figueiredo JL, Carrasco-Marín F (2010) Wet air oxidation of trinitrophenol with activated carbon catalysts: effect of textural properties on the mechanism of degradation. Appl Catal Environ 100:310–317
223. Gomes HT, Machado BF, Ribeiro A, Moreira I, Rosário M, Silva AMT, Figueiredo JL, Faria JL (2008) Catalytic properties of carbon materials for wet oxidation of aniline. J Hazard Mater 159:420–426
224. Apolinário AC, Silva AMT, Machado BF, Gomes HT, Araújo PP, Figueiredo JL, Faria JL (2008) Wet air oxidation of nitro-aromatic compounds: reactivity on single- and multi-component systems and surface chemistry studies with a carbon xerogel. Appl Catal Environ 84:75–86
225. Yang S, Zhu W, Li X, Wang J, Zhou Y (2007) Multi-walled carbon nanotubes (MWNTs) as an efficient catalyst for catalytic wet air oxidation of phenol. Catal Commun 8:2059–2063
226. Yang S, Li X, Zhu W, Wang J, Descorme C (2008) Catalytic activity, stability and structure of multi-walled carbon nanotubes in the wet air oxidation of phenol. Carbon 46:445–452
227. Rocha RP, Sousa JP, Silva AMT, Pereira MFR, Figueiredo JL (2011) Catalytic activity and stability of multiwalled carbon nanotubes in catalytic wet air oxidation of oxalic acid: the role of the basic nature induced by the surface chemistry. Appl Catal Environ 104:330–336
228. Sousa JPS, Silva AMT, Pereira MFR, Figueiredo JL (2010) Wet air oxidation of aniline using carbon foams and fibers enriched with nitrogen. Sep Sci Technol 45:1546–1554
229. Yang S, Cui Y, Sun Y, Yang H (2014) Graphene oxide as an effective catalyst for wet air oxidation of phenol. J Hazard Matter 280:55–62

230. Zhang Y, Peng F, Zhou Y (2016) Structure, characterization, and dynamic performance of a wet air oxidation catalyst Cu-Fe-La/γ-Al_2O_3. Chin J Chem Eng 24:1171–1177
231. Nakatsuji T, Kunishige M, Li J, Hashimoto M, Matsuzono Y (2013) Effect of CeO_2 addition into Pd/Zr–Pr mixed oxide on three-way catalysis and thermal durability. Catal Comm 35:88–94
232. Bernardi M, ML D, Dodouche I, Descorme C, Deleris S, Blanchet E, Besson M (2012) Selective removal of the ammonium-nitrogen in ammonium acetate aqueous solutions by catalytic wet air oxidation over supported Pt catalysts. Appl Catal Environ 128:64–71
233. Wang C, Wang GR, Wang JF (2014) A bi-component Cu catalyst for the direct synthesis of methylchlorosilane from silicon and methyl chloride. Chin J Chem Eng 22:299–304
234. Massa P, Ivorra F, Haure P, Fenoglio R (2011) Catalytic wet peroxide oxidation of phenol solutions over CuO/CeO_2 systems. J Hazard Mater 190:1068–1073
235. Fazlollahi F, Sarkari M, Gharebaghi H, Atashi H, Zarei MM, Mirzaei AA, Hecker WC (2013) Preparation of Fe–Mn/K/Al_2O_3 Fischer–Tropsch catalyst and its catalytic kinetics for the hydrogenation of carbon monoxide. Chin J Chem Eng 21:507–519
236. Wenbing M, Hongpeng L, Xuemei M (2013) Study on supercritical water oxidation of oily wastewater with ethanol. Res J Appl Sci Eng Technol 6(6):1007–1011
237. Bambang V, Jae-Duck K (2007) Supercritical water oxidation for the destruction of toxic organic wastewaters: a review. J Environ Sci 19:513–522
238. Fourcault A, Garcia-Jarana B, Sanchez-Oneto J, Mariasa F, Portela JR (2009) Supercritical water oxidation of phenol with air. Experimental results and modeling. Chem Eng J 152:227–233
239. Paraskeva P, Diamadopoulos E (2006) Technologies for olive mill wastewater (OMW) treatment: a review. J Chem Technol Biotechnol 81:1475–1485
240. Xu D, Wang S, Tang X, Gong Y, Guo Y, Wang Y, Zhang J (2012) Design of the first pilot scale plant of China for supercritical water oxidation of sewage sludge. Chem Eng Res Des 90:288–297
241. Han D, Zhang H, Fang L, Lin C (2015) Continuous monitoring of total organic carbon based on supercritical water oxidation improved by $CuSO_4$ catalyst. J Anal Bioanal Tech S13:002. doi:10.4172/2155-9872.S13-002
242. Marrone PA (2013) Supercritical water oxidation-current status of full-scale commercial activity for waste destruction. J Super Fluids 79:283–288
243. Li X, Li G (2015) A review: pharmaceutical wastewater treatment technology and research in China. Asia-Pacific energy equipment engineering research conference (AP3ER 2015), pp 345–348
244. Youngprasert B, Poochinda K, Ngamprasertsith S (2010) Treatment of acetonitrile by catalytic supercritical water oxidation in compact-sized reactor. J Water Resource Protect 2:222–226
245. Dong X, Gan Z, Lu X, Jin W, Yu Y, Zhang M (2015) Study on catalytic and non-catalytic supercritical water oxidation of p-nitrophenol wastewater. Chem Eng J 277:30–39
246. Yu L, Han M, He F (2017) A review of treating oily wastewater. Arabian J Chem 10:S1913–S1922
247. Medoll M (1982) Processing methods for the oxidation of organics in supercritical water. US Patent 4,338,199
248. Abelleira J, Sánchez-Oneto J, Portela JR, Martínez de la Ossa EJ (2013) Kinetics of supercritical water oxidation of isopropanol as an auxiliary fuel and co-fuel. Fuel 111:574–583
249. Ding ZY, Frisch MA, Li L, Gloyna EF (1996) Catalytic oxidation in supercritical water. Ind Eng Chem Res 35:3257–3279
250. Tomita K, Oshima Y (2004) Stability of manganese oxide in catalytic supercritical water oxidation of phenol. Ind Eng Chem Res 43:7740–7743

251. Silva CLD, Garlapalli RK, Trembly JP (2017) Removal of phenol from oil/gas produced water using supercritical water treatment with TiO_2 supported MnO_2 catalyst. J Environ Chem Eng 5:488–493
252. Arslan-Alaton I, Ferry JL (2002) $H_4SiW_{12}O_{40}$-catalyzed oxidation of nitrobenzene in supercritical water: kinetic and mechanistic aspects. Appl Catal B-Environ 38:283–293
253. Nunoura T, Lee G, Matsumura Y, Yamamoto K (2003) Reaction engineering model for supercritical water oxidation of phenol catalyzed by activated carbon. Ind Eng Chem Res 42:3522–3531
254. Krajnc M, Levec J (1997) Oxidation of phenol over a transition-metal oxide catalyst in supercritical water. Ind Eng Chem Res 36:3439–3445
255. Ding ZY, Aki SN, Abraham MA (1995) Catalytic supercritical water oxidation: phenol conversion and product selectivity. Environ Sci Technol 29:2748–2753
256. Zhang X, Savage PE (1998) Fast catalytic oxidation of phenol in supercritical water. Catal Today 40:333–342
257. Yu J, Savage PE (2000) Kinetics of catalytic supercritical water oxidation of phenol over TiO_2. Environ Sci Technol 34:3191–3198
258. Yu J, Savage PE (1999) Catalytic oxidation of phenol over MnO_2 in supercritical water. Ind Eng Chem Res 38:3793–3801
259. Angeles-Hernández MJ, Leeke GA, Santos RC (2008) Catalytic supercritical water oxidation for the destruction of quinoline over MnO_2/CuO mixed catalyst. Ind Eng Chem Res 48:1208–1214
260. Civan F, Özaltun DH, Kıpcak E, Akgün M (2015) The treatment of landfill leachate over Ni/Al_2O_3 by supercritical water oxidation. J Super Fluids 100:7–14
261. Aki SNVK, Abraham MA (1998) An economic evaluation of catalytic supercritical water oxidation: comparison with alternative waste treatment technologies. Environ Prog Sustain Energy 17(4):246–255
262. Lee G, Nunoura T, Matsumura Y, Yamamoto K (2002) Comparison of the effects of the addition of NaOH on the decomposition of 2-chlorophenol and phe-nol in supercritical water and under supercritical water oxidation conditions. J Super Fluids 24:239–250
263. Qi XH, Zhuan YY, Yuan YC, WX G (2002) Decomposition of aniline in supercritical water. J Hazard Mater B 90:51–62
264. Kazemi N, Tavakoli O, Seif S, Nahangi M (2015) High-strength distillery wastewater treatment using catalytic sub- and supercritical water. J Super Fluids 97:74–80
265. Chen JH, Ma CY, Xi DL, Li Q (2011) Study on catalytic supercritical water oxidation process for treating the perfume waste water. Environ Eng 29:36–39
266. Lin KS, Wang HP (2000) Supercritical water oxidation of 2-chlorophenol catalyzed by Cu^{2+} cations and copper oxide clusters. Environ Sci Technol 34:4849–4854

Impact on Disinfection Byproducts Using Advanced Oxidation Processes for Drinking Water Treatment

Brooke K. Mayer and Donald R. Ryan

Abstract Since the inception of drinking water treatment systems, ensuring the production of microbiologically safe drinking water has been a primary objective. While chemical oxidants are often successfully employed to mitigate microbial risks, the chemical reactions that occur between oxidants and the dissolved or particulate constituents present in source waters, e.g., natural organic matter (NOM), can produce byproducts associated with unintended health consequences. These disinfection byproducts (DBPs) are potentially carcinogenic, mutagenic, genotoxic, and/or teratogenic. Since the discovery of DBPs in the early 1970s, considerable effort has been afforded to develop regulations or guidelines striving to simultaneously control microbial pathogens and DBPs. As advanced oxidation processes (AOPs) gain traction as an integral part of advanced treatment trains in water, wastewater, and water reuse scenarios, their impact on DBPs, in terms of both formation and destruction, is an increasingly important consideration and is the focus of this chapter.

This chapter begins with a brief overview of major drinking water disinfection processes, followed by an introduction to common classes of disinfection byproducts (DBPs) and their precursors, and concludes with discussion of the influence of AOPs on DBP formation, formation potential, and removal.

Keywords Disinfection byproducts (DBP), Haloacetic acids (HAA), Natural organic matter (NOM), Oxidation, Trihalomethanes (THM)

B.K. Mayer (✉) and D.R. Ryan
Department of Civil, Construction and Environmental Engineering, Marquette University, Milwaukee, WI 53151, USA
e-mail: Brooke.Mayer@marquette.edu; Donald.Ryan@marquette.edu

Contents

1 Brief Introduction to Disinfection 346
   1.1 Common Disinfectants 347
   1.2 Advanced Oxidation Process-Based Disinfection 349
2 Disinfection Byproducts 349
   2.1 Disinfection Byproduct Regulations 354
   2.2 Conventional Disinfection Byproducts: Trihalomethanes, Haloacetic Acids, and Oxyhalides 355
   2.3 Emerging Disinfection Byproducts, Including Brominated, Iodinated, and Nitrogenous Species 357
   2.4 Non-halogenated Organic Disinfection Byproducts 359
3 Disinfection Byproduct Precursors 359
   3.1 Bulk Organic Measurements 360
   3.2 Natural Organic Matter Fractionation 361
   3.3 Structure and Reactivity of Natural Organic Matter (NOM) Fractions 364
   3.4 Synthetic Organic Disinfection Byproduct Precursors 366
4 Disinfection Byproduct Formation in Relation to Advanced Oxidation Processes 366
   4.1 Disinfection Byproduct Formation in Advanced Oxidation Processes 367
   4.2 The Impact of Advanced Oxidation Processes on Disinfection Byproduct (DBP) Formation Potential: Influence of Oxidation on Precursor Organic Matter and Subsequent DBP Formation 370
5 Mitigation of Disinfection Byproducts 372
   5.1 Alternative Disinfectants 372
   5.2 Removal of Precursor Natural Organic Matter Prior to Disinfection 374
   5.3 Mitigation of Preformed Disinfection Byproducts Using Advanced Oxidation Processes 376
6 Summary 378
References 379

1 Brief Introduction to Disinfection

The five most commonly used drinking water disinfection strategies are free chlorine, combined chlorine, chlorine dioxide, ozone, and ultraviolet (UV) light. Figure 1 shows the relative distribution of the use of these disinfectants in the United States based on surveys conducted by the American Water Works Association (AWWA).

UV inactivates microorganisms on the basis of electromagnetic radiation, which primarily disrupts nucleic acids. Each of the other four common disinfection processes involves additions of oxidizing chemicals. These oxidizers damage microbial proteins (amino acids) and genomes (nucleic acids) [3], thereby preventing microbial replication, i.e., causing inactivation. However, these oxidants are relatively nonselective, so they also react with other materials in the water (such as natural organic matter, NOM), and can produce disinfection byproducts (DBPs), which are potentially carcinogenic, mutagenic, genotoxic, and/or teratogenic. This is represented by the generalized reaction:

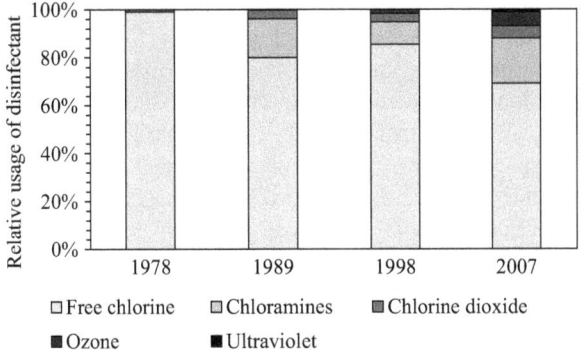

Fig. 1 Temporal summary of relative disinfectant use in municipal drinking water treatment in the United States. Adapted from [1, 2]. The cited surveys reported the use of multiple types of disinfectants for some systems (yielding totals in excess of 100% of the number of systems surveyed); here the relative distribution of type of disinfectant used is shown, resulting in a maximum of 100%

$$\text{Oxidant (e.g., HOCl)} + \text{Precursor (e.g., NOM)} \rightarrow \text{DBP (e.g., CHCl}_3\text{)}$$

The extent of DBP formation is a function of the type of disinfectant used, disinfectant dose, system parameters (e.g., contact time), and water quality (e.g., quantity and character of NOM, pH, and temperature).

1.1 Common Disinfectants

Free chlorine (primarily in the form of HOCl and OCl$^-$ [$E^o = 1.49$ V for HOCl/Cl$^-$]) is typically added as gaseous Cl$_2$ or NaOCl. It is the most commonly used disinfectant for water treatment as it is broadly effective against microbial pathogens and is comparatively inexpensive. Although free chlorine remains the dominant drinking water disinfectant, concerns over chlorine-resistant microbes (e.g., *Cryptosporidium* and *Giardia*) and formation of halogenated DBPs have increased interest in alternative disinfection strategies. Free chlorine is associated with the production of classically regulated DBPs such as trihalomethanes (THMs) and haloacetic acids (HAAs), as well as emerging DBPs, including chloral hydrate, chlorophenols, formaldehyde, haloketones, halogenated furanones, and haloacetonitriles. In general, the production of organochlorine compounds during chlorination is caused by reactions between chlorine and humic substances [4]. There are three general pathways through which free chlorine reacts with water constituents: oxidation, addition, and substitution [5]. When organic compounds have double bonds, the chlorine can undergo an addition reaction, but this reaction is often too slow to be of importance in water treatment. Thus, most chlorinated DBPs are formed via oxidation and substitution reactions, with reactions occurring much more rapidly at high pH than at low pH [4].

Combined chlorine is the sum of the species formed through the reaction of chlorine and ammonia: monochloramine (NH_2Cl), dichloramine ($NHCl_2$), and trichloramine (NCl_3). Combined chlorine, or chloramines, is not as effective as free chlorine for inactivation of pathogens and is thus not commonly used as a primary disinfectant. However, its longer residual makes it a common choice for secondary disinfection to avoid biological regrowth in the distribution system. Chloramines reduce the amount of THMs and HAAs formed in comparison to free chlorine; however, they introduce concerns for nitrosamine and cyanogen chloride DBP formation. Reactions of chloramines with humic materials and amino acids produce haloacetonitriles and non-halogenated acetonitriles, following a pathway similar to that for chlorine [4].

Chlorine dioxide (ClO_2) is a powerful disinfecting agent ($E° = 0.95$ V for ClO_2/ClO_2^-) but is volatile, typically requiring onsite generation. It is widely used as a disinfectant in continental Europe but is not as commonly used in the United States. At the dosages typically used in drinking water treatment, ClO_2 (which is more selective than free chlorine) does not react with NOM, so it avoids formation of THMs and HAAs, and produces almost no identifiable organic byproducts (although low levels of some aldehydes and ketones can result) [6]. Chlorine dioxide reacts only by oxidation, which explains the lack of organochlorine compound formation [4]. Although stable in pure water, ClO_2 decomposes in drinking water as it is photoreactive and can also undergo disproportionation to produce the inorganic DBPs chlorite (ClO_2^-) and chlorate (ClO_3^-), the kinetics and degree of which depend on ambient water quality parameters [7].

Of the common chemical disinfectants, ozone (O_3, $E° = 2.07$ V for O_3/O_2) is the strongest oxidizing agent and must be generated on-site as it is unstable. Ozone provides effective oxidation of many chemical contaminants as well as inactivation of microbial pathogens. Ozonation itself does not produce halogenated DBPs and has thus become increasingly common. However, when bromide is present in waters, ozone can produce brominated DBPs such as bromate (BrO_3^-). In general, bromate forms through a combination of molecular ozone attack and reactions of bromide with free radicals, which are formed as ozone decomposes during water treatment [4]. The radical pathway may play a more important role than the molecular ozone pathway [4]. Ozone can also react with bromide to form brominated organics such as bromoform, dibromoacetonitrile, and dibromoacetone [4, 8]. As ozone progressively degrades complex organics, non-halogenated DBPs such as formaldehyde (CH_2O) may also result.

In the case of UV disinfection, microorganisms are inactivated via disruption of their genetic material (DNA or RNA) rather than via chemical oxidation. For the fluences (i.e., UV dose, which is a function of intensity and exposure time) typically used in drinking water treatment (<200 mJ/cm^2), there is no evidence of DBP formation, nor are DBP levels exacerbated using post-UV disinfection [9].

1.2 Advanced Oxidation Process-Based Disinfection

Recalcitrant contaminants, such as endocrine-disrupting compounds (EDCs), pharmaceuticals and personal care products (PPCPs), pesticides, etc., are typically the primary targets for AOPs, and there are numerous reports of this application. However, the hydroxyl radicals (HO•, $E^o = 2.70$ V) or sulfate radicals (SO_4•$^-$, $E^o = 2.5–3.1$ V) generated from AOPs or sulfate-radical-based AOPs (SR-AOPs), respectively, may also provide some degree of disinfection [10]. The use of AOPs for disinfection is far less frequently reported in comparison to chemical degradation (e.g., for UV/TiO_2 [11–14]) but may be realized as a secondary outcome of the use of AOP treatments in drinking water. Disinfection using solar irradiation-based AOPs may be of particular interest for use in developing countries [15]. Hydroxyl radicals may oxidize and disrupt cell walls and membranes, thereby lysing the cell, or they may diffuse into the cells and react with intracellular components. AOP-based disinfection may be limited by mass transfer through the cell walls or membranes as HO• reacts with most biological molecules at diffusion-controlled rates. Although there may be some oxidative enhancement in virus inactivation due to HO•, UV appears to be primarily responsible for microbial inactivation using UV-based AOPs [10].

Since the oxidation pathway in AOPs relies on radicals rather than halogens, halogenated DBP production during AOP treatments is generally less of a concern in comparison to traditional oxidative disinfectants. However, ozone-associated DBPs may be generated during ozone-based AOPs, and non-halogenated organic DBPs can occur as complex organic matter is degraded. DBP formation during AOP treatments is discussed in greater detail in Sect. 4.1.

2 Disinfection Byproducts

Reactions between oxidizing disinfectants and the organic or inorganic precursor material found in water can lead to the generation of potentially harmful DBPs, as depicted in the generalized illustration in Fig. 2.

The range of DBPs produced via reactions between oxidants and organic precursors (NOM or anthropogenic organic pollutants) or inorganic precursors (i.e., bromide or iodide) include halogenated organics, organic oxidation byproducts, and inorganics. Table 1 lists several important classes of DBPs, their main causative agents, and examples of established drinking water regulations.

The formation of DBPs during drinking water disinfection has been recognized since the 1970s [23, 24], but advances in analytical techniques and risk assessment continue to facilitate discovery and better characterization of the more than 600 currently identified DBPs. Of note, the number of DBPs identified and quantified in water to date is only a small fraction of those potentially formed. As illustrated for

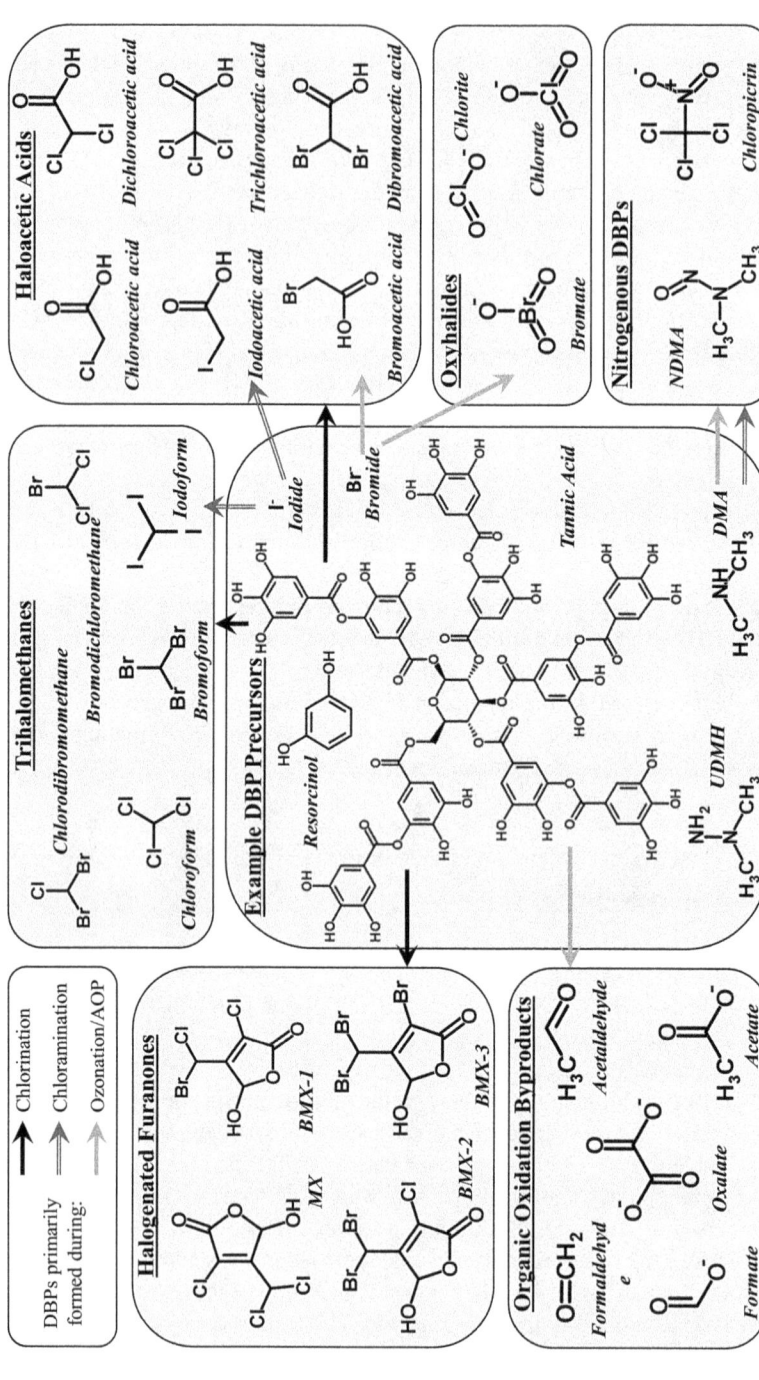

Fig. 2 Examples of disinfection byproducts (DBPs) produced from reactions between oxidizing disinfectants and precursor material, in accordance with the general reaction: Oxidant + Precursor → DBP. All chemical structures are from the ChemSpider website (http://www.chemspider.com/)

Table 1 Examples of classical and emerging drinking water disinfection byproducts (DBPs) produced by conventional oxidative disinfection processes and advanced oxidation processes (AOPs) (all concentrations in µg/L)

		Primarily a byproduct of						Selected drinking water regulations				
Species	Chemical formula	Free chlorine	Chloramines	ClO$_2$	O$_3$	O$_3$-based AOP	Non-O$_3$-based AOP	Australia[a]	European Union (EU)[b]	Health Canada (HC)[c]	US Environmental Protection Agency (USEPA)[d]	World Health Organization (WHO)[e]
Trihalomethanes (THM)	TTHM[f]							250	100	100	80	g
Chloroform	CHCl$_3$	X	X									300
Bromodichloromethane	CHBrCl$_2$	X	X									60
Dibromochloromethane	CHBr$_2$Cl	X	X									100
Bromoform	CHBr$_3$	X	X		X	X						100
Dichloroiodomethane	CHICl$_2$	X	X									
Chlorodiiodomethane	CHI$_2$Cl	X	X									
Bromochloroiodomethane	CHBrICl	X	X									
Bromoiodomethane	CHBr$_2$I		X									
Bromodiiodomethane	CHBrI$_2$		X									
Triiodomethane	CHI$_3$		X									
Haloacetic acids (HAA)	HAA5[h]									80	60	
Monochloroacetic acid	C$_2$H$_3$ClO$_2$	X	X					150				20
Dichloroacetic acid	C$_2$H$_2$Cl$_2$O$_2$	X	X					100				50
Trichloroacetic acid	C$_2$HCl$_3$O$_2$	X	X					100				200
Monobromoacetic acid	C$_2$H$_3$BrO$_2$	X	X									–
Dibromoacetic acid	C$_2$H$_2$Br$_2$O$_2$	X	X									
Bromochloroacetic acid	C$_2$H$_2$BrClO$_2$	X	X									
Dibromochloroacetic acid	C$_2$HBr$_2$ClO$_2$	X	X									
Bromodichloroacetic acid	C$_2$HCl$_2$O$_2$	X	X									
Tribromoacetic acid	C$_2$HBr$_3$O$_2$	X	X									
Oxyhalides												
Bromate	BrO$_3$$^-$				X	X		20	10	10	10	10
Chlorite	ClO$_2$$^-$			X				800		1,000	1,000	700
Chlorate	ClO$_3$$^-$			X						1,000		700
Haloketones (HK)												
1,1-Dichloro-2-propanone	C$_3$H$_4$Cl$_2$O	X	X									

(continued)

Table 1 (continued)

Species	Chemical formula	Primarily a byproduct of						Selected drinking water regulations				
		Free chlorine	Chloramines	ClO$_2$	O$_3$	O$_3$-based AOP	Non-O$_3$-based AOP	Australia[a]	European Union (EU)[b]	Health Canada (HC)[c]	US Environmental Protection Agency (USEPA)[d]	World Health Organization (WHO)[e]
1,1,1-Trichloro-2-propanone	C$_3$H$_5$Cl$_3$O	X										
Chlorophenols												
2-Chlorophenol	C$_6$H$_5$ClO	X	X					300				
2,4-Dichlorophenol	C$_6$H$_4$Cl$_2$O	X	X					200				
2,4,6-Trichlorophenol	C$_6$H$_3$Cl$_3$O	X	X					20		5		200
Chloral hydrate (trichloroacetaldehyde)	C$_2$H$_3$Cl$_3$O$_2$	X	X					100				
Haloacetonitriles (HANs)												
Dichloroacetonitrile	C$_2$HCl$_2$N	X	X									20
Dibromoacetonitrile	C$_2$HBr$_2$N	X	X									70
Trichloroacetonitrile	C$_2$Cl$_3$N	X	X									
Bromochloroacetonitrile	C$_2$HBrClN	X	X									
Nitrosamines												
N-Nitrosodimethylamine (NDMA)	C$_2$H$_6$N$_2$O		X		X	X		0.1		0.04		0.1
Trihalonitromethanes												
Trichloronitromethane (chloropicrin)	CCl$_3$NO$_2$	X	X									
Bromodichloronitromethane	CBrCl$_2$NO$_2$	X	X									
Dibromochloronitromethane	C$_2$Br$_2$ClNO$_2$	X	X									
Tribromonitromethane	CBr$_3$NO$_2$	X	X									
Cyanogen halides (CNX)												
Cyanogen chloride (as cyanogen)	CNCl		X					80				70
Cyanogen bromide	CNBr	X	X									
Aldehydes												
Formaldehyde	CH$_2$O	X			X	X	X	500				
Acetaldehyde	C$_2$H$_4$O	X			X	X	X					
Glyoxal	C$_2$H$_2$O$_2$	X			X	X	X					

Methylglyoxal	$C_3H_4O_2$	x		x		x	x		x
Carboxylates									
Formate	CHO_2^-					x	x		x
Acetate	$C_2H_3O_2^-$					x	x		x
Oxalate	$C_2O_4^{2-}$					x	x		x
Keto acids									
Glyoxylic acid	$C_2H_2O_3$					x	x		x
Pyruvic acid	$C_3H_4O_3$					x	x		x
Ketomalonic acid	$C_3H_2O_5$					x	x		x

Updated and adapted from [16, 17]
[a][18]
[b][19]
[c][20]
[d][21]
[e][22]
[f]THM = sum of four THMs: chloroform, bromodichloromethane, dibromochloromethane, and bromoform
[g]The sum of the ratio of the concentration of each to its respective guideline value should not exceed 1
[h]HAA5 = sum of five HAAs: monochloroacetic acid, dichloroacetic acid, trichloroacetic acid, monobromoacetic acid, and dibromoacetic acid

chlorinated drinking water in Fig. 3, more than half of the total organic halide (TOX) formed during chlorination has yet to be chemically identified [25, 26].

2.1 Disinfection Byproduct Regulations

Since their discovery in the 1970s, DBPs have been widely regulated as drinking water contaminants, as demonstrated by the list of select DBPs and applicable regulations/guidelines shown in Table 1. For historical context, a brief description of the regulatory basis for DBPs in the United States is provided here. Next, the most commonly regulated, or classical, DBPs are introduced, including trihalomethanes, haloacetic acids, chlorate, chlorite, and bromate. This section concludes with brief descriptions of several classes of emerging DBPs.

In the United States, DBPs were first regulated in 1979, beginning with the TTHM (total trihalomethanes) Rule, which set a maximum contaminant level (MCL) of 100 µg/L TTHM based on a running annual average of distribution system samples. In 1986, amendments to the Safe Drinking Water Act (SDWA) noted that disinfectants and DBPs should be regulated [28]. Thus, to simultaneously control microbial pathogens, residual disinfectants, and DBPs, three related rules were developed: the Information Collection Rule (ICR), the Disinfectants/DBP Rule (D/DBPR, implemented in two stages), and the Enhanced Surface Water Treatment Rule (ESWTR, implemented in stages, e.g., Long-Term 2 Enhanced Surface Water Treatment Rule or LT2). In Stage 1 of the D/DBPR, the USEPA reduced the existing TTHM standard to 80 µg/L and expanded regulations to include haloacetic acids (HAA5s = 60 µg/L), bromate (BrO_3^- = 10 µg/L), and chlorite (ClO_2^- = 1,000 µg/L). In Stage 2 of the D/DBPR, the DBP MCLs were maintained, but compliance was amended to a locational running annual average basis (rather than the previous approach of averaging concentrations across distribution system sampling points).

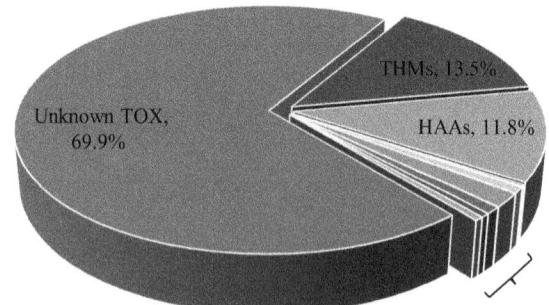

Fig. 3 Distribution of characterized total organic halides (TOX) in chlorinated drinking water, including unknown TOX, trihalomethanes (THMs), haloacetic acids (HAAs), and several classes of emerging halogenated disinfection byproducts (DBPs) which account for <5% of the total DBPs. Adapted from [27]

$\sum$(Iodo THMs, Halofuranones, Haloacetonitriles, Halogenated Aldehydes, Haloketones, Haloacetates, Halonitromethanes), 4.8%

In addition to these established federal regulations, the USEPA also periodically reviews the status of unregulated DBPs (and other drinking water contaminants) as part of the Contaminant Candidate List (CCL) and Unregulated Contaminant Monitoring Rule (UCMR). The CCL is a list of currently unregulated contaminants that are known or believed to occur in public water systems and are thus identified as research priorities in order to better inform determinations of risks and regulations. The USEPA's 4th version of the CCL (CCL4, announced in 2016) includes several additional DBPs: acetaldehyde (an O_3-DBP), chlorate (formed during ClO_2 and hypochlorite disinfection), formaldehyde (O_3 based), bromochloromethane, N-nitrosodiethylamine (NDEA), N-nitrosodimethylamine (NDMA), N-nitrosodi-n-propylamine (NDPA), N-nitrosodiphenylamine (NDPhA), and N-nitrosopyrrolidine (NPYR) [29]. The state of California's Department of Public Health has already set a notification level of 10 ng/L for NDEA, NDMA, and NDPA and a public health goal for NDMA of 3 ng/L [30]. Monitoring results from public water systems collected as part of the USEPA's UCMR program suggest that more than 10% of the US' chloraminated systems could be affected if a NDMA MCL was introduced at the California action level of 10 ng/L [31].

2.2 Conventional Disinfection Byproducts: Trihalomethanes, Haloacetic Acids, and Oxyhalides

The trihalomethanes (THMs) were the first DBPs to be discovered and are one of the most prevalent classes of DBPs resulting from chlorine disinfection [25]. Together with haloacetic acids, THMs account for approximately 25% of the halogenated DBPs from chlorination [32]. Chloramination can also generate THMs, albeit typically to a much lesser extent, as can ozone through production of bromoform. Many DBP regulations are based on total trihalomethanes (TTHMs), which are calculated as the sum of four THMs: chloroform, bromoform, bromodichloromethane, and chlorodibromomethane, of which chloroform is often found at the highest concentrations [25]. All of the TTHM species demonstrate carcinogenicity in rodents [25].

Surface water typically contains higher concentrations of precursor NOM and, as such, is associated with higher DBP production, as shown in Fig. 4. Other commonly used indicators of THM formation include chlorine dose, pH, temperature, bromide concentration, and disinfection contact time. In general, as these parameters increase, so does DBP formation (although DBP responses to increasing pH are mixed). A number of empirical and semi-mechanistic DBP formation models have been used to predict DBP concentrations based on these and other parameters, often using a multiple linear or nonlinear regression approach. When applied to the treatment scenarios for which they were specifically developed, these models can be helpful indicators of operation, risk assessment, etc. [33], although they tend to overpredict DBPs for conditions least conducive to formation while

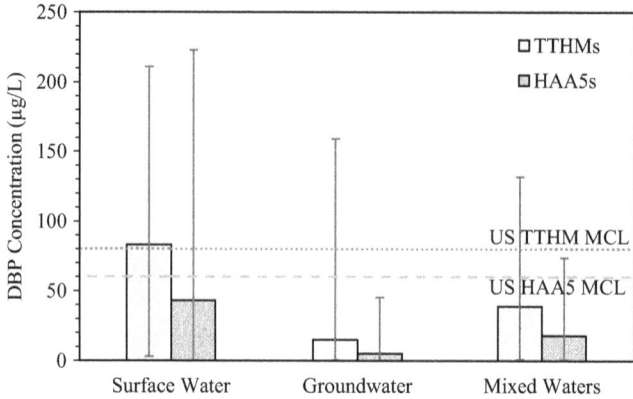

Fig. 4 Formation of TTHMs and HAA5s in finished water surveyed in 1992 at more than 100 US drinking water treatment plants with varying source waters. The bars represent average values, while the error bars illustrate the range in values reported, and the lines denote regulatory maximum contaminant levels (MCLs) in the US. Data from [36]

underpredicting for the conditions most conducive for DBPs [8, 16, 34, 35]. However, Mayer et al. [16] reported that TTHM and HAA5 models generally performed poorly when applied to DBP data not used to directly develop the model, regardless of the use and extent of AOP treatment and type of source water. This suggests that bulk indicators and/or models should be used cautiously as metrics for AOP mitigation of DBP formation potential.

The main haloacetic acids (HAAs) include nine different halogenated compounds, as shown in Table 1. When regulated, the HAAs are sometimes dealt with on the basis of individual compounds but are sometimes grouped together, for example, as HAA5, or the sum of five of the HAAs: bromoacetic acid, dibromoacetic acid, chloroacetic acid, dichloroacetic acid, and trichloroacetic acid, all five of which are mutagenic [25]. The remaining four main HAAs were more difficult to quantify when the US regulation was promulgated and so were excluded from regulation. Like THMs, chlorination generally produces the highest levels of HAAs, although lower levels can result from chloramination, ClO_2, and O_3 [37].

The oxyhalides chlorite (ClO_2^-) and chlorate (ClO_3^-) are the inorganic DBPs produced through reactions of NOM with ClO_2. Chlorine dioxide rapidly reacts with NOM and inorganic matter, degrading to chlorite, chlorate, and chloride [25].

When bromide is present in source water, brominated DBPs including inorganic bromate (BrO_3^-) and organic brominated DBPs may pose a concern. Bromate is of particular concern for ozonation processes when bromide is present in source waters at high levels (>50–100 µg/L) [25, 38]. Bromate is both genotoxic and carcinogenic, and of the DBPs regulated in the United States, it is the most potent carcinogen in laboratory animals [25]. It is formed by a series of oxidations mediated by O_3 or a combination of O_3 and HO• reacting with natural bromide, where O_3 sequentially oxidizes bromide to hypobromite (BrO^-), followed by bromite (BrO_2^-), and finally to bromate [39]. The HO• produced as part of the ozonation process also participates in the intermediate reactions [39].

2.3 Emerging Disinfection Byproducts, Including Brominated, Iodinated, and Nitrogenous Species

To date, the greatest emphasis in DBP research and mitigation has been on a subset of more conventionally regulated DBPs including THMs, HAAs, and BrO_3^-. However, there are a large number of "emerging" DBPs, including those not currently widely regulated. As the majority of TOX often consists of unidentified compounds (see Fig. 3), developing a better understanding of emerging DBPs is an important area of research. In this section, several major classes of emerging DBPs are introduced.

2.3.1 Organic Brominated and Iodinated Disinfection Byproducts

Bromide and iodide are naturally occurring inorganic DBP precursors for a variety of DBPs. These ions may be present, and of concern, in source waters impacted by seawater intrusion and natural salt deposits, e.g., coastal cities or areas affected by oil and gas brines. When bromide is present, organic brominated DBPs (B-DBPs) are primarily produced by chlorination in the same manner as the classical DBPs. In contrast, iodinated DBPs (I-DBPs) are primarily produced by chloramination of iodide-containing waters. Both B-DBPs and I-DBPs are more toxic and carcinogenic than their chlorine analogs, in order of greatest health risk: chlorinated < brominated < iodinated [40].

Although iodo-THMs and iodo-HAAs are among the most toxic unregulated DBPs, iodate (IO_3^-) is generally not problematic and is readily reduced to iodide (I^-) after consumption [41]. In processes that use strong oxidizing disinfectants, e.g., ozone or chlorine, I^- is readily oxidized to IO_3^- (as shown below), which results in very low amounts of organic I-DBPs remaining in solution.

$$I^- \rightarrow IO^-/HIO \rightarrow IO_2^- \rightarrow IO_3^-$$

For example, 90% of naturally occurring I^- was converted to IO_3^- during ozonation [42]. This stems from the higher activity of hypoiodous acid (HIO) with strong oxidizing disinfectants in comparison to reactivity with NOM. Thus, the oxidation reactions can be optimized as a sink to form IO_3^- rather than organic I-DBPs when I^- reacts with NOM [43].

The smaller oxidative potential of chloramines yields ineffective oxidation of HIO to IO_3^-, which can increase formation of I-DBPs during chloramination [42]. Iodide can be oxidized prior to ammonia additions during chloramination using prechlorination or preozonation; however, this approach could increase the formation of B-DBPs and chlorinated DBPs [40, 42].

2.3.2 Halogenated Furanones

Halogenated furanones, including MX (3-chloro-4-(dichloromethyl)-5-hydroxy-2 (5H)-furanone) and its brominated analogues (BMX), are highly mutagenic DBPs. Weinberg et al. [26] found that MX and BMX analogues are generally found at concentrations of 60 ng/L or less, but they have also been found at concentrations as high as 80 ng/L [44]. These species can account for more than 50% of the mutagenicity in bioassays of chlorinated drinking waters [45, 46]. The formation pathway for MX may be similar to that for THMs and HAAs as occurrence of MX analogues is positively correlated with occurrence of chloroform, and MX analogues have been found in effluent from chlorine and chlorine dioxide disinfection processes which also contained THMs and HAAs [47]. Chloramine disinfection processes yield comparatively smaller concentrations of MX and BMX analogues [47].

2.3.3 Nitrogenous Disinfection Byproducts

Nitrogenous DBPs (N-DBPs) are a subset of the classical carbonaceous DBPs (C-DBPs) [27]. They may result when sufficient levels of nitrogen are present for DBP incorporation, e.g., using source waters impaired by algal blooms, during chloramination, in the presence of certain polymers, or at water reuse facilities.

Nitrosamines are one of the most widely studied N-DBPs. They are a family of emerging DBPs often found at ng/L concentrations but associated with serious health risks as they have been reported to be carcinogenic, mutagenic, and/or teratogenic [48, 49]. N-Nitrosodimethylamine (NDMA) is the most common nitrosamine and is a probable human carcinogen, with 0.7 ng/L NDMA correlating to a lifetime cancer risk of 10^{-6} [50].

The NDMA formation potential varies as a function of source water and treatment process. It is primarily produced during chloramination of water [51], which directly adds nitrogen during disinfection. Low molar yields of NDMA may be generated by reacting precursors with either chloramines or ozone, but most precursors are more reactive with just one of the disinfectants (i.e., chlorine- or ozone-reactive NDMA precursors) [52]. Some studies have shown that preozonation increases NDMA formation potential in water [53], but others have found decreased NDMA formation potential [54].

In general, NDMA precursors comprise secondary, tertiary, or quaternary amines, with sources including PPCPs, pesticides, water or wastewater treatment chemicals, and industrial chemicals [52]. As wastewater effluent may contain more of these N-rich compounds in comparison to many drinking water sources, NDMA is of particular concern in wastewater effluent and water reuse scenarios. Naturally occurring bromide can also contribute to NDMA formation by catalyzing reactions with precursors [55]. Padhye et al. [56] found that NDMA formation tripled in the presence of 100 μM bromide.

In addition to NOM, polymers may serve as NDMA precursors. As shown in Fig. 5, cationic amine-based polymers, such as polyamine and poly-DADMAC, used to enhance coagulation and dewatering can contain dimethylamine (DMA) and other small amine moieties which serve as significant NDMA precursors [48, 51, 57]. When dosed at optimal levels, minimal polymer residual remains in solution following physicochemical treatment. Accordingly, it is unlikely that sufficient concentrations of polymer will be present to react with chloramines to form NDMA. Rather, the primary concern lies in degradation of the polymers and release of NDMA precursors. Polyamine is generally more susceptible to degradation and DMA release than poly-DADMAC and therefore is more closely linked to DBP formation [57]. Alternative polymers produced from natural products may perform as well as poly-DADMAC while limiting nitrosamine formation potential [58].

2.4 Non-halogenated Organic Disinfection Byproducts

When condensed aromatic compounds are oxidized, low molecular weight, non-halogenated organic DBPs may result. These products may include aldehydes, carboxylic acids, and keto acids, all of which can enhance biodegradability and can even be toxic. The most commonly observed organic byproducts are aldehydes such as formaldehyde and acetaldehyde [59]. Other common organic byproducts include formate, acetate, and oxalate [17].

3 Disinfection Byproduct Precursors

DBP precursors include both organic matter (e.g., NOM) and inorganic compounds (e.g., bromide and iodide). Increases in the ratio of bromide ion relative to chlorine or organic matter can shift speciation of THMs and HAAs toward more bromine-

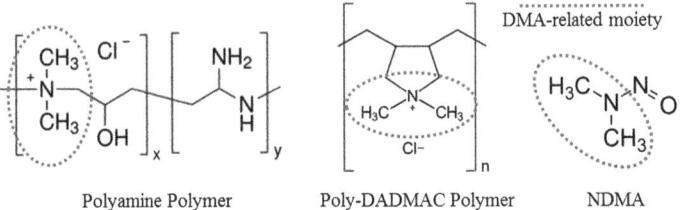

Fig. 5 *N*-Nitrosodimethylamine (NDMA) and common polymers acting as NDMA disinfection byproduct (DBP) precursors, with the dimethylamine (DMA)-related moiety (primary component serving as NDMA precursor) highlighted in each molecule

substituted species, which are associated with more significant health risks compared to chlorinated DBPs [60].

Natural organic matter is ubiquitous in drinking water sources and serves as the primary organic DBP precursor. Its enigmatic character derives from a complex mixture of numerous compounds ranging from aromatic phenolic compounds and aliphatic carboxylic acids to nitrogenous compounds such as proteins, sugars, amino acids, and large biopolymers, e.g., lignin [61]. Understanding the quantity and character of NOM is essential for water treatment design to effectively mitigate NOM-related problems such as DBP formation; greater coagulant dose requirements; increased bioavailability of organics in the water; and aesthetic concerns such as color, taste, and odor [62]. Multiple approaches can be used to provide indications of the quantity and quality of complex mixtures of NOM, including bulk organic measurements and fractionation on the basis of fluorescence excitation/emissions, size, or operational behavior (e.g., hydrophilic, acid, etc.).

3.1 Bulk Organic Measurements

Total organic carbon (TOC), dissolved organic carbon (DOC), ultraviolet absorbance at a wavelength of 254 nm (UV_{254}), and specific ultraviolet absorbance (SUVA) are bulk parameters often used as indicators of NOM quantity and general character. These parameters are also commonly used as surrogates for estimating DBP formation potential, although they do have limitations. They do not always correlate to DBP concentrations in finished drinking water since other factors such as disinfectant type and dose, as well as water pH, temperature, bromide concentrations, etc. all strongly influence DBP formation [16, 36]. However, bulk organic parameters are still commonly used as they offer an easy, rapid, and inexpensive approach to gauging NOM.

Natural waters used as drinking water sources typically contain low levels of synthetic organic contaminants, so TOC is often considered synonymous with NOM [61]. The DOC fraction is classified as the organic carbon that passes through a 0.45 μm filter, whereas the fraction that is retained is the particulate organic matter (POM, often accounting for <10% of TOC) [63, 64].

Spectrophotometric analysis in the wavelength range of 220–280 nm is also commonly used as an indicator of the presence of organics. Molar absorptivities vary widely due to the diversity of chromophores present in NOM; for example, carboxylic acids and aromatic compounds are associated with a wavelength of 220 nm [65], with maximum absorption at 254 nm for most aromatic groups. Accordingly, absorbance at 254 nm (UV_{254}) is primarily used to indicate the presence of the dense, aromatic, hydrophobic portion of NOM [66] and is also used as a rough indicator of overall NOM content [61].

SUVA is another helpful correlation parameter for DBP formation potential [67] as it is indicative of the relative aromaticity of the NOM. The SUVA value normalizes UV_{254} relative to the DOC concentration.

$$\text{SUVA} = \frac{\text{UV}_{254}}{\text{DOC}}, (\text{L/mg m})$$

High SUVA values (>4) indicate largely aromatic, hydrophobic, high molecular weight compounds, whereas low SUVA values (<4) represent small molecular weight, hydrophilic compounds [68].

These bulk organic parameters are useful in describing the quantity of NOM and providing an indication of its character, which can provide an indication of water quality through treatment stages as well as DBP formation potential (as bulk organic matter parameters increase, DBP formation potential generally increases). However, bulk parameters do not always correlate to DBPs, nor do they provide information regarding specific NOM constituents such as amino acids, sugars, and carbohydrates.

3.2 Natural Organic Matter Fractionation

To better characterize the array of diverse compounds present in NOM, fractionation techniques based on parameters such as absorbance, size, hydrophobicity, etc. may be employed.

3.2.1 Fluorescence Absorption and Emission-Based Fractionation

Analysis using a 3-D fluorescence excitation emission matrix (EEM) can be useful for understanding NOM fractions [69]. In water, fluorophores are generally divided into humic-like fluorophores and protein-like fluorophores [70, 71]. The NOM can be fractionated prior to analysis in order to express characteristic peaks, locations of which are illustrated in the EEM matrix in Fig. 6 [72–75]. Each fraction is characterized by its own potential for DBP production. The humic-like peak correlates strongly with TTHM formation potential, and the tryptophan-like peak correlates well with NDMA formation potential [76]. EEM analysis may provide better correlation to NDMA formation potential compared to UV_{254} and SUVA indicators [76–78].

3.2.2 Size-Based Fractionation

Size exclusion chromatography (SEC) and high-performance liquid chromatography-SEC (HPLC-SEC) can be used to characterize NOM samples on the basis of molecular size. Although molecular size can provide an indication of DBP formation potential, there is a great deal of variation in observed correlations between size of precursor NOM and resulting DBP formation. While some studies have found the hydrophobic NOM fraction with an apparent molecular weight (AMW) of 1–10 kDa

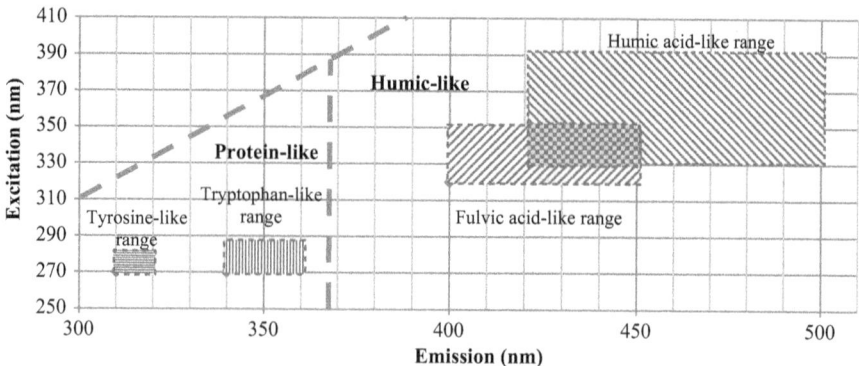

Fig. 6 Excitation emission matrix (EEM) wavelengths associated with natural organic matter (NOM) fractions. Adapted from [79, 80]

to be the primary DBP precursor [81], hydrophilic and lower molecular weight fractions may also generate significant DBPs [16, 82–84]. For example, the bulk portion of NDMA precursors consists of small molecular weight compounds (<3,000 Da) [85, 86]. Similarly, Zhao et al. [87] found that low molecular weight dissolved organic matter serves as a significant THM precursor.

Figure 7 shows several example datasets from disinfected surface waters illustrating the relative contribution of different NOM size fractions to C-DBP, N-DBP, and I-DBP formation potential. As exemplified here, there is no clear trend between NOM molecular weight fractions and DBP formation; therefore, size fractionation by itself is a less important predictor for DBP formation compared to chemical composition of the NOM and water quality parameters [88, 89].

Thus, while using UV detectors for HPLC-SEC provides helpful information for evaluating the potential for removal of high molecular weight organic matter via coagulation or other processes, their use for understanding DBP formation potential may still be limited. Alternatively, coupling UV detectors in tandem with DOC analyzers or fluorospectrometers can be used to provide more informative datasets by detecting aromatic and nonaromatic datasets as a function of molecular weight, both of which parameters are important for predicting DBP formation potential [90, 91]. Compound classes and their respective apparent molecular weights are shown in Table 2.

3.2.3 Resin Fractionation

To better understand complex organic matter, resin fractionation methods can be used to classify NOM in accordance with behavior during different water treatment processes. The predominant properties of organic matter can be characterized as hydrophobic, hydrophilic, acidic, basic, and neutral (although distinctions are not always clear, e.g., a great deal of dissolved organic matter is recognized as both

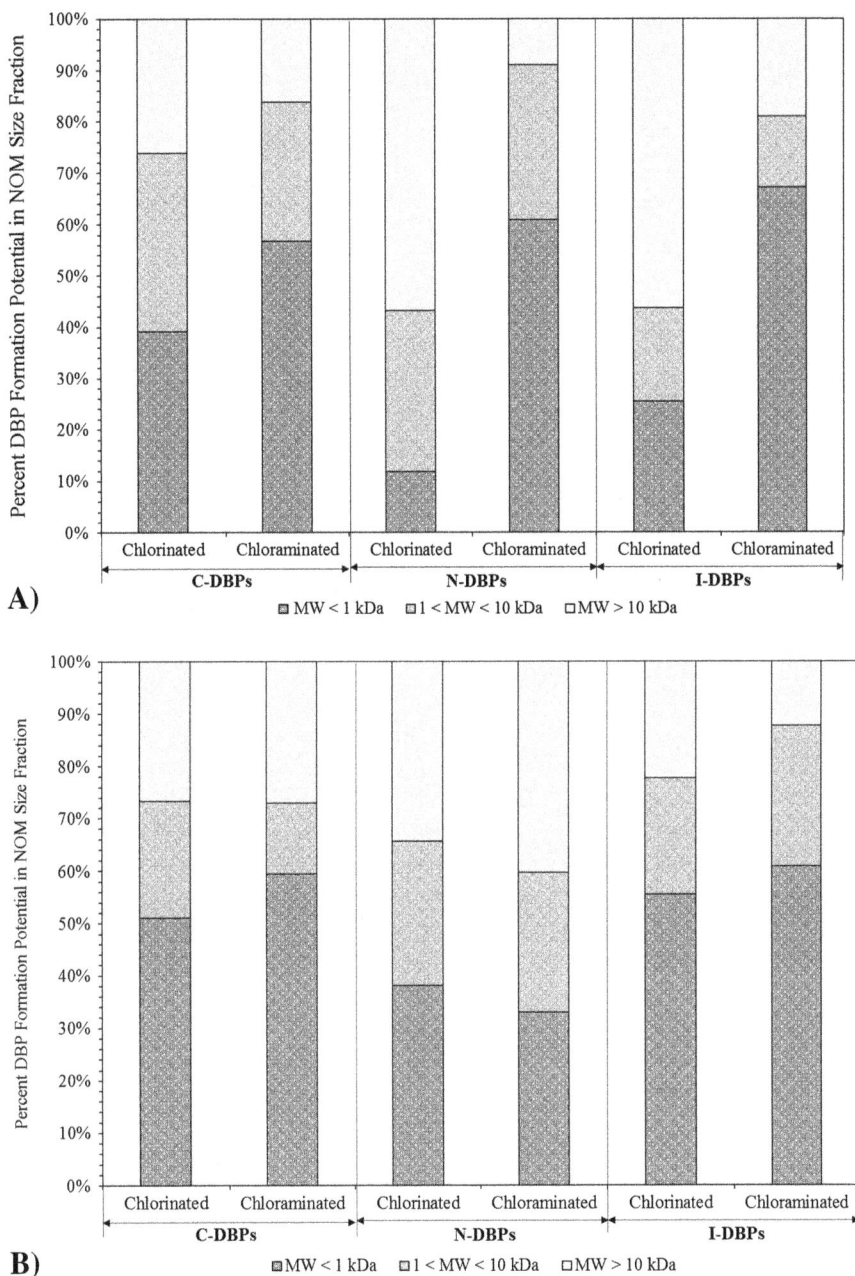

Fig. 7 Disinfection byproduct formation relative to natural organic matter (NOM) molecular weight (MW) size fraction for C-DBPs, N-DBPs, and I-DBPs in chlorinated and chloraminated water from the (**a**) Huangpu River, China, and (**b**) the Yangtze River, China. Data from [89]

Table 2 Size characterization using a combination of high-performance liquid chromatography-size exclusion chromatography (HPLC-SEC) with (ultraviolet) UV and online dissolved organic carbon (DOC) detectors to characterize natural organic matter (NOM) by apparent molecular weight (AMW)

Compound class	Apparent molecular weight (AMW)
Polysaccharide- and protein-like substances	High (>10,000 g/mol)
Highly aromatic and fulvic substances	Medium (1,000–5,000 g/mol)
Aliphatic-like substances	Low (<680 g/mol)

Adapted from [92]

amphiphilic and amphoteric) [61]. Each of these fractions, or subsets thereof, can be isolated using sequential resin-based separation.

The hydrophobic portion of NOM is composed of densely aromatic structures, conjugated double bonds, and high molecular weight compounds. The hydrophobic fraction is also characterized by a high specific surface charge, making it more amenable to removal via coagulation [93]. The hydrophobic fraction can account for more than half of the DOC in water [64, 93, 94], although there is great variation in fractionation among different source waters. This can be problematic for drinking water treatment since the hydrophobic portions of NOM serve as significant precursors for DBPs and produce greater amounts of unidentified total organic halogen products (TOX) [95–97].

The hydrophilic fraction of NOM contains low molecular weight polar compounds such as carboxylic acids, as well as nitrogenous compounds including sugars, peptides, and amino acids [61]. The hydrophilic NOM fraction is also a significant precursor for DBPs and generally forms more NDMA than the hydrophobic fractions [98]. The hydrophilic fraction has been observed to be more reactive with bromine and iodine than the hydrophobic NOM fraction [95], as illustrated in Fig. 8a, b, which shows DBP formation potential from different NOM fractions in several waters.

As shown in Fig. 8, although the presence of different fractions can provide an indication of DBP formation potential, actual DBP production can vary widely with source water and operational parameters in the disinfection process.

3.3 Structure and Reactivity of Natural Organic Matter (NOM) Fractions

The physicochemical properties (K_{ow}, pK_a, and molecular weight) of individual NOM species do not correlate well to DBP formation potential [101]. However, halogen substitution efficiency is an effective indicator of DBP formation potential. As shown in Fig. 9, compounds exhibiting high halogen substitution efficiency, and which produce substantial THMs and HAAs, include ferulic acid, L-tryptophan, and resorcinol [101]. These compounds are considered activated aromatics because the

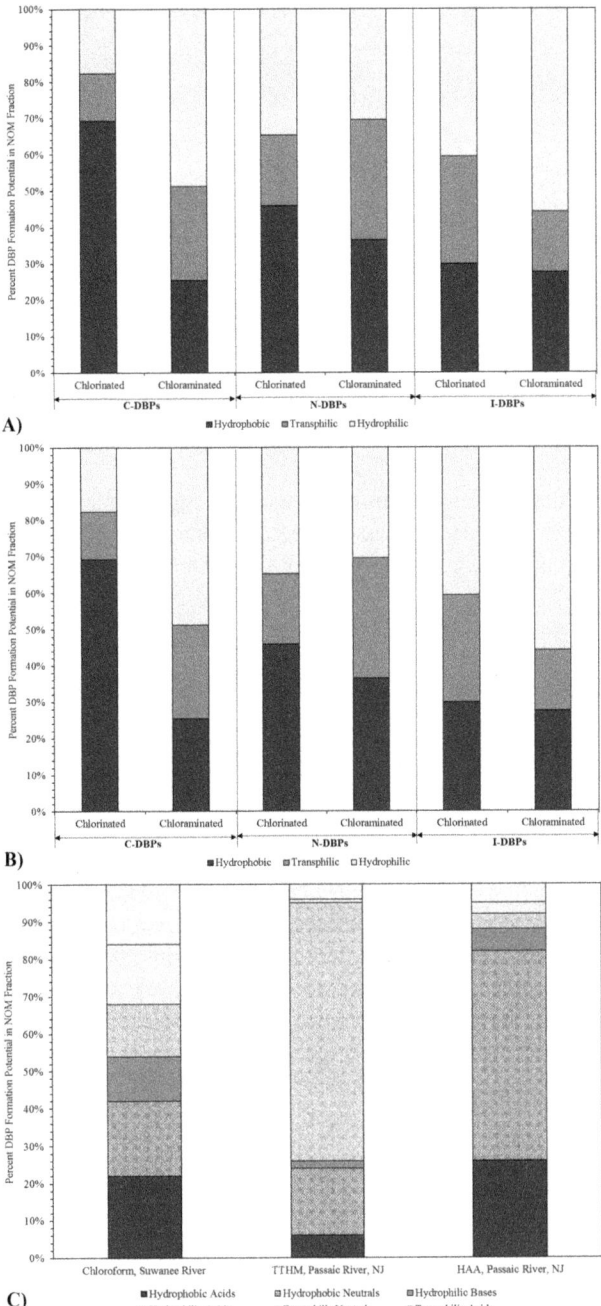

Fig. 8 Disinfection byproduct (DBP) formation potential associated with natural organic matter (NOM) fractions in variable source waters. C-DBPs, N-DBPs, and I-DBPs formed during chlorination or chloramination of water from (**a**) the Huangpu River, China, and (**b**) the Yangtze River, China [89]. (**c**) Shows chloroform [99], total trihalomethane (TTHM), and haloacetic acid (HAA) formation following chlorination [100]

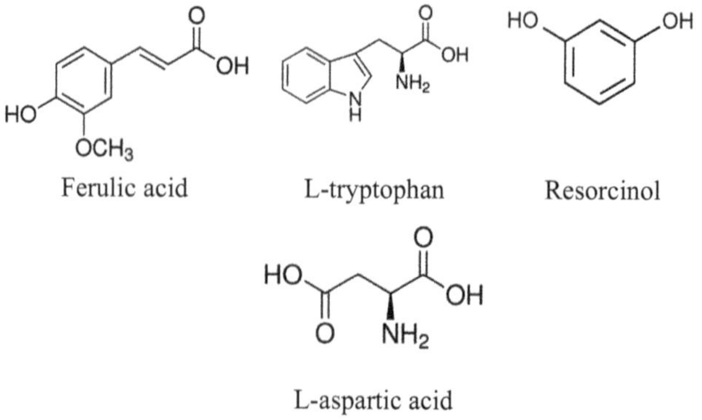

Ferulic acid   L-tryptophan   Resorcinol

L-aspartic acid

Fig. 9 Compounds exhibiting high halogen substitution efficiency and which produce substantial trihalomethanes (THMs) and/or haloacetic acids (HAAs)

constituents on the ring are electron donors. Although not an activated aromatic species, L-aspartic acid also has many electron donating functional groups, has a high substitution efficiency, and serves as a significant precursor for HAA species.

3.4 Synthetic Organic Disinfection Byproduct Precursors

In addition to NOM, synthetic organic compounds can serve as precursor material for DBP formation. For example, the active compounds in sunscreens can be transformed to halogenated DBPs when swimming pools are disinfected with chlorine [102]. Understanding of DBPs produced by oxidation of the activated aromatic rings in synthetic compounds is a relatively new undertaking in comparison to their NOM-derived relatives. However, research is progressing, often stemming from studies of oxidative degradation of compounds such as pesticides (e.g., S-triazine herbicides and isoproturon), pharmaceuticals (e.g., carbamazepine and acetaminophen), antibacterial agents (e.g., triclosan and carbadox), textile dyes (e.g., azo), bisphenol A, alkylphenol ethoxylate surfactants, etc. [25].

4 Disinfection Byproduct Formation in Relation to Advanced Oxidation Processes

Advanced oxidation processes primarily impact DBP formation in two different ways: directly through in situ DBP formation and indirectly by influencing the DBP formation potential of the organics in the water matrix. In the first case, DBPs may be generated during the operation of AOPs themselves, particularly when using AOPs with direct inputs of chemical oxidants, e.g., ozone or hydrogen peroxide.

Additionally, organic oxidation byproducts may result during AOPs as complex organics are progressively degraded. In addition to direct formation during AOP treatments, AOPs may influence subsequent DBP formation potential as the DBP precursor organic material undergoes dramatic transformations during AOPs, which can serve to either decrease or increase DBP formation during subsequent disinfection processes.

4.1 Disinfection Byproduct Formation in Advanced Oxidation Processes

Hydroxyl radicals can attack organic molecules by radical addition, hydrogen abstraction, electron transfer, and radical combination. Select kinetic rate constants for reactions of HO• with organic matter, with and without radical scavengers (i.e., dissolved organic matter [DOM]), are shown in Table 3. The rate constants are generally greater for benzene-based compounds (44–120 × 10^8/M s) in comparison to carboxylic acid-based compounds (1–120 × 10^8/M s; at 190/M s, cysteine appears to be an outlier, perhaps based on its thiol properties). However, there are no distinct correlations between structure, molecular weight, and reactivity [103].

Westerhoff et al. [105] analyzed the reactivity of ozone and HO• with hydrophobic organic acids isolated from a variety of source waters. Although NOM characteristics varied among samples, there were strong correlations between SUVA values and the rate constants for O_3 and HO• interactions, as shown in Fig. 10. The data demonstrate ozone's strong preference for densely conjugated aromatic NOM, whereas HO• reactions with NOM were rapid and nonselective.

The general progression of oxidation using AOPs follows the pathway:

$$\text{organic pollutant} \rightarrow \text{aldehydes} \rightarrow \text{carboxylic acids}$$
$$\rightarrow \text{carbon dioxide and mineral acids}$$

Accordingly, although AOPs can effectively mitigate organic precursors to limit downstream formation of DBPs, they can also directly generate their own suite of DBPs in the event of incomplete oxidation, wherein the products do not completely mineralize due to insufficient chemical and/or energy inputs. DBPs produced during AOP treatments may include BrO_3^-, NDMA, and small molecular weight organic compounds, e.g., acetaldehyde and formaldehyde [59, 106].

Table 3 Hydroxyl radical (HO•) kinetic rate constants with various organic compounds, with and without dissolved organic matter (DOM)

Compound	Chemical formula	Compound class	$k_{HO•}$ ($\times 10^8$/ M s)	$k_{HO•, DOM}$ ($\times 10^8$/ M_{carbon} s)
Benzaldehyde	C_7H_6O	Benzene-based	44	6
Hydroquinone	$C_6H_6O_2$	Benzene-based	52	9
Catechol	$C_6H_6O_2$	Benzene-based	110	18
Phthalic acid	$C_8H_6O_4$	Benzene-based, carboxylic acid	59	7
Salicylic acid	$C_7H_6O_3$	Benzene-based, carboxylic acid	120	17
Oxalic acid	$C_2H_2O_4$	Carboxylic acid	1	0.2
Citric acid	$C_6H_8O_7$	Carboxylic acid	3	1
Tartaric acid	$C_4H_6O_6$	Carboxylic acid	14	4
Cysteine	$C_3H_7NO_2S$	Carboxylic acid, thiol	190	63
Mean value for hydrophobic organic acids isolated from a variety of sources		Organic acids		3.6

Data from [103–105]

4.1.1 Bromate and Organic Brominated Disinfection Byproducts

Bromate (BrO_3^-) formation stems from ozonation, wherein O_3 directly oxidizes Br^- to BrO_3^-:

$$Br^- \rightarrow BrO^-/HBrO \rightarrow BrO_2^- \rightarrow BrO_3^-$$

Consequently, O_3-based AOPs such as O_3/UV and O_3/H_2O_2 can directly generate BrO_3^-, the extent of which depends on water quality and process operation. During O_3-based AOPs, the synergistic effect of O_3 and HO• may oxidize Br^- to BrO_3^- as HO• can participate in intermediate steps by producing radical species (e.g., Br•, BrO•) [107]. Relative to O_3-only processes, O_3/UV may produce similar amounts of BrO_3^- [108]. However, O_3/H_2O_2 can produce less BrO_3^- than O_3/UV, and O_3-only processes since H_2O_2 can reduce BrO^- to Br^-; hence optimized H_2O_2 doses can mitigate BrO_3^- formation [109, 110]. Although HO• is involved in BrO_3^- formation during ozonation processes, non-O_3-based AOPs such as UV/H_2O_2 and TiO_2 photocatalysis (UV/TiO_2), which rely on the production of HO• radicals, have not been shown to form significant amounts of BrO_3^- [109].

The production of organic B-DBPs can be much less prominent than BrO_3^- during ozonation and AOPs due to competing kinetics between HBrO and HO•.

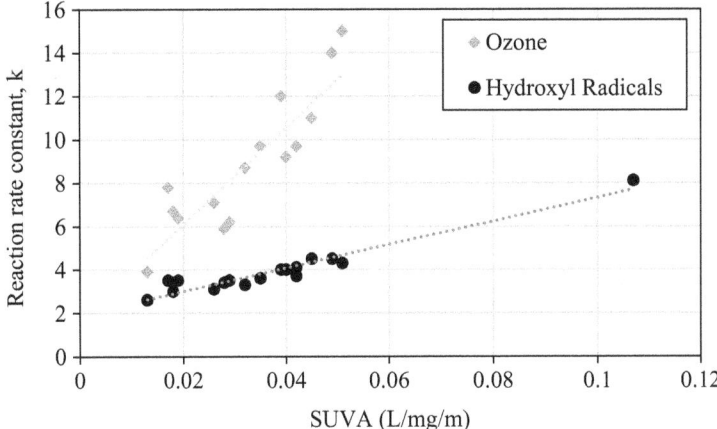

Fig. 10 Reactivity of natural waters with ozone (k in units of $\times 10^3$ per second) and hydroxyl radicals (k in units of $\times 10^{-8}$ L/mol C s). Data from [105]

Hypobromous acid reacts more readily with O_3 and HO• compared to the precursor NOM necessary to form B-DBP haloorganics [111].

4.1.2 N-Nitrosodimethylamine (NDMA)

The use of AOP treatments can have variable results in terms of NDMA formation, as a function of influent water quality parameters. This is demonstrated in Fig. 11, where Zhao et al. [112] analyzed seven different source waters and found that UV/H_2O_2 yielded the highest amount of NDMA for some waters but lower NDMA for other waters. Interestingly, there was no correlation with bulk organic precursors (DOC, UV_{254}, SUVA), although the two waters yielding the most nitrosamines using UV/H_2O_2 had high UV_{254} values. In contrast, the water with the highest TOC and UV_{254} yielded small amounts of NDMA relative to others [112].

4.1.3 Non-halogenated Organic Disinfection Byproducts

AOPs can produce non-halogenated organic DBPs, primarily due to oxidation of condensed aromatic compounds, which produces low molecular weight organics. These byproducts can enhance biodegradability and can even be toxic. Wert et al. [113] compared organic byproduct formation during O_3 treatment to O_3/H_2O_2 AOP treatment and showed that O_3/H_2O_2 produced more organic byproducts than ozonation alone. This indicated greater organic degradation using the AOP, leading to incomplete oxidation. Incomplete oxidation can be problematic in distribution systems as it increases assimilable organic carbon (AOC), which can enhance downstream biological growth. Moreover, incomplete oxidation can actually increase DBP formation potential, as described in the following section.

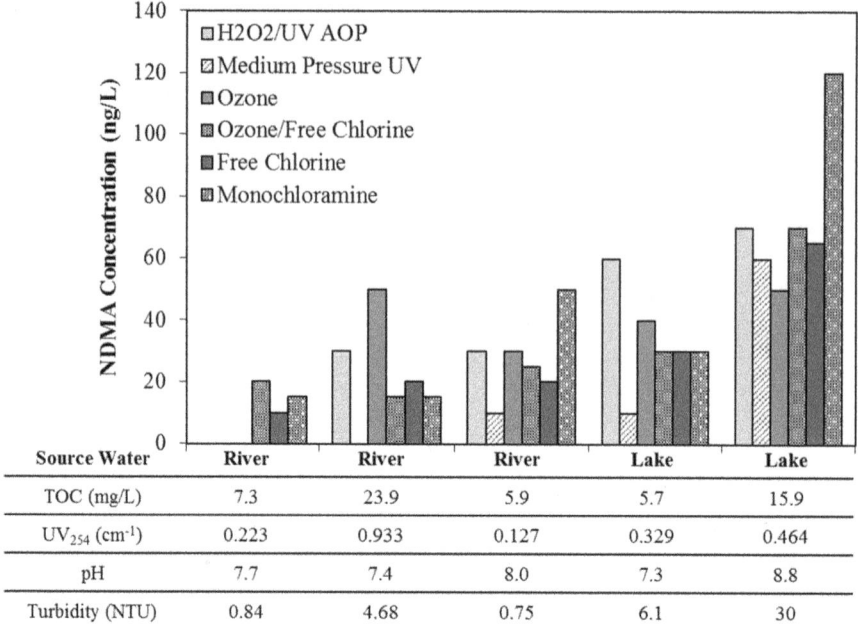

Fig. 11 Comparison of *N*-nitrosodimethylamine (NDMA) formation during disinfection of variable quality untreated source water. The disinfection strategies tested include conventional approaches as well as advanced treatments such as the UV/H_2O_2 advanced oxidation process (AOP). Data from [112]

4.2 The Impact of Advanced Oxidation Processes on Disinfection Byproduct (DBP) Formation Potential: Influence of Oxidation on Precursor Organic Matter and Subsequent DBP Formation

Although AOPs have the ability to mineralize NOM and recalcitrant synthetic organic compounds, variations in chemical and energy inputs as well as process configuration can lead to incomplete oxidation. Incomplete oxidation products can include small organics such as carboxylates, ketones, and aldehydes. These compounds are formed when HO• oxidizes the hydrophobic, densely conjugated portions of NOM, thus opening ring structures and decreasing aromaticity. This is illustrated for the photo(electro)catalytic degradation pathways of phenol shown in Fig. 12. As the ring structures open, the NOM becomes more hydrophilic, and the exposed ring structure can exhibit greater halogen substitution efficiency, thereby serving as an active site for halogenated DBP formation [114].

In many drinking water systems, an oxidizing disinfectant such as free or combined chlorine is added prior to releasing the water to the distribution system to provide residual disinfectant to inactivate pathogens, maintain water quality, and protect against biological regrowth. The use of this type of secondary disinfection

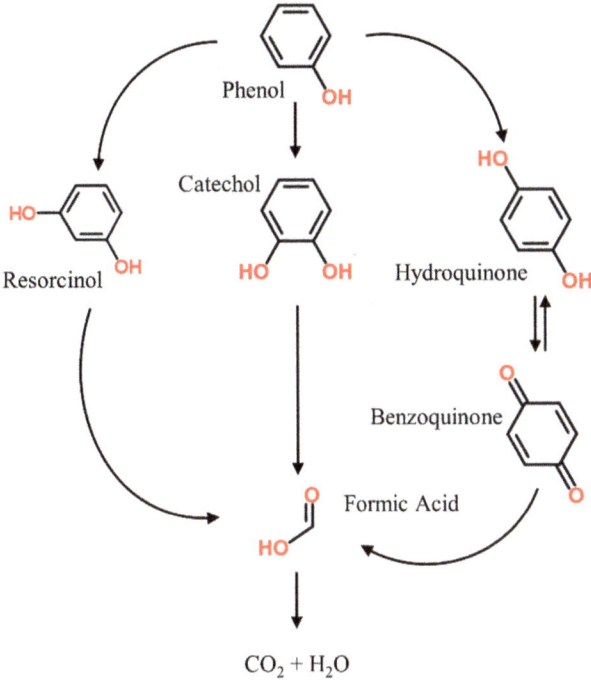

Fig. 12 Photo(electro)catalytic degradation pathways of phenol mineralization using $BiVO_4$ as the catalyst with visible light irradiation. Based on [115]

after AOP treatments wherein incomplete oxidation occurs can potentially exacerbate DBP production, depending on the type and dose of residual disinfectant. Incomplete oxidation leads to conversion of the humic fraction toward the fulvic fraction, which can increase THM and HAA formation potential as the structural properties of NOM change to resemble significant precursors, such as resorcinol. Structural changes in NOM due to reactions with HO• can particularly affect the formation of B-DBPs and I-DBPs because, in comparison to chlorine, bromine and iodine are more reactive with hydrophilic fractions of NOM produced by HO• [95].

AOP-derived chemical changes in NOM can also significantly impact NDMA formation as the hydrophilic NOM fraction produced during incomplete oxidation generally forms more NDMA than the hydrophobic fraction [98]. Additionally, AOPs can degrade residual polymer, which can potentially exacerbate the release of NDMA precursors, thereby greatly increasing NDMA formation when downstream chloramines are used as the residual disinfectant [56].

The extent to which DBP formation potential increases or decreases following AOP treatment is a function of numerous parameters and can thus vary widely across systems as well as within systems. The variation in AOPs' impact on DBP formation potential for selected studies is shown in Table 4.

As oxidation progresses toward NOM mineralization using greater energy and/or chemical AOP inputs, DBP formation potential decreases, as illustrated in Fig. 13, for TiO_2 photocatalysis AOP treatment followed by free chlorine addition.

Complete mineralization, and the resultant reduction in DBP formation potential, is energy and/or chemically intensive, perhaps making combinations of AOP treatment together with processes such as filtration more feasible for implementation in multi-barrier drinking water treatment trains. For example, dual barriers using AOPs followed by biological activated carbon (BAC) filtration can be used to more effectively mitigate DBP concerns using more feasible energy inputs. Toor and Mohseni [120] showed that using UV/H_2O_2 AOP followed by BAC significantly reduced DBP formation potential, whereas using the AOP as a stand-alone treatment required extremely high initial H_2O_2 concentrations and UV fluences to effectively reduce DBP formation (>23 mg/L H_2O_2 and >1,000 mJ/cm^2; for perspective, <200 mJ/cm^2 is typical for drinking water treatment).

5 Mitigation of Disinfection Byproducts

To mitigate DBPs, the recommended course of action is to prevent DBP formation in the first place. Several strategies are employed for this purpose, including employing alternative disinfectants and removing precursor material prior to disinfection. An additional option is to remove/destroy DBPs after they have formed.

5.1 Alternative Disinfectants

Free chlorine remains the most commonly used disinfectant, but concerns over DBP production have led to implementation of alternative disinfectants, including other chemical oxidants such as chloramines, chlorine dioxide, and ozone, as well as non-oxidant-based strategies. It is also possible to move the point of chlorination further downstream in the treatment process to allow reductions in NOM precursor material prior to disinfection [124].

As described in Sect. 2.1, each of the oxidizing disinfectants is characterized by varying degrees of effectiveness against different microbial pathogens and also has potential to produce DBPs, although they may impact the magnitude of DBP formation or the type of DBP produced. For example, monochloramine reduces production of THMs and HAAs; however, it can increase formation of iodo-acids, which are considered one of the most toxic DBPs [40]. Likewise, chlorine dioxide introduces concerns for chlorite and chlorate production, while ozonation can increase bromate formation.

Heat (e.g., boiling or pasteurization) and electromagnetic radiation (e.g., gamma and UV radiation) offer non-oxidant-based disinfection strategies. For water disinfection, the only one of these approaches in routine practice is UV radiation [17]. A

Table 4 Impact of advanced oxidation processes (AOPs) on disinfection byproduct (DBP) formation potential reported in select studies

AOP	Source water	Process description	Trend in DBP formation potential (FP) following AOP treatment	Reference
O_3	River Ruhr, Germany	1.5 mg O_3/mg DOC	↓ TTHMFP and AOXFP[a]	[116]
O_3	Indoor swimming pool	6 mg/L O_3	↓ TTHMFP, AOXFP, TOC, AOX	[117]
O_3/ H_2O_2	Indoor swimming pool	6 mg/L O_3, 1.5 mg/L H_2O_2	↑ TTHMFP ↓ AOXFP, TOC, AOX	[117]
O_3/ UV	Indoor swimming pool	6 mg/L O_3, 2.93 W low-pressure UV lamp	↑ TTHMFP ↓ AOXFP, TOC, AOX	[117]
O_3/ UV	Seymour reservoir, Canada	0.62±0.019 mg O_3/mL, UV fluence = 1.61 J/cm^2	↓ TTHMFP and HAA5FP	[118]
UV/ H_2O_2	Ohio river sample from water treatment plant	5–10 mg/L H_2O_2, medium- and high-pressure UV lamps	↑ TTHM yield following post-chlorination	[119]
UV/ H_2O_2	River Ruhr, Germany	15 W low-pressure UV lamp, 8 mg/L H_2O_2 initially	↑ TTHMFP prior to 1,050 min of irradiation ↓ TTHMFP after 1,050 min of irradiation and 5.6 mg/L H_2O_2 consumed	[116]
UV/ H_2O_2	Vancouver reservoir	5–15 mg/L H_2O_2, low-pressure high-output UV lamp	↓ TTHMFP for fluences >1,500 mJ/cm^2	[114]
UV/ H_2O_2	Raw surface water	0–23 mg/L H_2O_2, low-pressure UV fluence = 0–3,500 mJ/cm^2	↓ DBP formation potential for >1,000 mJ/cm^2 and initial H_2O_2 concentrations >23 mg/L	[120]
UV	Indoor swimming pool	2.93 W low-pressure UV lamp	↑ TTHMFP ↓ AOXFP, TOC, AOX	[117]
UV/ TiO_2	Arizona surface water	7 mW/cm^2 low-pressure UV, 400 mg/L TiO_2	↑ TTHMFP using 5–20 kWh/m^3 ↓ TTHMFP for energy inputs >20 kWh/m^3	[121]
UV/ TiO_2	Myponga reservoir, Australia	Blacklight blue fluorescent UV lamp, 0.1 g/L TiO_2	↓ TTHMFP ↑ HAA5FP before 30 min of irradiation ↓ HAA5FP after 30 min of irradiation	[122]

(continued)

Table 4 (continued)

AOP	Source water	Process description	Trend in DBP formation potential (FP) following AOP treatment	Reference
UV/ TiO$_2$	Hillsborough river water, Florida; Sacramento-San Joaquin Estuary, California	450–1,200 W high-pressure UV lamps, 1.0 g/L TiO$_2$	↓ THMFP	[123]

[a]AOX represents the organically bound halogens adsorbable on activated carbon

major advantage of UV disinfection is that at the fluences typically used in drinking water treatment (<200 mJ/cm^2), there is no evidence of DBP formation, nor are DBP levels exacerbated using post-UV disinfection [9]. However, UV provides no residual disinfection, so an oxidant such as chlorine or chloramines is typically applied prior to release of the water in the drinking water distribution system to protect against pathogen regrowth.

While AOPs can provide some degree of disinfection, they are unlikely to be used explicitly for this purpose. Moreover, increased reactivity of the NOM due to the production of small organic compounds during AOP-based incomplete oxidation can exacerbate DBP formation and also lead to challenges in the distribution system. Enhanced biodegradability of the small organics in the effluent can lead to increased corrosion, nitrification, taste and odor compound formation, and enhanced microbial growth that can impair the microbial safety of the drinking water [62, 114, 125].

To simultaneously mitigate concerns related to a wide spectrum of microbial pathogens, DBPs, and contaminants of emerging concern, water treatment plants may use dual barriers including an alternative disinfectant such as ozone or chloramination alongside advanced treatment (e.g., AOP-BAC) [126].

5.2 Removal of Precursor Natural Organic Matter Prior to Disinfection

Given the trade-offs associated with the use of alternative disinfectants, the most common strategy to mitigate DBPs is to avoid formation by removing precursor NOM prior to disinfection [127].

Physicochemical unit operations including enhanced coagulation or softening, granular activated carbon adsorption, and membrane processes are considered best available techniques for physical removal of NOM from water [83, 128]. These processes are often sufficient to control DBP formation, but alternatives such as ozonation or AOP treatments have also been investigated [16, 84, 121, 129].

Coagulation processes preferentially remove the humic and higher molecular weight portions of NOM. Removal of 15–50% TOC is commonly achieved using

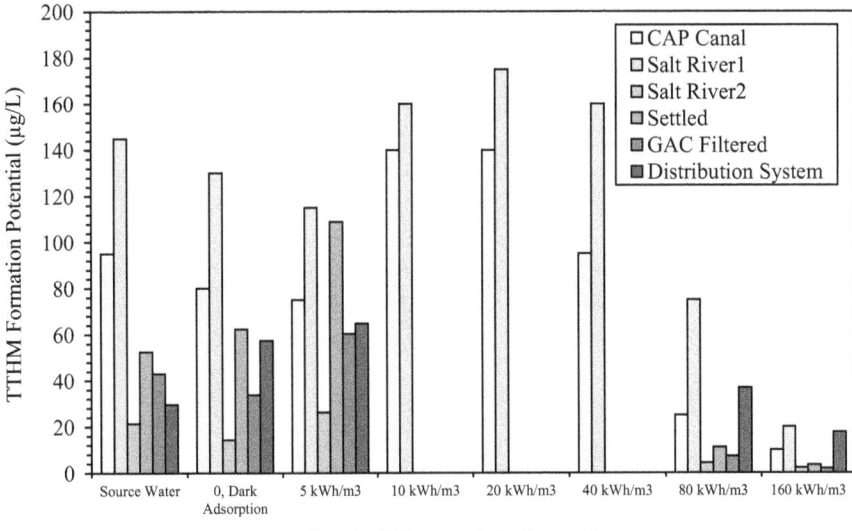

Fig. 13 Total trihalomethane (THM) formation potential in various source waters resulting from UV/TiO$_2$ advanced oxidation process (AOP) treatment followed by free chlorine addition. The data show that TTHM formation potentials increase in some source waters following incomplete oxidation with relatively low-energy inputs (≤ 5 kWh/m^3). In all waters tested, high-energy inputs approached more complete natural organic matter (NOM) mineralization, and TTHM formation potential decreased. Adapted from [84, 121]

enhanced coagulation (which relies on the use of higher coagulant doses and/or pH adjustment to target NOM removal) or enhanced softening [17]. Since the hydrophobic portion of NOM, which generally consists of humic substances, is primarily removed by coagulation, further treatment may be necessary to remove the hydrophilic, fulvic portion of NOM, which can also contribute to DBP formation.

Adsorption of NOM is possible using activated carbon (granular [GAC] or powdered [PAC]) or ion-exchange resins. However, the use of activated carbon for NOM can be costly as a significant fraction of the NOM, comprised of large molecular weight organics, is poorly adsorbed, meaning that large amounts of activated carbon are used to remove relatively small amounts of NOM. With more selective ion-exchange media, e.g., MIEX resin, much greater NOM removal can be achieved more quickly [17].

In some cases, membrane filtration can effectively remove DBP precursors, although results vary for different types of membranes. For example, ultrafiltration with a molecular weight cutoff (MWCO) of 100,000 Da was not effective for controlling DBP formation in pilot studies, whereas nanofiltration with a MWCO of 400–800 Da was effective when little-to-no bromide was present [130]. Microfiltration and ultrafiltration are not effective at removing NDMA precursors, but 57–98% of precursors were removed using nanofiltration and reverse osmosis [131]. Reverse osmosis has been effectively used for NOM removal, rejecting >90% NOM and DBP precursors [132].

The use of O_3 or AOP treatments in combination with BAC filtration can provide an effective means for removing NOM to both reduce DBP formation potential and decrease AOC in the distribution system. Using AOPs or ozonation can break carbon-carbon double bonds in NOM, which transforms the NOM into more readily biodegradable organic matter (higher AOC or biodegradable dissolved organic carbon – BDOC), which is particularly amenable to biofiltration. By combining ozonation with biologically active filtration, 35–40% DOC removal can be achieved [17]. This sequential AOP-biological process treatment train concept is depicted in Fig. 14.

5.3 Mitigation of Preformed Disinfection Byproducts Using Advanced Oxidation Processes

Another strategy to mitigate DBPs is to remove them from water following DBP generation during disinfection processes. While the use of phase-transfer processes or redox chemistry may offer options to accomplish this, DBP removal is typically a less effective and economical approach compared to avoiding DBP formation in the first place via removal of precursors prior to disinfection processes, as described in Sect. 5.2.

Phase-transfer processes such as activated carbon adsorption and air stripping can be used to remove DBPs from water. However, these approaches may introduce elevated operation and maintenance costs that limit feasibility of implementation. For example, GAC is characterized by low THM adsorption capacity, meaning that large amounts of GAC would be required for adequate DBP removal [134, 135]. Additionally, while volatile THMs will readily be removed by air stripping, HAAs will remain [136].

The hydroxyl radicals produced during AOPs can degrade organic DBPs, as illustrated by the kinetic rate constants of several different types of DBPs shown in Table 5.

AOPs have been used to mitigate DBPs in swimming pool water, where O_3/UV and O_3/H_2O_2 generally improved removal of TOC precursors and DBPs (quantified as AOX) beyond levels achieved by O_3 alone. For THM formation potential, O_3 provided a slight advantage over the AOPs due to O_3 oxidation selectivity in comparison to the small reactive molecules produced via HO• reactions, which are more easily transformed to THMs [117].

NDMA readily penetrates reverse osmosis membranes and is poorly adsorbed and stripped, so UV-AOPs have been used to meet regulatory guidelines for potable water reuse [137, 141]. Notably, NDMA is photodegradable using direct UV irradiation (λ<260 nm), so additions of H_2O_2, O_3, or TiO_2 to provide advanced oxidation would not be necessary if other recalcitrant compounds were not also

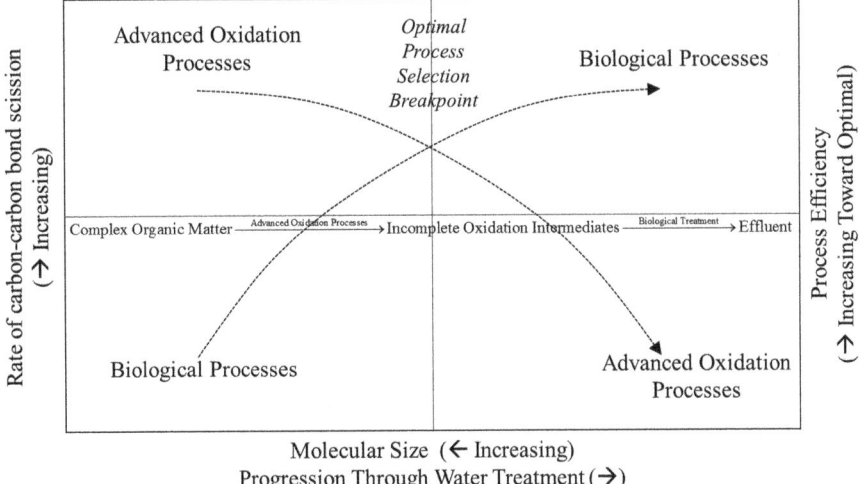

Fig. 14 Conceptual sequential advanced oxidation process (AOP)/biological process (e.g., BAC) approach to treatment, taking advantage of the efficiency of each process for degrading organic matter. The breakpoint between AOPs and biological processes is illustrated, where biological treatment becomes more attractive in terms of rate and efficiency as incomplete oxidation byproducts can be easily biodegraded. Modified after [133]

being targeted, e.g., 1,4-dioxane [141]. Degradation of NDMA during UV treatments can proceed via three pathways: hemolytic cleavage of N-NO bonds, heterolytic cleavage of N-NO bonds, and photooxidation [142]. During UV/H_2O_2 AOP treatment, HO•-based degradation of NDMA occurs by hydrogen atom abstraction from the methyl groups in NDMA [137, 143].

Using a UV dose of $\geq$1,000 mJ/cm^2 together with 3 mg/L H_2O_2, 1 log removal of NDMA can typically be achieved [141]. As noted previously, additions of H_2O_2 do not substantially improve NDMA degradation beyond UV alone. However, direct photolysis of NDMA can produce DMA as a major byproduct, and as DMA reacts with chloramines to produce NDMA, care must be taken to avoid regeneration of NDMA [141].

The electrical energy per order (EEO) of reduction is a metric used to evaluate electrical efficiency of processes, e.g., UV-AOPs. For one log reduction of NDMA using UV/H_2O_2, the EEO is approximately 21–265 kWh/m^3 using doses of 5–6 mg/L H_2O_2 (although this can vary greatly across different waters) [144]. The EEO for destruction of THMs using TiO_2 photocatalysis has been reported at 19–64 kWh/m^3 in several surface waters [121]. For perspective, EEO values less than 0.265 kWh/m^3 are considered favorable for water treatment [145], which demonstrates that AOPs are not an efficient means of destroying preformed DBPs.

Table 5 Hydroxyl radical (HO•) rate constants with various disinfection byproducts (DBPs)

DBP	Chemical formula	DBP class	$k_{HO•}$ ($\times 10^8$/ M s)
Chloroform	$CHCl_3$	Trihalomethane	0.11
Dichloromethane	CH_2Cl_2	Halomethane	0.22
Bromoform	$CHBr_3$	Trihalomethane	1.5
Dibromomethane	CH_2Br_2	Halomethane	0.99
Bromodichloromethane	$CHBrCl_2$	Trihalomethane	0.711
Chlorodibromomethane	$CHBr_2Cl$	Trihalomethane	0.831
N-Nitrosodimethylamine (NDMA)	$C_2H_6N_2O$	Nitrosamines	4.3
Chloronitromethane	CH_2ClNO_2	Halonitromethane	1.94
Dichloronitromethane	$CHCl_2NO_2$	Halonitromethane	5.12
Trichloronitromethane (chloropicrin)	CCl_3NO_2	Halonitromethane	0.497
Bromonitromethane	CH_2BrNO_2	Halonitromethane	0.836
Dibromonitromethane	$CHBr_2NO_2$	Halonitromethane	4.75
Tribromonitromethane	CBr_3NO_2	Halonitromethane	3.25
Bromochloronitromethane	$CHBrClNO_2$	Halonitromethane	4.2
Bromodichloronitromethane	$CBrCl_2NO_2$	Halonitromethane	1.02
Dibromochloronitromethane	CBr_2ClNO_2	Halonitromethane	1.80

Values from [137–140]

6 Summary

Although the premise of disinfection is easily understood, interactions between disinfectants and the target and nontarget constituents in complex drinking water matrices are incredibly complex. Accordingly, there is not a one-size-fits-all approach to minimizing the risks posed by microbial pathogens against those posed by chemical contaminants, e.g., DBPs. While a given disinfection process may significantly reduce one DBP, it may lead to substantial increases in another DBP (e.g., chloramination may reduce THMs and HAAs but increase NDMA). Likewise, shifts in precursor NOM composition during unit operations prior to disinfection can have variable impacts on subsequent DBP formation during disinfection processes. Overall, removal of this precursor material prior to disinfection offers the most effective strategy to mitigate DBPs.

The use of AOPs in drinking water treatment trains may impact DBPs via three different pathways: (1) direct formation during AOP treatment, (2) indirect influences on DBP formation potential in downstream disinfection, and (3) destruction of preformed DBPs. High AOP energy and/or chemical inputs are needed to destroy preformed DBPs, making it much more efficient to avoid DBP production in the first place.

The direct generation of DBPs during AOP treatments is generally less of a concern in comparison to disinfection strategies based on additions of halogenated oxidants such as chlorine since AOPs do not directly form halogenated DBPs. However, direct inputs of chemical oxidants, e.g., O_3 or H_2O_2, during AOPs can generate DBPs such as BrO_3^- or NDMA. Moreover, as complex organics are progressively degraded, AOPs can also directly generate their own suite of DBPs in the event of incomplete oxidation, which produces small molecular weight organic compounds, e.g., acetaldehyde and formaldehyde. These non-halogenated organic DBPs can be toxic or can lead to subsequent issues such as enhanced DBP formation potential, higher biodegradability (AOC or BDOC) in the distribution system, etc.

In addition to direct DBP formation, AOPs may also strongly influence subsequent DBP formation potential as DBP precursor organic material undergoes dramatic transformations during AOPs. Although AOPs can effectively mitigate organic precursors to limit downstream formation of DBPs given sufficient energy and/or chemical inputs to yield complete mineralization, lower inputs lead to incomplete oxidation, which can exacerbate DBP formation. Thus, AOPs may either decrease or increase DBP yields during subsequent disinfection processes depending on the water matrix and process operation.

To simultaneously mitigate concerns related to a wide spectrum of microbial pathogens, DBPs, and contaminants of emerging concern, drinking water treatment plants may use dual barriers including an alternative disinfectant such as ozone or chloramination alongside advanced treatment (e.g., AOP-BAC). The impact of these processes on DBP formation potential must be evaluated for each individual source water using selected operational parameters as changes in DBP precursors during treatment processes can have widely variable impacts on DBP formation, and care must be taken to avoid increasing DBP formation via incomplete oxidation during AOP treatments.

References

1. AWWA (2000) Committee report: disinfection at large and medium-size systems. J Am Water Work Assoc 92:32–43
2. AWWA (2010) Committee report: disinfection survey, part 1 – recent changes, current practices, and water quality. J Am Water Work Assoc 100:76–90
3. Mayer BK, Yang Y, Gerrity DW, Abbaszadegan M (2015) The impact of capsid proteins on virus removal and inactivation during water treatment processes. Microbiol Insights 8(Suppl 2):15–28
4. Amy G, Bull R, Craun GF, Siddiqui M (2000) Disinfectants and disinfection by-products. World Health Organization, Geneva
5. Johnson JD, Jensen JN (1986) THM and TOX formation: routes, rates, and precursors. J Am Water Work Assoc 78:156–162. doi:10.1111/j.1467-8330.1974.tb00606.x
6. Bull RJ, Gerba C, Trussell RR (1990) Evaluation of health risks associated with disinfection. Crit Rev Environ Control 20:77–114

7. WHO (2016) Chlorine dioxide, chlorate and chlorite in drinking-water: background document for development of WHO guidelines for drinking-water quality. World Health Organization
8. Amy G, Siddiqui M, Ozekin K et al (1998) Empirically based models for predicting chlorination and ozonation by-products: trihalomethanes, haloacetic acids, chloral hydrate, and bromate. USEPA, EPA 815-R-98, Cincinnati, OH
9. USEPA (2006) Ultraviolet disinfection guidance manual for the final long term 2 enhanced surface water treatment rule. Washington, DC
10. Mamane H, Shemer H, Linden KG (2007) Inactivation of E. coli, B. subtilis spores, and MS2, T4, and T7 phage using UV/H2O2 advanced oxidation. J Hazard Mater 146:479–486. doi:10.1016/j.jhazmat.2007.04.050
11. Matsunaga T, Tomoda R, Nakajima T et al (1988) Continuous-sterilization system that uses photosemiconductor powders. Appl Environ Microbiol 54:1330–1333
12. Ryu H, Gerrity D, Crittenden JC, Abbaszadegan M (2008) Photocatalytic inactivation of Cryptosporidium parvum with TiO2 and low-pressure ultraviolet irradiation. Water Res 42:1523–1530. doi:10.1016/j.watres.2007.10.037
13. Gerrity DW, Ryu H, Crittenden JC, Abbaszadegan M (2008) Photocatalytic inactivation of viruses using titanium dioxide nanoparticles and low-pressure UV light. J Environ Sci Health A Tox Hazard Subst Environ Eng 43:1261–1270
14. Cho M, Chung H, Choi W, Yoon J (2005) Different inactivation behaviors of MS-2 phage and Escherichia coli in TiO2 photocatalytic disinfection. Appl Environ Microbiol 71:270–275
15. Alrousan DMA, Dunlop PSM, McMurray TA, Byrne JA (2009) Photocatalytic inactivation of E. coli in surface water using immobilised nanoparticle TiO2 films. Water Res 43:47–54. doi:10.1016/j.watres.2008.10.015
16. Mayer BK, Daugherty E, Abbaszadegan M (2015) Evaluation of the relationship between bulk organic precursors and disinfection byproduct formation for advanced oxidation processes. Chemosphere 121:39–46. doi:10.1016/j.chemosphere.2014.10.070
17. Crittenden JC, Trussell RR, Hand DW et al (2012) Water treatment: principles and design, 3rd edn. Wiley, Hoboken, NJ
18. NHMRC, NRMMC (2011) Australian drinking water guidelines paper 6 national water quality management strategy. Commonwealth of Australia, Canberra
19. EU (1998) Council Directive. http://eur-lex.europa.eu/LexUriServ/LexUriServ.do?uri=OJ:L:1998:330:0032:0054:EN:PDF
20. Health Canada (2017) Guidelines for Canadian drinking water quality summary table. Ottawa, ON, Canada
21. USEPA (2009) National Primary Drinking Water Regulations. Washington, DC
22. WHO (2017) Guidelines for drinking-water quality: Fourth edition incorporating the first addendum
23. Bellar TA, Lichtenberg JJ, Kroner RC (1974) The occurrence of organohalides in chlorinated drinking water. J Am Water Work Assoc 66:703–706
24. Rook JJ (1974) Formation of halogens during the chlorination of natural water. Water Treat Exam 23:234–243
25. Richardson SD, Plewa MJ, Wagner ED et al (2007) Occurrence, genotoxicity, and carcinogenicity of regulated and emerging disinfection by-products in drinking water: a review and roadmap for research. Mutat Res 636:178–242. doi:10.1016/j.mrrev.2007.09.001
26. Weinberg HS, Krasner SW, Richardson SD, Thruston AD (2002) The occurrence of disinfection by-products (DBPs) of health concern in drinking water: results of a nationwide DBP occurrence study. Athens, GA
27. Mitch WA, Krasner SW, Paul W, Dotson A (2009) Occurrence and formation of nitrogenous disinfection by-products
28. Clark RM, Boutin BK (2001) Controlling disinfection by-products and microbial contaminants in drinking water. Washington, DC

29. USEPA (2016) Final CCL4 chemical contaminants. In: Contaminant candidate list (CCL) and regulatory determination. https://www.epa.gov/ccl/chemical-contaminants-ccl-4. Accessed 11 Jul 2017
30. Khiari D (2017) Water Research Foundation Focus Area: NDMA and other nitrosamines. J Am Water Work Assoc 109:38–42
31. Russell CG, Brown R, Porter K, Reckhow D (2017) Practical considerations for implementing nitrosamine control strategies. doi: 10.5942/jawwa.2017.109.0054
32. Krasner SW, Weinberg HS, Richardson SD et al (2006) Occurrence of a new generation of disinfection byproducts. Environ Sci Technol 40:7175–7185
33. Chowdhury S, Champagne P, McLellan PJ (2009) Models for predicting disinfection byproduct (DBP) formation in drinking waters: a chronological review. Sci Total Environ 407:4189–4206. doi:10.1016/j.scitotenv.2009.04.006
34. Sohn J, Amy G, Cho J et al (2004) Disinfectant decay and disinfection by-products formation model development: chlorination and ozonation by-products. Water Res 38:2461–2478. doi:10.1016/j.watres.2004.03.009
35. Amy GL, Chadik PA, Chowdhury ZK (1987) Developing models for predicting trihalomethane formation potential and kinetics. J Am Water Work Assoc 79:89–97
36. Arora H, LeChevallier MW, Dixon KL (1997) DBP occurrence survey. J Am Water Work Assoc 89:60–68
37. Richardson SD (1998) Drinking water disinfection byproducts. In: Encyclopedia of environmental analysis and remediation, pp 1398–1421
38. von Gunten U (2003) Ozonation of drinking water: part II. Disinfection and by-product formation in presence of bromide, iodide or chlorine. Water Res 37:1469–1487
39. von Gunten U, Bruchet A, Costentin E (1996) Bromate formation in advanced oxidation processes. J Am Water Work Assoc 88:53–65
40. Plewa MJ, Wagner ED, Richardson SD et al (2004) Chemical and biological characterization of newly discovered Iodoacid drinking water disinfection byproducts. Environ Sci Technol 38:4713–4722. doi:10.1021/Es049971v
41. Taurog A, Howells EM, Nachimson HI (1966) Conversion of iodate to iodide in vivo and in vitro. J Biol Chem 241:4686–4693
42. Bichsel Y, von Gunten U (2000) Formation of iodo-trihalomethanes during disinfection and oxidation of iodide-containing waters. 34:2784–2791. doi: 10.1021/es9914590
43. Hua G, Reckhow DA, Kim J (2006) Effect of bromide and iodide ions on the formation and speciation of disinfection byproducts during chlorination. Environ Sci Technol 40:3050–3056. doi:10.1021/es0519278
44. Wright JM, Schwartz J, Vartiainen T et al (2002) 3-Chloro-4-(dichloromethyl)-5-hydroxy-2 (5H)-furanone (MX) and mutagenic activity in massachusetts drinking water. Environ Health Perspect 110:157–164. doi:10.1289/ehp.02110157
45. Smeds A, Vartiainen T, Mäki-Paakkanen J, Kronberg L (1997) Concentrations of Ames mutagenic chlorohydroxyfuranones and related compounds in drinking waters. Environ Sci Technol 31:1033–1039. doi:10.1021/es960504q
46. Kronberg L, Vartiainen T (1988) Ames mutagenicity and concentration of the strong mutagen 3-chloro-4-(dichloromethyl)-5-hydroxy-2(5H)-furanone and of its geometric isomer E-2-chloro-3-(dichloromethyl)-4-oxo-butenoic acid in chlorine-treated tap waters. Mutat Res 206:177–182
47. Onstad GD, Weinberg HS, Krasner SW (2008) Occurrence of halogenated furanones in U.S. drinking waters. Environ Sci Technol 42:3341–3348. doi:10.1021/es071374w
48. Mitch WA, Gerecke AC, Sedlak DL (2003) A N-nitrosodimethylamine (NDMA) precursor analysis for chlorination of water and wastewater. Water Res 37:3733–3741. doi:10.1016/S0043-1354(03)00289-6
49. Linge KL, Kristiana I, Liew D et al (2017) Formation of N-nitrosamines in drinking water sources: case studies from Western Australia. J Am Water Work Assoc 109:E184–E196. doi:10.5942/jawwa.2017.109.0036

50. California Department of Health Services (2005) California drinking water activities related to NDMA and other nitrosamines. California Department of Health Services, Sacramento. http://www.waterboards.ca.gov/drinking_water/certlic/drinkingwater/NDMA.shtml
51. Choi J, Valentine RL (2002) Formation of N-nitrosodimethylamine (NDMA) from reaction of monochloramine: a new disinfection by-product. Water Res 36:817–824. doi:10.1016/S0043-1354(01)00303-7
52. Marti EJ, Dickenson ERV, Trenholm RA, Batista JR (2017) Treatment of specific NDMA precursors by biofiltration. J Am Water Work Assoc 109:E273–E286. doi:10.5942/jawwa.2017.109.0070
53. Park SH, Padhye LP, Wang P et al (2015) N-nitrosodimethylamine (NDMA) formation potential of amine-based water treatment polymers: effects of in situ chloramination, breakpoint chlorination, and pre-oxidation. J Hazard Mater 282:133–140. doi:10.1016/j.jhazmat.2014.07.044
54. Shah AD, Krasner SW, Lee CFT et al (2012) Trade-offs in disinfection byproduct formation associated with precursor preoxidation for control of N-nitrosodimethylamine formation. Environ Sci Technol 46:4809–4818. doi:10.1021/es204717j
55. von Gunten U, Salhi E, Schmidt CK, Arnold WA (2010) Kinetics and mechanisms of N-nitrosodimethylamine formation upon ozonation of N,N-dimethylsulfamide-containing waters: bromide catalysis. Environ Sci Technol 44(15):5762–5768. doi:10.1021/es1011862
56. Padhye L, Luzinova Y, Cho M et al (2011) PolyDADMAC and dimethylamine as precursors of N-nitrosodimethylamine during ozonation: reaction kinetics and mechanisms. Environ Sci Technol 45:4353–4359. doi:10.1021/es104255e
57. Park SH, Wei S, Mizaikoff B et al (2009) Degradation of amine-based water treatment polymers during chloramination as N-nitrosodimethylamine (NDMA) precursors. Environ Sci Technol 43:1360–1366. doi:10.1021/es802732z
58. Cornwell DC, Brown RA (2017) Replacement of alum and polyDADMAC with natural polymers – turbidity removal and residuals reduction impacts. J Am Water Work Assoc 109:E252–E264
59. Glaze WH, Weinberg HS (1993) Identification and occurrence of ozonation by-products in drinking water. American Water Works Association Foundation
60. Krasner SW, McGuire MJ, Jacangelo JG et al (1989) The occurrence of disinfection by-products in U.S. drinking water. J Am Water Work Assoc 81:41–53
61. Leenheer JA, Croue J-P (2003) Characterizing dissolved aquatic matter. Environ Sci Technol 37:18a–26a
62. Jacangelo JG, DeMarco J, Owen DM, Randtke SJ (1995) Selected processes for removing NOM: an overview. J Am Water Work Assoc. doi:10.2307/41295153
63. Danielsson LG (1982) On the use of filters for distinguishing between dissolved and particulate fractions in natural waters. Water Res 16:179–182
64. Thurman EM (1985) Organic geochemistry of natural waters 1st edn. Nijhoff/Junk, Dordrecht, The Nederlands
65. Korshin G, Chow CWK, Fabris R, Drikas M (2009) Absorbance spectroscopy-based examination of effects of coagulation on the reactivity of fractions of natural organic matter with varying apparent molecular weights. Water Res 43:1541–1548. doi:10.1016/j.watres.2008.12.041
66. Edzwald JK, Becker WC, Wattier KL (1985) Surrogate parameters for monitoring organic matter and THM precursors. J Am Water Work Assoc 77:122–132
67. Kitis M, Karanfil T, Kilduff JE, Wigton A (2001) The reactivity of natural organic matter to disinfection by-products formation and its relation to specific ultraviolet absorbance. Water Sci Technol 43:9–16
68. Edzwald J, Tobiason J (1999) Enhanced coagulation: US requirements and a broader view. Water Sci Technol 40:63–70. doi:10.1016/S0273-1223(99)00641-1
69. Coble PG (1996) Characterization of marine and terrestrial DOM in seawater using excitation-emission matrix spectroscopy. Mar Chem 51:325–346

70. Wu FC, Kothawala DN, Evans RD et al (2007) Relationships between DOC concentration, molecular size and fluorescence properties of DOM in a stream. Appl Geochem 22:1659–1667. doi:10.1016/j.apgeochem.2007.03.024
71. Baghoth SA, Dignum M, Grefte A et al (2009) Characterization of NOM in a drinking water treatment process train with no disinfectant residual. Water Sci Technol Water Supply 9:379–386. doi:10.2166/ws.2009.569
72. Seredyńska-Sobecka B, Baker A, Lead JR (2007) Characterisation of colloidal and particulate organic carbon in freshwaters by thermal fluorescence quenching. Water Res 41:3069–3076. doi:10.1016/j.watres.2007.04.017
73. Baker A, Tipping E, Thacker SA, Gondar D (2008) Relating dissolved organic matter fluorescence and functional properties. Chemosphere 73:1765–1772. doi:10.1016/j.chemosphere.2008.09.018
74. Hudson N, Baker A, Ward D et al (2007) Can fluorescence spectrometry be used as a surrogate for the biochemical oxygen demand (BOD) test in water quality assessment? An example from South West England. doi: 10.1016/j.scitotenv.2007.10.054
75. Spencer RGM, Bolton L, Baker A (2007) Freeze/thaw and pH effects on freshwater dissolved organic matter fluorescence and absorbance properties from a number of UK locations. Water Res 41:2941–2950. doi:10.1016/j.watres.2007.04.012
76. Yang L, Kim D, Uzun H et al (2015) Assessing trihalomethanes (THMs) and N-nitrosodimethylamine (NDMA) formation potentials in drinking water treatment plants using fluorescence spectroscopy and parallel factor analysis. Chemosphere 121:84–91. doi:10.1016/j.chemosphere.2014.11.033
77. Bieroza M, Baker A, Bridgeman J (2009) Relating freshwater organic matter fluorescence to organic carbon removal efficiency in drinking water treatment. Sci Total Environ 407:1765–1774. doi:10.1016/j.scitotenv.2008.11.013
78. Peiris RH, Hallé C, Budman H et al (2010) Identifying fouling events in a membrane-based drinking water treatment process using principal component analysis of fluorescence excitation-emission matrices. Water Res 44:185–194. doi:10.1016/j.watres.2009.09.036
79. Chen W, Westerhoff P, Leenheer JA, Booksh K (2003) Fluorescence excitation-emission matrix regional integration to quantify spectra for dissolved organic matter. Environ Sci Technol 37:5701–5710
80. Matilainen A, Gjessing ET, Lahtinen T et al (2011) An overview of the methods used in the characterisation of natural organic matter (NOM) in relation to drinking water treatment. Chemosphere 83:1431–1442. doi:10.1016/j.chemosphere.2011.01.018
81. Chow AT, Gao S, Dahlgren RA (2005) Physical and chemical fractionation of dissolved organic matter and trihalomethane precursors: a review. J Water Supply Res Technol 54:475–507
82. Hwang CJ, Sclimenti MJ, Bruchet A et al (2001) DBP yields of polar NOM fractions from low humic waters. Proceedings of water quality technology conference AWWA. Denver, CO
83. Liu S, Lim M, Fabris R, et al. (2008) TiO2 photocatalysis of natural organic matter in surface water: impact on trihalomethane and haloacetic acid formation potential. Environ Sci Technol 42:6218–6223
84. Mayer BK, Daugherty E, Abbaszadegan M (2014) Disinfection byproduct formation resulting from settled, filtered, and finished water treated by titanium dioxide photocatalysis. Chemosphere 117:72–78
85. Pehlivanoglu-Mantas E, Sedlak DL (2008) Measurement of dissolved organic nitrogen forms in wastewater effluents: concentrations, size distribution and NDMA formation potential. Water Res 42:3890–3898. doi:10.1016/j.watres.2008.05.017
86. Mitch WA, Sedlak DL (2004) Characterization and fate of N-nitrosodimethylamine precursors in municipal wastewater treatment plants. Environ Sci Technol 38:1445–1454. doi:10.1021/es035025n
87. Zhao Z-Y, Gu J-D, Fan X-J, Li H-B (2006) Molecular size distribution of dissolved organic matter in water of the Pearl River and trihalomethane formation characteristics with chlorine

and chlorine dioxide treatments. J Hazard Mater 134:60–66. doi:10.1016/j.jhazmat.2005.10.032
88. Kitis M, Karanfil T, Wigton A, Kilduff JE (2002) Probing reactivity of dissolved organic matter for disinfection by-product formation using XAD-8 resin adsorption and ultrafiltration fractionation. Water Res 36:3834–3848
89. Lin L, Xu B, Lin YL et al (2014) A comparison of carbonaceous, nitrogenous and iodinated disinfection by-products formation potential in different dissolved organic fractions and their reduction in drinking water treatment processes. Sep Purif Technol 133:82–90. doi:10.1016/j.seppur.2014.06.046
90. Her N, Amy G, Foss D et al (2002) Optimization of method for detecting and characterizing NOM by HPLC-size exclusion chromatography with UV and on-line DOC detection. Environ Sci Technol 36:1069–1076. doi:10.1021/es015505j
91. Her N, Amy G, Sohn J, Gunten U (2008) UV absorbance ratio index with size exclusion chromatography (URI-SEC) as an NOM property indicator. J Water Supply Res Technol 57:35. doi:10.2166/aqua.2008.029
92. Her N, Amy G, McKnight D et al (2003) Characterization of DOM as a function of MW by fluorescence EEM and HPLC-SEC using UVA, DOC, and fluorescence detection. Water Res 37:4295–4303. doi:10.1016/S0043-1354(03)00317-8
93. Sharp EL, Jarvis P, Parsons SA, Jefferson B (2006) Impact of fractional character on the coagulation of NOM. Colloids Surf A Physicochem Eng Asp 286:104–111. doi: 10.1016/j.colsurfa.2006.03.009
94. Swietlik J, Dabrowska A, Raczyk-Stanisławiak U, Nawrocki J (2004) Reactivity of natural organic matter fractions with chlorine dioxide and ozone. Water Res 38:547–558. doi:10.1016/j.watres.2003.10.034
95. Hua G, Reckhow DA (2007) Characterization of disinfection byproduct precursors based on hydrophobicity and molecular size. Environ Sci Technol 41:3309–3315
96. Chen C, Zhang X, Zhu L et al (2008) Disinfection by-products and their precursors in a water treatment plant in North China: seasonal changes and fraction analysis. Sci Total Environ 397:140–147. doi:10.1016/j.scitotenv.2008.02.032
97. Croue J-P, Korshin GV, Benjamin M (2006) Characterization of natural organic matter in drinking water. Denver, CO
98. Chen Z, Valentine RL (2007) Formation of N-nitrosodimethylamine (NDMA) from humic substances in natural water. Environ Sci Technol 41:6059–6065. doi:10.1021/es0705386
99. Croue J-P, Debroux FF, Amy GL, Aiken GR, Leenheer JA (1999) Natural organic matter: structural characteristics and reactive properties. In: Singer PC (ed) Formation and control of disinfection by-products in drinking water. American Water Works Association, Denver
100. Marhaba TF, Van D (2000) The variation of mass and disinfection by-product formation potential of dissolved organic matter fractions along a conventional surface water treatment plant. J Hazard Mater 74:133–147
101. Bond T, Henriet O, Goslan EH et al (2009) Disinfection byproduct formation and fractionation behavior of natural organic matter surrogates. Environ Sci Technol 43:5982–5989
102. Zwiener C, Richardson SD, de Marini DM et al (2007) Drowning in disinfection byproducts? Assessing swimming pool water. J Environ Sci Technol 41:363–372
103. Westerhoff P, Mezyk SP, Cooper WJ, Minakata D (2007) Electron pulse radiolysis determination of hydroxyl radical rate constants with Suwannee river fulvic acid and other dissolved organic matter isolates. Environ Sci Technol 41:4640–4646. doi:10.1021/es062529n
104. Buxton GV, Greenstock CL, Helman WP, Ross AB (1988) Critical review of rate constants for reactions of hydrated electrons, hydrogen atoms and hydroxyl radicals in aqueous solution. J Phys Chem Ref Data Monogr 17:513–886. doi:10.1063/1.555805
105. Westerhoff P, Aiken G, Amy G, Debroux J (1999) Relationships between the structure of natural organic matter and its reactivity towards molecular ozone and hydroxyl radicals. Water Res 33:2265–2276. doi: 10.1016/S0043-1354(98)00447-3

106. Richardson SD, Thruston AD, Caughran TV et al (1999) Identification of new ozone disinfection byproducts in drinking water. Environ Sci Technol 33:3368–3377. doi:10.1021/es981218c
107. von Gunten U, Hoigne J (1994) Bromate formation during ozonation of bromide-containing waters: interaction of ozone and hydroxyl radical reactions. Environ Sci Technol 28:1234–1242. doi:10.1021/es00056a009
108. Kishimoto N, Nakamura E (2012) Bromate formation characteristics of UV irradiation, hydrogen peroxide addition, ozonation, and their combination processes. Int J Photoenergy. doi:10.1155/2012/107293
109. Symons JM, Zheng MCH (1997) Technical note: does hydroxyl radical oxidize bromide to bromate? J AWWA 89:106–109
110. von Gunten U, Oliveras Y (1998) Advanced oxidation of bromide-containing waters: bromate formation mechanisms. Environ Sci Technol 32:63–70. doi:10.1021/es970477j
111. Haag WR, Hoigne J (1983) Ozonation of bromide-containing waters: kinetics of formation of hypobromous acid and bromate. Environ Sci Technol 17:261–267. doi:10.1021/es00111a004
112. Zhao Y-Y, Boyd JM, Woodbeck M et al (2008) Formation of N-nitrosamines from eleven disinfection treatments of seven different surface waters. Environ Sci Technol 42:4857–4862. doi:10.1021/es7031423
113. Wert EC, Rosario-Ortiz FL, Drury DD, Snyder SA (2007) Formation of oxidation byproducts from ozonation of wastewater. Water Res. doi:10.1016/j.watres.2007.01.020
114. Sarathy S, Mohseni M (2010) Effects of UV/H2O2 advanced oxidation on chemical characteristics and chlorine reactivity of surface water natural organic matter. Water Res 44:4087–4096. doi:10.1016/j.watres.2010.05.025
115. Bennani Y, Perez-Rodriguez P, Alani MJ et al (2016) Photoelectrocatalytic oxidation of phenol for water treatment using a BiVO4 thin-film photoanode. J Mater Res 31:2627–2639. doi:10.1557/jmr.2016.290
116. Kleiser G, Frimmel FH (2000) Removal of precursors for disinfection by-products (DBPs) – differences between ozone- and OH-radical-induced oxidation. Sci Total Environ 256:1–9. doi:10.1016/S0048-9697(00)00377-6
117. Glauner T, Kunz F, Zwiener C, Frimmel FH (2005) Elimination of swimming pool water disinfection by-products with advanced oxidation processes (AOPs). Acta Hydrochim Hydrobiol. doi:10.1002/aheh.200400605
118. Chin A, Bérubé PR (2005) Removal of disinfection by-product precursors with ozone-UV advanced oxidation process. Water Res 39:2136–2144. doi:10.1016/j.watres.2005.03.021
119. Dotson AD, Keen VOS, Metz D, Linden KG (2010) UV/H2O2 treatment of drinking water increases post-chlorination DBP formation. Water Res 44:3703–3713. doi:10.1016/j.watres.2010.04.006
120. Toor R, Mohseni M (2007) UV-H2O2 based AOP and its integration with biological activated carbon treatment for DBP reduction in drinking water. Chemosphere 66:2087–2095. doi:10.1016/j.chemosphere.2006.09.043
121. Gerrity D, Mayer B, Ryu H et al (2009) A comparison of pilot-scale photocatalysis and enhanced coagulation for disinfection byproduct mitigation. Water Res 43:1597–1610. doi:10.1016/j.watres.2009.01.010
122. Liu S, Lim M, Fabris R et al (2008) Removal of humic acid using TiO2 photocatalytic process – fractionation and molecular weight characterisation studies. doi:10.1016/j.chemosphere.2008.01.061
123. Hand DW, Perram DL, Crittenden JC (1995) Destruction of DBP precursors with catalytic oxidation. J Am Water Work Assoc 87:84–96
124. USEPA (1999) Microbial and disinfection byproduct rules: simultaneous compliance guidance manual. United States Environmental Protection Agency, Washington, EPA 815-R-99-015

125. Prest EI, Hammes F, van Loosdrecht MCM, Vrouwenvelder JS (2016) Biological stability of drinking water: controlling factors, methods, and challenges. Front Microbiol. doi:10.3389/fmicb.2016.00045
126. Liao X, Chen C, Yuan B et al (2017) Control of nitrosamines, THMs, and HAAs in heavily impacted water with O3-BAC. J Am Water Work Assoc 109:E215–E225
127. Kulkarni P, Chellam S (2010) Disinfection by-product formation following chlorination of drinking water: artificial neural network models and changes in speciation with treatment. Sci Total Environ 408:4202–4210. doi:10.1016/j.scitotenv.2010.05.040
128. USEPA (2006) National Primary Drinking Water Regulations: stage 2 disinfectants and disinfection byproducts rule
129. WHO (2000) Environmental health criteria 216: disinfectants and disinfectant by-products, Geneva
130. Laîné J-M, Jacangelo JG, Cummings EW et al (1993) Influence of bromide on low-pressure membrane filtration for controlling DBPs in surface waters. Am Water Work Assoc 85:87–99
131. Uzun H, Kim D, Karanfil T (2017) The removal of N-nitrosodimethylamine formation potential in drinking water treatment plants. J Am Water Works Assoc 109:15–28. doi:10.5942/jawwa.2017.109.0047
132. Fu P, Ruiz H, Thompson K, Spangenberg C (1994) Selecting membranes for removing NOM and DBP precursors. J Am Water Work Assoc 88:55–72
133. Comninellis C, Kapalka A, Malato S et al (2008) Advanced oxidation processes for water treatment: advances and trends for R & D. J Chem Technol Biotechnol 83:769–776. doi:10.1002/jctb
134. Speth TF, Miltner RJ (1990) Technical note: adsorption capacity of GAC for synthetic organics. JAWWA 82:72–75
135. Crittenden JC, Berrigan JK, Hand DW (1986) Design of rapid small-scale adsorption tests for a constant diffusivity. J Water Pollut Control Fed 58(4):312–319
136. Marhaba TF (2000) A new look at disinfection by-products in drinking water. Water Eng Manag 147:30–34
137. Mezyk SP, Cooper WJ, Madden KP, Bartels DM (2004) Free radical destruction of N-nitrosodimethylamine in water. Environ Sci Technol 38:3161–3167. doi:10.1021/es0347742
138. Mezyk SP, Helgeson T, Cole SK et al (2006) Free radical chemistry of disinfection-byproducts. 1. Kinetics of hydrated electron and hydroxyl radical reactions with halonitromethanes in water. J Phys Chem A 110:2176–2180. doi:10.1021/jp054962+
139. Cole SK, Cooper WJ, Fox RV et al (2007) Free radical chemistry of disinfection byproducts. 2. Rate constants and degradation mechanisms of trichloronitromethane (chloropicrin). Environ Sci Technol 41:863–869. doi:10.1021/es061410b
140. Haag RW, David Yao CC (1992) Rate constants for reaction of hydroxyl radicals with several drinking water contaminants. Environ Sci Technol 26:1005–1013
141. Fujioka T, Masaki S, Kodamatani H, Ikehata K (2017) Degradation of N-nitrosodimethylamine by UV-based advanced oxidation processes for potable reuse: a short review. Curr Pollut Rep 3:79–87. doi:10.1007/s40726-017-0052-x
142. Lee C, Choi W, Yoon J (2005) UV photolytic mechanism of N-nitrosodimethylamine in water: roles of dissolved oxygen and solution pH. Environ Sci Technol 39:9702–9709. doi:10.1021/es051235j
143. Sharpless CM, Linden KG (2003) Experimental and model comparisons of low- and medium-pressure Hg lamps for the direct and H2O2 assisted UV photodegradation of N-nitrosodimethylamine in simulated drinking water. Environ Sci Technol 37:1933–1940. doi:10.1631/jzus.A0820642
144. Tchobanoglous G, Burton FL, Stensel HD (2003) Wastewater engineering: treatment and reuse. McGraw Hill, New York
145. Bolton JR, Cater SR (1994) Homogeneous photodegradation of pollutants in contaminated water: an introduction. Aquatic Surf Photochem 467–490

Evolution of Toxicity and Estrogenic Activity Throughout AOP's Surface and Drinking Water Treatment

Tatjana Tišler and Albin Pintar

Abstract Nowadays, increasing pollution of surface and drinking water with hundreds of chemical compounds, lots of them toxic and persistent, presents a major problem in access to safe drinking water supply. Although low concentrations of pollutants such as natural and synthetic estrogens have been measured in surface and drinking water, they could be biologically active even at these concentration levels. Advanced oxidation processes (AOPs) are very powerful tools for removal of recalcitrant compounds from water streams as they are able to mineralize or degrade organic compounds. In this chapter, development of toxicity and estrogenic activity of water samples treated by AOPs, which depends on the oxidation process used and experimental conditions, is discussed. Evolution of species-specific toxicity was particularly evident during the photocatalytic oxidation of herbicide iodosulfuron, as greatly enhanced toxicity of treated samples was detected in the case of water fleas *Daphnia magna*. Therefore, it is clearly seen that the efficiency of applied AOPs for removal of pollutants from surface and drinking water should be checked by a comprehensive analysis including determination of remaining parent compounds, identification, and quantification of intermediates as well as assessment of possible biological adverse effects of treated water samples.

Keywords 17β-Estradiol, Biological activity, Bisphenols, Catalytic wet air oxidation, Iodosulfuron, Photocatalysis

Contents

1 Introduction 388
2 Treatment Efficiency Evaluations of AOPs 389

T. Tišler (✉) and A. Pintar (✉)
Department for Environmental Sciences and Engineering, National Institute of Chemistry, Ljubljana, Slovenia
e-mail: tatjana.tisler@ki.si; albin.pintar@ki.si

3 Endocrine Disrupting Compounds (EDCs) 390
  3.1 Evolution of Toxicity and/or Estrogenic Activity of E2- and BPA-Treated Samples
      During AOPs 391
4 Pesticides 394
  4.1 Evolution of Toxicity and/or Estrogenic Activity of Iodosulfuron Treated Samples
      During AOPs 394
5 Conclusions 399
References 399

Abbreviations

Acetyl-CoA	Acetyl coenzyme A
AHAS	Acetohydroxyacid synthase enzyme
AOPs	Advanced oxidation processes
BPA	Bisphenol A
BPAF	Bisphenol AF
BPF	Bisphenol F
CWAO	Catalytic wet air oxidation
D. magna	*Daphnia magna*
D. rerio	*Danio rerio*
D. subspicatus	*Desmodesmus subspicatus*
DAD	Diode array detector
DNA	Deoxyribonucleic acid
E1	Estrone
E2	17β-Estradiol
E3	Estriol
EDCs	Endocrine disrupting compounds
EE2	17α-Ethinylestradiol
ER Calux	Estrogen receptor-mediated chemical-activated luciferase gene expression
GFP	Green fluorescent protein
HPLC	High-performance liquid chromatography
ISO	International Organization for Standardization
MQ	Ultrapure water
MS	Mass spectrometry
MTT	3-(4,5-Dimethylthiazol-2-yl)-2,5-diphenyltetrazolium bromide
NMR	Nuclear magnetic resonance
TOC	Total organic carbon
UV	Ultraviolet
V. fischeri	*Vibrio fischeri*
WWTPs	Wastewater treatment plants
YES	Yeast estrogen screen assay

1 Introduction

In recent years, the presence of numerous chemicals such as biocides, pharmaceuticals, personal care products, and endocrine disrupting compounds (EDCs) in effluents, surface water, groundwater, and drinking water has become an issue of increasing international attention and concern. It is well known that the presence of these compounds in environmental samples could have adverse effects on wildlife [1]. Furthermore, if these compounds occur in drinking water supplies, they may pose threat to the human population by development or acceleration of different diseases [2]. Pollution of surface and ground water has contributed significantly to the lack of drinking water supply, most commonly in poor and water-short countries, which is mostly linked to faster population growth rate in these countries [3].

A major point source of these compounds in the aquatic environment are effluents from wastewater treatment plants (WWTPs); for this reason, an efficient treatment of effluents is necessary to remove these compounds before entering aquatic environment. Despite removal of a large portion of compounds from wastewater streams during various treatment processes, WWTPs do not remove all compounds completely. Microorganisms present in WWTPs are able to degrade biodegradable compounds, and the level of degradation depends on complexity of molecule and hydrophilicity. However, toxic and/or persistent pollutants could not be removed from wastewater streams by means of otherwise very effective conventional biological water treatment processes [4]. In such cases advanced oxidation processes (AOPs), which are based on generation of highly reactive oxidative radicals such as hydroxyl radicals, have demonstrated high level of removal efficiency of biocides, EDCs, pharmaceuticals, and personal care products from water samples [5–7]. Heterogeneous photocatalysis, Fenton-based oxidation, ultrasound oxidation, electrochemical oxidation, and ozone-based technologies are most frequently used for these purposes [8].

2 Treatment Efficiency Evaluations of AOPs

An ideal goal of AOPs is a complete mineralization of organic molecules into carbon dioxide, water, and associated mineral salts. Actually, degradation processes during AOPs are usually incomplete leading to production of numerous intermediates or end products. The removal efficiency of a tested (model) compound in the applied AOP is mostly followed by chemical analyses of parent organic matter and identification of produced intermediates or end products. For these purposes, liquid chromatography and gas chromatography methods combined with different techniques such as mass spectrometry are most frequently used [9–13]. However, intermediates and end products might exhibit adverse effects on aquatic organisms comparable to or even higher than that observed for the parent compounds. For these reasons, treated samples should be checked for possible baseline toxicity or specific toxicity (e.g., estrogenic activity, neurotoxicity, genotoxicity, etc.) induced by a target mode of toxic action, which is indicative of certain groups of chemicals [14].

Numerous bioassays, which are based on measuring the degree of response elicited in the organisms, exposed to a single chemical compound or mixtures of compounds (e.g., environmental samples), have been developed. Microorganisms, algae, invertebrates, and fishes are typically used taxonomic groups of aquatic organisms involved in bioanalytical testing [15]. Literature review shows that the efficiency of toxicity removal of aqueous samples after AOPs is mostly determined by means of the standardized toxicity test procedures using the luminescence inhibition test with bacteria *Vibrio fischeri* [16–18], the growth inhibition test with green algae *Desmodesmus subspicatus* and *Pseudokirchneriella subcapitata* [19, 20], the immobility acute test with water fleas *Daphnia magna* [19, 21, 22], and fish early-life stage toxicity test with zebrafish, rainbow trout, and bluegill sunfish [23, 24].

In addition to the standard procedures with above mentioned organisms, cell lines are used for toxicological analyses of AOP-treated aqueous samples. Mammalian cells (Chinese hamster ovary cells and human ductal breast epithelial tumor cells) were applied to detect cytotoxic (MTT test), genotoxic (alkaline comet assay), and mutagenic (Ames test) effects of aqueous samples with selected micropollutants before and after oxidative treatment [25]. Gou et al. [11] reported a novel approach, based on toxicogenomics toxicity assay with a GFP-fused whole cell array of *Escherichia coli* K12, which provides insights into dynamics of the nature of toxicity mechanisms during the course of electro-Fenton oxidative process of three selected contaminants of emerging concern.

The measurement of specific modes of action, i.e., estrogenic activity of treated samples, is mostly done by in vitro yeast-based estrogenic assays such as yeast estrogen screen (YES) assay [26–31], E-screen assay using estrogenic receptor-positive human breast adenocarcinoma cell line MCF-7 [14, 32], and ER Calux test with human ductal breast epithelial tumor cells [25]. The complexity of interactions as a consequence of toxicokinetic and toxicodynamic factors is covered by in vivo bioassays, which are mainly based on the induction of a sensitive biomarker, yolk precursor protein vitellogenin, confirming the exposure of fish males and juveniles to estrogens and xenoestrogens. Fish species like Japanese medaka fish *Oryzias latipes* [27], rainbow trout and brown trout [23, 33], and zebrafish *Danio rerio* [34] are frequently used for these purposes. Mboula et al. [13] reported that a genetically modified organism, transgenic medaka line, containing a green fluorescence protein gene regulated by the regulatory sequence of the choriogenin H (ChgH) gene was used to assess the estrogenic activity of photocatalytically treated BPA aqueous samples.

3 Endocrine Disrupting Compounds (EDCs)

The ubiquitous presence of EDCs in wastewater, surface, and drinking water has become a major concern, since it is evident that they have a potential to elicit adverse effects on endocrine systems of wildlife and human, including hormone-dependent cancers [2, 35]. EDCs, a large group of structurally diverse chemicals, include a variety of natural and synthetic compounds, e.g., estrogenic hormones and xenoestrogens such as industrial chemicals (surfactants, plasticizers), pesticides, heavy metals, and pharmaceuticals. The largest contribution to the estrogenic activity of water streams belongs to the natural estrogens such as estrone (E1), 17-β-estradiol (E2), estriol (E3), and synthetic estrogens like 17α-ethinylestradiol (EE2). These compounds are of a major concern as they induce reproductive and developmental disturbances in wild fish population at extremely low concentrations [36].

The main sources of surface water pollution with EDCs are considered to be municipal wastewater streams [37], mainly contaminated with estrogens (E1, E2, E3, and EE2) and industrial effluents as well as landfill leachates, where xenoestrogens (e.g., BPA, alkylphenols, phthalate esters) are frequently present [38, 39].

Bisphenol A (BPA) is an important industrial high-production-volume chemical, which is used in the production of epoxy resins and polycarbonate plastics. It is also present in thermographic papers, coatings, flame retardants, and dental sealants [40]. Due to its widespread use, it is commonly present in all segments of aquatic environment. BPA is considered to cause adverse effects on reproduction, development, growth, and behavior of wildlife and humans [41–43]. Therefore, stringent regulations on the production and application of BPA have been accepted in the European Union [44, 45]. Consequently, the restriction has forced intensive production and application of possible alternatives to BPA. Numerous bisphenol analogues such as bisphenol B, bisphenol E, bisphenol F (BPF), bisphenol S, bisphenol AF (BPAF), and bisphenol AP have been synthesized. Their presence has already been confirmed in the aquatic environment [46, 47].

3.1 Evolution of Toxicity and/or Estrogenic Activity of E2- and BPA-Treated Samples During AOPs

3.1.1 17β-Estradiol (E2)

Most natural and synthetic estrogens as well as xenoestrogens are chemicals with phenol moiety, which is related to the estrogenic activity of compounds due to formulation of hydrogen bonds between the OH groups and the estrogenic receptors. The breakage and oxidation of benzene rings occur rapidly due to generation of hydroxyl radicals during AOPs; therefore, a lack of ability of intermediates to activate the estrogenic receptors by binding to them is anticipated [26, 48]. Bistan et al. [49] reported a successful removal of E2 and estrogenic activity during catalytic wet air oxidation (CWAO) and photolytic/photocatalytic oxidation. The presence of catalysts in both processes clearly enhanced the degradation of E2, especially Ru/TiO_2 catalyst used in the CWAO process was found to be very effective as complete E2 removal was noticed. We found that no estrogenically active intermediates were formed during both oxidation processes, which was confirmed by removal of estrogenic activity of treated samples. Similar results with no estrogenic activity of intermediates produced during photocatalytic oxidation of E2 were also obtained in other studies [31, 48].

However, the presence of some estrogenically active intermediates of E2 generated during different AOPs was reported. Mboula et al. [13] reported that during photocatalytic oxidation of E2, the latter was completely removed from the aqueous solution under UV irradiation using the synthesized catalysts, but the estrogenic activity of intermediates was not removed probably due to the presence of phenol moiety in most determined intermediates. It appears that the removal efficiency depends on the light source and catalyst used. The residual estrogenic activity of treated aqueous samples was mostly found after oxidation processes of E2 such as ozonation [28, 32] combined by hydrogen peroxide-assisted oxidation [30] and chlorination [29, 32, 50, 51]. Estrogenic activity of end product solutions produced in various AOPs could be attributed to: (1) production of oxidized by-products, which are equally or even more estrogenically active in comparison to the initial solution, or (2) the presence of very low concentrations of residual E2, which could remain in treated samples despite very high removal efficiency. It should be noted that E2 is estrogenically active at very low concentrations (ng/L) [52].

According to the research findings, it is evident that diverse degradation pathways generate different intermediates of E2 during AOPs; a degradation mechanism depends on applied treatment process and operating conditions. This also applies to removal of estrogenic activity and toxicity of xenoestrogens from aqueous samples by means of oxidation processes.

3.1.2 Bisphenols

Many studies reported the efficient degradation of BPA. Among the most effective techniques are the application of O_3 and UV/H_2O_2 with low-pressure Hg UV lamps that resulted in high level of BPA degradation as well as removal of estrogenic activity and toxicity of treated aqueous samples [27, 53]. It was found that the composition of aqueous phase plays an important role in removal efficiency of BPA, its toxicity, and estrogenic activity. Richard et al. [25] investigated the efficiency of O_3 and UV/H_2O_2 oxidation of BPA in concentrations detected in surface water by combining chemical and toxicological analyses. They determined the formation of cytotoxic by-products after 60 min of UV/H_2O_2 treatment in pure water, but no toxic effects of spiked wastewater effluent were observed after ozonation.

Due to high removal efficiency of BPA from aqueous samples, several studies deal with photocatalytic oxidation of BPA [9, 18, 54, 55]. In our study [56] removal of BPA was studied in two oxidation processes, catalytic wet air oxidation (CWAO) and photolytic/photocatalytic oxidation. CWAO process was found to be most effective when Ru(3.0 wt. %)/TiO_2 catalyst was used at the operating temperature 230°C, as complete degradation of BPA and TOC conversion (>97%) were determined accompanied with complete removal of estrogenic activity and toxic effects to algae *D. subspicatus* and bacteria *V. fischeri*. On the other hand, remaining toxicity and estrogenic activity were determined in most photolytically/photocatalytically treated aqueous samples, even in the sample with complete removal of BPA performed in the presence of commercial TiO_2-P25 catalyst illuminated with UV light. Acute toxicity to *V. fischeri*, *D. magna*, and *D. rerio* as well as estrogenic activity was determined in these samples. However, in some cases even higher toxicity was detected in comparison to the initial BPA solution, which could be attributed to the occurrence of intermediates produced during photo (cata)lysis.

It is well known that production of hydroxyl radicals during oxidation processes with a TiO_2 photocatalyst enhanced the degradation of BPA in aqueous samples. A mixture of intermediates is produced during BPA oxidation processes before a complete mineralization is achieved, and some of them are able to elicit adverse biological effects such as genotoxicity, estrogenic activity, and/or toxicity. Degradation pathways of BPA are related to the applied oxidation process and operating/reaction conditions which result in production of various by-products. For example, Chiang et al. [16] reported that photocatalytic degradation of BPA depended on pH value of initial BPA aqueous solution, which in turn resulted in the formation of different intermediates during the oxidation process. At higher pH value (pH = 10) more stable, but less toxic, intermediates were produced. On the contrary, less stable and in the early stage of BPA oxidation more toxic intermediates were generated at lower pH (pH = 3.0). Mboula et al. [55] reported that production of some by-products suspected to exhibit estrogenic activity depends on catalyst nature and experimental conditions, i.e., illumination of a photocatalyst. Gou et al. [11] studied toxicity evolution during electro-Fenton oxidation of BPA by means of quantitative toxicogenomics assay and found that BPA and its intermediates could cause DNA damage and induce membrane stress.

Degradation of BPA takes place via different pathways generating various by-products, which involve (1) para-substituted phenols such as *p*-hydroxyacetophenone, *p*-isopropenylphenol, and *p*-isopropylphenol, (2) hydroxylated BPA such as monohydroxylated, dihydroxylated BPA and their quinones, and, finally, (3) aliphatic compounds such as formic, acetic, oxalic, succinic, and fumaric acids, which are produced after ring opening [9, 11, 12, 17, 55]. Some of them like *p*-hydroxyacetophenone exerted slight estrogenic activity detected by YES assays [57].

Erjavec et al. [58] studied removal of toxicity and estrogenic activity of BPA, BPF, and BPAF from aqueous solutions in a three-phase batch and continuous stirred-tank reactor using glass fiber supported TiO_2 photocatalyst under UV light. In the batch reactor, toxic effects of all photocatalytically treated bisphenols were diminished and completely eliminated after 4 and 6 h of photocatalytic oxidation in the case of *V. fischeri* and *D. magna*, respectively. Furthermore, complete removal of estrogenic activity of aqueous samples containing BPA and BPF was also determined after 4 h. BPAF is less prone to photocatalytic oxidation as 57.0, 30.1, 26.4, and 14.7% of the remaining TOC were detected after 2, 4, 6, and 8 h of degradation process in the batch reactor, respectively. Although various intermediates were produced during the photocatalytic oxidation of BPAF, no acute toxicity of the aqueous samples after 4 h of treatment was obtained to luminescence of *V. fischeri* and mobility of water fleas *D. magna*. Furthermore, it was confirmed by in vitro yeast estrogen screen assay using modified yeast strain *Saccharomyces cerevisiae* (John P. Sumpter, Brunel University, UK) that photocatalytic oxidation of BPAF produced intermediates with no estrogenic activity of treated samples after 4 h. However, the initial aqueous solutions of BPA, BPF, and BPAF demonstrated estrogenic activity in a dose-response relationship; the highest estrogenic potential was obtained in the case of BPAF.

4 Pesticides

Increasing use of pesticides has become a major concern worldwide due to pollution of surface and ground water through leaching and runoffs. The presence of herbicides in water poses a potential risk to aquatic organisms as well as humans. Such example is atrazine, a triazine-based herbicide, which was considered as a potential carcinogen and endocrine disruptor. In 2004 atrazine was banned from the EU market due to ubiquitous and persistent groundwater contamination and its persistence in the environment [59]. New products have appeared on the market, and one of the recently used has been iodosulfuron-methyl-ester (referred to as iodosulfuron). It is a systemic sulfonylurea herbicide, widely used to control annual grasses and broad-leaved weeds mostly in cereal crops [60]. Due to its relatively high solubility in water (up to 65 g/L, depending on the pH value), iodosulfuron could be highly mobile in the environment and contaminates surface and drinking water [61].

A mode of toxic action of iodosulfuron, like other sulfonylureas, involves inhibition of the enzyme acetohydroxyacid synthase (AHAS), and consequently synthesis routes of essential amino acids for algal growth are blocked [62]. Branched-chain amino acids such as valine, leucine, and isoleucine are such examples. AHAS is involved in the first step of biosynthesis process of valine (from pyruvate alone), leucine (from pyruvate and acetyl-CoA), and isoleucine (from pyruvate and 2-ketobutyrate). Contrary to microorganisms and plants, which are able to synthesize fats, amino acids, vitamins, etc., animals are not able to produce them and are therefore tied to a complex diet [62].

4.1 Evolution of Toxicity and/or Estrogenic Activity of Iodosulfuron Treated Samples During AOPs

In many cases, mineralization of pollutants during photocatalytic oxidation processes is less successful, which could lead to production of more toxic intermediates in comparison to the parent chemicals. Such an example is shown below in the case of photocatalytic oxidation of iodosulfuron.

Removal of iodosulfuron from aqueous solution was studied in photocatalytic runs carried out at different experimental conditions. Iodosulfuron-methyl-ester was dissolved in tap or ultrapure water (initial concentration was 26.5 mg/L) and oxidized in a three-phase slurry reactor under UV or visible light irradiation in the presence of synthesized sol-gel Nd- or nitrogen-doped TiO_2 catalysts [63]. After 3 h of photocatalytic oxidation, toxicity of treated samples was determined by means of toxicity tests with organisms from different taxonomic groups in order to detect possible toxic potential of intermediates generated during incomplete degradation of iodosulfuron. The luminescence of freeze-dried bacteria *Vibrio fischeri*, the immobility of water fleas *Daphnia magna*, and the growth of unicellular green algae *Desmodesmus subspicatus* were determined following the exposure to tested samples. Toxicity tests were performed according to the protocols and ISO standards, which were described in detail previously [24, 43].

The measured conversions and corresponding residual concentrations of parent compound obtained during treatment of aqueous solutions of iodosulfuron by means of photolytic/photocatalytic oxidation carried out at various experimental conditions are listed in Table 1, while the toxicity of treated samples to bacteria, algae, and water fleas is plotted in Figs. 1 and 2.

Table 1 Experimental conditions of photolytic/photocatalytic oxidation runs during treatment of aqueous solutions of iodosulfuron

Sample	Catalyst	Light	Water	Gas	Conversion (%)	Residual concentration (mg/L)
0	Initial solution	/	/	/	/	26.5
1	No catalyst	UV	Tap	Air	47.4	13.9
2	No catalyst	UV	MQ	N_2	62.2	10.0
3	No catalyst	Visible	MQ	N_2	0.19	26.4
4	No catalyst	Visible	MQ	Air	0.00	26.5
5	TiO_2 P-25	UV	MQ	N_2	99.1	0.25
6	TiO_2 P-25	Visible	MQ	N_2	87.4	3.32
7	TiO_2 P-25	UV	MQ	Air	98.8	0.32
8	TiO_2 P-25	UV	Tap	Air	99.3	0.20
9	TiO_2 P-25	Visible	Tap	Air	36.0	16.9
10	TiO_2P-25	Visible	MQ	Air	90.1	2.62
11	TiO_2-Nd10	UV	MQ	Air	99.0	0.28
12	TiO_2-Nd10	UV	Tap	Air	56.3	11.6
13	TiO_2-Nd10	Visible	MQ	Air	18.8	21.5
14	TiO_2-Nd10	Visible	Tap	Air	5.79	24.9
15	TiO_2-N (1.1)	UV	MQ	Air	/	/
16	TiO_2-N (1.1)	UV	MQ	Air	91.0	2.39
17	TiO_2-N (1.1)	Visible	MQ	Air	10.5	23.7
18	TiO_2-N (1.2)	UV	MQ	Air	97.2	0.75
19	TiO_2-N (1.2)	Visible	MQ	Air	19.9	21.2
20	TiO_2-PcN2	UV	MQ	Air	97.1	0.76
21	TiO_2-PcN2	Visible	Tap	Air	7.10	24.6
22	TiO_2-PcN3	Visible	MQ	Air	10.8	23.6

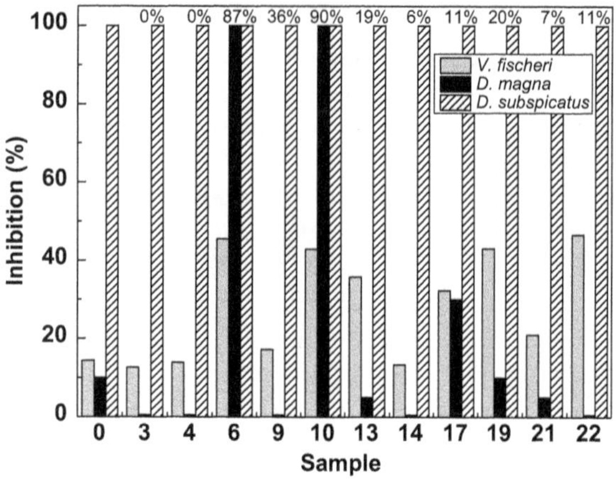

Fig. 1 Toxicity of initial aqueous solution of iodosulfuron and treated samples under visible light to bacteria *V. fischeri*, water fleas *D. magna*, and algae *D. subspicatus*. Numbers listed above bars represent the % of iodosulfuron conversion during the treatment by means of photolytic/photocatalytic oxidation. See Table 1 for sample notation

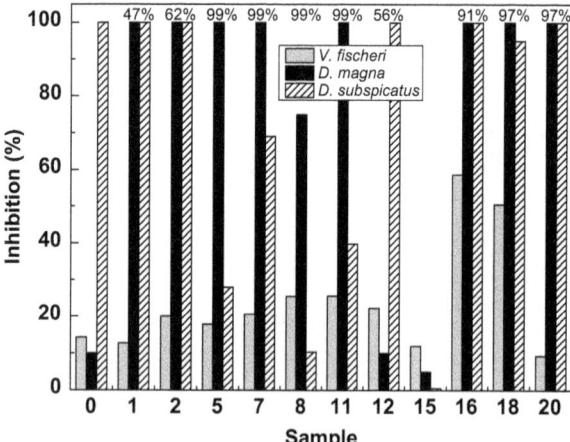

Fig. 2 Toxicity of initial aqueous solution of iodosulfuron and treated samples under UV light to bacteria *V. fischeri*, water fleas *D. magna*, and algae *D. subspicatus*. Numbers listed above bars represent the % of iodosulfuron conversion during the treatment by means of photolytic/photocatalytic oxidation. See Table 1 for sample notation

The results showed that the highest removal of iodosulfuron (>95%) was mostly determined in photocatalytic experiments using ultrapure (MQ) water under UV illumination. Contrary, the removal efficiency of iodosulfuron was significantly lower in the photocatalytic experiments in which visible light illumination was utilized (Table 1). It is obvious that the degradation level of iodosulfuron dissolved in tap water is lower than in ultrapure water due to the ionic content such as hydrogen carbonates, chlorides, nitrates, etc., which can adsorb to the catalyst surface and consequently decrease the catalyst activity [63].

In the experiments without catalysts (samples 3 and 4) and utilizing visible light, no photolytic degradation of iodosulfuron was determined, and consequently the toxicity of both samples to bacteria, algae, and water fleas remained similar to that of the initial aqueous solution of iodosulfuron (sample 0). The presence of commercial Degussa TiO_2 P25 catalyst enhanced the degradation of iodosulfuron dissolved in ultrapure water (samples 6 and 10) under visible light, which increased the toxicity to bacteria (up to 47% of inhibition) and particularly to water fleas. The presence of Nd- and N-doped TiO_2 catalysts in photocatalytic experiments caused low conversion of iodosulfuron (up to 20%) under visible light; higher degradation level was achieved using ultrapure water as a solvent of iodosulfuron. These samples showed slightly higher toxicity to *Vibrio fischeri*; toxicity threshold (20% of inhibition) was exceeded. However, toxicity has not changed significantly in the case of *Daphnia magna*. Because of high residual iodosulfuron concentration in photocatalytically treated samples under visible light, 100% of inhibition of algal growth was determined in all tested samples (Fig. 1).

In the photolytic experiments without catalysts and utilizing UV light illumination, photolysis exhibited 47% (sample 1) and 62% (sample 2) of degradation of iodosulfuron; no increased toxicity was observed for luminescence, but immobility of all exposed water fleas and 100% inhibition of algal growth were determined in photolytically treated samples. In photocatalytically treated solutions illuminated with UV light, degradation of iodosulfuron in ultrapure water (samples 5, 7, and 11) and tap water (sample 8) was almost complete (99%), and the toxicity of these samples to algae considerably decreased. However, intensive toxic effects (100% immobility) to water fleas were determined in most of end product solutions in which very high removal of iodosulfuron from aqueous solutions was achieved. No clear correlation between the measured conversion levels and toxicity data toward *Vibrio fischeri* was obtained (Fig. 2).

To summarize, UV light enhanced the conversion of iodosulfuron as the acquired percentages of conversion were mostly higher than 90% in the presence of different catalysts, except in the case of run 12, where tap water was used as a reaction medium. Consequently, noticeable increase of toxicity to water fleas was detected in most end product solutions, but diminished toxic effects to algae were observed in the samples with the highest % of iodosulfuron removal (Table 1). In the case of *Vibrio fischeri*, no clear correlation between % of iodosulfuron removal and inhibition of luminescence was obtained. However, the use of UV light mostly results in lower toxicity of end product solutions to *Vibrio fischeri* in comparison to runs in which visible light irradiation was utilized (Figs. 1 and 2).

It is clearly shown that low degradation level of iodosulfuron in aqueous samples after photocatalytic oxidation is well correlated with high toxicity obtained to algae (Fig. 1). We assume that the residual iodosulfuron in treated samples inhibited the enzyme acetohydroxyacid synthase (AHAS). As these amino acids are crucial for algal growth and development, the growth of algae was completely inhibited in the aqueous samples with residual iodosulfuron.

We found a clear correlation between the conversion of iodosulfuron and toxic effects on water fleas. When high percentage of conversion (equal or higher than 87%) was attained during photocatalytic oxidation (samples 5, 6, 7, 8, 10, 11, 16, 18, and 20), toxicity to *Daphnia magna* highly increased in comparison to the initial aqueous solution (sample 0). It was determined that only about 25% of TOC was removed after 180 min under given operational conditions. We confirmed by

HPLC analyses that by-products were produced during photocatalytic oxidation of iodosulfuron (Fig. 3) in significant quantities relating to the TOC data.

Sleiman et al. [10] studied intermediates produced during iodosulfuron degradation by TiO_2 photocatalysts under UV light irradiation. Based on HPLC-MS and HPLC-NMR analyses, more than 20 intermediates were identified, which were classified into two groups according to the presence/absence of sulfonylurea bridge. In our study, these two groups of intermediates were also detected by HPLC analysis. Immediately after starting the experiment, group A intermediates appeared in the solution due to iodosulfuron degradation and then progressively disappeared during the process of photocatalytic oxidation. Simultaneously, group B intermediates that were produced after the cleavage of sulfonylurea bridge appeared and remained in the aqueous solution until the end of the experiment (Fig. 3). Sleiman et al. [10] also found that the ultimate organic product of the degradation process of iodosulfuron was cyanuric acid, which was very resistant to further oxidation. Furthermore, after the aromatic (benzene) ring opening, several short carboxylic acids arose, and they were determined by means of HPLC-DAD. Among them most important are oxalic, formic, acetic, glycolic, and glyoxylic acids; the major one was acetic acid, which reached a maximum concentration at 49 μmol/L (3 mg/L) after 120 min. According to the results obtained in our previous study, acetic and formic acids were acutely toxic to water fleas [64], and the presence of these

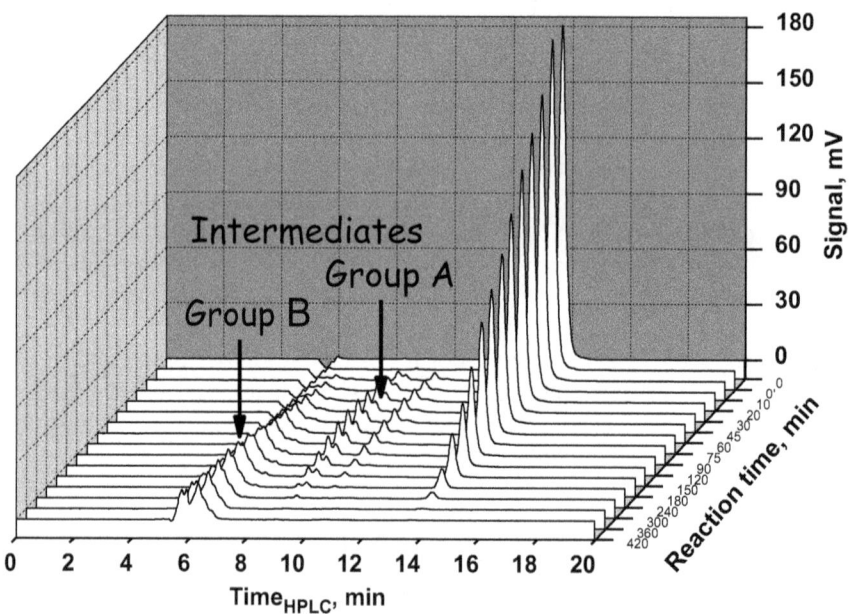

Fig. 3 3D HPLC chromatograms of aqueous-phase samples withdrawn during photocatalytic oxidation of aqueous solution of iodosulfuron ($c_0 = 26.5$ mg/L) in a slurry reactor in the presence of TiO_2-N (1.1) catalyst under UV high pressure lamp (Reproduced from Kralchevska et al. [63] with permission from Elsevier)

compounds is a probable reason for the significantly increased toxicity of photocatalytically treated samples to *Daphnia magna* with almost complete degradation of iodosulfuron.

From the obtained results, we conclude that in the process of heterogeneous photocatalytic oxidation, removal efficiency of iodosulfuron from aqueous solution and toxicity of treated samples are attributed to the use of different operating and reaction conditions of the concerned oxidative treatment, i.e., type of water matrix, source of illumination, gas, and catalyst present in the aqueous suspension.

5 Conclusions

Application of AOPs for water treatment purposes is undoubtedly very perspective tool for efficient removal of persistent organic pollutants from surface and drinking water, because in many cases these processes are able to completely mineralize water dissolved organic pollutants. However, the main concern relates to incomplete degradation of compounds, leading to production of intermediates or end products that can be formed during utilization of AOPs. Furthermore, degradation products could be even more biologically active in comparison to parent compounds. For instance, production of more toxic intermediates during photocatalytic oxidation of iodosulfuron was clearly observed as significantly enhanced toxicity to water fleas *D. magna* was determined in treated samples exhibiting the highest iodosulfuron removal efficiency. The toxicity of photocatalytically treated samples was found to be species-specific as the increased toxicity was not observed to algae *D. subspicatus* and bacteria *V. fischeri*. Therefore, chemical analytical measurements of remaining original compounds in treated water streams and identification of degradation products should be accompanied with a comprehensive assessment of potential biological activity, covering non-specific toxicity and different modes of action (e.g., estrogenic/androgenic activity in the case of EDCs), in order to determine suitability of the applied advanced oxidation process for surface and drinking water treatment purposes.

Acknowledgments The authors gratefully acknowledge funding by the Ministry of Education, Science and Sport of the Republic of Slovenia through Research program No. P2-0150. The authors thank Mr. Andraž Premru for conducting the experimental work.

References

1. Schwarzenbach RP, Escher BI, Fenner K, Hofstetter TB, Johnson CA, von Gunten U, Wehrli B (2006) The challenge of micropollutants in aquatic systems. Science 313:1072–1077
2. Woodruff TJ (2011) Bridging epidemiology and model organisms to increase understanding of endocrine disrupting chemicals and human health effects. J Steroid Biochem Mol Biol 127:108–117

3. Jury WA, Vaux HJ (2007) The emerging global water crisis: managing scarcity and conflict between water users. Adv Agron 95:1–76
4. Alvarez-Corena JR, Bergendahl JA, Hart FL (2016) Advanced oxidation of five contaminants in water by UV/TiO$_2$: reaction kinetics and byproducts identification. J Environ Manage 181: 544–551
5. Esplugas S, Bila DM, Krause LGT, Dezotti M (2007) Ozonation and advanced oxidation technologies to remove endocrine disrupting chemicals (EDCs) and pharmaceuticals and personal care products (PPCPs) in water effluents. J Hazard Mater 149:631–642
6. Silva CP, Otero M, Esteves V (2012) Processes for the elimination of estrogenic steroid hormones from water: a review. Environ Pollut 165:38–58
7. Ribeiro AR, Nunes OC, Pereira MFR, Silva AMT (2016) An overview on the advanced oxidation processes applied for the treatment of water pollutants defined in the recently launched Directive 2013/39/EU. J Environ Manage 181:544–551
8. Klavarioti M, Mantzavinos D, Kassinos D (2009) Removal of residual pharmaceuticals from aqueous systems by advanced oxidation processes. Environ Int 35:402–417
9. Ohko Y, Ando I, Niwa C, Tatsuma T, Yamamura T, Nakashima T, Kubota Y, Fujishima A (2001) Degradation of bisphenol A in water by TiO$_2$ photocatalyst. Environ Sci Technol 35: 2365–2368
10. Sleiman M, Conchon P, Ferronato C, Chovelon J-M (2007) Iodosulfuron degradation by TiO$_2$ photocatalysis: kinetic and reactional pathway investigations. Appl Catal B Environ 71:279–290
11. Gou N, Yuan S, Lan J, Gao C, Alshawabkeh AN, Gu AZ (2014) A quantitative toxicogenomics assay reveals the evolution and nature of toxicity during the transformation of environmental pollutants. Environ Sci Technol 48:8855–8863
12. Kondrakov AO, Ignatev AN, Frimmel FH, Bräse S, Horn H, Revelsky AI (2014) Formation of genotoxic quinones during bisphenol A degradation by TiO$_2$ photocatalysis and UV photolysis: a comparative study. Appl Catal B Environ 160–161:106–114
13. Mboula VM, Héquet V, Andrès Y, Gru Y, Colin R, Dona-Rodríguez JM, Pastrana-Martínez LM, Silva AMT, Leleu M, Tindall AJ, Mateos S, Falaras P (2015) Photocatalytic degradation of estradiol under simulated solar light and assessment of estrogenic activity. Appl Catal B Environ 162:437–444
14. Macova M, Escher BI, Reungoat J, Carswell S, Lee Chue K, Keller J, Mueller JF (2010) Monitoring the biological activity of micropollutants during advanced wastewater treatment with ozonation and activated carbon filtration. Water Res 44:477–492
15. Rizzo L (2011) Bioassays as a tool for evaluating advanced oxidation processes in water and wastewater treatment. Water Res 45:4311–4340
16. Chiang K, Lim TM, Tsen L, Lee CC (2004) Photocatalytic degradation and mineralization of bisphenol A by TiO$_2$ and platinized TiO$_2$. Appl Catal A Gen 261:225–237
17. Rodríguez EM, Fernández G, Klamerth N, Maldonado MI, Álvarez PM, Malato S (2010) Efficiency of different solar advanced oxidation processes on the oxidation of bisphenol A in water. Appl Catal B Environ 95:228–237
18. Repousia V, Petala A, Frontistisa Z, Antonopoulou M, Konstantinou I, Kondarides DI, Mantzavinos D (2017) Photocatalytic degradation of bisphenol A over Rh/TiO$_2$ suspensions in different water matrices. Catal Today 284:59–66
19. Arslan-Alaton I, Aytac E, Kusk KO (2014) Effect of Fenton treatment on the aquatic toxicity of bisphenol A in different water matrices. Environ Sci Pollut Res 21:12122–12128
20. Brienza M, Mahdi Ahmed M, Escande A, Plantard G, Scrano L, Chiron S, Bufo SA, Goetz V (2016) Use of solar advanced oxidation processes for wastewater treatment: follow-up on degradation products, acute toxicity, genotoxicity and estrogenicity. Chemosphere 148: 473–480
21. Lu N, Lu Y, Liu F, Zhao K, Yuan X, Zhao Y, Li Y, Qin H, Zhu J (2013) H$_3$PW$_{12}$O$_{40}$/TiO$_2$ catalyst-induced photodegradation of bisphenol A (BPA): kinetics, toxicity and degradation pathways. Chemosphere 91:1266–1272

22. Rozas O, Vidal C, Baeza C, Jardim WF, Rossner A, Mansilla HD (2016) Organic micropollutants (OMPs) in natural waters: oxidation by UV/H_2O_2 treatment and toxicity assessment. Water Res 98:109–118
23. Stalter D, Magdeburg A, Weil M, Knacker T, Oehlmann J (2010) Toxication or detoxication? In vivo toxicity assessment of ozonation as advanced wastewater treatment with the rainbow trout. Water Res 44:439–448
24. Tišler T, Erjavec B, Kaplan R, Şenilă M, Pintar A (2015) Unexpected toxicity to aquatic organisms of some aqueous bisphenol A samples treated by advanced oxidation processes. Water Sci Technol 72:29–37
25. Richard J, Boergers A, vom Eyser C, Besterd K, Tuerk J (2014) Toxicity of the micropollutants bisphenol A, ciprofloxacin, metoprolol and sulfamethoxazole in water samples before and after the oxidative treatment. Int J Hyg Environ Health 217:506–514
26. Coleman HM, Routledge EJ, Sumpter JP, Eggins BR, Byrne JA (2004) Rapid loss of estrogenicity of steroid estrogens by UVA photolysis and photocatalysis over an immobilised titanium dioxide catalyst. Water Res 38:3233–3240
27. Chen P-J, Linden KG, Hinton DE, Kashiwada S, Rosenfeldt EJ, Kullman SW (2006) Biological assessment of bisphenol A degradation in water following direct photolysis and UV advanced oxidation. Chemosphere 65:1094–1102
28. Bila D, Montalvao AF, Azevedo DA, Dezotti M (2007) Estrogenic activity removal of 17β-estradiol by ozonation and identification of by-products. Chemosphere 69:736–746
29. Rosenfeldt EJ, Chen PJ, Kullman S, Linden KG (2007) Destruction of estrogenic activity in water using UV advanced oxidation. Sci Total Environ 377:105–113
30. Maniero MG, Bila DM, Dezotti M (2008) Degradation and estrogenic activity removal of 17β-estradiol and 17α-ethinylestradiol by ozonation and O_3/H_2O_2. Sci Total Environ 407:105–115
31. Benotti MJ, Stanford BD, Wert EC, Snyder SA (2009) Evaluation of a photocatalytic reactor membrane pilot system for the removal of pharmaceuticals and endocrine disrupting compounds from water. Water Res 43:1513–1522
32. Alum A, Yoon Y, Westerhoff P, Abbaszadegan M (2004) Oxidation of bisphenol A, 17-β-estradiol, and 17α-ethynyl estradiol and byproduct estrogenicity. Environ Toxicol 19:257–264
33. Henneberg A, Triebskorn R (2015) Efficiency of advanced wastewater treatment technologies for the reduction of hormonal activity in effluents and connected surface water bodies by means of vitellogenin analyses in rainbow trout (*Oncorhynchus mykiss*) and brown trout (*Salmo trutta f. fario*). Environ Sci Eur 27:1–12
34. Wang WJ, Wang W, Tian H, Zhang X, Ru S (2017) A novel enzyme-linked immunosorbent assay based on anti-lipovitellin monoclonal antibodies for quantification of zebrafish (*Danio rerio*) vitellogenin. Ecotoxicol Environ Saf 136:78–83
35. Gross TS, Arnold BS, Sepulveda MS, McDonald K (2003) Endocrine disrupting chemicals and endocrine active agents. In: Hoffman DJ, Rattner BA, Burton A, Cairns J (eds) Handbook of ecotoxicology. CRC Press, Boca Raton, pp 1033–1098
36. Jobling S, Nolan M, Tyler CR, Brighty G, Sumpter JP (1998) Widespread sexual disruption in wild fish. Environ Sci Technol 32:2498–2506
37. Körner W, Bolz U, Süßmuth W, Hiller G, Schuller W, Hanf V, Hagenmaier H (2000) Input/output balance of estrogenic active compounds in a major municipal sewage plant in Germany. Chemosphere 40:1131–1142
38. Yasuhara A, Shiraishi H, Nishikawa M, Yamamoto T, Nakasugi O, Okumura T, Kenmotsu K, Fukui H, Nagase M, Kawagoshi Y (1999) Organic components in leachates from hazardous waste disposal sites. Waste Manage Res 17:186–197
39. Zhao J-L, Chen X-W, Yan B, Wei C, Jiang Y-X, Ying G-G (2015) Estrogenic activity and identification of potential xenoestrogens in a coking wastewater treatment plant. Ecotoxicol Environ Saf 112:238–246

40. Staples CA, Dorn PB, Klecka GM, O'Block ST, Harris LR (1998) A review of the environmental fate, effects, and exposures of bisphenol A. Chemosphere 36:2149–2173
41. Chen M-Y, Ike M, Fujita M (2002) Acute toxicity, mutagenicity, and estrogenicity of bisphenol-A and other bisphenol. Environ Toxicol 17:80–86
42. Mihaich EM, Friederich U, Caspers N, Hall AT, Klecka GM, Dimond SS, Staples CA, Ortego LS, Hentges SG (2009) Acute and chronic toxicity testing of bisphenol A with aquatic invertebrates and plants. Ecotoxicol Environ Saf 72:1392–1399
43. Tišler T, Krel A, Gerzelj U, Erjavec B, Sollner Dolenc M, Pintar A (2016) Hazard identification and risk characterization of bisphenols A, F and AF to aquatic organisms. Environ Pollut 212:472–479
44. Directive 2011/8/EU (2011) Amending Directive 2002/72/EC as regards the restriction of use of bisphenol A in plastic infant feeding bottles. Off J Eur Union L 26/11, Jan 29
45. Directive 2011/10/EU (2011) Regulation on plastic materials and articles intended to come into contact with food. Off J Eur Union L 12/1, Jan 15
46. Liao C, Liu F, Moon H-B, Yamashita N, Yun S, Kannan K (2012) Bisphenol analogues in sediments from industrialized areas in the United States, Japan, and Korea: spatial and temporal distributions. Environ Sci Technol 46:11558–11565
47. Song S, Ruan T, Wang T, Liu R, Jiang G (2012) Distribution and preliminary exposure assessment of bisphenol AF (BPAF) in various environmental matrices around a manufacturing plant in China. Environ Sci Technol 46:13136–13143
48. Ohko Y, Iuchi K-I, Niwa C, Tatsuma T, Nakashima T, Iguchi T, Kubota Y, Fujishima A (2002) 17β-estradiol degradation by TiO_2 photocatalysis as a means of reducing estrogenic activity. Environ Sci Technol 36:4175–4181
49. Bistan M, Tišler T, Pintar A (2012) Conversion and estrogenicity of 17β-estradiol during photolytic/photocatalytic oxidation and catalytic wet-air oxidation. Acta Chim Slov 59:389–397
50. Hu J, Cheng S, Aizawa T, Terao J, Kunikane S (2003) Products of aqueous chlorination of 17β-estradiol and their estrogenic activities. Environ Sci Technol 37:5665–5670
51. Li M, Xu B, Liungai Z, Hu H-Y, Chen C, Qiao J, Lu Y (2016) The removal of estrogenic activity with UV/chlorine technology and identification of novel estrogenic disinfection by-products. J Hazard Mater 307:119–126
52. Bistan M, Podgorelec M, Marinšek Logar R, Tišler T (2012) Yeast estrogen screen assay as a tool for detecting estrogenic activity in water bodies. Food Technol Biotechnol 50:427–433
53. Rosenfeldt EJ, Linden KG (2004) Degradation of endocrine disrupting chemicals bisphenol A, ethinyl estradiol, and estradiol during UV photolysis and advanced oxidation processes. Environ Sci Technol 38:5476–5483
54. Wang R, Ren D, Xia S, Zhang Y, Zhao J (2009) Photocatalytic degradation of bisphenol A (BPA) using immobilized TiO_2 and UV illumination in a horizontal circulating bed photocatalytic reactor (HCBPR). J Hazard Mater 169:926–932
55. Mboula VM, Héquet V, Andrès Y, Pastrana-Martínez LM, Dona-Rodríguez JM, Silva AMT, Falaras P (2013) Photocatalytic degradation of endocrine disruptor compounds under simulated solar light. Water Res 47:3997–4005
56. Bistan B, Tišler T, Pintar A (2012) Catalytic and photocatalytic oxidation of aqueous bisphenol A solutions: removal, toxicity, and estrogenicity. Ind Eng Chem Res 51:8826–8834
57. Ike M, Chen M-Y, Jin C-S, Fujita M (2002) Acute toxicity, mutagenicity, and estrogenicity of biodegradation products of bisphenol-A. Environ Toxicol 17:457–461
58. Erjavec B, Hudoklin P, Perc K, Tišler T, Sollner Dolenc M, Pintar A (2016) Glass fiber-supported TiO_2 photocatalyst: efficient mineralization and removal of toxicity/estrogenicity of bisphenol A and its analogs. Appl Catal B Environ 183:149–158
59. Sass JB, Colangelo A (2006) European Union bans atrazine, while the United States negotiates continued use. Int J Occup Environ Health 12:260–267
60. Brigante M, Emmelin C, Previtera L, Baudot R, Chovelon JM (2005) Abiotic degradation of iodosulfuron-methyl-ester in aqueous solution. J Agric Food Chem 53:5347–5352

61. European Commission, Health and Consumer Protection Directorate (2003) Review report for the active substance iodosulfuron. SANCO/10166/2003-Final
62. Duggleby RG, McCourt JA, Guddat LW (2008) Structure and mechanism of inhibition of plant acetohydroxyacid synthase. Plant Physiol Biochem 46:309–324
63. Kralchevska R, Milanova M, Tišler T, Pintar A, Tyuliev G, Todorovsky D (2012) Photocatalytic degradation of the herbicide iodosulfuron by neodymium or nitrogen doped TiO_2. Mater Chem Phys 133:1116–1126
64. Pintar A, Batista J, Tišler T (2008) Catalytic wet-air oxidation of aqueous solutions of formic acid, acetic acid and phenol in a continuous-flow trickle-bed reactor over Ru/TiO_2 catalysts. Appl Catal B Environ 84:30–41

Chemometric Methods for the Optimization of the Advanced Oxidation Processes for the Treatment of Drinking and Wastewater

Messias Borges Silva, Cristiano Eduardo Rodrigues Reis,
Fabrício Maciel Gomes, Bruno dal Rovere Contesini,
Ana Paula Barbosa Rodrigues de Freitas, Hélcio José Izário Filho,
Leandro Valim de Freitas, and Carla Cristina Almeida Loures

Abstract Advanced Oxidative Processes (AOP) have been successfully employed as efficient water treatment methods. The utilization of AOP on drinking and wastewater represents currently an alternative to costly, hazardous, and slow processes. In order to further establish the ground basis for AOP in water safety and security, reliable and consistent methods of analysis are required. As an alternative to basic statistical methods, which may not successfully describe and forecast the application of a given treatment methodology of water, the use of chemometrics has increased significantly over the past decades. Chemometric analyses are an intersection between analytical chemistry and applied statistical models in order to predict and extract information from a given condition. This chapter introduces the concepts of chemometrics in environmental engineering issues and the utilization of experimental design to efficiently analyze experimental data in environmental samples. Two case studies are presented to demonstrate the importance of chemometrics

---

M.B. Silva (✉), F.M. Gomes, A.P.B.R. de Freitas, L.V. de Freitas, and C.C.A. Loures
School of Engineering of Lorena - EEL – USP, University of São Paulo, Lorena, Brazil

Faculdade de Engenharia do Campus de Guaratinguetá -UNESP - FEG, Universidade Estadual Paulista - Julio de Mesquita Filho, Guaratinguetá, Brazil
e-mail: messias@dequi.eel.usp.br; fmgomes@usp.br; anapaulabrfreitas@gmail.com; leandrovalimdefreitas@gmail.com; carla.loures@cefet-rj.br

C.E.R. Reis and B.R. Contesini
School of Engineering of Lorena - EEL – USP, University of São Paulo, Lorena, Brazil
e-mail: cristianorodreis@gmail.com; bruno.contesini@usp.br

H.J. Izário Filho
School of Engineering of Lorena – EEL – USP, University of São Paulo, Lorena, Brazil
e-mail: helcio@dequi.eel.usp.br

in water analyses: (1) considering a Taguchi L_{16} experimental design, and an optimization study using Response Surface Methodology, to evaluate photo-Fenton and ozone AOP-based treatment on an effluent with high concentration of organic matter; (2) using a Taguchi L_9 array to evaluate the combination of photocatalytic degradation and AOP of an industrial effluent. The results showed in this chapter demonstrate how a given statistical method can be successfully employed within the intersection of environmental analyses and water issues.

Keywords Advanced oxidative processes, Chemometric analyses, Chemometric methods, Taguchi methods

Contents

1 Introduction 406
2 Chemometrics in Environmental Studies 408
3 AOP for Treating Drinking Water 410
4 Design of Experiments for the Optimization of AOP for the Treatment of Drinking Water 410
5 Case 1: AOP Using RSM and Taguchi Orthogonal Array (Adapted from de Freitas et al.) 412
    5.1 Characteristics of the Design of Experiments 412
    5.2 Results of the RSM (Fe^{2+}/H_2O_2) for Removing the Total Organic Carbon (TOC) . 415
    5.3 Conclusion Case 1 416
6 Case 2: Taguchi Method in AOP (Adapted from de Freitas et al.) 417
    6.1 Conclusion Case 2 420
References 421

1 Introduction

Water scarcity is one of the main challenges of the twenty-first century that many societies around the world are already facing. Throughout the last century, use and consumption of water grew at twice the rate of population growth and, although there is not water scarcity globally, the number of regions with chronic levels of water shortages is increasing. Water scarcity is not only a natural phenomenon, but is also caused by human action. Nowadays, there is enough freshwater to supply people who inhabit the planet, however, it is distributed unevenly, wasted, polluted, and unsustainably managed [1]. Water recycling and remediation of industrial effluents are key topics of applied research, in order to reduce the environmental impact of the human activity, and to revert many of the social consequences of misuse of waterbodies [2]. The growing concern of providing fresh and safe water systems to the growing society is a concern of all the United Nations Sustainable development goals (SDG) signatory countries. Among the over 21 SDG, several of those are directly or indirectly related to water issues, either by the production of nutritional food systems or by providing freshwater systems to the world as a public health concern. Water management and water engineering have been intensive areas of research over the past decades, both of which encompass a few operational steps. As a rule, the first step for a successful water management program relies on water monitoring. Water monitoring is, then, followed by elements of assessment, restoration, and ultimate protection of stream water quality. As one can imagine, assessing the aforementioned steps require a strong and robust system of monitoring of physical, biological, and chemical parameters. As streams and waterbodies are usually not in a steady state, the monitoring processes usually require a collection of temporal data points. As one can imagine, the collection of a large number of data points within monitoring stations can be understood in a larger-scale level would generate a complicated data matrix aimed at the evaluation of water quality. The required expertise to extract the most information from such a complex data collection may not be omnipresent though.

The use of chemometric methods, or multivariate statistical techniques, has strongly grown over the past years. The urge to pursue the United Nations SDG has given an additional boost for research and development in water quality and safety area. The basis for chemometric methods relies on the identification of the natural clustering pattern and variables in group for the analysis of similarities and discrepancies among different samples. The most common techniques for classification of water quality parameters are known as principal component analysis (PCA) with factor analysis (FA), and cluster analysis (CA). This chapter reviews direct applications of such methods, analyzing how the combination of experimental design optimization techniques with chemometrics can enhance the interpretation of the most important information one could obtain from a complex matrix.

Chemometrics have been increasingly accepted as standard methods, which provide several avenues for exploratory assessment of water quality data sets and classification of water qualities. Chemometric methods identify the natural clustering pattern and group variables on the basis of similarities between the samples. The most common chemometric methods for classification are namely, cluster analysis (CA) and principal component analysis (PCA) with factor analysis (FA). The discriminant analysis (DA) is used to confirm the groups found by means of the CA and PCA. These multidimensional data analysis methods are increasingly in use for environmental studies dealing with measurements and monitoring. The importance of crosschecking the group analyses between different methods (e.g., CA and PCA) is also relevant in order to avoid type-II statistical errors. The use of confirmatory techniques is found, for example, with discriminant analysis (DA), which will also be further discussed along this chapter. The use of chemometric methods does not hold as objective to add complexity to datasets. Chemometric methods aim, on the other hand, to reduce the natural complexity of large datasets, by offering enhanced interpretation and visualization alternatives. As a general rule, environmental data do not hold a normal distribution pattern. Since most traditional statistical design analyses rely on normality of data, the use of FA and DA, for example, can lead to the accumulation of a significant number of outliers and can greatly impair the result analysis. Statistical transformation methods, which could normalize positively skewed data patterns, as Box-Cox, Power, and Logarithmic transformations. The use of each transformation method depends greatly on the type of data sets. Particularly in environmental applications, the use of Box-Cox transformation has been widely used to transform data into linear models.

2 Chemometrics in Environmental Studies

There has been a significant increase in use of Chemometrics in environmental studies over the past decades, especially due to the increase in regulation from governmental agencies and the enhancement of analytical measurement techniques. Among the different areas of interest in environmental chemometrics, one could divide them into three major categories: quantitative chemical analytic methods, prediction of toxicity and toxicological effects, and monitoring and assessment of environmental quality [3]. Though there are many robust analytical techniques, two main challenges are still found on the quantitative chemical analysis of organic compounds classified as pollutants in environmental samples: (1) the natural complexity of natural components, often not easily identifiable by traditional methods that are not coupled with mass spectrometry, and (2) the low concentrations of analytes, which are often near or under the detection limit for the established methods [4]. Though there is lack of selectivity and the presence of interfering chemicals in complex matrices, the use of advanced statistical analyses increases the signal response from non-chromatographic methods. The use of multivariate calibration methods in complex environmental samples focuses on the application of mathematical models that relate multivariate signals and noises with analyte concentrations or chemical properties of the interest compound [5]. The use of univariate calibration relies on a single numerical scalar value per sample, e.g. intensity of signal, being recorded and analyzed. On the other hand, multivariate analysis relates complex data arrays on a sample basis and allows further reading on analytes that lack selectivity. The use of first-order calibration methods identifies a single response as a vector of values, e.g. time profiles, spectra, and voltammograms, on a sample basis. As a result of first-order analysis, one could quantify an analyte in an interfering environment, since the interfering compounds are also present in the calibration steps and the model development [5]. The use of second-order calibration methods, however, provides an instrumental response that can be analyzed as a data matrix on a sample basis, as is the case for an HPLC-Diode Array Detection chromatogram [6]. The statistical advantage of a second-order method relies on the power of the sample signal that can be identified in the presence of interfering compounds, though these may not be present in the calibration standards. Though there are higher-order data sets, their related calibration strategies and analysis are often more complex due to the need of higher-order tensors for calibration, therefore even higher costs are associated.

Multivariate analyses in environmental samples, thus, stand out as an alternative that provides sufficient analyte concentration in complex matrices. Over the last 30–40 years, the increased use of second and third order instrumental data coupled with different areas of analytical chemistry has improved and been thoroughly studied.

Second-order calibration algorithms, such as parallel factor analysis (PARAFAC) [7], resolution-alternating least squares (MCRALS), and partial least-squares with residual bilinearization (PLS-RBL), have been applied to different environmental analysis, and have been reviewed by Mas et al. [5]. Monitoring of environmental samples yields large sets of data, e.g. chemical concentrations, physical parameters, geospatial data, etc. The assumption of chemometric methods is that the variation of each parameter coming from a single sample can be explained through a reduced number of different contributions from independent environmental sources, on a sense that diffuse and point sources can be characterized according to their composition and attributed to a single factor (e.g., industrial, agricultural, anthropogenic), which can, then, help to discriminate among the different temporal and geographical patterns [5].

The use of PCA, as briefly explained in the Introduction, is based on the extraction of information from the main orthogonal matrix, i.e. from the principal components, widely used in exploratory analyses. The number of degrees of freedom in PCA often allows the analyzer to increase interpretability by rotating the principal components by a number of methods, such as the varimax rotation. The varimax rotation allows a change of coordinates that maximizes the vector sum of the factors. Despite being a powerful statistical tool, PCA has some drawbacks that often do not predict variance in the data matrix, and in environmental complex samples, the attribution of individual factors to single specific sources is not usually common [8]. Other alternatives not yet mentioned to PCA include UNMIX and positive matrix factorization (PMF), as well as MCRALS. Recent use of UNMIX and PMF includes air quality measurement performed by the United States Environmental Protection Agency (US-EPA) [9].

The use of chemometrics based on the assessment of toxicity of chemicals often relies on in vitro and in vivo methods to provide indicators about the reactions of a given chemical into a biological system. As many other analytical procedures, measuring with accuracy the toxicological effects of a certain compound can involve high expenditures, and the cost may be inhibitive. A common alternative to efficiently estimate the toxicity of a given compound is through the application of quantitative structure-activity relationships (QSAR), which are data tools relating indirect toxicity through the structure of a chemical compound [10]. QSAR derives mathematical models between the backbone of a molecule and its toxicological effect associated with a library database from previous studies. The use of QSAR relies, then, on efficient methods to select and predict the most relevant parameters related with the expected ranges of toxicity levels. Coupling QSAR with other regression methods, as stepwise and PCA, has been proposed [10], and it has been known to be a source of robust predictive models of toxicity in water and environmental samples. The use of QSAR modeling techniques can be, then, summarized according to different groups, according to the statistical method being used to support the chemical modeling, which are, for example, (1) linear regression, (2) non-linear regression (e.g., artificial neural networks, and support vector machines), and (3) methods for classification [10]. The intent of this chapter is not to provide an extensive literature review on analytical methods, nor statistical approaches, on chemometrics for environmental approaches. Therefore, further details regarding these particular methods are not described herein and can be found in the cited literature.

3 AOP for Treating Drinking Water

The use of AOP to treat drinking water, i.e., one of the purest types of water, with extremely low levels of toxic chemicals and organic matter, has been an intensive subject of study for over half a century now. The use of AOP yields the removal or the degradation of natural organic matter by a series of chemical reactions. As mentioned in the previous sections, the very first step one should have when treating environmental samples, as is the case of drinking water, is the full characterization of such sample. AOPs can be divided in a series of oxidants, radiation, and different catalysts (e.g., UV, O_3, H_2O_2, TiO_2, Fe^{2+}); AOPs usually operate at temperatures close to 25°C, and rely on chemical reactions to completely oxidize or mineralize organic contaminants. The use of OH radical is widespread in AOP, and though OH-based degradation is not selective, it has high chemical reaction rate. OH radical reacts with organic matter in three different ways: (1) by an electron addition to OH from an organic substituent, (2) by H abstraction, yielding carbon radicals, which react with oxygen to produce peroxyl radicals, and subsequently ketones/aldehydes and carbon dioxide, and (3) by addition of OH radicals to double bonds [11]. Among the many constituents of organic matter, the presence of chromophores, i.e., visible-light absorbing and emitting aromatic rings, is particularly of interest in drinking water. The capacity to degrade chromophoric compounds is, therefore, a desirable trait to technologies of treating drinking water. AOP methods based on TiO_2 and UV, and combinations thereof, are particularly known to be highly effective in this case. As a natural result of the application of AOPs in drinking water, if a complete mineralization is not achieved, a wide range of low-molecular weight compounds may be found in treated drinking water, and further physical treatment (e.g., filtration) is often required. Nitrous oxide (N_2O) is an important pollutant, which is emitted during the biological nutrient removal (BNR) processes of wastewater treatment. Since it has a greenhouse effect which is 265 times higher than that of carbon dioxide, even relatively small amounts can result in a significant carbon footprint [12].

4 Design of Experiments for the Optimization of AOP for the Treatment of Drinking Water

Aiming at the adequate treatment of water, the use of statistical techniques and the application of Design of Experiments (DOE) can be determinant for the efficient understanding of the influence of the factors involved in the process; thus contributing to the improvement of its results. A DOE method is a test or a series of tests, in which deliberate changes are made to the input variables of a process or system, making it possible to observe and identify the reasons for changes in output variables or responses [13].

A usually complex and costly experimental situation is easily resolved with DOE. All factors are considered in a minimal number of experiments, and the results are verified with recognized statistical methods. The analysis of variance (ANOVA) is a standard statistical technique that is commonly used in order to determine the significance of the independent variables on the output responses. It does not analyze the data directly, but determines the percentage of contribution of each factor in determining the variability (variance) of data [14].

A direct application of DOE methods is based on orthogonality of arrays, fully explored by Box-Cox, Placket Burmann, and Taguchi methods. The aim of Taguchi method is to minimize the variability of the product, identifying the means of giving it robustness during the process. By reducing the variability, by means of the ideal adjustment of the input variables, aiming at a certain behavior of the output variable, the method also allows to reduce the impact of noise sources, reduce costs, reduce product development time, and achieve robustness [15]. The fundamental principle is that to ensure a consistent quality, one must look for processes that are insensitive to defects from fluctuations, that is, the process must be designed so that its performance is the least sensitive of all types of noise.

Response surface methodology (RSM) is a collection of statistical and mathematical techniques useful for developing, improving, and optimizing processes. It has also important applications in the design, development, and formulation of new products, as well as in the improvement of existing product designs [16]. Experiments for fitting a predictive model involving several continuous variables are known as response surface experiments. The objectives of RSM include the determination of variable settings for which the mean response is optimized and the estimation of the response surface is in the vicinity of this optimum location [17]. The purpose of mathematical modeling as part of the RSM is described in three main steps:

1. Establish a relationship, albeit approximate, between y (output variables) and x_1, x_2, ..., x_k (input variables) that can be used to predict the response values for given settings of the control variables.
2. Determine, through hypothesis testing, the significance of the factors whose levels are represented by $x_1, x_2, ..., x_k$.
3. Determine the optimum settings of $x_1, x_2, ..., x_k$ that result in the optimized response over a certain region of interest [18].

RSM has several advantages compared to the classical experimental or optimization methods in which one variable at a time technique is used. Firstly, RSM offers a large amount of information from a small number of experiments. Indeed, classical methods are time consuming and a large number of experiments are needed to explain the behavior of a system. Secondly, it is possible to observe the interaction effect of the independent parameters on the response using RSM [19]. Engineering problems usually have more than one output variable, often the interests of input variable adjustments are divergent for each response. Thus, it is necessary to define a method that unifies, in the most appropriate way, the statistical results obtained for all output variables. Multiresponse optimization problems often involve incommensurate and conflicting responses. In order to overcome the trade-offs involved in conflicting multivariate optimization, one could explicit the preferential parameter to be optimized. In order to illustrate the selection of an optimization parameter, one could hypothesize a process in which the optimum temperature and pH levels have been determined experimentally. However, in such hypothetical case, adjusting pH would potentially damage or be too costly for the overall process, therefore, the optimization would be done only by adjusting the temperature factor.

The object of multi-response optimization is to determine the conditions on the independent variables that lead to optimal or nearly optimal values of the response variables. Desirability function appears to have been first proposed as a criterion for response optimization by Harrington, and popularized by Derringer and Suich [20].

The two case studies presented in Sects. 5 and 6 were developed by members of the research group responsible for this chapter. Though these cases were focused on the utilization of AOP to treat industrial effluents, the streamline is similar to the cases in drinking water when it comes to the basic chemometric studies. The results discussed on Sects. 5 and 6 are analyzed by applications of Taguchi and RSM methods utilizing different combinations of AOP.

5 Case 1: AOP Using RSM and Taguchi Orthogonal Array (Adapted from de Freitas et al. [11])

De Freitas et al. employed different sorts of AOP to reduce the chemical load of an effluent rich in phenolic compounds and with high organic load [8]. Initially, an L_{16} Taguchi Orthogonal Array was utilized to evaluate the factors of Photo-Fenton and Ozone AOP methods on the removal of organic matter. Initial results indicated removal levels of 29.07% of Total Organic Carbon (TOC). The process was then optimized with a Photo-Fenton RSM. With such optimization, the highest percentage of TOC removal was of 54.68%. This condition is associated with a mass ratio of hydrogen peroxide and ferrous ions equal to 8, which corresponds to 47.8 g of H_2O_2 and 5.95 g of Fe^{2+}.

5.1 Characteristics of the Design of Experiments

Table 1 presents each variable with its respective selected levels. The distribution of factors was made according to symbol in it, with the intention of obtaining interactions among them, as allowed in Taguchi method.

Table 2 shows the percentage of variation in the TOC for the experiments conducted in the design of experiments L_{16}. Experiments 13 and 15 had a higher percentage of TOC removal for the AOP (O_3/UV and Photo-Fenton).

The statistical analysis of the L_{16} Taguchi design (Fig. 1) showed the most significant parameters for degrading organic matter in the effluent. These corresponded to uncoded factor pH = 3, which was set on a low level, and the adjusted

Table 1 Control factors and levels for the exploratory study on the phenolic effluent treatment (reproduced from [11] with permission)

Factor	Symbol	Level 1	Level 2
[a]Hydrogen Peroxide (g)	A	38.7	46.5
[a]Ferrous ions (g)	B	1.26	1.55
pH	C	3	5
Ozone (L/h)	D	3	5
Temperature	E	30	35
Lamp power	F	16	28

[a]Fenton reagent

Table 2 Taguchi orthogonal array L_{16} with TOC removal percentage (reproduced from [11] with permission)

Row	A	B	AB	C	AC	BC	E	AE	F	AF	D	AD	AF	TOC reduction (%)
1	1	1	1	1	1	1	1	1	1	1	1	1	1	18.28
2	1	1	1	1	1	1	2	2	2	2	2	2	2	23.82
3	1	1	1	1	1	2	2	2	2	2	2	2	2	9.38
4	1	1	1	2	2	2	2	2	2	2	1	1	1	15.74
5	1	2	2	1	1	2	1	1	2	2	1	2	2	23.72
6	1	2	2	1	1	2	2	2	1	1	2	1	1	22.48
7	1	2	2	2	2	1	1	1	2	2	2	1	1	15.89
8	1	2	2	2	2	1	2	2	1	1	1	2	2	15.55
9	2	1	2	1	2	2	1	2	1	2	1	1	2	25.71
10	2	1	2	1	2	2	2	1	2	1	2	2	1	23.74
11	2	1	2	2	1	1	1	2	1	2	2	2	1	22.88
12	2	1	2	2	1	1	2	1	2	1	1	1	2	21.07
13	2	2	1	1	2	1	1	2	2	1	1	2	1	29.07
14	2	2	1	1	2	1	2	1	1	2	2	1	2	26.78
15	2	2	1	2	1	2	1	2	2	1	2	1	2	28.68
16	2	2	1	2	1	2	2	1	1	2	1	2	1	21.82

maximum level factors were: 46.5 g hydrogen peroxide, 1.55 g ferrous ions, and UV lamp power of 28 W. In the utilized DOE, the interactions among factors were also evaluated, in which those that were the most significant for the process were hydrogen peroxide and pH adjusted for level 1, and hydrogen peroxide and temperature for level 2.

The statistical significance of the factors and their interactions towards the TOC reduction for the treatment of the phenolic effluent was confirmed by ANOVA, as shown in Table 3. The statistical analysis, at 95% of confidence, showed the most significant factors for removing the organic load. According to Array F, whose critical value was 18.51, and for a p-value which was less than 5%, the most significant factors to the degradation of the organic matter of the phenolic effluent were H_2O_2, Fe^{2+}, pH, UV, and interactions H_2O_2/pH and H_2O_2/temperature.

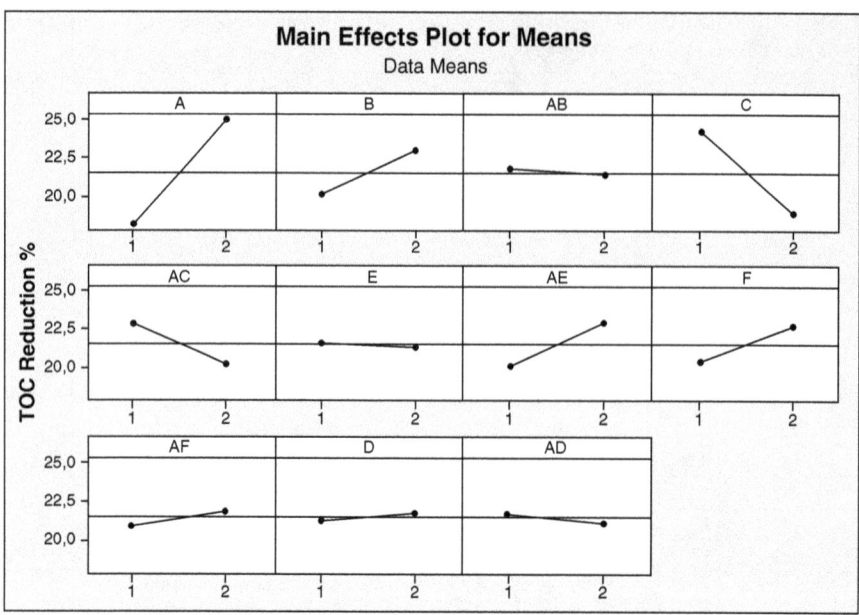

Fig. 1 Main effects of percentage variation measures of TOC in the treatment of phenolic effluent from the L_{16} design (reproduced from [11] with permission)

Table 3 Analysis of variance of Taguchi's orthogonal array L_{16} (reproduced from [11] with permission)

Source	DF	SS	Adj SS	Adj MS	F	P
A	1	185.232	185.232	185,232	178.97	0.00
B	1	32.833	32.833	32.833	31.72	0.005
AB	1	0.555	0.555	0.555	0.54	0.505
C	1	110.986	110.986	110.986	107.23	0.000
AC	1	26.112	26.112	26.112	25.23	0.007
E	1	0.585	0.585	0.585	0.57	0.494
AE	1	32.490	32.490	32.490	31.39	0.005
F	1	21.160	21.160	21.160	20.44	0.011
AF	1	3.667	3.667	3.667	3.54	0.133
D	1	0.616	0.616	0.616	0.60	0.483
AD	1	1.103	1.103	1.103	1.07	0.360
Residual error	4	4.140	4.140	1.035		
Total	15	419.480				

Hydrogen peroxide was the most significant factor, with $F = 195.6$ and a p-value of 0.5%, and the other factors are pH ($F = 117.2441$ and p-value $= 0.84\%$), ferrous ions ($F = 34.6842$ and p-value $= 2.76\%$), lamp power ($F = 22.3531$ and p-value $= 4.19\%$) and the interactions H_2O_2/pH ($F = 27.5844$ and p-value $= 3.44\%$) and H_2O_2/temperature ($F = 34.3219$ and p-value $= 2.79\%$). The values obtained by

ANOVA confirmed the significance shown by the main effects chart, displayed as Fig. 1.

The best experiment for the degradation of the organic load of the phenolic effluent was experimental condition 13, with 29.07% removal. It has 46.5 g of hydrogen peroxide, 1.55 g of ferrous ions, pH = 3, ozone flow of 3 L/h, temperature of 30°C and 28 W ultraviolet lamp power.

5.2 Results of the RSM (Fe^{2+}/H_2O_2) for Removing the Total Organic Carbon (TOC)

The percentage of TOC removal for the experiments conducted by the RSM (Fe^{2+}/H_2O_2) is reported in Table 4. Experiment 10 presents the greatest percentage of organic load removal, i.e. 54.68%.

Table 5 presents the ANOVA for the linear and quadratic model proposed to explain the response of the TOC reduction, based on significant terms of the Pareto chart and on the variance analysis of the effects. The ANOVA with 95% confidence, F critical equals to 6.61 and a p-value which is less than 5% showed that the

Table 4 Photo-Fenton response surface methodology (reproduced from [11] with permission)

Experiment	Fe^{2+}	H_2O_2	TOC (%)
1	−1	−1	12.14
2	1	−1	15.28
3	−1	1	33.66
4	1	1	42.35
5	0	0	49.58
6	0	0	51.26
7	0	0	52.69
8	−(2)1/2	0	47.6
9	0	(2)1/2	44.53
10	(2)1/2	0	54.68
11	0	−(2)1/2	1.44

Table 5 Analysis of variance of RSM Photo-Fenton (reproduced from [11] with permission)

Factors	[a](SS)	[a](df)	[a](SMQ)	F
A – H_2O_2 (Q)	1,612.300	1	1,612.300	10.8436
B – H_2O_2 (L)	661.906	1	661.906	4.4517
C – Fe^{2+} (L)	429.284	1	429.284	2.8872
D – Fe^{2+} (L)	44.894	1	44.894	0.3019
CB	7.701	1	7.701	0.0518
Error	743.432	5	148.686	
Total SS	3,483.202	10		

[a]SS sum of squares, df degrees of freedom, SMQ average sum of squares

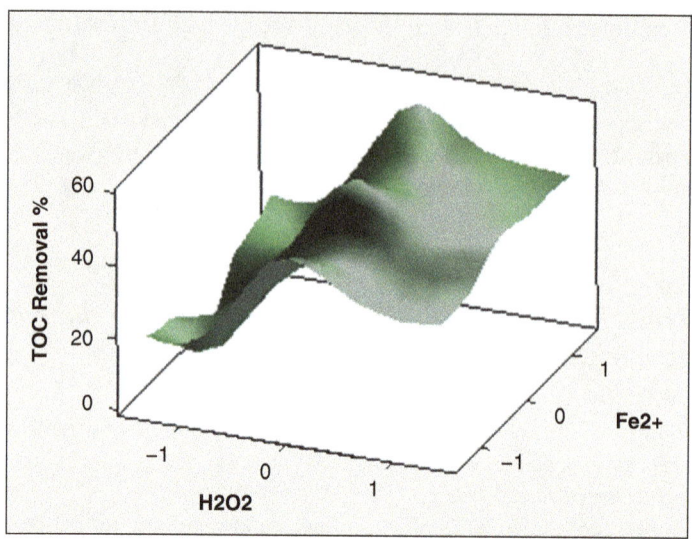

Fig. 2 Response surface results for TOC removal (%) evaluating the concentrations of ferrous ion and hydrogen peroxide (reproduced from [11] with permission)

significant factor for the removal of TOC was the addition of hydrogen peroxide to the quadratic model ($F = 10.8436$ and p-value $= 2.16\%$).

Figure 2 illustrates how the percentage of TOC removal increases as the mass amount of H_2O_2 and Fe^{2+} increases. It is possible to observe a significant drop in the quadratic model developed from the RSM when the Fenton reagent is at a low level (-1).

The greatest reduction of TOC (54.68%) was observed for level 0 of hydrogen peroxide and ferrous ions at a high level (+1.5), which corresponds to experimental condition number 10 of the response surface chart. In Fig. 2, it is observed that the majority of experimental data is distributed in the optimal region, which is indicated in light and dark green color. The parameters used in the optimization of organic matter degradation of the phenolic effluent for a better experimental condition were 48.3 g of hydrogen peroxide and 5.95 g of ferrous ions. Experimental condition number 10, with 54.68% removal of the organic load of the effluent, was the most significant for the process.

5.3 Conclusion Case 1

The case study 1 illustrated how AOP (Photo-Fenton and O_3/UV) were evaluated with Taguchi and RSM designs. Taguchi results demonstrated that Freitas et al. [11] selected initial factorial levels that were not significant for the removal of organic load. In this DOE, it was found that the greatest reduction of TOC (%) was related to an increase in the concentration of Fe^{2+} and H_2O_2, and to the absence of ozone in the reaction medium. The Photo-Fenton RSM has demonstrated that optimum conditions of treatment were within the studied region, since the conducted experiment had reached a significant curvature, making it necessary for the accomplishment of the star, i.e. rotational, experiment. The obtained results were significant for the removal of the TOC (%).

In the process optimization for RSM related to the Photo Fenton method, the highest percentage of TOC removal 54.68% was achieved, which corresponded to experimental condition number 10. This condition was associated with the mass ratio of hydrogen peroxide and ferrous ions being equal to eight, which corresponded to 47.8 g of H_2O_2 and 5.95 g of Fe^{2+}. It is concluded that the Photo-Fenton is the most effective process for the phenolic effluent treatment used in this work.

6 Case 2: Taguchi Method in AOP (Adapted from de Freitas et al. [16])

Case study 2 was developed upon the use of a polyester resin effluent, which was supplied by Valspar industry, located in São Bernardo do Campo, State of São Paulo, Brazil.

An experimental design Taguchi L_9 with AOP and heterogeneous photocatalysis was used for 1 L of fresh effluent diluted three times. The prepared effluent sample was previously homogenized and conditioned at room temperature. Semiconductor titanium dioxide (TiO_2) and H_2O_2 (30% w/v) were added during the initial 50 min of 1 h of total reaction. The temperature of the reaction medium during the whole period of the photocatalytic process was controlled at 25 °C by using a thermostatic bath. A centrifugal pump was used for conducting the effluent from the tubular reactor to the storage tank. Ultraviolet lamps of 15 and 21 W were used. The statistical planning performed is represented by Taguchi L_9 orthogonal array to which the response variable was TOC and independent variables as factors proposed for this stage were: pH, titanium dioxide, hydrogen peroxide and UV radiation power. Table 6 shows the treatment variables with their levels for selected AOP.

Table 6 Control variables and their levels (reproduced from [16] with permission)

Control variables (factors)	Level 1	Level 2	Level 3
A – pH	3.0	5.0	7.0
B – TiO_2 (g/L)	0.083	0.167	0.250
C – H_2O (g)[a]	120.0	151.0	182.0
D – UV (W)	Not present	15	21

[a]$[H_2O_2] = 30\%$ w/w

Table 7 Results of replica 2 of the percentage reduction obtained in experiments, initial TOC of 7,920 mg/L (reproduced from [16] with permission)

Experiment	pH factor A	TiO_2 factor B	H_2O_2 factor C	UV factor D	Reduction of TOC (%)
1	1	1	1	1	31.28
2	1	2	2	2	34.5
3	1	3	3	3	36.4
4	2	1	2	3	35.72
5	2	2	3	1	31.65
6	2	3	1	2	30.02
7	3	1	3	2	30.74
8	3	2	1	3	33.01
9	3	3	2	1	27.48

Initially, the mass of H_2O_2 (30% w/w) was calculated by a stoichiometric ratio that depended on the organic load of the effluent. This led a mass of H_2O_2 of 50 g per liter of effluent.

The amount of TOC of the effluent *in natura* had a mean value of 7,920 mg/L that was subjected to pre-treatment. For each experimental amount of TOC, a sample in a 60-min reaction was determined. Table 7 shows the arrangement of orthogonal Taguchi L_9 for the treatment of effluent polyester resin using AOP.

Statistical analysis of Taguchi L_9, in Fig. 3, showed the most significant parameters for the degradation of organic matter in the wastewater, the latter reflecting of pH = 3, adjusted at a low level and factors set at a maximum level of 182 g hydrogen peroxide and ultraviolet lamp power of 21 W. According to the plan performed, the level of titanium dioxide added to the process can be adjusted at low or medium level, i.e. with values 0.083 g/L and 0.167 g/L.

Statistical analysis at a level of 95% showed the most significant factors for the removal of organic load. According to the distribution F, which critical value under these conditions corresponded to 4.26 and a p-value less than 5%, the most important factors for the degradation of organic matter in the effluent were H_2O_2, pH, and UV. The most significant factor was the ultraviolet lamp power with F of 60.65201 and a p-value less than 0.001%, and then the remaining factors were the pH ($F = 30.11586$, p-value $= 0.10\%$) and H_2O_2 ($F = 4.67497$; p-value $= 4.053\%$). The values obtained by ANOVA confirmed the significance shown in the graph of main effects. The ANOVA with $F = 2$ demonstrated that the factor TiO_2 was significant for TOC removal. An F value greater than 2 was considered as a relevant effect (factor). The statistical significance factor in TOC reduction in the effluent treatment was confirmed by ANOVA, as shown in Table 8. Multiple linear regressions provided another statistical approach to evaluate variables in a quantitative approach. Significant parameters to the regression analysis are shown in Table 9, where pH, UV, and H_2O_2 are relevant for the degradation process of the organic load of the polyester resin effluent.

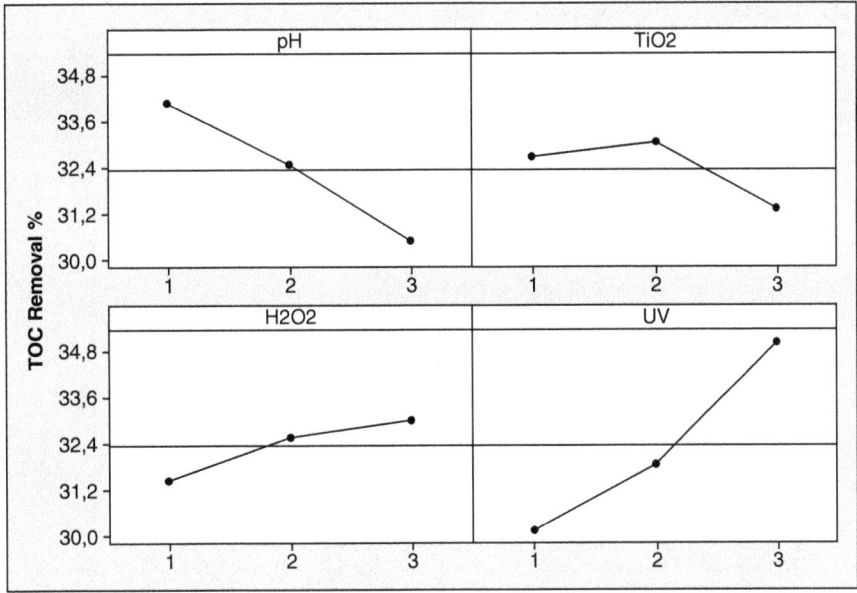

Fig. 3 Main effects in TOC percentage variation measurements in the effluent treatment of L_9 planning (reproduced from [16] with permission)

Table 8 Analysis of variance Taguchi L_{16} orthogonal array obtained for TOC (%) removal (reproduced from [16] with permission)

Source of variation	SQ	GL	SMQ	F	P-value
pH	53.0700	2.00	26.53499	30.11586	0.00010
TiO_2	3.7711	2.00	1.88554	2.13999	0.17366
H_2O_2	8.2382	2.00	4.11909	4.67497	0.04053
UV	106.8806	2.00	53.44030	60.65201	<0.00001
Error	7.9299	9.00	0.88110		

Table 9 Regression parameters

Factor	Coefficient	t-value	Beta coeff.	Probability
pH	−1.04933	−4.573584	−0.542040	0.0003
TiO_2	−5.73042	−1.042771	−0.123584	0.1580
H_2O_2	0.0266726	1.627560	0.192891	0.0638
UV	0.245688	5.791488	0.686380	0.0000
Constant	32.2247			

Multiple linear regression showed a coefficient of determination (R^2) of 0.817404, which demonstrates the efficiency of the degradation of effluent using the polyester resin of the experimental design. An ANOVA (Table 10) was performed in order to validate the multiple linear regression equation.

Table 10 ANOVA of multiple linear regression

Sources of variation	df	SQ	MSQ	F	P-value
Due to regression	4	147.0425	36.7603	14.55	0.0001
Independent	13	32.84718	2.526706		

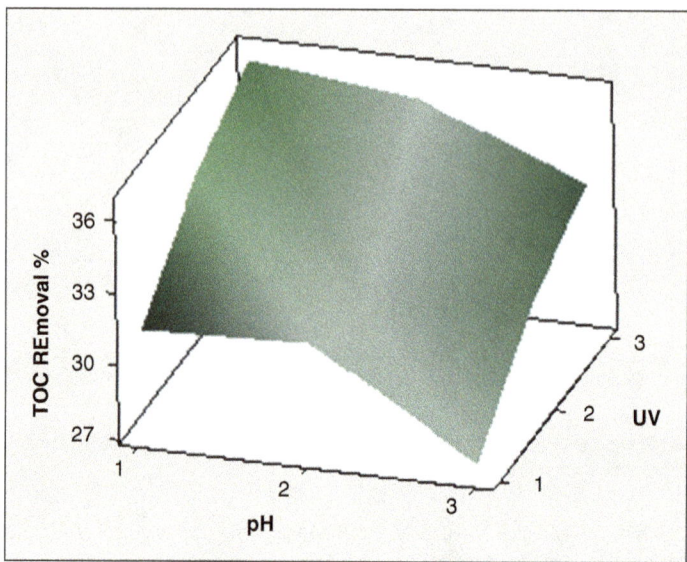

Fig. 4 Graph of the two most influential factors in the process

The most significant factors in the process show how the percentage removal of TOC is influenced by pH, as seen in Fig. 4. An increasing degradation of the organic load is observed on the surface. This is achieved by independent variables: pH and power of the ultraviolet lamp. The percentage of increased removal of organic load occurs when there is an increase of the power of the lamp and a decrease of pH. The greatest percentage reduction of the organic load is equal to 39.489%, which parameters used in this experimental condition were H_2O_2 = 182 g, pH = 3, TiO_2= 0.250 g/L and the lamp power of 21 W. The response variable was significant for the degradation of organic matter in the effluent.

6.1 Conclusion Case 2

Taguchi planning was applied to the degradation of effluent organic load. The experimental design showed that further reduction of TOC (%) is related to an increase in pH and ultraviolet intensity. The results obtained were significant for the removal of TOC (%) from polyester resin effluent treated by AOP, and heterogeneous photocatalysis.

The Taguchi orthogonal array shows that the optimal removal corresponds to experimental condition number 3. This condition was inclusive of the weight ratio of hydrogen peroxide at 183 g, pH = 3, TiO_2 = 0.250 g/L and the lamp intensity = 21 W. It can be concluded that the process of heterogeneous photocatalysis is optimally suitable for the treatment of the effluent studied in this work.

References

1. Ruiz-Aguirre A, Polo-López M, Fernández-Ibáñez P, Zaragoza G (2017) Integration of membrane distillation with solar photo-Fenton for purification of water contaminated with bacillus sp. and clostridium sp. spores. Sci Total Environ 595(1):110–118. https://doi.org/10.1016/j.scitotenv.2017.03.238
2. Batista LMB, dos Santos AJ, da Silva DR, de Melo Alves AP, Garcia-Segura S, Martinez-Huitle CA (2017) Solar photocatalytic application of NbO_2OH as alternative photocatalyst for water treatment. Sci Total Environ 596:79–86. https://doi.org/10.1016/j.scitotenv.2017.04.019
3. Slutsky B (1998) Handbook of chemometrics and qualimetrics: part A by D.L. Massart, B.G.M. Vandeginste, L.M.C. Buydens, S. De Jong, P.J. Lewi, and J. Smeyers-Verbeke. Data handling in science and technology volume 20a. Elsevier: Amsterdam. J Chem Inf Comput Sci 38(6):1254–1254. https://doi.org/10.1021/ci980427d
4. Hopke PK (2015) Chemometrics applied to environmental systems. Chemometr Intell Lab Syst 149:205–214. https://doi.org/10.1016/j.chemolab.2015.07.015
5. Mas S, de Juan A, Tauler R, Olivieri AC, Escandar GM (2010) Application of chemometric methods to environmental analysis of organic pollutants: a review. Talanta 80(3):1052–1067. https://doi.org/10.1016/j.talanta.2009.09.044
6. Barnard TE, Booksh K, Brereton R, Coomans D, Deming S, Hayashi Y, Mallet Y, Matsuda R, Nocerino J, Olivero R (2013) Chemometrics in environmental chemistry-statistical methods, vol 2. Springer, Berlin
7. Murphy KR, Bro R, Stedmon CA (2014) Chemometric analysis of organic matter fluorescence. In: Coble P, Baker A, Lead J, Reynolds D, Spencer R (eds) Aquatic organic matter fluorescence. Cambridge University Press, New York, pp 339–375
8. Pinto VS, Gambarra-Neto FF, Flores IS, Monteiro MR, Lião LM (2016) Use of 1H NMR and chemometrics to detect additives present in the Brazilian commercial gasoline. Fuel 182:27–33
9. Pancras JP, Vedantham R, Landis MS, Norris GA, Ondov JM (2011) Application of epa unmix and nonparametric wind regression on high time resolution trace elements and speciated mercury in Tampa, Florida aerosol. Environ Sci Technol 45(8):3511–3518
10. Kar S, Roy K, Leszczynski J (2017) On applications of QSARs in food and agricultural sciences: history and critical review of recent developments. In: Roy K (ed) Advances in QSAR modeling. Springer, Cham, pp 203–302
11. de Freitas APBR, de Freitas LV, Loures CCA, da Silva AF, Gonçalves LG, Silva MB (2014) Taguchi orthogonal array combined with Monte Carlo simulation in the optimization of wastewater treatment. AJTAS 3(6-1):19–22. http://dx.doi.org/10.11648/j.ajtas.s.2014030601.12
12. Massara TM, Malamis S, Guisasola A, Baeza JA, Noutsopoulos C, Katsou E (2017) A review on nitrous oxide (N_2O) emissions during biological nutrient removal from municipal wastewater and sludge reject water. Sci Total Environ 596:106–123
13. Montgomery DC (2013) Design and analysis of experiments design and analysis of experiments 8th edn. Wiley, New York

14. Myers RH, Montgomery DC, Anderson-Cook CM (2016) Response surface methodology: process and product optimization using designed experiments. Wiley, New York
15. Carbas RJC (2008) Estudo paramétrico de juntas adesivas pelo método de Taguchi. Ms. Degree Thesis in Mechanical Engineering, Universidade do Porto, Portugal
16. de Freitas APBR, de Freitas LV, Samanamud GL, Marins FAS, Loures CCA, Salman F, dos Santos TL, Silva MB (2013) Multivariate analysis in advanced oxidation process. In: The statistical significance of the factors and the sciences. InTech, Rijeka
17. Dean A, Voss D, Draguljic D (2017) Response surface methodology. In: Design and analysis of experiments. Springer, New York, pp 565–614
18. Bas D, Boyaci H (2007) Modeling and optimization I: usability of response surface methodology. J Food Eng 78(3):836–845
19. Chabbi A, Yallese MA, Meddour I, Nouioua M, Mabrouki T, Girardin F (2017) Predictive modeling and multi-response optimization of technological parameters in turning of polyoxymethylene polymer (POM C) using RSM and desirability function. Measurement 95:99–115
20. Derringer G, Suich R (1980) Simultaneous optimization of several response variables. J Qual Technol 12(4):214–219

Index

A

Acetaldehyde, 126, 352, 355, 359, 367, 379
Acetaminophen, 119
Acetic acid, 74, 224, 238, 393, 398
Acetohydroxyacid synthase (AHAS), 394
Acetone, 74, 126, 224
Acetylaminoantipyrine, 148
Acid Orange II, 80, 81, 222, 231, 241, 242
Activated carbon (AC), 15, 28, 34, 44, 69, 77, 103, 165, 204, 214, 230, 271, 315–329
   biological (BAC), 90, 202, 372
   granular activated carbon (GAC), 44, 136, 181, 374, 375
   metal-functionalized, 78, 89
   powdered (PAC), 44, 45, 72, 181, 375
Adsorption, 7, 71–92, 105, 315, 325, 374
Advanced oxidation processes (AOPs), 1, 21, 53, 73, 267
   disinfection, 267
   Fenton-like, 69
   NOMs, 57
Al/Fe-pillared clay, 69
Akaganeite, 79
Alachlor, 142, 144, 148
Aldehydes, 28, 59, 74, 116, 125, 180, 348, 352, 354, 359, 370, 410
Algae, 13, 42, 55, 143, 275, 276, 390–399
Algal blooms, 6, 8, 358
Alkylphenol ethoxylate, 366
Alkylphenols, 391
Amino acids, 28, 46, 56, 73, 102, 105, 107, 135, 262, 276, 280, 346, 348, 360, 364, 394, 397

Ammonia, 219, 348, 357
Ammonium, 32, 61, 75
AMPA, 143
Ampicillin, 161, 167, 168
Antibiotic-resistant bacteria (ARB), 161
Antibiotics, 28, 139, 143, 161, 270
Aquifers, 1–12, 17, 31, 159
Aromatic acids, 73, 102, 364
Arsenic, 13, 24, 25, 33, 41
Artificial recharge, 1, 17
Ascaris lumbricoides, 165, 260, 274
Ascaris suum, 272
Aspartic acid, 366
Assimilable organic carbon (AOC), 29
Atrazine, 142, 144, 148, 172, 394
Azo dyes, 78, 80, 128, 149, 231, 235, 242, 317

B

Bacteria, 10, 42, 55, 159, 257–284, 395
   regrowth, 71, 135
   removal, 42
Batch reactors, 211, 215, 271, 393
Benzene, 140, 367, 398
Benzoic acid, 108, 116
Benzothiazole, 180
Bicarbonates, 142
Biocides, 172, 261, 266, 388
Biological activity, 387
Biologically activated carbon (BAC), 90, 202, 372
Bis(2-chloroisopropyl) ether, 202
Bisphenols, 387, 392
   bisphenol A, 366, 391

Bluegill sunfish, 390
Boreholes, 7, 12
Brachionus calyciflorus, 147
Bromate, 46, 60, 65, 145, 196, 202, 348, 351, 354, 356, 368, 372
Bromine, 57, 214, 264, 359, 364, 371
Bromochloromethane, 355
Bromodichloromethane, 355
Bromophenols, 180
Butanal, 126
Butyl butyrate, 193

C
Camphor, 193
Carbadox, 366
Carbamazepine, 146, 366
Carbapenem, 161
Carbohydrates, 28, 56, 102, 126, 361
Carbonates, 12, 30, 109, 115, 122, 135, 142, 158, 198, 280, 396
Carbon dioxide, 30, 32, 73, 74, 82, 88, 109, 389, 410
Carbon nanotubes, 69, 90, 214, 241, 320, 324, 327
Carbon xerogels, 327
Carboxylic acids, 102
Catalase (CAT), 162, 260
Catalysts, 211, 309
   clay, 77, 315, 324
   cost-effective, 309
   load, 237
   physicochemical properties, 241
   sludge-based (SBC), 321
Catalytic cracking, 220
Catalytic supercritical water oxidation (CSCWO), 328–330
Catalytic wet peroxide oxidation (CWPO), 69, 214, 325, 387
Chemical oxidation, 183, 213, 257, 348, 366
Chemometrics, methods/analyses, 405
Chloramines, 357
Chlorfenvinphos, 143
Chlorination, 14, 72, 86, 155, 257, 264, 281, 347, 354–357, 372, 392
   byproducts, 193
Chlorine, 46, 264, 300
Chlorine dioxide, 183, 264, 300, 346–348, 356, 358, 372
Chlorobenzene, 224
p-Chloro-*m*-cresol (PCMC), 81
Chlorodibromomethane, 355
Chloroform, 62, 74
Chlorophenols, 78, 139, 143, 180, 347, 352

Chromium, 13, 26
Ciprofloxacin, 119, 167, 168
Citric acid, 173, 368
Clarification, 35
Clarithromycin, 161, 167
Clay catalysts, 77, 315, 324
Climate change, 3
Coagulation, 35, 72
   enhanced, 41, 374
Cobalt doped red mud (Co/RM), 321
Colloids, stability, 36
Coloured dissolved organic matter (CDOM), 119
Compound parabolic collector, 155
Contaminant Candidate List (CCL), 355
Continuous reactors (flow reactors), 211, 217
Continuous stirred-tank reactor (CSTR), 221, 231
Copper, 78, 83, 229, 327, 330
Copper oxide, 321
Cost, 297
m-Cresol, 81, 229
Cryptosporidium sp., 165, 257, 261, 297
Cyanides, 266
Cyanobacteria, 180, 186, 202
Cyclocitral, 180
Cysteine, 368

D
Danio rerio, 390
Daphnia magna, 143, 147, 387, 390, 397
DBP formation potential (DBPFP), 59
Dechlorination, 81
Desmodesmus subspicatus, 390
Dichloramine, 348
Dichloroacetic acid, 148
2,6-Dichloro-1,4-benzoquinone, 200
Diclofenac, 27, 146
Diethyl disulfide, 202
Dimethylamine (DMA), 359
Dimethyl-decahydro-naphthalen-1-ol, 195
Dimethyl disulfide, 202
Diseases, 259
Disinfectants, 14, 25, 46, 59, 71, 260, 300, 347, 374
   alternative, 372
   oxidizing, 349, 350, 357, 372
   secondary, 300
Disinfection, 21, 34, 46, 257, 259, 297
   AOP, 349
Disinfection byproducts (DBP), 55, 60, 116, 345, 349

brominated (B-DBPs), 357, 368
iodinated (I-DBPs), 357
mitigation, 372
nitrogenous (N-DBPs), 358
precursors, 359
Dissolved air flotation, 44
Dissolved organic carbon (DOC), 28, 102, 109, 360
Dissolved organic matter (DOM), 102
Diuron, 148
DNA, damage, 46, 260, 263, 272, 276, 303, 305, 348, 393
repair, 263
Drinking water, 1, 179, 257
disinfection methods, 264
sources, 4, 34
treatment, 21, 24, 55, 72, 84, 140, 179, 202, 300, 345
Dunaliella tertiolecta, 172
Dune filtrate, 47
Dyes, 139, 226, 236, 299, 366

E

Electrochemical oxidation, 137, 189, 196, 281, 312, 389
Electrocoagulation, 73
Electrospray ionization (ESI), 104, 111
Emerging contaminants (ECs), 133, 136, 141, 147, 148
Endocrine-disrupting compounds (EDCs), 42, 136, 141, 147, 213, 230, 310, 349, 390
Endosulfan, 42
Energy efficiency, 133, 147
Entamoeba histolytica, 260, 264, 274
Enterococcus faecalis, 169, 274
Enzymes, 28, 46, 162, 263, 265, 282
ESI-FT-ICRMS, 111
Estradiol, 27, 387, 391
Estriol, 391
Estrogenic assays, 147
2-Ethyl-cyclohexanone, 195
Ethinylestradiol, 27, 391
Ethylenediamine-N,N'-disuccinic acid (EDDS), 173
Ethylenediaminetetraacetic acid (EDTA), 173
Ethylenethiourea (ETU), 42
2-Ethyl-1-hexanol, 193
Eutrophication, 8, 11
Evaporation, 2, 3, 5

F

Feed flow rate, 238
Fenton, 59, 155, 211, 313
heterogeneous, 211, 316
intracellular, 269
microorganisms, inactivation, 269
oxidation, 222
temperature, 232
Feroxyhyte, 163
Ferric chloride, 39, 43, 73
Ferric hydroxides, 230
Ferric oxyhydroxides, 122, 158
Ferritin, 163
Ferulic acid, 364, 366
Fe-ZSM-5, 82
Filtration, 13, 27, 33, 42, 47, 72, 103, 181, 197, 202, 204
Fixed-bed reactors, 217
Flavoproteins, 163
Flocculation, 35, 72
Floc removal, 35, 43
Fluidized-bed reactors (FBR), 219
Fluorescence excitation emission matrix (FEEM), 29, 110
Fluorescent regional integration (FRI), 110
Fluorine, 214
Formaldehyde, 116, 126, 347, 352, 355, 359, 367, 379
Formic acid, 116, 238, 393, 398
Formylaminoantipyrine, 148
FT-ICR-MS, 111
FT-IR spectroscopy, 110
Fullerene, 91
Fulvic acids, 28, 56, 109, 135, 186, 362
Fungi, 160, 169, 173, 186, 197, 262, 272, 275
Furanones, halogenated, 358
Furfuryl alcohol (FFA), 121
Fusarium solani, 272

G

Geosmin (GSM), 53, 60–65, 179–186
Giardia sp., 257, 261, 264, 274, 275, 347
Glyoxal, 116, 352
Glyoxylic acids, 353, 398
Glyphosate, 143, 145
Goethite, 158, 163, 172, 271, 315, 316
Granular activated carbon (GAC), 44, 136, 181, 374, 375
Graphene oxide (GO), 316, 327

Groundwater, 6, 11, 21, 122
  chlorinated organics, 136
  drinking water, 1
  Fenton/photo-Fenton, 122, 155
  microbial contamination, 159
  overexploitation, 9
  pesticides, 394

H

Haloacetic acid formation potential (HAAFP), 58
Haloacetic acids (HAA), 57, 58, 103, 117, 345, 347, 355, 366
Haloacetonitriles, 117
Heavy metals, 25, 31, 33, 41, 57, 71, 84, 103, 391
Helminths, 25, 173, 260, 272, 274
Hematite, 83, 158, 172, 225, 242, 271
Hepatitis, 260, 276
Herbicides, 10, 172, 366, 394
cis-3-Hexen-1-ol, 180
Hormone disruptors, 27
Hospital wastewater, 143, 161
Humic acids, 28, 37, 41, 46, 47, 87, 173, 186, 374
Humins, 28, 56
Hydrodechlorination, 81
Hydrodynamic cavitation, 74
Hydrogen peroxide, 58, 60, 76, 123, 138, 214, 266, 317, 392, 412
  ozonation, 312
  photolysis, 139
Hydrogen sulfide, 32, 88, 180
Hydroperoxyl radicals, 104, 111, 184, 214, 223, 234, 266, 267
Hydroquinone, 368
Hydrotalcite, 241
p-Hydroxyacetophenone, 393
Hydroxy-cyclohexadienyl (HCHD•), 123
Hydroxyl radicals, 60, 76, 140, 265, 302, 310, 349, 367, 389
2-Hydroxy-3-methylbutyric acid methyl ester, 200
Hydroxy-4a-methyloctahydronaphthalen-2-one, 195
Hypobromite, 62, 65, 356
Hypobromous acid, 369
Hypochlorite, 14, 117, 264, 355
Hypochlorous acid, 117, 214

I

Inactivation, 297
Infiltration, 3
Iodine, 264
Iodosulfuron, 387, 394
Ion exchange, 13, 15, 29, 55, 75, 87, 88, 104, 375
β-Ionone, 180
Iron, 14, 21, 32, 269, 327
Iron acetate, 242
Iron chelates, 173
Iron chloride, 90
Iron molybdate, 315
Iron nitrate, 242
Iron oxides, 32, 41, 78, 158, 163, 242, 317
Iron sludge, 136, 217, 224, 225, 311, 313
Iron sulfate, 242
Iron sulfide, 272
Isobutyl isobutyrate, 193
2-Isobutyl-3-methoxypyrazine (IBMP), 179, 180, 200
Isooctyl alcohol, 193
p-Isopropenylphenol, 393
2-Isopropyl-3-methoxypyrazine (IPMP), 179–182, 184, 185, 200
2-Isopropyl-3-methylpyrazine, 200
p-Isopropylphenol, 393
Isoproturon, 142, 145

K

Ketones, 59, 74, 125, 195, 266, 348, 370, 410
  cyclic, 195

L

Lakes, 5, 11
Lepidocrocite, 79, 163
Lipids, 46, 161, 260, 266, 276, 305
  peroxidation, 162
Logarithmic reduction values (LRV), 25

M

Magnetite, 83, 163, 225
Maleic acid, 74, 123, 138
Manganese, 14, 21, 32, 33, 48, 316, 317
Mepanipyrim, 148
Mercaptans, 180
Methane, 32

Methicillin-resistant *Staphylococcus aureus*, 161
Methylaminoantipyrine, 148
2-Methyl-2-bornene, 193
3-Methylbutanoic acid, 200
Methyl glyoxal, 116, 353
2-Methylisoborneol (2-MIB), 60, 62, 179, 180, 186
Methyl isobutyrate, 193
Methyl methacrylate, 193
Methyl orange, 78, 79
3-Methyl-2-oxobutanoic acid, 200
3-Methyl-2-oxobutanoic acid methyl ester, 200
4-Methyl-2-oxovaleric acid, 200
2-Methylpropanoic acid, 200
Methyl propionate, 193
Metolachlor, 53, 60, 62
Microbial contamination, 159
Microbiological standards, 25
Microorganisms, removal, 42
Microplastics, 27
Micropollutants, 21, 24, 31, 42, 55, 74, 85, 135, 136, 297, 390
Monochloramine, 264, 300, 348, 372
Montmorillonite, 80, 88, 124, 321, 325
MTBE (Methyl tert-butyl ether), 112
Municipal water supply, 1
Myponga Reservoir, Adelaide, Australia, 126

N

Natural organic matter (NOM), 28, 53, 69, 133, 345
  fractionation, 361
  intermediates, 99
  oxidation, 99
  separation, 99
  structure/reactivity, 364
Nickel, 321
Nitrates, 10, 32, 141
Nitrilotriacetic acid, 173
Nitrite, 141
Nitrobenzene, 139
N-Nitrosodiethylamine (NDEA), 355
N-Nitrosodimethylamine (NDMA), 355, 359, 369
N-Nitrosodiphenylamine (NDPhA), 355
N-Nitrosodi-*n*-propylamine (NDPA), 355
N-Nitrosopyrrolidine (NPYR), 355
Nonanal, 193
Nonanoic acid, 193
Non-purgeable organic carbon (NPOC), 109

O

Odor, 179
Orange II dye, 222
Oryzias latipes, 390
Oxalic acid, 74, 148, 173, 224, 238, 321, 368, 393, 398
Oxidation, 111, 259, 345
  chemical, 183, 213, 257, 348, 366
  electrochemical, 137, 189, 196, 281, 312, 389
  photocatalytic, 59
  sonochemical, 136
Oxidative stress, 257
Oxyhalides, 351, 355
Oxyhydroxides, 122, 158, 163, 321
Ozonation, 58, 86, 112, 135, 144–148, 186, 195, 265, 281, 309, 318, 348, 392
  photocatalytic, 139
Ozone, 14, 28, 46, 53, 58, 60, 112, 135, 139, 169, 197, 265, 300, 318, 413

P

Paracetamol, 317
Parasites, 46, 160, 257, 299
Particle size, 243
Pathogens, 2, 10, 24, 73, 155, 161, 259, 298, 347, 370
PCR (polymerase chain reaction), 160
Pentachlorophenol, 143
Pentafluorobenzyl hydroxylamine hydrochloride (PFBHA), 126
1-Penten-3-one, 180
Peptides, 102, 364
Peroxidase, 260, 316
Peroxone, 136, 140
Peroxyl radicals, 74, 104, 112, 138, 410
Persistent organic pollutants (POPs), 213
Personal care products, 24, 388
Persulfate, 172, 188–201, 205
Pesticides, 10, 26, 136, 142
pH, 157, 225
Pharmaceutical and personal care products (PPCPs), 143, 230, 349
Pharmaceuticals, 26, 136
Pharmaceutical wastewater, 226
Phenol, 79, 148, 226, 228
Phosphate, 41, 107, 280
Photocatalysis, 59, 137, 141, 387
  heterogeneous, 124, 137, 275, 322
  homogenous, 116, 138, 271
Photodegradation, 119
Photo-Fenton, 60, 155, 165, 271

Photo-Fenton (*cont.*)
  microorganisms, inactivation, 271
  solar driven, 165
Photolysis, 31, 47, 58, 65, 76, 117, 139, 187, 200, 377, 397
Photosensitizing agent, 119
Phytophthora capsici, 272, 273
Plastics, 27
Polarity rapid assessment method (PRAM), 29
Poliovirus, 264
Pollutants, concentration, 236
  removal, 21
Polyoxometalates, 195
Polysaccharides, 28
Potassium permanganate, 183
Powdered activated carbon (PAC), 44, 45, 72, 181, 375
Practical calcium carbonate precipitation potential (PCCPP), 31
Propanal, 126
Protection, 1
Proteins, 28, 73, 102, 105, 107, 161, 260, 276, 346, 360
Protozoans, 25, 46, 160, 173, 186, 257–279, 298
Pseudokirchneriella subcapitata, 147, 390
Pseudomonas aeruginosa, 265
Pseudo-nitzschia delicatissima, 172
Pyrite, 225

Q
Quality deterioration, 1
Quantitative microbial risk assessment (QMRA), 25

R
Rainbow trout, 390
REACH, 27
Reactive chlorine species (RCS), 118, 260
Reactive oxygen species (ROS), 104, 111, 119, 162, 260, 267, 277, 302, 304, 389
Recharge, artificial 17
Removal, 21, 42, 53, 179
Resin fractionation, 362
Resorcinol, 364
Reversed-phase liquid chromatography (RP-LC), 107
River bank filtrate, 47

River basin management plans (RBMPs), 16
Rivers, 6, 11

S
Safe Drinking Water Act (SDWA), 24
Salicylic acid, 368
Saline water, 2, 5
Salmonella typhimurium, 268
Sand filtration, 13, 27, 33, 42, 47, 72, 103, 181, 197, 202, 204
Saturation index, SI (Langelier saturation index, LSI), 30
Sedimentation, 5, 34, 36, 42, 43, 72, 203, 204
Sepiolite, 241
Singlet oxygen, 104, 111, 121, 136, 184, 196, 266, 267, 274
Size-exclusion chromatography (SEC), 107, 116, 361
Sludge-based catalyst (SBC), 321
Solar disinfection/treatment, 155, 303, 349
  photo-Fenton, 160, 163, 165
Soluble microbiological products (SMPs), 28
Specific ultraviolet absorbance (SUVA), 29, 56, 85, 104, 108, 117, 360, 367
Springs, 5–7, 12
Streams, 6, 11
Sulfamethoxazole, 146, 167
Sulfanilamide, 80
Sulfate radical anion, 117, 193, 198, 349
Sulfates, 158, 172, 196, 242, 268, 281
Sulfonylureas, 394, 398
Sunscreens, 366
Superoxide radical, 104, 111, 121, 136, 184, 200, 266, 269, 275, 302, 322
Superoxygen dismutase (SOD), 162, 260
Surface run-off, 34
Surface water, 2, 11, 21, 34
  Fenton/photo-Fenton, 155
  microbial contamination, 159
Surfactants, 139
Sweep coagulation, 41

T
Taguchi methods, 406
Tannins, 102
Tartaric acid, 156, 173, 223, 313, 368
Taste, 179
Taste and odor (T&O) compounds, 179–186, 197, 202, 205

Index

Tetrachloroethylene, 139, 140
Tetracycline, 167, 168
Textile effluents/wastewater, 219–226, 234–236
Theoretical calcium carbonate precipitation potential at 90°C (TCCPP90), 31
Thioethers, 202
Titanium dioxide (TiO_2), 137, 275, 302
  / UV, 124
Toluene, 224
Total organic carbon (TOC), 28, 55, 83, 89, 108, 109, 135, 317, 360, 412, 415
Total organic halide (TOX), 354, 364
Toxicity tests, 143, 148, 390, 395
Transpiration, 3
Treatment, efficiency evaluations, 389
  groundwater, 14
  surface water, 13
S-Triazines, 366, 394
Triazole, 42
Trichloramine, 348
Trichloroacetic acid, 351, 356
2,4,6-Trichloroanisole (TCA), 179–185, 197
Trichloroethane, 139, 224
Trichloroethylene (TCE), 139, 140, 149
Trichloronitromethane, 117, 378
2,4,6-Trichlorophenol (TCP), 200, 352
Triclosan, 366
Trihalomethane formation potential (THMFP), 58, 61
Trihalomethanes (THM), 103, 135, 180, 265, 345–355, 375, 378
2,4,6-Trinitrotoluene, 222
Tryptophan, 115, 361, 364, 366
Tyrosol, 229

U
Ultrasound, 73, 81, 84, 103, 126, 135, 149, 165, 196, 221, 312, 314, 318, 320
  oxidation, 389
Unregulated Contaminant Monitoring Rule (UCMR), 355
UV/C_2, 116
UV, disinfection, 46, 47, 53, 59, 124, 262, 348, 374
  VUV, 46, 57, 140, 312
UV/H_2O_2, 116

UV-Vis, 56, 61, 79, 88, 104, 108, 121, 193

V
Vacuum ultraviolet (VUV), 46, 57, 140, 312
Vanadium, 321
Vancomycin-resistant *Enterococcus* spp., 161
Vibrio fischeri, 143, 147, 148, 390–399
Viruses, 10, 25, 42, 46, 160, 166, 173, 260–282, 299, 349
Volatile chlorinated organic carbons (VCOCs), 139

W
Wastewater treatment, 211
  plants (WWTPs), 26, 389
Water, cycle, 3
  disinfection, 155, 259
    resistance, 260
  microbiology, 257
  pollutants, 133
  quality, chemical, 25
    conditioning, 30
    NOM, 28
  resource management/protection, 15
  treatment, 1, 69
    conventional, 21
Wells, 7, 12
Wet hydrogen peroxide catalytic oxidation (WHPCO), 214, 317

X
Xenoestrogens, 390

Y
Yeast estrogen screen (YES) assay, 390

Z
Zebrafish, 143, 390
Zeolites, 15, 69, 77, 88, 92, 189, 195, 214, 231–235, 271, 315–325
  USY, 241
Zinc, 30, 314, 320, 328
ZnO, 36, 59, 137, 277, 312, 320–329

CPSIA information can be obtained
at www.ICGtesting.com
Printed in the USA
LVHW02*1614150718
583816LV00003B/262/P